X International Conference on Kaon Physics 2016

AA006418

Journal of Physics: Conference Series Volume 800

Birmingham, United Kingdom
14 – 17 September 2016

ISBN: 978-1-5108-3843-7
ISSN: 1742-6588

Printed from e-media with permission by:

Curran Associates, Inc.
57 Morehouse Lane
Red Hook, NY 12571

Some format issues inherent in the e-media version may also appear in this print version.

Copyright© (2016) by the Institute of Physics
All rights reserved. The material featured in this book is subject to
IOP copyright protection, unless otherwise indicated.

Printed by Curran Associates, Inc. (2017)

For permission requests, please contact the Institute of Physics
at the address below.

Institute of Physics
Dirac House, Temple Back
Bristol BS1 6BE UK

Phone: 44 1 17 929 7481
Fax: 44 1 17 920 0979

techtracking@iop.org

Additional copies of this publication are available from:

Curran Associates, Inc.
57 Morehouse Lane
Red Hook, NY 12571 USA
Phone: 845-758-0400
Fax: 845-758-2633
Email: curran@proceedings.com
Web: www.proceedings.com

Table of contents

Volume 800

X International Conference on Kaon Physics
14–17 September 2016, Birmingham, United Kingdom

Accepted papers received: 21 December 2016
Published online: 23 March 2017

Preface

Paper

Invited lectures

Posters

The International Conference on Kaon Physics 2016 took place at the University of Birmingham (United Kingdom) on 14–17 September 2016. This conference continued the KAON series, offering an opportunity for theorists and experimentalists from the high-energy physics community to discuss all aspects of kaon physics. The 2016 edition saw a strong participation from theory and phenomenology and the first kaon results from the LHCb experiment at CERN, as well as updates from several experiments around the world including NA62 and KOTO.

All papers published in this volume of KAON2016 have been peer reviewed through processes administered by the proceedings Editors. Reviews were conducted by expert referees to the professional and scientific standards expected of a proceedings journal published by IOP Publishing.

The organizers and the participants wish to thank the University of Birmingham, the European Research Council (advanced grant 268062, starting grant 336581), CERN, the UK Science and Technology Facility Council and the UK Institute for Particle Physics Phenomenology for their support in the organization of this successful edition.

Content from this work may be used under the terms of the Creative Commons Attribution 3.0 licence. Any further distribution of this work must maintain attribution to the author(s) and the title of the work, journal citation and DOI.
Published under licence by IOP Publishing Ltd

Scientific programme

- CP and T violation
- CKM matrix and flavour mixing
- Rare decays
- Precision tests of the Standard Model
- CPT and quantum mechanics
- Lepton universality and flavour violation
- Lattice gauge theory
- Chiral perturbation theory
- Physics beyond the Standard Model
- Future opportunities in kaon physics

International Advisory Committee

Andrzej Buras – Munich University
Sergio Bertolucci – CERN
Brigitte Bloch-Devaux – University of Turin
Fabio Bossi – INFN Frascati
Douglas Bryman – University of British Columbia
Augusto Ceccucci – CERN
Gilberto Colangelo – Bern University
Peter Cooper – FNAL
Giancarlo D'Ambrosio – INFN Napoli
Paolo Franzini – University of Rome
Bob Hsiung – National Taiwan University
Enrico Iacopini – INFN Firenze
Gino Isidori – INFN Frascati
Takashi Kaneko – KEK
Konrad Kleinknecht – Mainz University
Steve Kettell – BNL
Marc Knecht – CNRS Marseille
Takeshi Komatsubara – KEK
Chris Sachrajda – University of Southampton
Monica Tecchio – University of Michigan
Bob Tschirhart – FNAL
Taku Yamanaka – Osaka University
Heinrich Wahl – University of Ferrara

Local Organizing Committee

Martin Gorbahn – University of Liverpool
Evgueni Goudzovski – University of Birmingham
Cristina Lazzeroni – University of Birmingham (chair)
Alexander Lenz – University of Durham

IOP Conf. Series: Journal of Physics: Conf. Series **800** (2017) 011001 doi:10.1088/1742-6596/800/1/011001

Proceedings Referees

Marc-Olivier Bettler – CERN
Patrizia Cenci – INFN Perugia
John Fry – University of Liverpool
Martin Gorbahn – University of Liverpool
Evgueni Goudzovski – University of Birmingham
Cristina Lazzeroni – University of Birmingham
Alexander Lenz – University of Durham
Diego Martinez Santos – Universidade de Santiago de Compostela

IOP Conf. Series: Journal of Physics: Conf. Series **800** (2017) 011002 doi:10.1088/1742-6596/800/1/011002

Peer review statement

All papers published in this volume of *Journal of Physics: Conference Series* have been peer reviewed through processes administered by the proceedings Editors. Reviews were conducted by expert referees to the professional and scientific standards expected of a proceedings journal published by IOP Publishing.

Content from this work may be used under the terms of the Creative Commons Attribution 3.0 licence. Any further distribution of this work must maintain attribution to the author(s) and the title of the work, journal citation and DOI.
Published under licence by IOP Publishing Ltd

Kaon Flavour Physics Strikes Back

Andrzej J. Buras

TUM Institute for Advanced Study, Lichtenbergstr. 2a, D-85748 Garching, Germany

E-mail: aburas@ph.tum.de

Abstract. In this short presentation I emphasize the increased importance of kaon flavour physics in the search for new physics (NP) that we should witness in the rest of this decade and in the next decade. The main actors will be the branching ratios for the rare decays $K^+ \to \pi^+\nu\bar{\nu}$ and $K_L \to \pi^0\nu\bar{\nu}$, to be measured by NA62 and KOTO, and their correlations with the ratio ε'/ε on which recently progress by lattice QCD and large N dual QCD approach has been made implying a new flavour anomaly. Further correlations of $K^+ \to \pi^+\nu\bar{\nu}$, $K_L \to \pi^0\nu\bar{\nu}$ and ε'/ε with ε_K, ΔM_K, $K_L \to \mu^+\mu^-$ and $K_L \to \pi^0\ell^+\ell^-$ will help us to identify indirectly possible NP at short distance scales. This talk summarizes the present highlights of this facinating field including some results from concrete NP scenarios.

1. Introduction

In three recent reports [1, 2, 3] I have stressed the increased importance of kaon flavour physics in the search for new physics (NP) which we should witness in the near future. Indeed after years of silence I expect that kaon flavour physics will strike back providing new insights in the dynamics at very short distance scales. The following pages can be considered as an express review of this fascinating field. Further details and in particular numerous references can be found in [1, 2, 3, 4].

2. Important Messages

2.1. ε'/ε

Presently in kaon flavour physics the most exciting appears to be the anomaly in ε'/ε. The present status of ε'/ε in the SM can be summarized as follows. The RBC-UKQCD lattice collaboration calculating hadronic matrix elements of all contributing operators but not including isospin breaking effects finds [5, 6]

$$(\varepsilon'/\varepsilon)_{\rm SM} = (1.38 \pm 6.90) \times 10^{-4}, \qquad ({\rm RBC-UKQCD}). \tag{1}$$

Using the hadronic matrix elements of QCD-penguin (Q_6) and EW-penguin (Q_8) $(V-A)\otimes(V+A)$ operators from RBC-UKQCD lattice collaboration but extracting the matrix elements of penguin $(V-A)\otimes(V-A)$ operators from the data on CP-conserving $K \to \pi\pi$ amplitudes and including isospin breaking effects one finds [7]

$$(\varepsilon'/\varepsilon)_{\rm SM} = (1.9 \pm 4.5) \times 10^{-4}, \qquad ({\rm BGJJ}) \tag{2}$$

that is confirmed within the errors by the recent analysis in [8]

$$(\varepsilon'/\varepsilon)_{\rm SM} = (1.1 \pm 5.1) \times 10^{-4}, \qquad ({\rm KNT}). \tag{3}$$

Content from this work may be used under the terms of the Creative Commons Attribution 3.0 licence. Any further distribution of this work must maintain attribution to the author(s) and the title of the work, journal citation and DOI.
Published under licence by IOP Publishing Ltd

All these results are significantly below the experimental world average from NA48 [9] and KTeV [10, 11] collaborations,

$$(\varepsilon'/\varepsilon)_{\rm exp} = (16.6 \pm 2.3) \times 10^{-4}, \tag{4}$$

suggesting that models providing enhancement of ε'/ε are favoured.

These results are based on NLO calculations of the Wilson coefficients of the relevant operators [12, 13, 14, 15, 16, 17]. Partial NNLO calculations have been performed in [18, 19, 20]. Complete NNLO result from Maria Cerda-Sevilla, Martin Gorbahn, Sebastian Jäger and Ahmet Kokulu should be available soon.

While these results, based on the hadronic matrix elements from RBC-UKQCD lattice collaboration, suggest some evidence for NP in ε'/ε, the large uncertainties in the hadronic matrix elements in question do not yet preclude that eventually the SM will agree with data. Therefore the upper bounds on the relevant hadronic matrix elements of Q_6 and Q_8 from large N dual QCD approach [21] are important as they give presently the strongest support to the anomaly in question, certainly stronger than present lattice results.

In the strict large N limit [22, 23, 24] the parameters $B_6^{(1/2)}$ and $B_8^{(3/2)}$ that represent the relevant hadronic matrix elements of the QCD penguin operator Q_6 and the electroweak penguin operator Q_8, respectively, are simply given by

$$B_6^{(1/2)} = B_8^{(3/2)} = 1, \qquad \text{(large N Limit)}\,. \tag{5}$$

But RBC-UKQCD results [6, 5] imply [7, 25]

$$B_6^{(1/2)} = 0.57 \pm 0.19\,, \qquad B_8^{(3/2)} = 0.76 \pm 0.05\,, \qquad \text{(RBC-UKQCD)}, \tag{6}$$

and the suppression of $B_6^{(1/2)}$ below unity is the main origin of the strong suppression of ε'/ε below the data within the SM. Yet in view of the large error in $B_6^{(1/2)}$ one could be sceptical about any claims that there is NP in ε'/ε. Future lattice results could in principle raise $B_6^{(1/2)}$ towards its large N value and above $B_8^{(3/2)}$ bringing the SM result for ε'/ε close to its experimental value.

However, the analyses of $B_6^{(1/2)}$ and $B_8^{(3/2)}$ within the dual QCD approach in [21, 26] show that such a situation is rather unlikely. Indeed, in this approach going beyond the strict large N limit one can understand the suppression of $B_6^{(1/2)}$ and $B_8^{(3/2)}$ below the unity as the effect of the meson evolution from scales $\mu = \mathcal{O}(m_\pi, m_K)$ at which (5) is valid to $\mu = \mathcal{O}(1\,\text{GeV})$ at which Wilson coefficients of Q_6 and Q_8 are evaluated [21]. This evolution has to be matched to the usual perturbative quark evolution for scales higher than 1 GeV and in fact the supressions in question and the property that $B_6^{(1/2)}$ is more strongly suppressed than $B_8^{(3/2)}$ are consistent with the perturbative evolution of these parameters above $\mu = \mathcal{O}(1\,\text{GeV})$. Thus we are rather confident that [21]

$$B_6^{(1/2)} < B_8^{(3/2)} < 1 \qquad \text{(dual QCD)}. \tag{7}$$

For further details, see [21, 2].

Additional support for the small value of ε'/ε in the SM comes from the recent reconsideration of the role of final state interactions (FSI) in ε'/ε [26]. Already long time ago the chiral perturbation theory practitioners put forward the idea that both the amplitude ReA_0, governed by the current-current operator $Q_2 - Q_1$ and the Q_6 contribution to the ratio ε'/ε could be enhanced significantly through FSI in a correlated manner (see e.g. [27] and other reference in [2]). However, as shown recently in [26] FSI are likely to be important for the $\Delta I = 1/2$ rule, in agreement with these papers but much less relevant for ε'/ε.

It should finally be noted that even without lattice results, varying all input parameters, the bound in (7) implies the upper bound on ε'/ε in the SM

$$(\varepsilon'/\varepsilon)_{\rm SM} < (8.6 \pm 3.2) \times 10^{-4}, \qquad ({\rm BG}). \tag{8}$$

On the other hand employing the lattice value for $B_8^{(3/2)}$ in (6) and $B_6^{(1/2)} = B_8^{(3/2)} = 0.76$, one obtains $(6.0 \pm 2.4) \times 10^{-4}$ instead of (8), well below the data.

All these findings give strong motivation for searching for NP which could enhance ε'/ε above its SM value. We will summarize the present efforts in this direction below.

2.2. Tensions between ε_K and $\Delta M_{s,d}$ in the SM and CMFV Models

In [28] we have pointed out a significant tension between ε_K and $\Delta M_{s,d}$ within the SM and models with constrained MFV (CMFV) implied by new lattice QCD results from Fermilab Lattice and MILC Collaborations [29] on $B^0_{s,d} - \bar{B}^0_{s,d}$ hadronic matrix elements. Even if this tension is certainly not as large as is the case of the ε'/ε anomaly the plots in [28], in particular in Fig. 5 of that paper, show that there is a clear tension between ε_K and $\Delta M_{s,d}$ in the SM and CMFV models. Moreover this tension persists independently of the values of CKM parameters. For smaller (exclusive) values of $|V_{cb}|$ one finds $\Delta M_{s,d}$ to agree well with the data, while ε_K is roughly 25% below its experimental value. For $|V_{cb}|$ in the ballpark of inclusive determinations one finds ε_K to agree with the data, while $\Delta M_{s,d}$ are then typically by 15% larger than their experimental values. These numbers are for the SM, in all other CMFV models the situation gets worse.

The improved $\Delta B = 2$ hadronic matrix elements from other lattice collaborations and improved values of $|V_{cb}|$ and $|V_{ub}|$ will tell us one day whether this tension persists and if this will turn out to be the case, whether there is a ε_K anomaly and/or a $\Delta M_{s,d}$ anomaly.

2.3. $K^+ \to \pi^+ \nu\bar{\nu}$ and $K_L \to \pi^0 \nu\bar{\nu}$ in the SM

These two rare decays allow to test the short distance scales far beyond the reach of the LHC. Even scales of $\mathcal{O}(100)$ TeV can be probed in this manner [30]. The present status of $K^+ \to \pi^+ \nu\bar{\nu}$ and $K_L \to \pi^0 \nu\bar{\nu}$ within the SM has been presented in [25] with the result

$$\mathcal{B}(K^+ \to \pi^+ \nu\bar{\nu}) = (8.4 \pm 1.0) \times 10^{-11}, \tag{9}$$

$$\mathcal{B}(K_L \to \pi^0 \nu\bar{\nu}) = (3.4 \pm 0.6) \times 10^{-11}. \tag{10}$$

But the most important outcome of this paper are parametric expressions for the branching ratios of these two decays in terms of the CKM input and the correlations between $K^+ \to \pi^+ \nu\bar{\nu}$ and $B_s \to \mu^+\mu^-$ and between $K^+ \to \pi^+ \nu\bar{\nu}$ and ε_K in the SM. These formulae should allow to monitor the numerical values for these branching ratios within the SM when the CKM input improves. Interesting correlations between $K^+ \to \pi^+ \nu\bar{\nu}$ and $K_L \to \pi^0 \nu\bar{\nu}$ and various observables are found in simplified models with flavour violating couplings of the SM Z and of a heavy Z' [31].

2.4. Strategy for ε'/ε and Lessons

In order to investigate the implications of ε'/ε anomaly on rare decays $K^+ \to \pi^+ \nu\bar{\nu}$ and $K_L \to \pi^0 \nu\bar{\nu}$ in a systematic fashion a strategy has been proposed in [32]. While ε'/ε plays the dominant role in this strategy it was useful to assume that there is also a modest ε_K anomaly.

Then ε'/ε and ε_K in the presence of NP contributions are given by

$$\frac{\varepsilon'}{\varepsilon} = \left(\frac{\varepsilon'}{\varepsilon}\right)^{\rm SM} + \left(\frac{\varepsilon'}{\varepsilon}\right)^{\rm NP}, \qquad \varepsilon_K \equiv e^{i\varphi_\varepsilon}\,\left[\varepsilon_K^{\rm SM} + \varepsilon_K^{\rm NP}\right] \tag{11}$$

with NP contributions parametrized as follows:

$$\left(\frac{\varepsilon'}{\varepsilon}\right)^{\rm NP} = \kappa_{\varepsilon'} \cdot 10^{-3}, \qquad 0.5 \le \kappa_{\varepsilon'} \le 1.5, \qquad \varepsilon_K^{\rm NP} = \kappa_\varepsilon \cdot 10^{-3}, \qquad 0.1 \le \kappa_\varepsilon \le 0.4\,. \tag{12}$$

The ranges for $\kappa_{\varepsilon'}$ and κ_ε indicate the required size of this contribution but can be kept as free parameters. They will be determined one day when the theory on ε'/ε and the CKM input improve.

In the simplest NP scenarios with tree-level Z and Z' exchanges, the imaginary parts of flavour-violating Z or Z' couplings to quarks are then determined as functions of $\kappa_{\varepsilon'}$. As ε_K is governed by the product of imaginary and real parts of these couplings, invoking it allows then to determine the corresponding real parts as functions of $\kappa_{\varepsilon'}$ and κ_ε.

Having fixed the flavour violating couplings of Z or Z' in this manner, one can express NP contributions to the branching ratios for $K^+ \to \pi^+\nu\bar\nu$, $K_L \to \pi^0\nu\bar\nu$, $K_L \to \mu^+\mu^-$ and to ΔM_K in terms of $\kappa_{\varepsilon'}$ and κ_ε. Explicit formulae can be found in [32]. In this manner one can directly study the impact of ε'/ε and ε_K anomalies in Z and Z' scenarios on these four observables.

In [32] numerous plots for the ratios

$$R_+^{\nu\bar\nu} \equiv \frac{\mathcal{B}(K^+ \to \pi^+\nu\bar\nu)}{\mathcal{B}(K^+ \to \pi^+\nu\bar\nu)_{\rm SM}}, \qquad R_0^{\nu\bar\nu} \equiv \frac{\mathcal{B}(K_L \to \pi^0\nu\bar\nu)}{\mathcal{B}(K_L \to \pi^0\nu\bar\nu)_{\rm SM}} \tag{13}$$

as functions of $\kappa_{\varepsilon'}$ and κ_ε within the models with tree-level Z and Z' exchanges have been presented. We will list the most important lessons from this study that depend on the flavour violating couplings $\Delta_{L,R}^{sd}(Z)$ and $\Delta_{L,R}^{sd}(Z')$ [33]. Moreover, we will use the abbreviations: (LHS $\equiv$ left − handed scenario) and (RHS $\equiv$ right − handed scenario) for NP scenarios in which only left-handed (LH) or right-handed (RH) flavour-violating couplings are present. The first six lesson deal with tree-level Z exchanges, the remaining four with Z' tree-level exchanges.

Lesson 1: In the LHS, a given request for the enhancement of ε'/ε determines the coupling ${\rm Im}\Delta_L^{sd}(Z)$.

Lesson 2: In LHS an enhanced ε'/ε implies uniquely *suppression* of $\mathcal{B}(K_L \to \pi^0\nu\bar\nu)$. This property is known from NP scenarios in which NP to $K_L \to \pi^0\nu\bar\nu$ and ε'/ε enters dominantly through the modification of Z-penguins.

Lesson 3: The imposition of the $K_L \to \mu^+\mu^-$ constraint in LHS determines the range for ${\rm Re}\Delta_L^{sd}(Z)$ which with the already fixed ${\rm Im}\Delta_L^{sd}(Z)$ allows to calculate the shifts in ε_K and ΔM_K. They are very small.

Lesson 4: With fixed ${\rm Im}\Delta_L^{sd}(Z)$ and the allowed range for ${\rm Re}\Delta_L^{sd}(Z)$, the range for $\mathcal{B}(K^+ \to \pi^+\nu\bar\nu)$ can be obtained. Both an enhancement and a suppression of $\mathcal{B}(K^+ \to \pi^+\nu\bar\nu)$ are possible. $\mathcal{B}(K^+ \to \pi^+\nu\bar\nu)$ can be enhanced by a factor of 2 at most.

Lesson 5: Analogous pattern is found in RHS, although the numerics is different. See Fig. 1 in [32]. In particular the suppression of $\mathcal{B}(K_L \to \pi^0\nu\bar\nu)$ for a given $\kappa_{\varepsilon'}$ is smaller. Moreover, an enhancement of $\mathcal{B}(K^+ \to \pi^+\nu\bar\nu)$ up to a factor of 5.7 is possible.

Lesson 6: In a general Z scenario with LH and RH flavour-violating couplings the pattern of NP effects changes because LR operators dominate NP contributions to ε_K and ΔM_K. One can then enhance simultaneously ε'/ε, ε_K, $\mathcal{B}(K^+ \to \pi^+\nu\bar\nu)$ and $\mathcal{B}(K_L \to \pi^0\nu\bar\nu)$ which is not possible in LHS and RHS. The correlations between ε'/ε and $K^+ \to \pi^+\nu\bar\nu$ and $K_L \to \pi^0\nu\bar\nu$ depend sensitively on the ratio of real and imaginary parts of the flavour-violating couplings involved. Moreover large departures from SM predictions for $K^+ \to \pi^+\nu\bar\nu$ and $K_L \to \pi^0\nu\bar\nu$ are possible and ε'/ε anomaly can be explained.

Z' models exhibit quite different pattern of NP effects in the K meson system than the LH and RH Z scenarios. In Z scenarios only electroweak penguins (EWP) can contribute to ε'/ε in an important manner because of flavour dependent diagonal Z coupling to quarks. But in Z'

models the diagonal quark couplings can be flavour universal so that QCD penguin operators (QCDP) can dominate NP contributions to ε'/ε. Interestingly, the pattern of NP in rare K decays depends on whether NP in ε'/ε is dominated by QCDP or EWP operators.

As demonstrated in [32] there is a large hierarchy between real and imaginary parts of the flavour violating couplings implied by ε'/ε anomaly in QCDP and EWP scenarios. In the case of QCDP imaginary parts dominate over the real ones, while in the case of EWP this hierarchy is opposite unless the ε_K anomaly is absent. Because of these different patterns there are striking differences in the implications of the ε'/ε anomaly for the correlation between $K^+ \to \pi^+\nu\bar{\nu}$ and $K_L \to \pi^0\nu\bar{\nu}$ in these two NP scenarios if significant NP contributions to ε'/ε are required. The plots in [32] and in particular analytic derivations presented there illustrate these differences in a spectacular manner. The main lessons are as follows.

Lesson 7: In the case of QCDP scenario the correlation between $\mathcal{B}(K_L \to \pi^0\nu\bar{\nu})$ and $\mathcal{B}(K^+ \to \pi^+\nu\bar{\nu})$ takes place along the branch parallel to the Grossman-Nir bound [34].

Lesson 8: In the EWP scenario the correlation between $\mathcal{B}(K_L \to \pi^0\nu\bar{\nu})$ and $\mathcal{B}(K^+ \to \pi^+\nu\bar{\nu})$ is very different from the one of the QCDP case. NP effects in rare K decays turn out to be modest in this case unless the diagonal quark couplings are $\mathcal{O}(10^{-2})$ and then the requirement of shifting upwards ε'/ε implies large effects in $K^+ \to \pi^+\nu\bar{\nu}$ and $K_L \to \pi^0\nu\bar{\nu}$ also in the EWP scenario.

Lesson 9: For fixed values of the neutrino and diagonal quark couplings in ε'/ε the predicted enhancements of $\mathcal{B}(K_L \to \pi^0\nu\bar{\nu})$ and $\mathcal{B}(K^+ \to \pi^+\nu\bar{\nu})$ are much larger when NP in QCDP is required to remove the ε'/ε anomaly than it is the case of EWP.

Lesson 10: In QCDP scenario ΔM_K is *suppressed* and this effect increases with increasing $M_{Z'}$ whereas in the EWP scenario ΔM_K is *enhanced* and this effect decreases with increasing $M_{Z'}$ as long as real couplings dominate. Already on the basis of this property one could differentiate between these two scenarios when the SM prediction for ΔM_K improves.

3. Results in specific NP models

3.1. Preliminaries

The latest analyses of NP contributions to ε'/ε in models with tree-level Z and Z' exchanges like 331 models, Littlest Higgs model with T-parity can be found in [35, 31, 36, 32, 37]. The analyses in supersymmetric models can be found in [38, 39, 40]. In view of space limitations we will only briefly summarize the results in 331 models and models with vector-like quarks.

3.2. 331 Flavour News

The 331 models are based on the gauge group $SU(3)_C \times SU(3)_L \times U(1)_X$ [41, 42, 43, 44, 45]. In these models new contributions to ε'/ε and other flavour observables are dominated by tree-level exchanges of a Z' with non-negligible contributions from tree-level Z exchanges generated through the $Z - Z'$ mixing. The size of these NP effects depends on $M_{Z'}$, on a parameter β, which distinguishes between various 331 models, on fermion representations under the gauge group and a parameter $\tan\bar{\beta}$ present in the $Z - Z'$ mixing. Extensive recent analyses in these models can be found in [46, 47, 48, 36, 37]. References to earlier analyses of flavour physics in 331 models can be found there and in [49, 50].

A detailed analysis of 331 models with different values of β, $\tan\bar{\beta}$ for two fermion representations F_1 and F_2, with the third SM quark generation belonging respectively to an antitriplet and a triplet under the $SU(3)_L$, has been presented in [48]: 24 models in total. Requiring that these models perform at least as well as the SM, as far as electroweak tests are concerned, seven models have been selected for a more detailed study of FCNC processes. Recent updated analyses of these seven models, that address the ε'/ε anomaly, have been presented in [36, 37] and we summarize the main results of these two papers. The main findings of [36, 37] for $M_{Z'} = 3\,\text{TeV}$ are as follows:

- Among seven 331 models singled out through electroweak precision study only three (M8, M9, M16) can provide significant shift of ε'/ε but not larger than 6×10^{-4}, that is $\kappa_{\varepsilon'} \leq 0.6$.
- The tensions between $\Delta M_{s,d}$ and ε_K can be removed in these models.
- Two of them (M8 and M9) can simultaneously suppress $B_s \to \mu^+\mu^-$ and bring the theory within 1σ range of the combined result from CMS and LHCb. The most recent result from ATLAS [51], while not accurate, appears to confirm this picture. On the other hand these models do not really help in the case of $B_d \to K^*\mu^+\mu^-$ anomalies [52, 53].
- In M16 the situation is opposite. The rate for $B_s \to \mu^+\mu^-$ can be reduced for $M_{Z'} = 3\,\text{TeV}$ by only a small amount but the anomaly in $B_d \to K^*\mu^+\mu^-$ can be significantly reduced.
- For higher values of $M_{Z'}$ the effects in $B_s \to \mu^+\mu^-$ and $B_d \to K^*\mu^+\mu^-$ are small. NP effects in rare K decays and $B \to K(K^*)\nu\bar{\nu}$ remain small in all 331 models even for $M_{Z'}$ of a few TeV. This could be challenged by NA62, KOTO and Belle II experiments in this decade.

All these results are valid for $|V_{ub}| = 0.0036$. For its inclusive value of $|V_{ub}| = 0.0042$, we find that for $|V_{cb}| = 0.040$ the maximal shifts in ε'/ε are increased to 7.7×10^{-4} and 8.8×10^{-4} for $M_{Z'} = 3\,\text{TeV}$ and $M_{Z'} = 10\,\text{TeV}$, respectively. Renormalization group effects are responsible for this enhancement of ε'/ε for increased $M_{Z'}$. A recent analysis in the MSSM in [39] identifies this effect as well. But as explained in [36] eventually for very high $M_{Z'}$, NP effects in ε'/ε will be suppressed.

Thus the main message from [36, 37] is that NP contributions in 331 models can simultaneously solve $\Delta F = 2$ tensions, enhance ε'/ε and suppress either the rate for $B_s \to \mu^+\mu^-$ or C_9 Wilson coefficient without any significant NP effects on $K^+ \to \pi^+\nu\bar{\nu}$ and $K_L \to \pi^0\nu\bar{\nu}$ and $b \to s\nu\bar{\nu}$ transitions. While sizable NP effects in $\Delta F = 2$ observables and ε'/ε can persist for $M_{Z'}$ outside the reach of the LHC, such effects in $B_s \to \mu^+\mu^-$ will only be detectable provided Z' will be discovered soon.

3.3. Models with vector-like quarks(VLQs)

A detailed analysis of flavour violation patterns in the K and $B_{s,d}$ sectors in eleven models with VLQs has been presented in [54]. The simplest (five of them) are the ones in which the gauge group is the SM one and the only new particles are VLQs in a single complex representation under the SM gauge group. A general classification of such models and references to the rich literature can be found in [54, 55]. In these models $\Delta F = 1$ FCNCs are dominated by tree-level Z exchanges, while $\Delta F = 2$ transitions by box diagrams with VLQs and scalars provided $M_{\text{VLQ}} \geq 5\,\text{TeV}$. Otherwise tree-level Z contributions cannot be neglected.

The summary of patterns of flavour violation in these models can be found in three DNA tables (Tables 5, 6, 10 in [54]) and the numerical results in Tables 8 and 9 of that paper. Our extensive numerical analysis has shown that NP effects in several of these models can still be very large and that simultaneous consideration of several flavour observables should allow to distinguish between these models. In particular models with left-handed and right-handed flavour violating currents can be distinguished from each other in this manner. Here we list most important results of this paper.

- All tensions between $\Delta M_{s,d}$ and ε_K can be easily removed in these models because the usual CMFV correlations between $\Delta M_{s,d}$ and ε_K are not valid in them. The box diagrams with VLQs and Higgs scalar exchanges are dominantly responsible for it.
- Tree-level Z contributions to ε'/ε can be large so that significant upward shift in ε'/ε can easily be obtained bringing the theory to agree with data.
- Simultaneously the branching ratio for $K^+ \to \pi^+\nu\bar{\nu}$ can be significantly enhanced over its SM prediction, but only in models with flavour violating RH currents. In models with

only LH currents $K^+ \to \pi^+\nu\bar{\nu}$ branching ratio can have at most its SM value because of the $K_L \to \mu\bar{\mu}$ constraint. On the other hand the positive shift in ε'/ε implies uniquely suppression of the $K_L \to \pi^0\nu\bar{\nu}$ branching ratio with the suppression being smaller in models with RH currents. The fact that in models with RH currents $K^+ \to \pi^+\nu\bar{\nu}$ can be enhanced, while $K_L \to \pi^0\nu\bar{\nu}$ suppressed is a clear signal of non-MFV sources at work. But also in models with pure LH currents the correlations between the branching ratios of these two decays differ from the MFV one.

- These features distinguish VLQ-models from 331 models, discussed above, in which NP effects are dominated by Z' exchanges with the maximal shift in ε'/ε amounting to 0.8×10^{-3} and NP effects in rare K decays being very small.
- Significant suppressions of the branching ratio for $B_s \to \mu^+\mu^-$, in particular in models with LH currents are possible. While such effects are also possible in 331 models, they cannot be as large as in VLQ models.
- On the other hand while 331 models can provide solutions to some LHCb anomalies, this is not possible in the VLQs models with SM gauge group and future confirmation of these anomalies could turn out to be a problem for the latter models.

Having the latter possibility in mind we have considered also six VLQ models with a heavy Z' related to $\mathrm{U(1)}_{\mathrm{L}_\mu-\mathrm{L}_\tau}$ symmetry and extended scalar sector. Some of such models, considered already in [56], can explain LHCb anomalies but NP effects in other observables are in my view less interesting than in models based on the SM gauge group. We refer to [54] for details. Future experimental results on $K^+ \to \pi^+\nu\bar{\nu}$, $K_L \to \pi^0\nu\bar{\nu}$, $B_s \to \mu^+\mu^-$ and LHCb anomalies and improved theoretical results on ε'/ε will tell us which of these VLQ models, if any, is selected by nature.

While the discovery of VLQs at the LHC would give a strong impetus to the models considered by us, non-observation of them at the LHC would not preclude their importance for flavour physics. In fact we have shown that large NP effects in flavour observables can be present for $M_{\rm VLQ} = 10$ TeV and in the flavour precision era one could even be sensitive to higher masses. In this context we have pointed out that the combination of $\Delta F = 2$ and $\Delta F = 1$ observables in a given meson system allows to determine the masses of VLQs in a given representation independently of the size of Yukawa couplings.

In summary the future of kaon flavour physics looks great and the coming years should be very exciting.

Acknowledgements

It is a pleasure to thank Monika Blanke, Fulvia De Fazio and Jean-Marc Gérard for very efficient studies discussed in this write-up and Christoph Bobeth, Alejandro Celis and Martin Jung for a very extensive analysis of VLQ models. I would also like to thank the organizers of Kaon 2016 for inviting me to this very interesting conference and for an impressive hospitality. This research was done and financed in the context of the ERC Advanced Grant project "FLAVOUR"(267104). It was also partially supported by the DFG cluster of excellence "Origin and Structure of the Universe".

References

[1] Buras A J 2015 *PoS* **EPS-HEP2015** 602 (*Preprint* 1510.00128)
[2] Buras A J 2016 (*Preprint* 1606.06735)
[3] Buras A J 2016 *8th International Workshop on QCD - Theory and Experiment (QCD@Work 2016) Martina Franca, Italy, June 27-30, 2016* (*Preprint* 1609.05711)
[4] Buras A J 2015 *PoS* **FWNP** 003 (*Preprint* 1505.00618)
[5] Blum T *et al.* 2015 *Phys. Rev.* **D91** 074502 (*Preprint* 1502.00263)
[6] Bai Z *et al.* (RBC, UKQCD) 2015 *Phys. Rev. Lett.* **115** 212001 (*Preprint* 1505.07863)

[7] Buras A J, Gorbahn M, Jäger S and Jamin M 2015 *JHEP* **11** 202 (*Preprint* 1507.06345)
[8] Kitahara T, Nierste U and Tremper P 2016 (*Preprint* 1607.06727)
[9] Batley J *et al.* (NA48) 2002 *Phys. Lett.* **B544** 97–112 (*Preprint* hep-ex/0208009)
[10] Alavi-Harati A *et al.* (KTeV) 2003 *Phys. Rev.* **D67** 012005 (*Preprint* hep-ex/0208007)
[11] Abouzaid E *et al.* (KTeV) 2011 *Phys. Rev.* **D83** 092001 (*Preprint* 1011.0127)
[12] Buras A J, Jamin M, Lautenbacher M E and Weisz P H 1992 *Nucl. Phys.* **B370** 69–104
[13] Buras A J, Jamin M, Lautenbacher M E and Weisz P H 1993 *Nucl. Phys.* **B400** 37–74 (*Preprint* hep-ph/9211304)
[14] Buras A J, Jamin M and Lautenbacher M E 1993 *Nucl. Phys.* **B400** 75–102 (*Preprint* hep-ph/9211321)
[15] Ciuchini M, Franco E, Martinelli G and Reina L 1993 *Phys. Lett.* **B301** 263–271 (*Preprint* hep-ph/9212203)
[16] Buras A J, Jamin M and Lautenbacher M E 1993 *Nucl. Phys.* **B408** 209–285 (*Preprint* hep-ph/9303284)
[17] Ciuchini M, Franco E, Martinelli G and Reina L 1994 *Nucl. Phys.* **B415** 403–462 (*Preprint* hep-ph/9304257)
[18] Buras A J, Gambino P and Haisch U A 2000 *Nucl. Phys.* **B570** 117–154 (*Preprint* hep-ph/9911250)
[19] Gorbahn M and Haisch U 2005 *Nucl. Phys.* **B713** 291–332 (*Preprint* hep-ph/0411071)
[20] Brod J and Gorbahn M 2010 *Phys. Rev.* **D82** 094026 (*Preprint* 1007.0684)
[21] Buras A J and Gerard J M 2015 *JHEP* **12** 008 (*Preprint* 1507.06326)
[22] Buras A J and Gérard J M 1986 *Nucl. Phys.* **B264** 371
[23] Bardeen W A, Buras A J and Gérard J M 1986 *Phys. Lett.* **B180** 133
[24] Buras A J and Gérard J M 1987 *Phys. Lett.* **B192** 156
[25] Buras A J, Buttazzo D, Girrbach-Noe J and Knegjens R 2015 *JHEP* **11** 033 (*Preprint* 1503.02693)
[26] Buras A J and Gerard J M 2016 (*Preprint* 1603.05686)
[27] Pallante E, Pich A and Scimemi I 2001 *Nucl. Phys.* **B617** 441–474 (*Preprint* hep-ph/0105011)
[28] Blanke M and Buras A J 2016 *Eur. Phys. J.* **C76** 197 (*Preprint* 1602.04020)
[29] Bazavov A *et al.* (Fermilab Lattice, MILC) 2016 *Phys. Rev.* **D93** 113016 (*Preprint* 1602.03560)
[30] Buras A J, Buttazzo D, Girrbach-Noe J and Knegjens R 2014 *JHEP* **1411** 121 (*Preprint* 1408.0728)
[31] Buras A J, Buttazzo D and Knegjens R 2015 *JHEP* **11** 166 (*Preprint* 1507.08672)
[32] Buras A J 2016 *JHEP* **04** 071 (*Preprint* 1601.00005)
[33] Buras A J, De Fazio F and Girrbach J 2013 *JHEP* **1302** 116 (*Preprint* 1211.1896)
[34] Grossman Y and Nir Y 1997 *Phys. Lett.* **B398** 163–168 (*Preprint* hep-ph/9701313)
[35] Blanke M, Buras A J and Recksiegel S 2016 *Eur. Phys. J.* **C76** 182 (*Preprint* 1507.06316)
[36] Buras A J and De Fazio F 2016 *JHEP* **03** 010 (*Preprint* 1512.02869)
[37] Buras A J and De Fazio F 2016 *JHEP* **08** 115 (*Preprint* 1604.02344)
[38] Tanimoto M and Yamamoto K 2016 (*Preprint* 1603.07960)
[39] Kitahara T, Nierste U and Tremper P 2016 (*Preprint* 1604.07400)
[40] Endo M, Mishima S, Ueda D and Yamamoto K 2016 (*Preprint* 1608.01444)
[41] Singer M, Valle J W F and Schechter J 1980 *Phys. Rev.* **D22** 738
[42] Pisano F and Pleitez V 1992 *Phys. Rev.* **D46** 410–417 (*Preprint* hep-ph/9206242)
[43] Frampton P H 1992 *Phys. Rev. Lett.* **69** 2889–2891
[44] Foot R, Hernandez O F, Pisano F and Pleitez V 1993 *Phys. Rev.* **D47** 4158–4161 (*Preprint* hep-ph/9207264)
[45] Montero J C, Pisano F and Pleitez V 1993 *Phys. Rev.* **D47** 2918–2929 (*Preprint* hep-ph/9212271)
[46] Buras A J, De Fazio F, Girrbach J and Carlucci M V 2013 *JHEP* **1302** 023 (*Preprint* 1211.1237)
[47] Buras A J, De Fazio F and Girrbach J 2014 *JHEP* **1402** 112 (*Preprint* 1311.6729)
[48] Buras A J, De Fazio F and Girrbach-Noe J 2014 *JHEP* **1408** 039 (*Preprint* 1405.3850)
[49] Diaz R A, Martinez R and Ochoa F 2005 *Phys. Rev.* **D72** 035018 (*Preprint* hep-ph/0411263)
[50] Carcamo Hernandez A, Martinez R and Ochoa F 2006 *Phys. Rev.* **D73** 035007 (*Preprint* hep-ph/0510421)
[51] Aaboud M *et al.* (ATLAS) 2016 (*Preprint* 1604.04263)
[52] Altmannshofer W and Straub D M 2015 *Eur. Phys. J.* **C75** 382 (*Preprint* 1411.3161)
[53] Descotes-Genon S, Hofer L, Matias J and Virto J 2015 (*Preprint* 1510.04239)
[54] Bobeth C, Buras A J, Celis A and Jung M 2016 (*Preprint* 1609.04783)
[55] Ishiwata K, Ligeti Z and Wise M B 2015 *JHEP* **10** 027 (*Preprint* 1506.03484)
[56] Altmannshofer W, Gori S, Pospelov M and Yavin I 2014 *Phys. Rev.* **D89** 095033 (*Preprint* 1403.1269)

IOP Conf. Series: Journal of Physics: Conf. Series **800** (2017) 012002 doi:10.1088/1742-6596/800/1/012002

LHCb searches for the strong CP-violating decays $\eta \to \pi\pi$ and $\eta' \to \pi\pi$

L Capriotti, on behalf of the LHCb Collaboration

School of Physics and Astronomy, The University of Manchester, Oxford Road, Manchester M13 9PL, United Kingdom

E-mail: lorenzo.capriotti@postgrad.manchester.ac.uk

Abstract. A search for the CP-violating strong decays $\eta \to \pi^+\pi^-$ and $\eta'(958) \to \pi^+\pi^-$ has been performed using approximately 2.5×10^7 events of each of the decays $D^+ \to \pi^+\pi^+\pi^-$ and $D_s^+ \to \pi^+\pi^+\pi^-$, recorded by the LHCb experiment. The data set corresponds to an integrated luminosity of $3.0\,\text{fb}^{-1}$ of pp collision data recorded during LHC Run 1 and $0.3\,\text{fb}^{-1}$ recorded in Run 2. No evidence is seen for $D^+_{(s)} \to \pi^+\eta^{(\prime)}$ with $\eta^{(\prime)} \to \pi^+\pi^-$, and upper limits at 90% confidence level are set on the branching fractions, $\mathcal{B}(\eta \to \pi^+\pi^-) < 1.6 \times 10^{-5}$ and $\mathcal{B}(\eta' \to \pi^+\pi^-) < 1.8 \times 10^{-5}$. The limit for the η decay is comparable with the existing one, while that for the η' is a factor of three smaller than the previous limit.

1. Introduction

The strength of *CP* violation in weak interactions in the quark sector is well below what would be required to serve as an explanation for the observed imbalance between the amounts of matter and antimatter in the universe. The QCD Lagrangian could contain a term, the θ term [1], that would give rise to *CP* violation in strong interactions; however, no strong *CP* violation has been observed. This apparent absence of *CP* violation in QCD is known as the "strong *CP* problem".

2. The strong *CP* problem

The strong *CP* problem arises from the structure of the degeneracy of the vacuum in QCD [2, 3]. The QCD vacuum has, in fact, a periodic structure; the true vacuum state, called θ-vacuum, is obtained as a superposition of all the vacuum states:

$$|\theta\rangle = e^{in\theta}\,|n\rangle\,.$$

When calculating vacuum to vacuum transitions, using the θ vacuum as true vacuum of the theory, an extra term in the QCD Lagrangian appears, called the "θ term",

$$\mathcal{L}_\theta = \theta \frac{g_s^2}{64\pi^2}\epsilon_{\mu\nu\alpha\beta}F_a^{\mu\nu}F_a^{\alpha\beta},$$

which violates *CP*. The experimental upper limit on the neutron electric dipole moment (nEDM) implies a limit $\theta \lesssim 10^{-10}$ [4]. The closeness of the value of θ to zero is seen as a fine-tuning problem of the theory.

Content from this work may be used under the terms of the Creative Commons Attribution 3.0 licence. Any further distribution of this work must maintain attribution to the author(s) and the title of the work, journal citation and DOI.
Published under licence by IOP Publishing Ltd

3. The $\eta^{(\prime)} \to \pi^+\pi^-$ decays

The decay modes $\eta \to \pi^+\pi^-$ and $\eta'(958) \to \pi^+\pi^-$ would both violate *CP* symmetry. In the Standard Model (SM) these decays could happen via the *CP*-violating weak interaction, through mediation by a virtual K^0_S meson, with expected branching fractions $\mathcal{B}(\eta \to \pi^+\pi^-) < 2 \times 10^{-27}$ and $\mathcal{B}(\eta' \to \pi^+\pi^-) < 4 \times 10^{-29}$ [5]. Based on the limit from the nEDM measurements, strong decays mediated by the θ term would have branching fractions below about 3×10^{-17} [5]. Any observation of larger branching fractions would indicate a new source of *CP* violation in the strong interaction. The current best limit for the $\eta \to \pi^+\pi^-$ decay mode, $\mathcal{B}(\eta \to \pi^+\pi^-) < 1.3 \times 10^{-5}$ at 90% confidence level (CL), is obtained from the KLOE experiment [6], which looked for $\eta \to \pi^+\pi^-$ in the decay $\phi(1020) \to \eta\gamma$. The limit for η', $\mathcal{B}(\eta' \to \pi^+\pi^-) < 5.5 \times 10^{-5}$ at 90% CL, is from the BESIII experiment [7], based on searches for $\eta' \to \pi^+\pi^-$ in radiative $J/\psi \to \eta'\gamma$ decays. In the study presented here, a new method is introduced to search for the decays $\eta \to \pi^+\pi^-$ and $\eta' \to \pi^+\pi^-$, exploiting the large sample of charm mesons collected by LHCb; any contribution would show as narrow peaks, consistent with mass resolution, in the inclusive $\pi^+\pi^-$ mass spectra.

4. Analysis strategy at LHCb

In the analysis [8], the decays $D^+ \to \pi^+\pi^+\pi^-$ and $D^+_s \to \pi^+\pi^+\pi^-$ are used to look for the presence of η and η' resonances in the $\pi^+\pi^-$ mass spectra, which could come from the known decays $D^+_{(s)} \to \pi^+\eta^{(\prime)}$ (inclusion of charge-conjugate modes is implied throughout). The data samples comprise about 25 million each of $D^+ \to \pi^+\pi^+\pi^-$ and $D^+_s \to \pi^+\pi^+\pi^-$ decays, from integrated luminosities of 3.0 fb^{-1} of *pp* collision data recorded by LHCb in LHC Run 1 and 0.3 fb^{-1} recorded in 2015 during Run 2. The data from Run 2 correspond to a 10% increase in integrated luminosity, but a 30% increase in number of signal candidates. They are processed through the Turbo stream, which performs an online reconstruction of candidates with offline quality, removing the need for any additional offline reconstruction [9].

For $N(\eta^{(\prime)})$ observed $\eta^{(\prime)}$ signal decays in the $\pi^+\pi^-$ mass spectrum from a total of $N(D^+_{(s)})$ mesons reconstructed in the $\pi^+\pi^+\pi^-$ final state, the measured branching fraction would be

$$\mathcal{B}\left(\eta^{(\prime)} \to \pi^+\pi^-\right) = \frac{N(\eta^{(\prime)})}{N(D^+_{(s)})} \times \frac{\mathcal{B}(D^+_{(s)} \to \pi^+\pi^+\pi^-)}{\mathcal{B}(D^+_{(s)} \to \pi^+\eta^{(\prime)})} \times \frac{1}{\epsilon(\eta^{(\prime)})}, \tag{1}$$

where $\epsilon(\eta^{(\prime)})$ is a small correction that accounts for any variation of efficiency with $\pi^+\pi^-$ mass. The values of $N(D^+_{(s)})$ and $N(\eta^{(\prime)})$ and their uncertainties are obtained from fits to the $\pi^+\pi^-\pi^+$ and $\pi^+\pi^-$ mass spectra of the selected $D^+_{(s)} \to \pi^+\pi^+\pi^-$ candidates; the branching fractions $\mathcal{B}(D^+_{(s)} \to \pi^+\pi^+\pi^-)$ and $\mathcal{B}(D^+_{(s)} \to \pi^+\eta^{(\prime)})$ and their uncertainties are taken from Ref. [10]; and the relative efficiency factors, ϵ, are obtained from simulations. Since the analysis starts from a given number of selected $D^+_{(s)} \to \pi^+\pi^+\pi^-$ decays, there are no normalisation channels.

5. Event selection

The event selection comprises an initial stage in which relatively loose criteria are applied to select samples of candidate $D^+_{(s)} \to \pi^+\pi^+\pi^-$ decays. A boosted decision tree (BDT) [11] is then used to further suppress backgrounds. Candidate $D^+_{(s)} \to \pi^+\pi^+\pi^-$ decays are required to have three good quality, high-p_T tracks consistent with coming from a vertex that is displaced from any primary vertex (PV) in the event, by typically 150 μm (300 μm) for D^+ (D^+_s). Loose particle identification criteria are applied, requiring the tracks to be consistent with the pion hypothesis. The three-track system is required to have total charge $\pm e$, its invariant mass must be in the range 1820–2020 MeV/c^2, and its combined momentum vector must be consistent with the direction from a PV to the decay vertex. The invariant mass of opposite-sign candidate pion pairs is required to be in the range 300–1650 MeV/c^2; this removes backgrounds where a random pion is associated with a vertex from either a $\gamma \to e^+e^-$ conversion, in which both electrons are misidentified as pions, or from a $D^0 \to K^-\pi^+$ decay, where the kaon is misidentified as a pion. The BDT has six input variables for each of the tracks, together with three variables related to the quality of the decay vertex and the association of the $D^+_{(s)}$ candidate with the PV. The track variables are related to track fit quality, particle identification probabilities and the quality of the track association to the decay vertex. The cut value of the BDT and the signal mass window are then selected by a bidimensional optimisation of the statistical significance of the $D^+_{(s)}$ signal. Figure 1 shows the $\pi^+\pi^+\pi^-$ mass spectra for Runs 1 and 2, before and after the BDT selection. The discontinuity in the Run 2 spectrum comes from the fact that the trigger has two separate output streams and there are different BDT cuts for D^+ and D^+_s. The yield per fb^{-1} is larger in Run 2 than in Run 1 by a factor 3.3, arising from the larger cross-section [12], and from a higher trigger efficiency for charm.

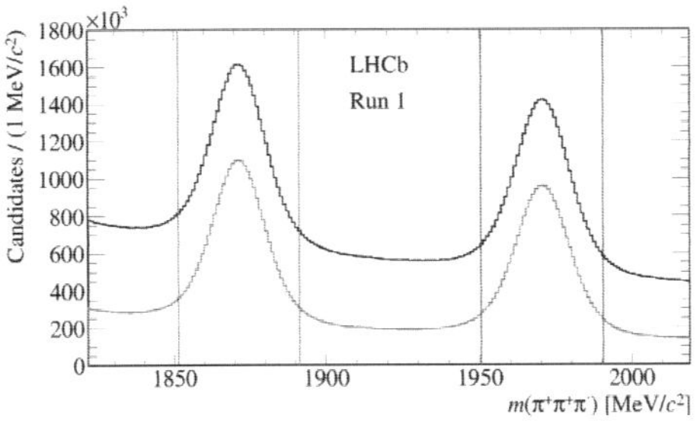

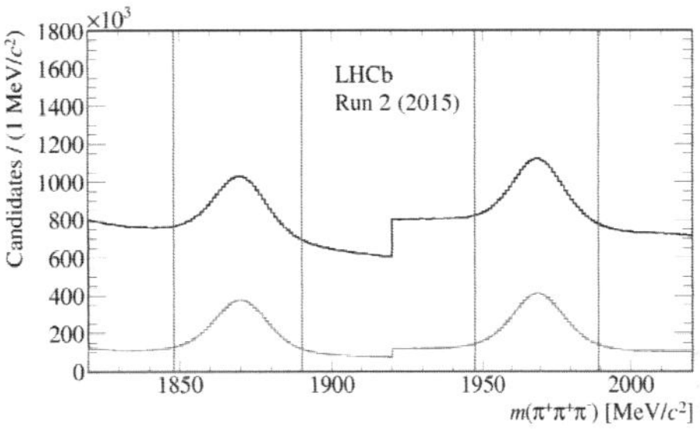

Figure 1: Three pions invariant mass before (black) and after (red) the BDT selection, for Run 1 (left) and Run 2 (right). The vertical lines define the optimised mass windows. Figures from [8].

6. Fits to the $\pi^+\pi^+\pi^-$ mass spectra

The BDT-selected events are fitted to extract the number of D^+ and D^+_s candidates. The background model comprises a fourth order Chebyshev polynomial for the combinatorial background and six shapes to account for sources of irreducible background, taken from high statistics simulations: one for $D^+_s \to K^+\pi^+\pi^-$, where the kaon is misidentified as a pion, one for $D^+_s \to \pi^+\pi^+\pi^-\pi^0$, where the π^0 is missing and four for $D^+_{(s)} \to (\eta^{(\prime)} \to \pi^+\pi^-\gamma)\pi^+$, where the photon is missing. The signal model is a sum of a Gaussian function and a double sided Crystal Ball function [13] for each of the two peaks. The fit results are shown in Figure 2. The total $D^+_{(s)} \to \pi^+\pi^+\pi^-$ signal yields in the optimised mass windows, summed over Run 1 and Run 2 data, are 2.49×10^7 for D^+ and 2.37×10^7 for D^+_s, with backgrounds of 1.38×10^7 and 1.08×10^7, respectively, within the same mass windows. Uncertainties of $\pm 2\%$ are assigned to each total yield to account for imperfections in the fits to the mass spectra.

7. Variation of efficiency as a function of $\pi^+\pi^-$ mass

The relative efficiency factors in Eq. 1 are obtained from simulation. Fully simulated $\pi^+\pi^-$ mass spectra from $D^+ \to \pi^+\pi^+\pi^-$ decays for Run 1 are divided by the generated spectra to give the relative efficiency as a function of the $\pi^+\pi^-$ mass. The efficiency is highest at large $\pi^+\pi^-$ masses, mainly due to the effects of the hardware and software triggers. The relative efficiencies in Run 1 data are found to be $\epsilon(\eta) = 0.85 \pm 0.01$ and $\epsilon(\eta') = 1.01 \pm 0.01$ for D^+ and $\epsilon(\eta) = 0.80 \pm 0.01$ and $\epsilon(\eta') = 1.03 \pm 0.01$ for D^+_s, where the uncertainties come from the simulation sample size. The relative efficiencies for Run 2 are found to be statistically compatible with those for Run 1, through a comparison of the $\pi^+\pi^-$ mass spectra from the D^+ and D^+_s signal candidates in the data. An additional systematic uncertainty of 2% is assigned to the Run 2 relative efficiencies, reflecting the statistical precision of the comparison.

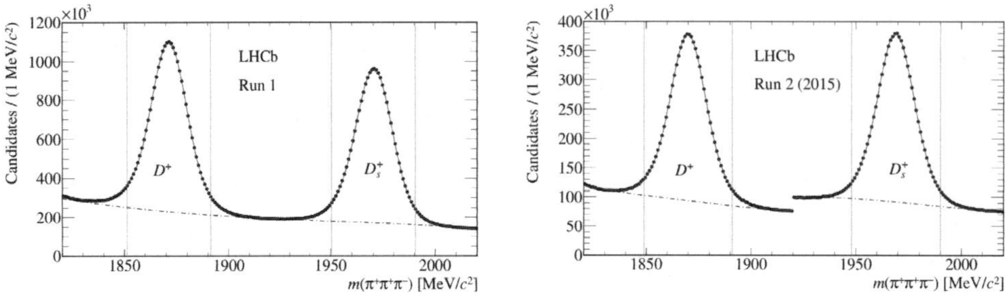

Figure 2: Fits to the $\pi^+\pi^+\pi^-$ invariant mass for Run 1 (left) and Run 2 (right). The black dashed line represents the sum of all background contributions. The vertical lines define the optimised mass windows. Figures from [8].

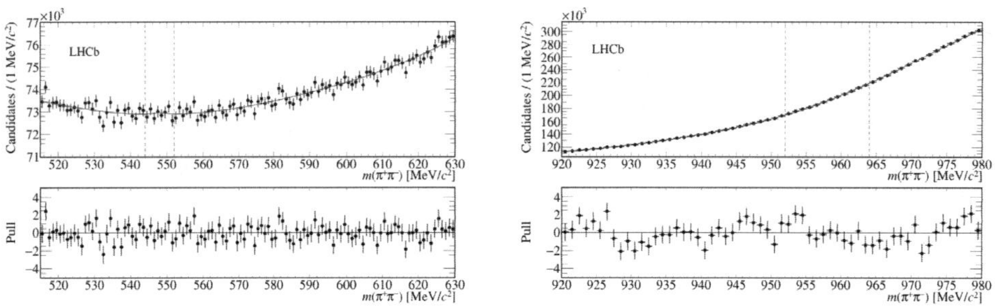

Figure 3: The $\pi^+\pi^-$ invariant mass distribution in the η (left) and η' (right) mass fitting region from the sum of the four samples, showing also the sum of the fitted curves and the pulls. The vertical dashed lines indicate the respective signal regions. Figures from [8].

8. Mass spectra for $\pi^+\pi^-$

For each of the η and η' resonances there are four separate $\pi^+\pi^-$ mass spectra, from the D^+ and the D_s^+ for each of Runs 1 and 2. Figure 3 shows the sums of the four $\pi^+\pi^-$ mass spectra for the η and η' mass fitting ranges. The vertical dashed lines indicate the signal regions, which cover the intervals 544–552 MeV/c^2 for the η and 952–964 MeV/c^2 for the η', in each case approximately ± 2 times the $\pi^+\pi^-$ mass resolution; they have been kept blind until the end of the analysis optimisation. The resolution is calculated from a dedicated $D^+ \to (\eta^{(\prime)} \to \pi^+\pi^-)\pi^+$ Monte Carlo sample, and cross-checked using $K_S^0 \to \pi^+\pi^-$ decays in data. The resolution from Monte Carlo and the one from data are found to be consistent and the difference is assigned as a systematic error. The mass resolutions are found to be $\sigma_\eta = 2.2 \pm 0.1$ MeV/c^2 and $\sigma_{\eta'} = 2.7 \pm 0.1$ MeV/c^2.

9. Systematic uncertainties

Several sources of systematic uncertainties are considered: errors on the branching fractions used in Equation 1, errors on the fitted numbers of D^+ and D_s^+ (dominated by the variation of the residuals), uncertainty in the variation of the relative efficiency, uncertainty in the extracted background PDF and in the dipion mass resolution. They are found to have no effect on the results, which are dominated by the statistical uncertainties; the systematic uncertainties together account for about 1% of the total uncertainty.

10. Limit setting

Since there is no evidence for any signal, the CL_s method [14] is used to obtain observed upper limits on the branching fraction. The signal PDFs are extracted by fitting a dedicated MC

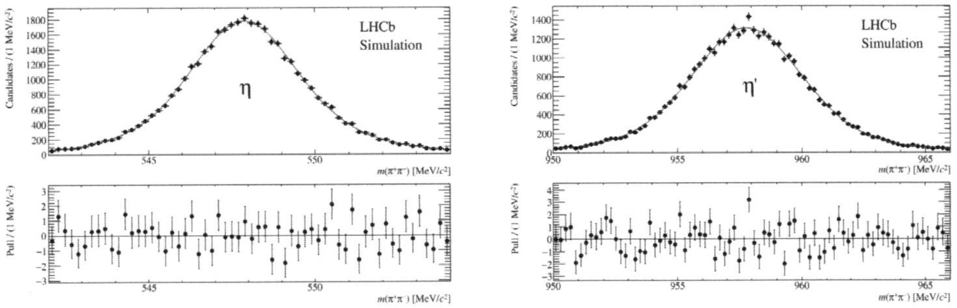

Figure 4: Mass spectra for $\eta \to \pi^+\pi^-$ (left) and $\eta' \to \pi^+\pi^-$ (right) from simulation, with the result of a fit to a double Gaussian function. Figures from [8].

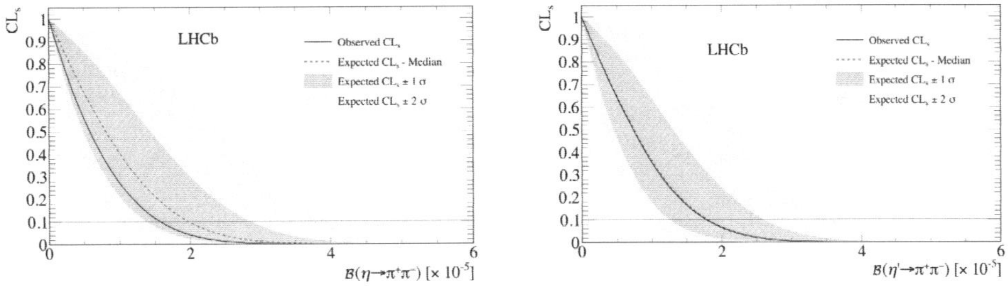

Figure 5: Values of CL$_s$ as function of $\mathcal{B}(\eta \to \pi^+\pi^-)$ (left) and $\mathcal{B}(\eta' \to \pi^+\pi^-)$ (right). The horizontal line indicates the 90% confidence level. Figures from [8].

sample, and are shown in Figure 4. The background PDFs come from the fits shown in Figure 3. The profile likelihood ratio is used as test statistic. The results of the CL$_s$ method are shown in Figure 5, along with the expected distribution. The observed limits at 90% CL are found to be $\mathcal{B}(\eta \to \pi^+\pi^-) < 1.6 \times 10^5$ and $\mathcal{B}(\eta' \to \pi^+\pi^-) < 1.8 \times 10^5$.

11. Conclusions

A new method is introduced to search for the decays $\eta \to \pi^+\pi^-$ and $\eta'(958) \to \pi^+\pi^-$, which would violate *CP* symmetry in the strong interaction. The method relies on the copious production of charm mesons at LHCb, and will improve in sensitivity as more data are collected at the LHC. With the LHC Run 1 data and data from the first year of Run 2, the limit obtained on the branching fraction for the decay $\eta \to \pi^+\pi^-$ is comparable to the existing limit, while that for $\eta' \to \pi^+\pi^-$ is a factor three better than the previous limit.

References

[1] H.-Y. Cheng, *The strong CP problem revisited*, Phys. Rept. **158** (1988) 1.
[2] R. Jackiw and C. Rebbi, *Vacuum periodicity in a yang-mills quantum theory*, Phys. Rev. Lett. **37** (1976) 172.
[3] C. G. Callan, Jr. R. F. Dashen, and D. J. Gross, *The Structure of the Gauge Theory Vacuum*, Phys. Lett. **B63** (1976) 334.
[4] J. Kuckei *et al.*, *Strong CP violation and the neutron electric dipole form factor*, Phys. Atom. Nucl. **70** (2007) 349, `arXiv:hep-ph/0510116`.
[5] C. Jarlskog and E. Shabalin, *How large are the rates of the CP violating* $\eta, \eta' \to \pi\pi$ *decays?*, Phys. Rev. **D52** (1995) 248.
[6] KLOE collaboration, F. Ambrosino *et al.*, *Upper limit on the* $\eta \to \pi^+\pi^-$ *branching ratio with the KLOE detector*, Phys. Lett. **B606** (2005) 276, `arXiv:hep-ex/0411030`.
[7] BESIII collaboration, M. Ablikim *et al.*, *Search for CP and P violating pseudoscalar decays into* $\pi\pi$, Phys. Rev. **D84** (2011) 032006, `arXiv:1106.5118`.

IOP Conf. Series: Journal of Physics: Conf. Series **800** (2017) 012002 doi:10.1088/1742-6596/800/1/012002

[8] LHCb collaboration, R. Aaij *et al.*, *Search for the CP-violating strong decays* $\eta \to \pi^+\pi^-$ *and* $\eta'(958) \to \pi^+\pi^-$, arXiv:1610.03666.

[9] R. Aaij *et al.*, *Tesla : an application for real-time data analysis in High Energy Physics*, Comput. Phys. Commun. (2016) , arXiv:1604.05596.

[10] Particle Data Group, K. A. Olive *et al.*, *Review of particle physics*, Chin. Phys. **C38** (2014) 090001, and 2015 update.

[11] B. P. Roe *et al.*, *Boosted decision trees as an alternative to artificial neural networks for particle identification*, Nucl. Instrum. Meth. **A543** (2005) 577, arXiv:physics/0408124.

[12] LHCb collaboration, R. Aaij *et al.*, *Measurements of prompt charm production cross-sections in pp collisions at* $\sqrt{s} = 13$ *TeV*, JHEP **03** (2016) 159, arXiv:1510.01707.

[13] T. Skwarnicki, *A study of the radiative cascade transitions between the Upsilon-prime and Upsilon resonances*, PhD thesis, Institute of Nuclear Physics, Krakow, 1986, DESY-F31-86-02.

[14] A. L. Read, *Presentation of search results: The* CL_s *technique*, J. Phys. G: Nucl. Part. Phys. **28** (2002) 2693.

Kaon 2016, 14 - 17 September 2016, Birmingham, UK

Status and Prospect of ϵ'/ϵ from a Lattice perspective

Nicolas Garron

Theoretical Physics Division, Department of Mathematical Sciences, University of Liverpool, Liverpool L69 7ZL, UK

E-mail: nicolas.garron@liverpool.ac.uk

Abstract. An ab-initio determination of the hadronic matrix elements relevant for a theoretical determination of $K \to \pi\pi$ decays has been a goal for the lattice QCD community for about three decades. These matrix elements are key to the theoretical determination of the CP violation parameters ε and ε'. This is not only relevant to challenge the Standard Model but also to provide stringent constraints on New-Physics scenarios. Although such a computation is very challenging, impressive results have been reported over the last few years. The most spectacular is the first realistic computation of ϵ'/ϵ with physical kinematics by RBC-UKQCD [1]. The novelty is the $\Delta I = 1/2$ channel, whereas the $\Delta I = 3/2$ contribution is now computed at several values of the lattice spacing and extrapolated to the continuum limit. After a short overview, I will report on the status and the prospect of direct and indirect violation in kaon decays from a lattice perspective.

Direct CP violation in kaon decays is a very small effect which is well measured experimentally, and as such provides a crucial test for the SM. Besides, since it comes mainly from virtual loops, one expects an important sensitivity to new physics. However the relevant weak matrix elements receive huge QCD contributions, which have to be computed non-perturbatively in a reliable way. This is why lattice QCD has a crucial role to play.

1. $K \to \pi\pi$ decays and CP violation

Various nice reviews are available on the subject, see for example [2]. I just recollect here some basic facts about $K \to \pi\pi$ phenomenology. Assuming isospin symmetry, the decays $K \to \pi\pi$ can be written in terms of the amplitudes $A\left[K \to (\pi\pi)_I\right] = A_I e^{i\delta_I}$, where I denotes the isospin of the two-pion state, either 0 or 2, and δ_I is the corresponding strong phase. The parameters of indirect (resp. direct) CP violation, ε (resp. ε') are given by

$$\varepsilon = \frac{A\left[K_L \to (\pi\pi)_0\right]}{A\left[K_S \to (\pi\pi)_0\right]}\,, \quad \frac{\varepsilon'}{\varepsilon} = \frac{1}{\sqrt{2}}\left(\frac{A\left[K_L \to (\pi\pi)_2\right]}{A\left[K_L \to (\pi\pi)_0\right]} - \frac{A\left[K_S \to (\pi\pi)_2\right]}{A\left[K_S \to (\pi\pi)_0\right]}\right)\,. \quad (1)$$

The first measurement of ε is the well-known discovery of indirect CP violation due to Christenson, Cronin, Fitch and Turlay [3] in 1964, for which Cronin and Fitch were awarded a Nobel prize. ε' has a long experimental history as it took tremendous

Content from this work may be used under the terms of the Creative Commons Attribution 3.0 licence. Any further distribution of this work must maintain attribution to the author(s) and the title of the work, journal citation and DOI.
Published under licence by IOP Publishing Ltd

efforts to measure direct CP violation. The final measurements are due to KTeV at Fermilab and NA48 at CERN [4, 5], the averages read

$$|\varepsilon| = 2.228(11) \times 10^{-3} , \qquad Re\left(\frac{\varepsilon'}{\varepsilon}\right) = 16.6(2.3) \times 10^{-4} . \tag{2}$$

In a theoretical approach, the standard framework to study $K \to \pi\pi$ decay is the $\Delta S = 1$ effective Hamiltonian obtained after integrating out the heavy degrees of freedom [6, 7]. In the three-flavour theory, it reads (see for example [8] for a review)

$$H_W = \frac{G_F}{\sqrt{2}} V_{us}^* V_{ud} \sum_{i=1}^{10} \left[z_i(\mu) + \tau y_i(\mu) \right] Q_i(\mu) . \tag{3}$$

2. Matching Lattice QCD with phenomenology

The physical amplitudes are given by $\langle\pi\pi|H_W|K\rangle$. In Eq. 3, the short-distance effects, computed in perturbation theory are factorised into the so-called Wilson coefficients, y_i, z_i, whereas V_{ij} are CKM matrix elements, $\tau = V_{ts}^* V_{td}/V_{us}^* V_{ud}$ and μ is an energy scale which can be thought as a cut-off. G_F is the Fermi constant. If we neglect the operators which emerge from the electric and magnetic dipole part of the electromagnetic and QCD penguins (see the talk by V.Lubicz at Lattice'14 and [9] for a recent lattice study by the ETM collaboration.), naively we find that there are ten four-quark operators Q_i which contribute. However, in four dimensions and in the three-flavour theory, they do not form a basis of the $\Delta S = 1$ four-quark operators as they are not linearly independent. Following for example [10], one can build a seven-operator basis, we denote them by Q_i'. Obtaining a reliable evaluation of the matrix elements $\langle\pi\pi|Q_i'|K\rangle$ is the most difficult part of the computation. Many attempts have been in the last decades using either effective theories or lattice simulations (or combinations of both), see for example [11, 12, 13, 14, 15, 16, 17] and reference therein.

Error propagation. The four-quark operators and the Wilson coefficients entering Eq 3 have to be computed in the same scheme and at the same scale. The Wilson coefficients are traditionally computed in $\overline{\mathrm{MS}}$ in the ten-operator pseudo-basis Q_i. The lattice results given [1] have been obtained in the seven-operator basis Q_i' and - for convenience - have been converted to the ten-operators pseudo-basis. However, one has to keep in mind that the errors given for the latter are obviously highly correlated, therefore a naive propagation of the lattice errors in quadrature will over-estimate the errors. Instead, since the hadronic errors dominate, I would advise to convert the Wilson coefficients to the basis used for the lattice computation. Eventually, when the determination of the hadronic matrix elements becomes more precise, one would have to take into account the correlations between the various operators: although the Q_i' are linearly independent in the algebraic sense, they are clearly correlated - mainly because the operators mix under renormalisation - but also because they come from the same Monte-Carlo simulation (they have been evaluated on the same gauge ensembles).

3. Lattice computation of $K \to \pi\pi$ amplitudes

From the lattice point of view, the computation of these amplitudes is very challenging for various reasons. First, chiral symmetry is crucial in order to prevent unwanted mixing with lower-dimension operators. Such a mixing comes with coefficients which diverge in inverse powers of the lattice spacing (eg $1/a^2$) and would need to be subtracted exactly before the continuum limit can be taken. However it is well known that maintaining both chiral and flavour symmetry at finite lattice spacing is numerically very expensive. With the RBC-UKQCD collaboration, we use the so-called Domain-Wall formulation, which respects chiral-flavour symmetry "almost exactly" but with numerical cost much higher than other discretisations of the Dirac operator. Furthermore, simulating light quarks is also very demanding, their propagators are difficult to compute and require large physical volumes , because finite volume artefacts are dominated by light pions. One possibility is to work on relatively coarse lattice lattice spacing relatively coarse. But if one wants to simulate finer lattices (and eventually take the continuum limit) one needs a lot of point in each space-time direction. Hence, simulating physical light quarks with a discretisation which preserves chiral-flavour symmetry is already a challenge by itself.

However, besides these technical difficulties, the main obstacle is to simulate the kinematic situation, in particular the final state made of two hadrons with non-vanishing momenta (because the phase-shift is not easily accessible from traditional Euclidean simulations, this is known as the Maiani-Testa no-go theorem [18]). The RBC-UKQCD collaboration follows the method proposed by Lellouch and Lüscher [19] which circumvent in a very elegant way this problem and allows us to access the physical amplitude. However, there is another technical problem: the physical state of interest appears as an excited state, which is much more difficult to extract because an excited state is statistically more noisy. Since we are already fighting with the noise in this project, we want to "eliminate" the unphysical states. We use two different techniques, depending on the isospin of the $\pi\pi$ state

- $K \to (\pi\pi)_2$, ie $\Delta I = 3/2$: taking advantage of the Wigner-Eckart theorem and using a peculiar choice of boundary condition, we compute the "unphysical process" $K^+ \to \pi^+\pi^+$, which is related to the physical one for pure QCD in the isospin limit,
- $K \to (\pi\pi)_{I=0}$, ie $\Delta I = 1/2$: we implement G-parity boundary conditions [20, 21], see for also [22, 23].

In practice this implies that for $\Delta I = 3/2$ we can use already existing gauge configurations, but for $\Delta I = 1/2$ we have to generate dedicated ensembles. $\Delta I = 3/2$ is also much easier for other reasons: only three operators contribute and there is no "disconnected diagram" (in which no quark line connects the initial kaon and the final two-pion states; these diagrams are numerically very demanding). In 2012, we performed the computation of the $\Delta I = 3/2$ channel [24, 25], with three flavours of dynamical quarks, physical kinematics and nearly-physical pion mass. However we had only one rather coarse lattice spacing $a \sim 1.4$ fm. Therefore, we renormalised the relevant four-quark operators at a rather low energy scale of $\mu \sim 1.4$ GeV and run non-perturbatively to $\mu = 3$ GeV ‡ using a step-scaling technique [27, 28] At $\mu = 3$ GeV, we match to $\overline{\rm MS}$ using perturbation theory, using [29]. Meanwhile there has been a very important breakthrough for Domain-Wall fermions, with a new formulation which preserves the same amount of chiral symmetry for a much lower cost [30]. Thanks to this formulation, the RBC-UKQCD collaboration could simulate new ensembles, still with three flavours and physical quark masses, but with finer lattice spacing (and

‡ The renormalisation is done with the Rome-Southampton method [26], where the renormalisation scale is the momentum given to the quarks. On the lattice, non-zero momenta imply discretisation artefacts, typically of order $(ap)^2$, therefore one keeps the scale below the cutoff: $\mu^2 = p^2 \ll (\pi/a)^2$.

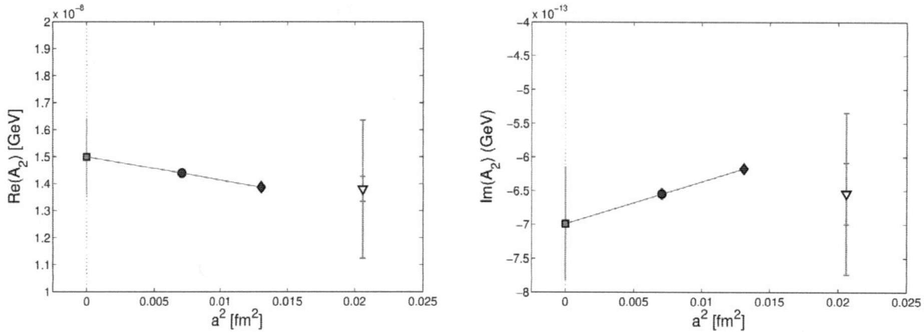

Figure 1. Real and Imaginary part of $A_2 = A[K \to (\pi\pi)]_{I=2}$. The triangle represents the 2012 computation on the coarse lattice and the blue points the 2014 determinations on the new ensembles (statistical error only), from which a continuum limit is extracted and shown in magenta (statistical and systematic errors combined). For the 2012 computation, we show both the statistical and the systematic error, largely dominated by the discretisation artefacts.

roughly the same physical volume) [31], namely $a \sim 0.11$ and $a \sim 0.84$ fm. In [32], these new lattices have been used to improve the determination of A_2: by performing a continuum extrapolation we were able to eliminate the main source of uncertainty of the previous computation, reducing the systematic error by roughly a factor 2 for the real part and a factor 1.5 for the imaginary part. Thanks to these new lattice determinations, the current errors on the theoretical determination of A_2 are of the order of 10% (see figure 1), we find :

$$Re(A_2) = 1.50(4)_{\rm stat}(14)_{\rm syst} \times 10^{-8}\text{GeV} \,, \tag{4}$$

$$Im(A_2) = -6.99(20)_{\rm stat}(84)_{\rm syst} \times 10^{-13}\text{GeV} \,. \tag{5}$$

The situation is more complicated for the $\Delta I = 1/2$ channel, as discussed above. First, to develop the numerical techniques needed to extract the signal, we did a pilot study with unphysical kinematics (at threshold and with heavy quark masses) but with all the diagrams included [33], so we had a first estimation for ImA_0. Recently we have reported on the first complete computation with physical quark masses and physical kinematics in [1]. From a more technical point of view, this computation requires the evaluation of all-to-all propagators and noise reduction techniques. As it is the first computation of this kind and because it requires new (and expensive) ensembles, the computation is performed at a single value of the lattice spacing, again at $a \sim 0.14$fm. The results read

$$Re(A_0) = 4.66(1.00)_{\rm stat}(1.21)_{\rm syst} \times 10^{-7}\text{GeV} \tag{6}$$

$$Im(A_0) = -1.90(1.23)_{\rm stat}(1.04)_{\rm syst} \times 10^{-11}\text{GeV} \tag{7}$$

and the corresponding theoretical value for ε'/ε $Re(\varepsilon'/\varepsilon) = 1.38(5.15)(4.43) \times 10^{-4}$, which is an approximate agreement($\sim 2.1\sigma$) with the experimental value. The final result is compatible with 0 but we show in [1] that we are able to extract a non-vanishing number for 6 of the 7 relevant operators. Rather than concluding that a significant deviation of the Standard Model prediction has been found, we note that the total error is much larger than the experimental one. From a phenomenological point of view, at this level of precision, these results do not invalidate the Standard Model, neither do they rule out the need for new-physics in $K \to \pi\pi$ decays. The important point is that for the first time ε'/ε has been computed with a full error budget, all the different contributions of the seven linearly independent operators are computed with controlled errors and with a precision which can be systematically improved. Furthermore, the individual contribution of each four-quark operator can be used in phenomenological applications. For example, the authors of [34] suggest to combine these contributions in a different way in the evaluation of ε'/ε in order to reduce the hadronic uncertainty, see also [35, 36, 37] at this conference. From a lattice perspective, now that the technology has been developed, reaching a precision of, say, 10% should be possible in the close future. In addition to reducing the statistical errors, the simulation can be done on finer lattices and extrapolated to the continuum limit, as done for A_2 (right now for A_0 we only have the points lying at the right-hand side of the plots shown in Fig. 1). Another dominant systematic error is due to the truncation of the perturbative series (needed to compute the Wilson coefficients). The renormalisation was performed at a scale of $\mu \sim 1.5$ GeV in order to keep the discretisation effects under control. Clearly this can be improved by running non-perturbatively to a higher scale, as done for the $\Delta I = 3/2$ channel, or again by performing the computation on finer lattice spacings. We should also improve on the normalisation of the weak matrix elements. In phenomenology it is a tradition to use the bag parameters, but it might not be the best choice since they are very sensitive to the quark mass and is not uniquely defined. An alternative could be to take the ratio of matrix elements, for example to divide $Q'_{i>1}$ by Q_1. On the lattice one expects some systematic errors to cancel in the ratios and milder continuum extrapolation. In the future we will also include the matrix elements of the four-quark operators which are not present in the SM but appear in generic BSM theories. It is also worth noting that another computation (done at threshold) has been done with Wilson fermions [38].

4. Conclusions and outlook

This is an exciting time for Kaon physics (see for example [39]); to a great extent this is due to the impressive progress achieved recently by the lattice community. The computation of $K \to (\pi\pi)_{I=2}$ is reaching a mature stage and a first computation $K \to (\pi\pi)_{I=0}$ with physical kinematic and complete error budget has recently been reported by the RBC-UKQCD collaboration. The results of these computations have an important role to play in particle physics phenomenology. Most of the $\Delta I = 1/2$ enhancement (the $\Delta I = 1/2$ puzzle) seems to come from non-perturbative effects [40]. Although this has to be confirmed with a precise computation of A_0, it certainly shows the importance of non-perturbative techniques. Regarding indirect CP violation, B_K is now known with an impressive precision. The various investigations of the $\Delta S = 2$ BSM operators are converging, the discrepancies observed by several collaborations are likely to be due to systematic errors affecting the non-perturbative renormalisation procedure in RI-MOM schemes [41]. Although a careful study is required, the solution could be provided by the SMOM schemes, which have a much better behaviour. Future improvements will also require the matching between lattice and phenomenology to be done at a much higher scale in order to decrease the error due to perturbation theory. I have presented here the new determinations of $K \to \pi\pi$ decay amplitudes and I have mentioned neutral kaon mixing matrix elements, but there are other new interesting and promising developments such as rare kaon decays and the $K_L - K_s$

mass difference (see [42] and [43]).

Acknowledgements- I would like to thank the organisers for a very pleasant conference and for inviting me to present this work. Most of the material I have presented is the result of the hard work done by RBC-UKQCD, it is a pleasure to thank my colleagues for such an enjoyable and fruitful collaboration. I acknowledge support from the Leverhulme Research grant RPG-2014-118.

[1] Z. Bai et al., (2015), 1505.07863.
[2] L. Lellouch, (2011) 629, 1104.5484.
[3] J.H. Christenson et al., Phys. Rev. Lett. 13 (1964) 138.
[4] NA48, V. Fanti et al., Phys. Lett. B465 (1999) 335, hep-ex/9909022.
[5] KTeV, A. Alavi-Harati et al., Phys. Rev. Lett. 83 (1999) 22, hep-ex/9905060.
[6] G. Buchalla, A.J. Buras and M.E. Lautenbacher, Rev. Mod. Phys. 68 (1996) 1125, hep-ph/9512380.
[7] M. Ciuchini et al., Nucl.Phys. B415 (1994) 403, hep-ph/9304257.
[8] V. Cirigliano et al., Rev. Mod. Phys. 84 (2012) 399, 1107.6001.
[9] M. Constantinou et al., PoS LATTICE2014 (2014) 390, 1412.1351.
[10] RBC Collaboration, T. Blum et al., Phys.Rev. D68 (2003) 114506, hep-lat/0110075.
[11] A.J. Buras, J.M. Grard and W.A. Bardeen, Eur. Phys. J. C74 (2014) 2871, 1401.1385.
[12] V. Cirigliano et al., Phys. Rev. Lett. 91 (2003) 162001, hep-ph/0307030.
[13] J. Bijnens and A. Celis, Phys. Lett. B680 (2009) 466, 0906.0302.
[14] J. Bijnens and J. Prades, JHEP 06 (2000) 035, hep-ph/0005189.
[15] T. Hambye, S. Peris and E. de Rafael, JHEP 05 (2003) 027, hep-ph/0305104.
[16] W.A. Bardeen, A.J. Buras and J.M. Gerard, Phys. Lett. B192 (1987) 138.
[17] E. de Rafael, Theoretical Advanced Study Institute in Elementary Particle Physics (TASI 94): CP Violation and the limits of the Standard Model Boulder, Colorado, May 29-June 24, 1994, 1995, hep-ph/9502254.
[18] L. Maiani and M. Testa, Phys. Lett. B245 (1990) 585.
[19] L. Lellouch and M. Luscher, Commun.Math.Phys. 219 (2001) 31, hep-lat/0003023.
[20] C.h. Kim and N.H. Christ, Nucl. Phys. Proc. Suppl. 119 (2003) 365, hep-lat/0210003, [,365(2002)].
[21] U.J. Wiese, Nucl. Phys. B375 (1992) 45.
[22] C. Kelly, LATTICE2015 (2015).
[23] A. Jüttner, LATTICE2015 (2015).
[24] T. Blum et al., Phys.Rev.Lett. 108 (2012) 141601, 1111.1699.
[25] T. Blum et al., Phys. Rev. D86 (2012) 074513, 1206.5142.
[26] G. Martinelli et al., Nucl. Phys. B445 (1995) 81, hep-lat/9411010.
[27] RBC and UKQCD Collaborations, R. Arthur et al., Phys.Rev. D85 (2012) 014501, 1109.1223.
[28] RBC Collaboration, UKQCD Collaboration, R. Arthur and P. Boyle, Phys.Rev. D83 (2011) 114511, 1006.0422.
[29] C. Lehner and C. Sturm, Phys.Rev. D84 (2011) 014001, 1104.4948.
[30] R.C. Brower, H. Neff and K. Orginos, (2012), 1206.5214.
[31] RBC, UKQCD, T. Blum et al., (2014), 1411.7017.
[32] T. Blum et al., Phys. Rev. D91 (2015) 074502, 1502.00263.
[33] T. Blum et al., (2011), 1106.2714.
[34] A.J. Buras et al., JHEP 11 (2015) 202, 1507.06345.
[35] A.J. Buras, KAON2016 (2016).
[36] M.C. Sevilla, KAON2016 (2016).
[37] S. Jaeger, KAON2016 (2016).
[38] N. Ishizuka et al., Phys. Rev. D92 (2015) 074503, 1505.05289.
[39] A.J. Buras, Proceedings, 2015 European Physical Society Conference on High Energy Physics (EPS-HEP 2015), 2015, 1510.00128.
[40] RBC, UKQCD, P.A. Boyle et al., Phys. Rev. Lett. 110 (2013) 152001, 1212.1474.
[41] N. Garron, R.J. Hudspith and A.T. Lytle, (2016), 1609.03334.
[42] RBC, UKQCD, N.H. Christ et al., Phys. Rev. D92 (2015) 094512, 1507.03094.
[43] N.H. Christ et al., Phys. Rev. D91 (2015) 114510, 1504.01170.

Recent results on strangeness from ALICE at LHC

Benjamin Dönigus (for the ALICE Collaboration)

Institut für Kernphysik, Goethe-Universität Frankfurt, Max-von-Laue-Str. 1, 60438 Frankfurt, Germany

E-mail: b.doenigus@gsi.de

Abstract. Recent measurements performed in high-multiplicity proton-proton (pp) and proton-lead (p-Pb) collisions have shown features that are similar to those observed in lead-lead (Pb-Pb) collisions. These observations warrant a comprehensive measurement of the production of identified particles as a function of multiplicity in all systems. The production of different strange and multi-strange particle species at mid-rapidity measured as a function of multiplicity in pp collisions at $\sqrt{s}$ = 7 TeV with the ALICE setup are conducted. Spectral shapes studied both for individual particles and via particle ratios such as (Λ/K^0_S) as a function of p_T exhibit an evolution with event multiplicity and the production rates of hyperons are observed to increase more strongly than those of non-strange hadrons. These phenomena are qualitatively similar to the ones observed in p-Pb and Pb-Pb collisions. The values in high-multiplicity pp and p–Pb collisions approach the ones in Pb–Pb.

1. Introduction

Ultra-relatvistic heavy-ion collisions at the LHC offer a unique way to study QCD matter at very high temperatures. Lattice QCD calculations [1,2] suggest that a transition from confined hadrons to a state of deconfined matter, the so-called quark-gluon plasma, is happening at temperatures above $T_c = (154\pm9)$ MeV and/or energy densities above $\epsilon_c = 0.34\pm0.16$ GeV/fm^3. These are clearly reached in these collisions as one can deduce for instance from the measurement of direct photon transverse momentum spectra. These lead to effective temperatures of $T_{eff} = (297 \pm 12(stat) \pm 41(syst))$ MeV averaged over the collision, extracted through the slope of the spectra [3]. Comparisions to models lead to initial temperatures of up to 740 MeV [4].

The evolution of the collisions themselves are imagined usually as a sequence of the following stages: two Lorentz contracted nuclei approach each other, they collide, and after a short time (less than 1fm/c) a quark-gluon plasma is formed which eventually expands, cools down and hadronizes. Finally the hadrons rescatter and freeze out. The temperature where the hadrons stop being produced is called chemical freeze-out temperature T_{ch} and the temperature when the hadrons stop scattering is denoted as kinetic freeze-out temperature T_{fo}.

If one uses an analogy between a light source and a particle source one can extract also a temperature from the measured multiplicities of the different particle species. This is possible using an approach based on a grand canonical ensemble with the main ingredients: (chemical freeze-out) temperature T_{ch}, volume V and baryo-chemical potential μ_B. The latter is basically zero at the LHC, namely baryons and anti-baryons are produced with equal amounts [5]. This approach is called (statistical) thermal model and the measurement of the production yield of different particle species, such as π, K, p, etc. can be used to extract a temperature of about 156 MeV at the LHC [6]. If this is done for the different available energies at different

Content from this work may be used under the terms of the Creative Commons Attribution 3.0 licence. Any further distribution of this work must maintain attribution to the author(s) and the title of the work, journal citation and DOI.
Published under licence by IOP Publishing Ltd

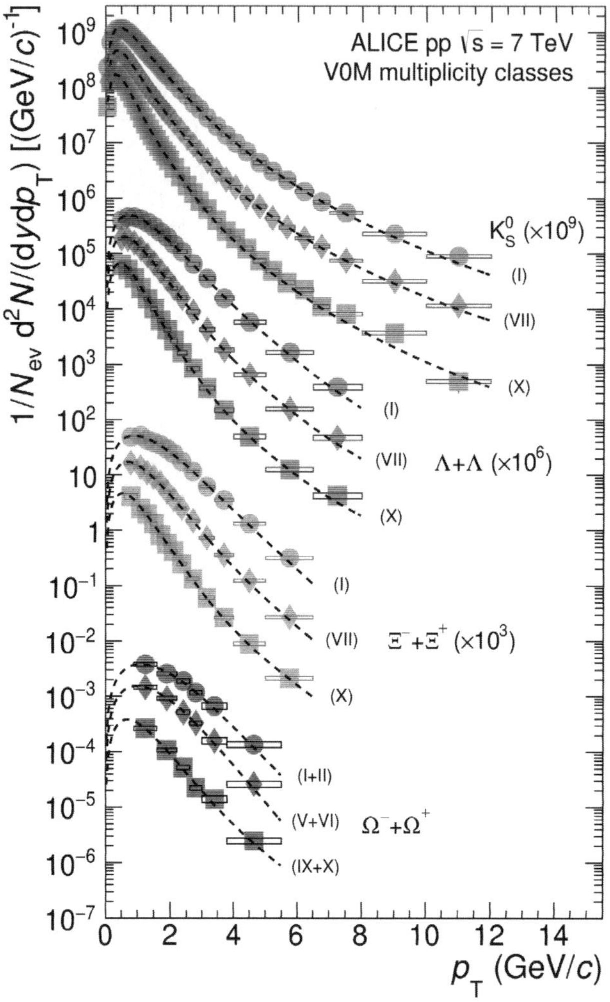

Figure 1. p_T-differential yields of K^0_S, $\Lambda + \overline{\Lambda}$, $\Xi^- + \overline{\Xi}^+$ and $\Omega^- + \overline{\Omega}^+$ measured in $|y| < 0.5$ for a selection of event classes, indicated by roman numbers in brackets, where I denotes the highest and X the lowest multiplicity. The data are scaled by different factors to improve the visibility. The dashed curves represent Tsallis-Lévy fits to each individual distribution to extract integrated yields. Figure taken from [15].

laboratories one sees that there is an increase up to energies reached at the SPS at CERN and then the temperature stays constant. This is another hint that the quark-gluon plasma is formed around a temperature of 159 MeV, which is the average of the extracted temperatures in the aforementioned plateau (T_{lim}). This leads to the assumption $T_{lim} \approx T_{ch} \approx T_c$ [7].

Before the observation of this interesting fact many signatures of the quark-gluon plasma have been proposed. One of the earliest ones was the possible enhancement of strangeness in ultra-relativistic heavy-ion collisions, relative to production in elementary collisions [8,9]. This was observed at the SPS [10] but is nowadays often interpreted in the opposite direction, as a lifting of the suppression when moving from elementary collisions such as pp to Pb–Pb for instance [11,12]. This can be understood also as the necessity to treat strangeness canonically and not as introduced before in a grand canonical ensemble [13,14].

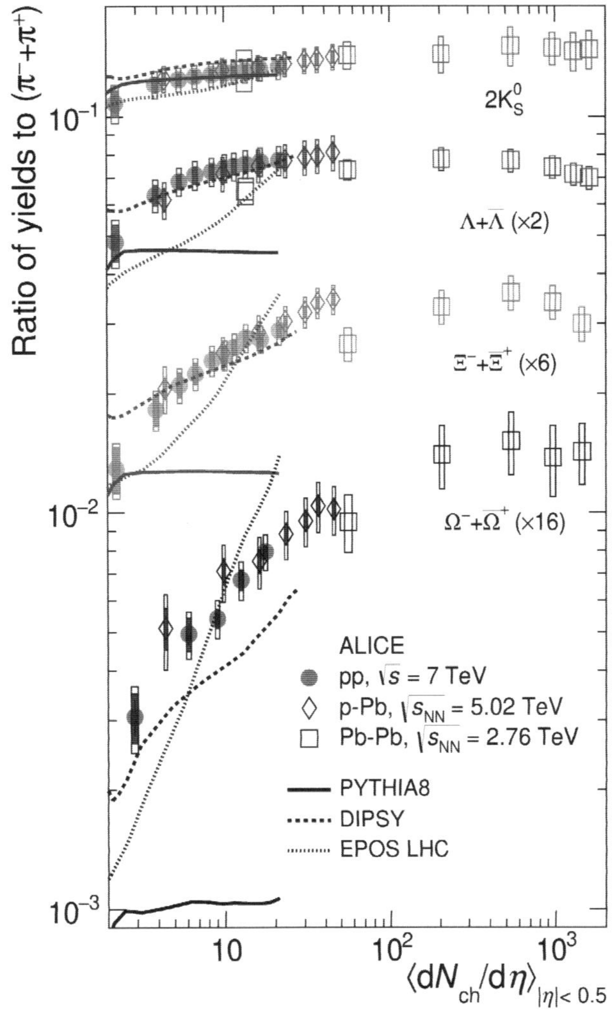

Figure 2. p_T-integrated yield ratios of strange and multi-strange hadrons to $(\pi^+ + \pi^-)$ as a function of $\langle dN_{ch}/d\eta \rangle$ measured in the rapidity interval $|y| < 0.5$ The empty and dark-shaded boxes show the total systematic uncertainty and the contribution uncorrelated across multiplicity bins, respectively The values are compared to calculations from MC models [17,18,19] and to results obtained in Pb–Pb and p–Pb collisions at the LHC . Figure from [16].

Measurement of strangeness production in heavy-ion collisions is nevertheless of high importance not only because the yields are needed for the determination of the chemical freeze-out temperature but also because they revealed some interesting points when the measurements where extended towards p–Pb collisions and in particular in multiplicity bins [15]. There it was observed that the ratios of hyperon production yields to those of charged pions increase as a function of multiplicity until the thermal model expectation is reached, nameley the ratios in high multiplicity p–Pb approaching those in Pb–Pb. As shown in [15], this trend is qualitatively reproduced within a thermal model with additional local conservation of strangeness. The high statistics and precision data taken by the ALICE Collaboration in pp collisions at $\sqrt{s} = 7$ TeV allows also the splitting in multiplicity classes and thus a comparison with the p–Pb and Pb–Pb measurements.

2. Results

The measured transverse momentum spectra for K^0_S, $\Lambda+\overline{\Lambda}$, $\Xi^-+\overline{\Xi}^+$ and $\Omega^-+\overline{\Omega}^+$ are depicted in Figure 1 for three different exemplary multiplicity classes. The multiplicity classes are numbered using roman figures from I to X, whereas I denotes the highest multiplicity and X the lowest (for details see [16]). One can already see here that the spectra become harder at higher multiplicities. A separate study shows in addition that the hardening is more pronounced for baryons than for mesons.

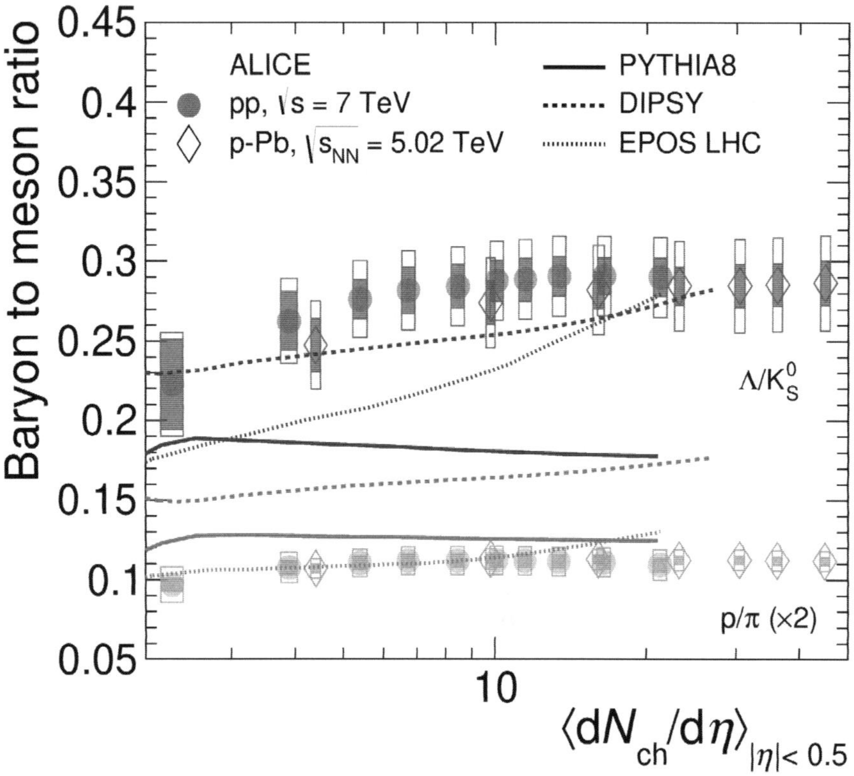

Figure 3. Particle production yield ratios $\Lambda/K^0_S = (\Lambda+\overline{\Lambda})/2K^0_S$ and $p/\pi = (p+\overline{p})/(\pi^+ + \pi^-)$ as a function of $\langle dN_{ch}/d\eta \rangle$ measured in the rapidity interval $|y| < 0.5$ The empty and dark-shaded boxes show the total systematic uncertainty and the contribution uncorrelated across multiplicity bins, respectively. The values are compared to calculations from MC models [17,18,19] in pp collisions at $\sqrt{s} = 7$ TeV and to results obtained in p–Pb collisions at the LHC. Figure taken from [16], where also more details can be found.

One can now extract the integrated production yield for a given multiplicity bin and build the ratio to the yield of charged pions to estimate the relative strangeness production. This is shown as a function of the mean charged particle multiplicity $\langle dN_{ch}/d\eta \rangle$ in Figure 2. Here a significant enhancement of strange and multi-strange particle production can be seen, where the rise depends strongly on the strangeness content. In addition one can see that no MC prediction describes the observed trend satisfactorily, whereas the pp data follows the trend observed in p–Pb nicely and this is not expected because both systems are expected to have different initial states. In particular it is visible that the particle ratios reach values that are similar to those observed in central Pb–Pb collisions.

To cancel the effect of strangeness the baryon to meson ratio is build as displayed for Λ/K^0_S and p/π in Figure 3. No increase for protons (non-strange) is observed, which is contrary to model predictions, e.g. DIPSY [18]. The ratios do not change significantly as a function of multiplicity, which shows that the observed enhanced production rates of strange hadrons with respect to pions is not due to the difference in the hadron masses but clearly connected to the strangeness content.

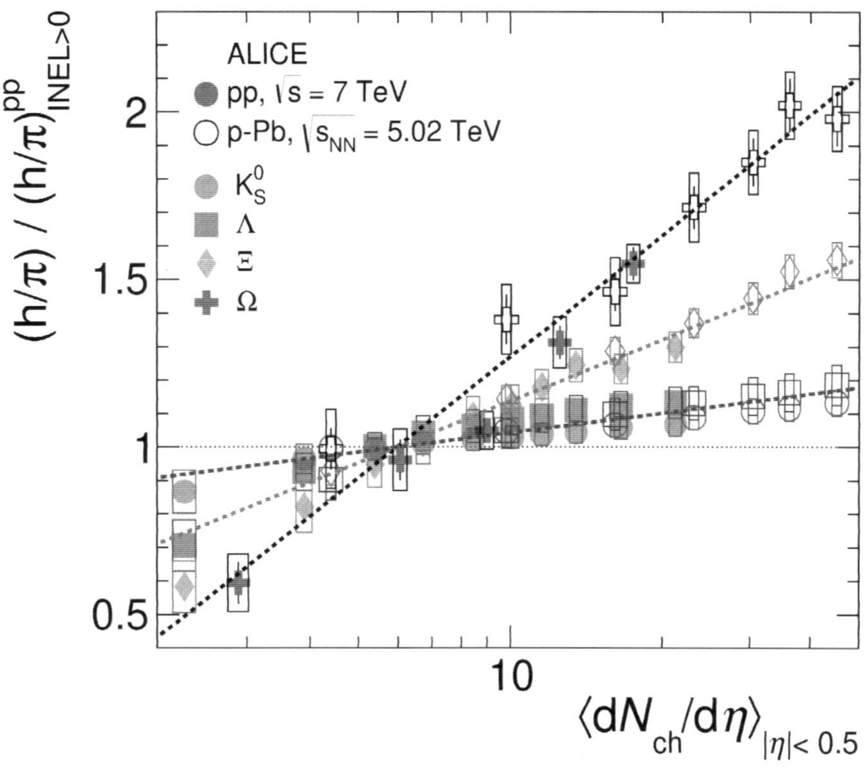

Figure 4. Particle production yield ratios of strange and multi-strange hadrons to pions normalised to the values measured in the inclusive INEL>0 pp sample, both in pp and in p–Pb collisions. The common systematic uncertainties cancel in the double-ratio. The empty boxes represent the remaining uncorrelated uncertainties. The lines represent a simultaneous fit of the results with the empirical scaling formula as discussed in [16] where also this figure is taken from.

The dynamical evolution is illustrated in Figure 4 as the ratio to pions for the previously shown data normalised by values measured in the inclusive INEL>0 pp sample, both for pp and p–Pb results. INEL>0 denotes here the requirement that the analysed events contain at least one charged particle produced with $p_T > 0$ in the pseudorapidity interval $|\eta| < 1$. The observed multiplicity-dependent enhancement with respect to the INEL>0 sample follows a hierarchy connected to the strangeness content of the hadrons. The observed increase is clearly more pronounced for baryons with higher strangeness content.

3. Conclusion

The presented multiplicity dependence of the production of primary strange (K^0_S, Λ, $\overline{\Lambda}$) and multi-strange (Ξ^-, $\overline{\Xi}^+$, Ω^-, $\overline{\Omega}^+$) hadrons in pp collisions at $\sqrt{s}$ = 7 TeV shows clearly that the smaller systems as pp and p–Pb reveal interesting results and deserve more attention. The measured p_T spectra become harder as the multiplicity increases. The spectral shapes as well as the multiplicity and mass dependence are showing similar behaviour as observed in p–Pb and Pb–Pb collisions at the LHC, this can be interpreted as a collective expansion of the system in the final state. The p_T-integrated yields of strange and multi-strange particles relative to pions increase significantly with multiplicity. There is a remarkable consistency across systems of this observable as a function of multiplicity. The observed enhancement increases with strangeness content rather than with mass or baryon number of the hadron and this trend can not be reproduced by the investigated MC generators. Further studies extending to higher multiplicity in small systems are essential, as they would give inside to the question if strangeness production saturates at the thermal equilibrium values such as predicted by the grand canonical statistical model or if it continues increasing. In addition it will be interesting to see if in pp collisions at higher energy, i.e. $\sqrt{s}$ = 13 TeV, the same trends will be observed.

4. Acknowledgments

The author acknowledges the support from the German Bundesministerium fur Bildung, Wissenschaft, Forschung und Technologie (BMBF) though the project grant 05P2015 - ALICE at High Rate (BMBF-FSP202).

5. References

[1] Bazavov A et al. (hotQCD), 2014 *Phys. Rev. D* **90** 094503
[2] Borsanyi S et al. (Budapest-Wuppertal group), JHEP 09 (2010) 073
[3] Adam J et al. (ALICE Collaboration), 2015 *Phys. Lett. B* **754** 235
[4] Chatterjee R et al., 2013 *Phys. Rev. C* **88** 034901
[5] Andronic A et al., 2011 *J. Phys. G* **38** 124081
[6] Floris M, 2014 *Nucl. Phys. A* **931** 103
[7] Andronic A et al., 2009 *Phys. Lett. B* **673** 142, erratum ibid 2009 **678** 516
[8] Rafelski J and Müller B, 1982 *Phys. Rev. Lett.* **48** 1066, erratum ibid 1986 **56** 2334
[9] Müller B, Koch P and Rafelski J, 1986 *Phys. Rept.* **142** 167
[10] Antinori F et al., 2006 *J. Phys. G* **32** 427
[11] Blume C and Markert C, 2011 *Prog. Part. Nucl. Phys.* **66** 834
[12] Hamieh S, Redlich K and A. Tounsi, 2000 *Phys. Lett. B* **486** 61
[13] Redlich K and Tounsi A, 2002 *Eur. Phys. J. C* **24** 589
[14] Tounsi A, Mischke A and Redlich K, 2003 *Nucl. Phys. A* **715** 565c
[15] Adam J et al. (ALICE Collaboration), 2016 *Phys. Lett. B* **758** 389
[16] Adam J et al. (ALICE Collaboration), 2016 arXiv:1606.07424
[17] Sjöstrand T et al., 2008 *Comput. Phys. Commun.* **178** 852
[18] Flensburg C et al., 2011 *JHEP* **08** 103
[19] Pierog T et al., 2015 *Phys. Rev. C* **92** 034906

IOP Conf. Series: Journal of Physics: Conf. Series **800** (2017) 012005 doi:10.1088/1742-6596/800/1/012005

Electromagnetic Corrections to Hadronic Decays from Lattice QCD

V. Lubicz[a,b], G. Martinelli[c,d], C.T. Sachrajda[e], F. Sanfilippo[e], S. Simula[f], N. Tantalo[g]

[a]Dipartimento di Fisica, Università Roma Tre, Via della Vasca Navale 84, I-00146 Rome, Italy, [b]INFN, Sezione di Roma Tre, Via della Vasca Navale 84, I-00146 Rome, Italy, [c]Physics Department and INFN Sezione di Roma La Sapienza Piazzale Aldo Moro 5, 00185 Roma, Italy, [d]TH Department, CERN, CH-1211, Geneva 23, Switzerland,[e]Department of Physics and Astronomy, University of Southampton, Southampton SO17 1BJ, UK, [f]Dipartimento di Fisica and INFN, Università di Roma "Tor Vergata", Via della Ricerca Scientifica 1, I-00133

E-mail: guido.martinelli@roma1.infn.it

Abstract. A new method, recently introduced for the lattice calculation of electromagnetic and isospin corrections to weak decays [1], is discussed. Using this method, for the first time, the electromagnetic effects in the leptonic decay rates $\pi^- \to \mu^- \bar{\nu}_\mu$ and $K^- \to \mu^- \bar{\nu}_\mu$ have been evaluated in lattice QCD. Preliminary results for the electromagnetic corrections to charged (neutral) pion and kaon masses and leptonic decay rates are presented.

1. Introduction

In Flavor Physics an accurate determination of the parameters of the Cabibbo-Kobayashi-Maskawa (CKM) matrix represents a crucial test of the Standard Model (SM). Moreover, improving the accuracy on the CKM parameters is at the heart of many searches for New Physics (NP), where we look for rather small effects. The Unitarity Triangle (UT) analysis, for example, allows a rather precise determination of the Wolfenstein parameters $\bar{\rho}$ and $\bar{\eta}$, obtained from a plethora of flavour measurements. This analysis has been performed by several groups, among which the **UT***fit* collaboration, see for example `http://www.utfit.org` (the results presented here are an update on the "Summer 2016" analysis and they will appear in the web-page as "Winter 2017"). The experimental measurements are mostly taken from `http://www.slac.stanford.edu/xorg/hfag`, while the non-perturbative QCD parameters come from the most recent lattice QCD averages [2]. Using the above inputs a global fit gives $\bar{\rho} = 0.153 \pm 0.013$ and $\bar{\eta} = 0.343 \pm 0.011$, see Figure 1.

For many physical quantities relevant for studies of flavour physics, recent improvements in lattice computations have led to such a precision that electromagnetic effects and isospin breaking contributions cannot be neglected anymore (see e.g. Ref. [2] and references therein). For light-quark flavours, important examples include the calculations of the leptonic decay constants f_K and f_π and of the form factor $f^+(0)$ in semileptonic $K_{\ell 3}$ decays. These are used to determine the CKM matrix element $|V_{us}|$ and the ratio $|V_{us}|/|V_{ud}|$ at high precision. For such quantities, which have been computed with a precision at the sub-percent level, the uncertainty due to the explicit breaking of isospin symmetry (of the order of $(m_u - m_d)/\Lambda_{QCD} \sim 0.01$) and to

Content from this work may be used under the terms of the Creative Commons Attribution 3.0 licence. Any further distribution of this work must maintain attribution to the author(s) and the title of the work, journal citation and DOI.
Published under licence by IOP Publishing Ltd

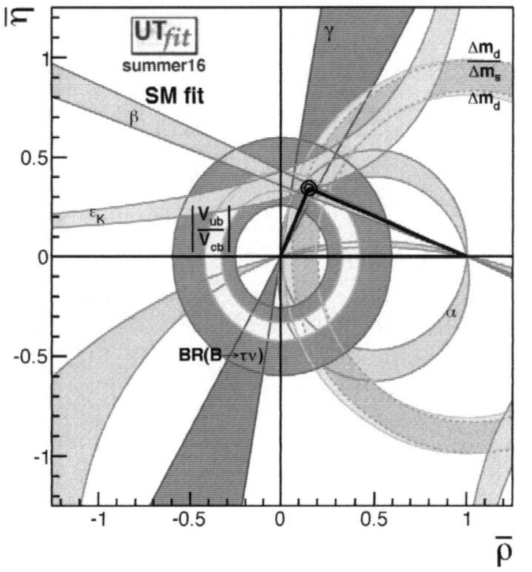

Figure 1: *$\bar{\rho}-\bar{\eta}$ plane showing the result of the SM fit. The black contours display the 68% and 95% probability regions selected by the given global fit. The 95% probability regions selected by the single constraints are also shown.*

electromagnetic corrections (of the order of $\alpha \sim 0.007$) is similar to, or even larger than, the quoted QCD errors [2]. Several collaborations have recently obtained remarkably accurate results in the calculation of the electromagnetic effects in the hadron spectrum and in the determination of quark masses with ab-initio lattice calculations see [3]-[5] for reviews on the subject.

In a recent paper, a new proposal to include electromagnetic and isospin-breaking effects in the non-perturbative calculation of hadronic decays was presented [1] and fully developed in [6]. As an example of the new method, the procedure to compute $O(\alpha)$ corrections to leptonic decays of pseudoscalar mesons was described in detail. This can then be used to determine the corresponding CKM matrix elements.

Whereas in the calculation of the hadron spectrum there are no infrared divergences, in the case of electromagnetic corrections to the hadronic amplitudes the infrared divergences are present and only cancel for well defined, measurable physical quantities. This requires the development of a new strategy, different than the usual approaches followed to compute the electromagnetic corrections to the spectrum. We proposed such a strategy in Ref. [1]. There we envisaged that at $O(\alpha)$ the physical observable is the inclusive decay rate of the pseudoscalar meson into a final state consisting of either $\ell^- \bar{\nu}_\ell$ or $\ell^- \bar{\nu}_\ell \gamma$, with the energy of the emitted photon in the rest frame of the pion smaller than an imposed cut-off ΔE. Here ℓ^- is a charged lepton and ν_ℓ the corresponding neutrino. The cut-off ΔE on the energy of the final-state (real) photon should be sufficiently small that for photons with such an energy we can neglect the structure of the meson and treat it as an elementary point-like particle, neglecting the structure-dependent corrections of $O(\alpha\,\Delta E/\Lambda_{\rm QCD})$. At $O(\alpha)$ the inclusive width can be written in the form

$$\Gamma(\Delta E) = \Gamma_0^{\rm tree} + \frac{\alpha}{4\pi}\lim_{L\to\infty}\Big(\Gamma_0(L) + \Gamma_1^{\rm pt}(\Delta E, L)\Big)\,, \tag{1}$$

where the suffix 0 or 1 indicates the number of photons in the final state; $\Gamma_0^{\rm tree}$ is the rate at $O(\alpha^0)$ given in Eq. (9) below; the superscript "pt" on Γ_1 denotes *point-like* and we have exhibited the dependence on L, the spacial extent of the box in which the lattice calculation is to be performed ($V = L^3$). It is now convenient to write

$$\lim_{L\to\infty}\Big(\Gamma_0(L) + \Gamma_1^{\rm pt}(\Delta E, L)\Big) = \lim_{L\to\infty}\Big(\Gamma_0(L) - \Gamma_0^{\rm pt}(L)\Big) + \lim_{L\to\infty}\Big(\Gamma_0^{\rm pt}(L) + \Gamma_1^{\rm pt}(\Delta E, L)\Big)\,. \tag{2}$$

The second term on the right-hand side of Eq. (2),

$$\Gamma^{\rm pt}(\Delta E) = \lim_{L\to\infty}\Big(\Gamma_0^{\rm pt}(L) + \Gamma_1^{\rm pt}(\Delta E, L)\Big) \tag{3}$$

can be evaluated in perturbation theory directly in infinite volume and the result has been presented in Ref. [1]. $\Gamma^{\rm pt}(\Delta E)$ is infrared finite and independent of the scheme used to regulate the divergences which are present separately in $\Gamma_0^{\rm pt}(L)$ and $\Gamma_1^{\rm pt}(\Delta E, L)$; its explicit expression is given by

$$\begin{aligned}\Gamma^{\rm pt}(\Delta E) \;&=\; \Gamma_0^{\rm tree}\times\left(1 + \frac{\alpha}{4\pi}\left\{3\log\left(\frac{m_P^2}{M_W^2}\right) + \log\left(r_\ell^2\right) - 4\log(r_E^2)\right.\right. \\ &+\frac{2-10r_\ell^2}{1-r_\ell^2}\log(r_\ell^2) - 2\frac{1+r_\ell^2}{1-r_\ell^2}\log(r_E^2)\log(r_\ell^2) - 4\frac{1+r_\ell^2}{1-r_\ell^2}{\rm Li}_2(1-r_\ell^2) - 3 \\ &+\left[\frac{3+r_E^2-6r_\ell^2+4r_E(-1+r_\ell^2)}{(1-r_\ell^2)^2}\log(1-r_E) + \frac{r_E(4-r_E-4r_\ell^2)}{(1-r_\ell^2)^2}\log(r_\ell^2)\right. \\ &\left.\left.\left. - \frac{r_E(-22+3r_E+28r_\ell^2)}{2(1-r_\ell^2)^2} - 4\frac{1+r_\ell^2}{1-r_\ell^2}{\rm Li}_2(r_E)\right]\right\}\right)\,,\end{aligned} \tag{4}$$

where $r_\ell = m_\ell/m_P$, $r_E = 2\Delta E/m_P$ and $0 \le r_E \le 1 - r_\ell^2$. $\Gamma_0(L)$ is infrared divergent and depends on the infrared regularisation. Since all momentum modes of the virtual photon contribute to Γ_0, it depends on the structure of the meson and is necessarily non-perturbative. It should therefore be computed in a lattice simulation. In Ref. [1] we stressed that the infrared divergence cancels in the difference $\Gamma_0(L) - \Gamma_0^{\rm pt}(L)$. In Ref. [6] we showed that the $1/L$ finite-volume (FV) corrections are also *universal*, that is they are independent of the structure of the pseudoscalar meson and hence cancel in the difference $\Gamma_0(L) - \Gamma_0^{\rm pt}(L)$. This allows us to calculate $\Gamma_0^{\rm pt}(L)$ in perturbation theory with a point-like pseudoscalar meson up to and including the $1/L$ corrections expanded in inverse powers of L

$$\Gamma_0^{\rm pt}(L) = C_0(r_\ell) + \tilde{C}_0(r_\ell)\log\left(m_P L\right) + \frac{C_1(r_\ell)}{m_P L} + \ldots\,, \tag{5}$$

m_P and m_ℓ are the masses of the pseudoscalar meson and the lepton respectively. The explicit expression for $\Gamma_0^{\rm pt}(L)$ has the form

$$\Gamma_0^{\rm pt}(L) = \Gamma_0^{\rm tree}\left\{1 + 2\frac{\alpha}{4\pi}Y(L)\right\}\,, \tag{6}$$

where

$$Y(L) = \left(1+r_\ell^2\right)\left[2(K_{31}+K_{32}) + \frac{\left(\gamma_E + \log\left[\frac{L^2 m_P^2}{4\pi}\right]\right)\log\left[r_\ell^2\right]}{\left(1-r_\ell^2\right)} + \frac{\log^2\left[r_\ell^2\right]}{2\left(1-r_\ell^2\right)}\right] + \\ + \frac{\left(1-3\,r_\ell^2\right)\log\left[r_\ell^2\right]}{\left(1-r_\ell^2\right)} - \log\left[\frac{M_W^2}{m_P^2}\right] + \log[m_P^2 L^2] - \frac{1}{2}K_P + \frac{1}{12} + \\ + \frac{1}{m_P L}\left(\frac{2r_\ell^2}{1-r_\ell^2}\left(K_{21}+K_{22}-2\pi\left(\frac{1}{1+r_\ell^2}+\frac{1}{r_\ell}\right)\right) - \frac{\pi(1+r_\ell^2)}{(1-r_\ell^2)}\left(K_{11}+K_{12}-3\right)\right) . \tag{7}$$

Note that in this expression we did not include the contribution of the muon wave-function renormalisation which is also not computed in $\Gamma_0(L)$ since it cancels exactly in the difference $\Gamma_0(L) - \Gamma_0^{\rm pt}(L)$ [1]. The dimensionless constants K_{ij} only depends on the ratio r_ℓ, and their explicit expression can be found in Ref. [?].

The coefficients $C_0(r_\ell)$, $\tilde{C}_0(r_\ell)$ and $C_1(r_\ell)$ are universal, although $C_0(r_\ell)$ and $C_1(r_\ell)$ depend on the infrared regulator. $\tilde{C}_0(r_\ell)$ is universal and does not depend on the regularisation. $C_0(r_\ell)$, $\tilde{C}_0(r_\ell)$ and $C_1(r_\ell)$ cancel the corresponding terms contained in $\Gamma_0(L)$. In this way $\Gamma_0(L)-\Gamma_0^{\rm pt}(L)$ is infrared finite and independent of the infrared regularisation up to terms of $O(1/L^2)$. Higher order FV terms are not universal and thus cannot be corrected with an analytic computation.

2. Calculation of the relevant amplitudes

The inclusive decay rate for $P^- \to \ell^- \bar{\nu}_\ell[\gamma]$ of Eq. (2) can be written as

$$\Gamma(\Delta E) = \Gamma_0^{(tree)} \times \left[1 + \delta R_P\right] , \tag{8}$$

where $\Gamma_0^{(tree)}$ is the tree-level decay rate given by

$$\Gamma_0^{(tree)} = \frac{G_F^2}{8\pi}|V_{q_1 q_2}|^2 m_\ell^2 \left(1 - \frac{m_\ell^2}{m_P^2}\right)^2 \left[f_P^{(0)}\right]^2 m_P , \tag{9}$$

and

$$\delta R_P = \alpha_{em}\frac{2}{\pi}\log\left(\frac{M_Z}{M_W}\right) + 2\frac{\delta A_P}{A_P^{(0)}} - 2\frac{\delta m_P}{m_P^{(0)}} + \Gamma^{(pt)}(\Delta E) . \tag{10}$$

The quantity $\Gamma^{(pt)}(\Delta E)$ can be read off from Eq. (4), while the term containing $\log(M_Z/M_W)$ comes from short-distance electroweak corrections, m_P is the physical pseudo scalar meson mass including both e.m. and strong Isospin Breaking (IB) corrections, and $A_P^{(0)}$ is the QCD axial amplitude

$$A_P^{(0)} \equiv Z_A \langle 0|\bar{q}_2 \gamma_0 \gamma_5 q_1|P^{(0)}\rangle \equiv f_P^{(0)} m_P^{(0)} \tag{11}$$

where $f_P^{(0)}$ defines the pseudo scalar meson decay constant known in pure QCD, and Z_A is the (lattice) renormalization constant of the axial current. Throughout this contribution the superscript $^{(0)}$ indicates quantities that do not contain e.m. and strong IB corrections.

The evaluation of δA_P and δm_P is obtained from the diagrams shown in Figs. 5 and 6 of Ref. [1]. It is in δA_P that the subtraction procedure described in Eq. (2), see also (7), is applied. We adopt the quenched QED approximation, which neglects the sea-quark electric charges and corresponds to consider only the connected diagrams shown in Fig. 2. In addition, with Wilson like fermions, one should include also the contributions coming from the tadpole operator, the e.m. shift in the critical mass and the insertion of the isovector scalar density (see Refs. [7, 8]).

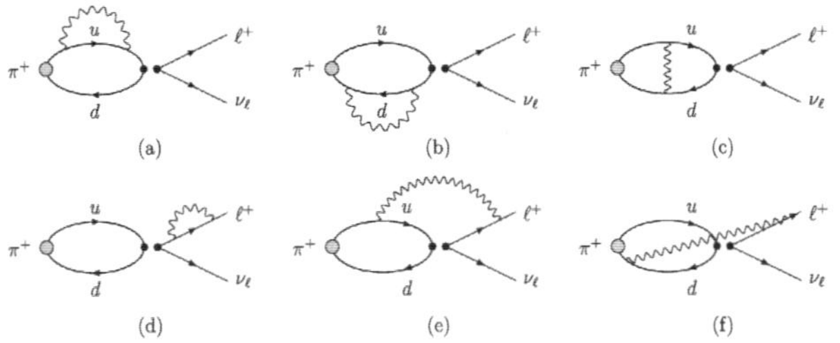

Figure 2: *Connected diagrams contributing at $O(\alpha_{em})$ to the amplitude for the decay $\pi^+ \to \ell^+ \nu$. The labels (a)-(f) identify the individual diagrams.*

3. Results for charged pion and kaon decays

We now insert the various ingredients described in the previous Sections in the master formula (10) in the case of the decays $\pi^- \to \mu^- \bar{\nu}_\mu[\gamma]$ and $K^- \to \mu^- \bar{\nu}_\mu[\gamma]$. It is understood that m_P represents the charged pseudo scalar mass, which includes e.m. and strong IB corrections. The latter are calculated as in Refs. [7, 8] and are generated by the mass difference $\delta m = m_d - m_u$ previously determined.

Preliminary results for the corrections δR_π and $\delta R_{K\pi} \equiv \delta R_K - \delta R_\pi$ are shown in Fig. 3, with the caveat that all photon energies (i.e. $\Delta E = \Delta E^{max} = m_P(1 - m_\mu^2/m_P^2)/2$) are included, since the experimental data on $\pi_{\ell 2}$ and $K_{\ell 2}$ decays are fully inclusive. The "universal" finite volume corrections up to order $O(1/L)$ are subtracted from the lattice data and the combined chiral, continuum and infinite volume extrapolations are performed using the fitting function

$$\delta R_P = A_0 + \frac{3}{16\pi^2}\log(\xi) + A_1\xi + A_2\xi^2 + Da^2 + \frac{K_2}{m_P^2 L^2} + \frac{K_2^\ell}{E_\ell^2 L^2} + \Gamma^{pt}(\Delta E^{max}) , \qquad (12)$$

where $\xi \equiv m_P^2/(4\pi f_0)^2$, E_ℓ is the lepton energy in the meson rest frame, $A_{0,1,2}$, D, K_2 and K_2^ℓ are free parameters, and the coefficient of the chiral log is taken from Ref. [9].

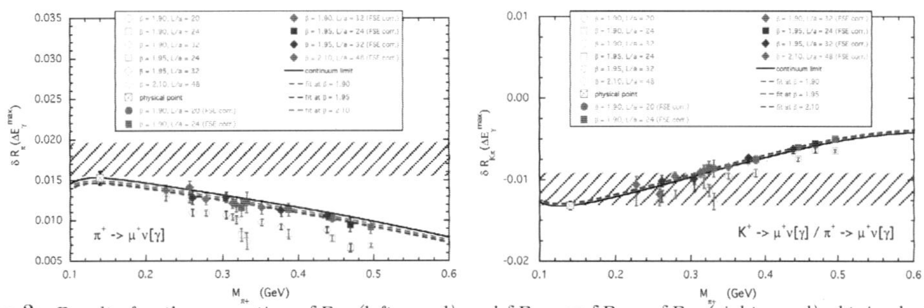

Figure 3: *Results for the corrections δR_π (left panel) and $\delta R_{K\pi} \equiv \delta R_K - \delta R_\pi$ (right panel) obtained after the subtraction of the "universal" FV terms computed in perturbation theory (7). The full markers correspond to the lattice data corrected by the residual FV corrections obtained using the fitting function (12) including the chiral log. The dashed lines represent the results in the infinite volume limit at each value of the lattice spacing, while the solid lines are the results in the continuum limit. The crosses represent the values δR_π^{phys} and $\delta R_{K\pi}^{phys}$ at the physical point. The shaded areas correspond respectively to the values 0.0176(21) and −0.0112(21) at 1-sigma level, obtained using ChPT (see Refs. [10, 11]).*

Adopting different fitting functions and FSE subtractions in order to estimate systematic errors and averaging the corresponding results, we finally get at the physical point

$$\delta R_\pi^{phys} = +0.0169\ (8)_{stat+fit}\ (11)_{chiral}\ (7)_{FSE}\ (2)_{a^2} = +0.0169\ (15) , \qquad (13)$$

$$\delta R_{K\pi}^{phys} = -0.0137\ (11)_{stat+fit}\ (6)_{chiral}\ (1)_{FSE}\ (1)_{a^2} = -0.0137\ (13) , \qquad (14)$$

where the errors do not include the QED quenching effects. Our findings (13)-(14) can be compared with the corresponding ChPT predictions 0.0176(21) and −0.0112(21) [10, 11].

Acknowledgements
We are particularly grateful to Z. Davoudi, L. Del Debbio and A. Patella for helpful discussions on finite volume corrections. Work partially supported by the ERC-2010 DaMESyFla Grant Agreement Number: 267985, by the MIUR (Italy) under a contract PRIN10 and by STFC Grant ST/L000296/1.

[1] N. Carrasco, V. Lubicz, G. Martinelli, C. T. Sachrajda, N. Tantalo, C. Tarantino and M. Testa, Phys. Rev. D **91** (2015) no.7, 074506 [arXiv:1502.00257 [hep-lat]].
[2] S. Aoki *et al.*, arXiv:1607.00299 [hep-lat].
[3] N. Tantalo, PoS LATTICE **2013** (2014) 007 [arXiv:1311.2797 [hep-lat]].
[4] A. Portelli, PoS LATTICE **2014** (2015) 013 [arXiv:1505.07057 [hep-lat]].
[5] A. Patella at Lattice 2016, https://conference.ippp.dur.ac.uk/event/470/session/1/contribution/400.
[6] V. Lubicz, G. Martinelli, C. T. Sachrajda, F. Sanfilippo, S. Simula and N. Tantalo, arXiv:1611.08497 [hep-lat].
[7] G. M. de Divitiis *et al.*, Phys. Rev. D **87** (2013) no.11, 114505 [arXiv:1303.4896 [hep-lat]].
[8] G.M. de Divitiis *et al.*, JHEP **1204** (2012) 124 [arXiv:1110.6294 [hep-lat]].
[9] M. Knecht *et al.*, Eur. Phys. J. C **12** (2000) 469 [hep-ph/9909284].
[10] J. L. Rosner *et al.*, Submitted to: Particle Data Book [arXiv:1509.02220 [hep-ph]].
[11] V. Cirigliano and H. Neufeld, Phys. Lett. B **700** (2011) 7 [arXiv:1102.0563 [hep-ph]].

IOP Conf. Series: Journal of Physics: Conf. Series **800** (2017) 012006 doi:10.1088/1742-6596/800/1/012006

Current status of ε_K in lattice QCD

Weonjong Lee

Director, Lattice Gauge Theory Research Center, CTP, and FPRD,
Department of Physics and Astronomy,
Seoul National University, Seoul 08826, South Korea

E-mail: wlee@snu.ac.kr

Abstract. We present the current status of ε_K evaluated directly from the standard model using lattice QCD inputs. The lattice QCD inputs include $\hat{B}_K$, ξ_0, ξ_2, $|V_{us}|$, $m_c(m_c)$, and $|V_{cb}|$. Recently, FLAG has updated $\hat{B}_K$, exclusive $|V_{cb}|$ has been updated with new lattice data in the $\bar{B} \to D\ell\bar{\nu}$ decay mode, and RBC-UKQCD has updated ξ_0 and ξ_2. We find that the standard model evaluation of ε_K with exclusive $|V_{cb}|$ (lattice QCD inputs) is 3.2σ lower than the experimental value, while that with inclusive $|V_{cb}|$ (heavy quark expansion) shows no tension.

1. Introduction

Since 2012, we have been monitoring ε_K, the indirect CP violation parameter in neutral kaons using lattice QCD inputs. The parameter ε_K is, in particular, very attractive to the particle physics community, since it is very precisely measured in experiment, and it provides a direct probe of CP violation in the standard model and in physics models beyond the standard model (BSM). In this paper, we present results of ε_K evaluated directly from the standard model with lattice QCD inputs and compare them with the experimental value. This paper is an update of our previous paper [1, 2].

2. Input parameters

In the standard model, the indirect CP violation parameter of the neutral kaon system ε_K can be expressed as follows,

$$\begin{aligned}\varepsilon_K &\equiv \frac{\mathcal{A}(K_L \to \pi\pi(I=0))}{\mathcal{A}(K_S \to \pi\pi(I=0))} \\ &= e^{i\theta}\sqrt{2}\sin\theta\Big(C_\varepsilon X_{\rm SD}\hat{B}_K + \frac{\xi_0}{\sqrt{2}} + \xi_{\rm LD}\Big) + \mathcal{O}(\omega\varepsilon') + \mathcal{O}(\xi_0\Gamma_2/\Gamma_1)\,. \end{aligned} \quad (1)$$

Here, the short distance contribution proportional to $\hat{B}_K$ occupies about 105% of ε_K, the long distance effect from the absorptive part ξ_0 gives about -5% correction, and the long distance effect from the dispersive part $\xi_{\rm LD}$ gives about $\pm 1.6\%$ correction. The details on C_ε, $X_{\rm SD}$, ξ_0, and $\xi_{\rm LD}$ are described in Ref. [1]. In order to determine ε_K directly from the standard model, we need 18 input parameters, and 6 of them can, in principle, be obtained from lattice QCD: V_{us}, V_{cb}, $\hat{B}_K$, ξ_0, $\xi_{\rm LD}$, and $m_c(m_c)$. Here, we address recent progress on determining those input parameters.

Content from this work may be used under the terms of the Creative Commons Attribution 3.0 licence. Any further distribution of this work must maintain attribution to the author(s) and the title of the work, journal citation and DOI.
Published under licence by IOP Publishing Ltd

Table 1: Results for $|V_{cb}|$

Decay mode	$\|V_{cb}\|$	Ref.
$\bar{B} \to D^*\ell\bar{\nu}$	39.04(49)(53)(19)	[3]
$\bar{B} \to D\ell\bar{\nu}$	40.7(10)(2)	[4]
ex-combined	39.62(60)	this paper
$\bar{B} \to X_c\ell\bar{\nu}$	42.00(64)	[5]

Table 2: Results for $|V_{ub}|$

Decay mode	$\|V_{ub}\|$	Ref.
$\bar{B} \to \pi\ell\bar{\nu}$	3.72(16)	[6]
$\bar{B} \to \pi\ell\bar{\nu}$	3.61(32)	[7]
ex-combined	3.70(14)	this paper
$\bar{B} \to X_u\ell\bar{\nu}$	4.45(16)(22)	[8]

2.1. $|V_{cb}|$

Recent results for $|V_{cb}|$ and $|V_{ub}|$ are presented in Tables 1 and 2. Recently, DeTar has collected the results of $\bar{B} \to D\ell\bar{\nu}$ decay at non-zero recoil from both lattice QCD [9, 10] and experiments of Babar [11] and Belle [12], and has made a combined fit of all the data simultaneously to determine $|V_{cb}|$ [4]. We have obtained the "ex-combined" result in Table 1 by taking a weighted average of the V_{cb} results from the $\bar{B} \to D^*\ell\bar{\nu}$ and $\bar{B} \to D\ell\bar{\nu}$ decay channels. Similarly, we have obtained the "ex-combined" result in Table 2 by taking a weighted average of the two V_{ub} results from $\bar{B} \to \pi\ell\bar{\nu}$ decay. In Fig. 1, we show all the results in various colors.[1] The inclusive results are about 3σ away from those from exclusive B meson decays respectively as well as from the LHCb results for $|V_{ub}/V_{cb}|$, which corresponds to the magenta band in Fig. 1.

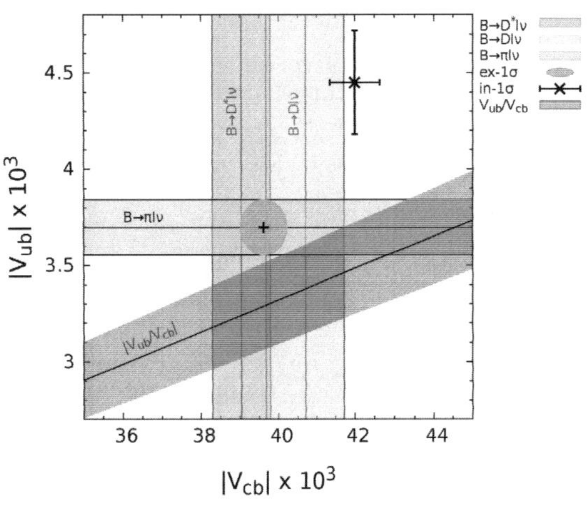

Figure 1: $|V_{cb}|$ versus $|V_{ub}|$. The sky-blue band represents $|V_{cb}|$ determined from the $B \to D^*\ell\bar{\nu}$ decay mode. The yellow-green band represents $|V_{cb}|$ determined from the $B \to D\ell\bar{\nu}$ decay mode. The yellow band represents $|V_{ub}|$ determined from the $B \to \pi\ell\bar{\nu}$ decay mode. The magenta band represents $|V_{ub}/V_{cb}|$ determined from the LHCb data of the $\Lambda_b \to \Lambda_c\ell\bar{\nu}$ and $\Lambda_b \to p\ell\bar{\nu}$ decay modes. The orange circle represents the combined results for exclusive $|V_{cb}|$ and $|V_{ub}|$ from the B meson decays within 1.0σ. The black cross ($\times$) represents the inclusive $|V_{cb}|$ and $|V_{ub}|$ from the heavy quark expansion. The details are given in Tables 1 and 2.

2.2. ξ_0 *and* $\xi_{\rm LD}$

There are two independent methods to determine ξ_0 in lattice QCD: the indirect and direct methods. In the indirect method, we determine ξ_0 from the experimental values of $\mathrm{Re}(\varepsilon'/\varepsilon)$, ε_K, ω and the lattice QCD input ξ_2. They are related to one another as follows,

$$\xi_0 = \frac{\mathrm{Im}A_0}{\mathrm{Re}A_0}, \qquad \xi_2 = \frac{\mathrm{Im}A_2}{\mathrm{Re}A_2}, \qquad \mathrm{Re}\left(\frac{\varepsilon'}{\varepsilon}\right) = \frac{\omega}{\sqrt{2}|\varepsilon_K|}(\xi_2 - \xi_0)\,. \tag{2}$$

[1] The plot is based on that by Andreas Kronfeld in Ref. [4].

Table 3: ξ_0 and $\xi_{\rm LD}$.

Input	Method	Value	Ref.
ξ_0	indirect	$-1.63(19) \times 10^{-4}$	[13]
ξ_0	direct	$-0.57(49) \times 10^{-4}$	[14]
$\xi_{\rm LD}$	—	$(0 \pm 1.6)\,\%$	[15]

Table 4: δ_0

Collaboration	δ_0	Ref.
RBC-UK-2016	23.8(49)(12)°	[14]
KPY-2011	39.1(6)°	[16]
CGL-2001	39.2(15)°	[17, 18]

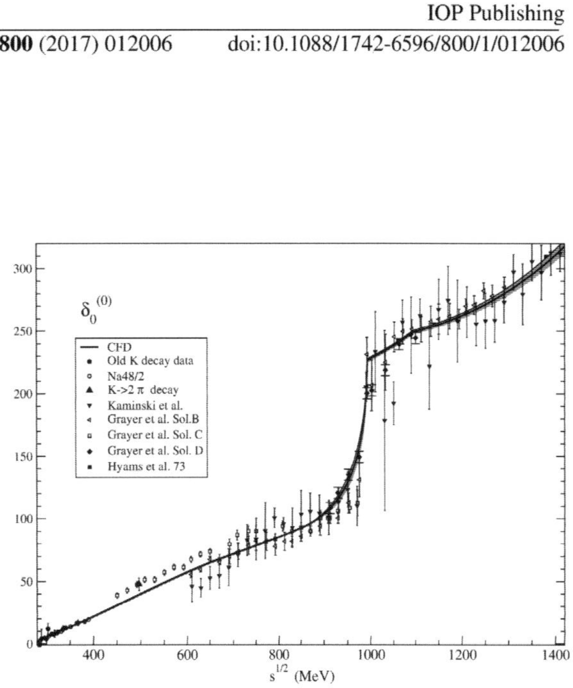

Figure 2: δ_0 from KPY-2011.

Recently, RBC-UKQCD reported updated results for ξ_2 [13]. The results for ξ_0 from the indirect method are given in Table 3.

Recently, RBC-UKQCD has reported new lattice QCD results for ImA_0 [14]. Combining their results with the experimental value of ReA_0, we can determine ξ_0 directly from the lattice input ImA_0. This is the direct method. In Ref. [14], RBC-UKQCD has also reported the S-wave $\pi - \pi$ scattering phase shift with isospin $I = 0$: $\delta_0 = 23.8(49)(12)$. This value has 3.0σ tension with the conventional determination of δ_0 in Refs. [16] (KPY-2011) and [17, 18] (CGL-2001). They used a singly subtracted Roy-like equation (KPY-2011) or a doubly subtracted Roy equation (CGL-2001) to do the interpolation around $\sqrt{s} = m_K$ (kaon mass). The values for δ_0 are summarized in Table 4. In Fig. 2, we show the fitting results of KPY-2011. Their fits to the experimental data work well from the $\pi - \pi$ threshold to $\sqrt{s} = 800\,\text{MeV}$, and are highly consistent with CGL-2001 in the interpolating region around $\sqrt{s} = m_K \approx 500\,\text{MeV}$.

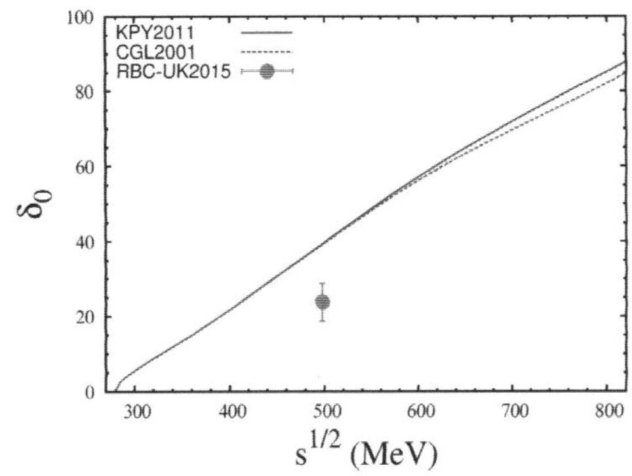

Figure 3: Comparison of δ_0.

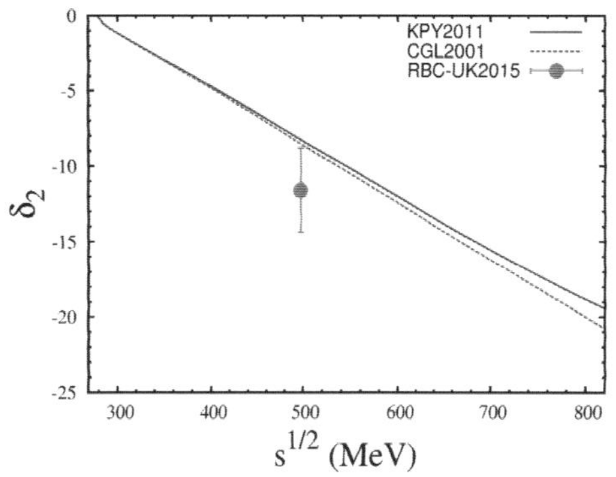

Figure 4: Comparison of δ_2.

In Fig. 3, we present the results of RBC-UKQCD together with the fitting results of KPY-2011 and CGL-2001. There is essentially no difference between KPY-2011 and CGL-2001 in the region near $\sqrt{s} = m_K \approx 500\,\text{MeV}$. Here, we observe the 3.0σ gap between RBC-UKQCD and KPY-2011. In contrast, for δ_2 (S-wave, I=2), there is no difference between RBC-UKQCD and KPY-2011 within statistical uncertainty, as one can see in Fig. 4.

Therefore, we conclude that the results of the indirect method are more reliable than those of the direct method for ξ_0, since the direct calculation of $\text{Im}A_0$ by RBC-UKQCD might have unresolved issues. Hence, we use the indirect method to determine ξ_0 in this paper.

Regarding ξ_{LD}, the long distance effect in the dispersive part, the theoretical master formula in the continuum is given in Ref. [1]. A theoretical framework for calculating it on the lattice is well established in Ref. [15]. There has been an on-going attempt to calculate it on the lattice [19]. However, this attempt [20], at present, is not mature and belongs to the category of exploratory study rather than to that of precision measurement. Hence, we use a rough estimate of ξ_{LD} given in Ref. [19] in this paper. It is summarized in Table 3.

2.3. $\hat{B}_K$

In Table 5, we present results for $\hat{B}_K$ calculated using lattice QCD tools with $N_f = 2+1$ flavors. Here, FLAG-2016 represents the global average of the results of BMW-2011 [21], Laiho-2011 [22], RBC-UK-2016 [23], and SWME-2016 [24], which is summarized in Ref. [25]. SWME-2014 represents the $\hat{B}_K$ result reported in Ref. [26]. RBC-UK-2016 represents that reported in Ref. [23].

The results of SWME-2016 are obtained using fitting based on staggered chiral perturbation theory (SChPT) in the infinite volume limit, and those of SWME-2014 are obtained using fitting based on SChPT with finite volume corrections incorporated at the NLO level. Here we use the FLAG-2016 result for $\hat{B}_K$.

Table 5: $\hat{B}_K$.

Collaboration	Value	Ref.
FLAG-2016	0.7625(97)	[25]
SWME-2014	0.7379(47)(365)	[26]
RBC-UK-2016	0.7499(24)(150)	[23]

Table 6: Wolfenstein parameters.

	CKMfitter	UTfit	AOF
λ	0.22548(68)/[27]	0.22497(69)/[28]	0.2253(8)/[29]
$\bar{\rho}$	0.145(13)/[27]	0.153(13)/[28]	0.139(29)/[30]
$\bar{\eta}$	0.343(12)/[27]	0.343(11)/[28]	0.337(16)/[30]

2.4. Other input parameters

For the Wolfenstein parameters λ, $\bar{\rho}$, and $\bar{\eta}$, both CKMfitter and UTfit updated their results in Refs. [27, 28], but the angle-only-fit has not been updated since 2015. The global unitarity triangle (UT) fits of both CKMfitter and UTfit use ε_K and $|V_{cb}|$ as input parameters to determine the apex $\bar{\rho}$ and $\bar{\eta}$. Hence, using them to evaluate ε_K introduces unwanted correlations through ε_K and $|V_{cb}|$. In contrast, the angle-only-fit (AOF) results are independent of ε_K and $|V_{cb}|$. Hence, we use the AOF results in this paper.

For the QCD corrections η_{cc}, η_{ct}, and η_{tt}, we use the same values as in Ref. [1], which are given in Table 7. In particular, we use the SWME value of η_{cc} reported in Ref. [1] instead of that in Ref. [31]. This issue is well explained in Ref. [1]. One of the reasons is that the size of the NNLO correction is already a conservative estimate for the truncation error of the NNNLO level in perturbation theory. Another reason is that the SWME result is highly consistent with that of Ref. [32].

In Table 8, we summarize other input parameters. They are the same as those in Ref. [1] except for the charm quark mass $m_c(m_c)$. For the charm quark mass, we use the HPQCD result reported in Ref. [35].

Table 7: QCD corrections.

Input	Value	Ref.
η_{cc}	1.72(27)	[1]
η_{tt}	0.5765(65)	[33]
η_{ct}	0.496(47)	[34]

Table 8: Other input parameters.

Input	Value	Ref.
G_F	$1.1663787(6) \times 10^{-5}$ GeV^{-2}	[29]
M_W	80.385(15) GeV	[29]
$m_c(m_c)$	1.2733(76) GeV	[35]
$m_t(m_t)$	163.3(2.7) GeV	[36]
θ	43.52(5)°	[29]
m_{K^0}	497.614(24) MeV	[29]
ΔM_K	$3.484(6) \times 10^{-12}$ MeV	[29]
F_K	156.2(7) MeV	[29]

3. Current status of ε_K

In Fig. 5, we show the results for ε_K evaluated directly from the standard model with the lattice QCD inputs described in the previous section. Here, the blue curve represents the theoretical evaluation of ε_K with the FLAG-2016 $\hat{B}_K$, AOF for the Wolfenstein parameters, and exclusive $|V_{cb}|$ that corresponds to ex-combined in Table 1. The red curve represents the experimental value of ε_K. In Fig. 6, the blue curve represents the same as that in Fig. 5 except for using the inclusive $|V_{cb}|$ in Table 1. Our preliminary results are, in units of 1.0×10^{-3},

$$|\varepsilon_K| = 1.69 \pm 0.17 \qquad \text{for exclusive } V_{cb} \text{ (lattice QCD)} \tag{3}$$

$$|\varepsilon_K| = 2.10 \pm 0.21 \qquad \text{for inclusive } V_{cb} \text{ (QCD sum rules)} \tag{4}$$

$$|\varepsilon_K| = 2.228 \pm 0.011 \qquad \text{(experimental value)} \tag{5}$$

We find that there is 3.2σ tension in the exclusive V_{cb} channel (lattice QCD), and no tension in the inclusive V_{cb} channel (heavy quark expansion; QCD sum rules).

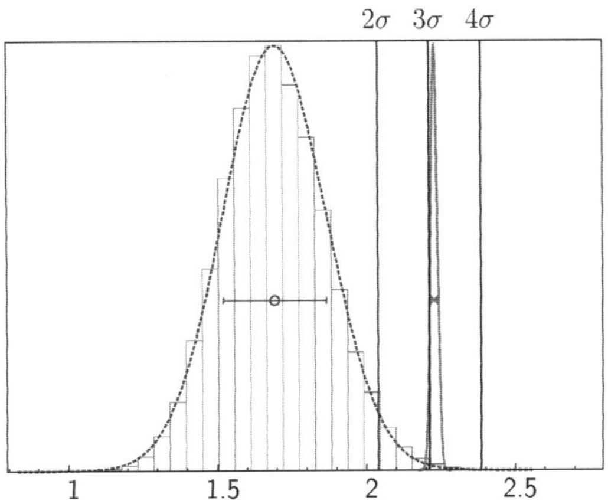

Figure 5: ε_K with exclusive V_{cb}.

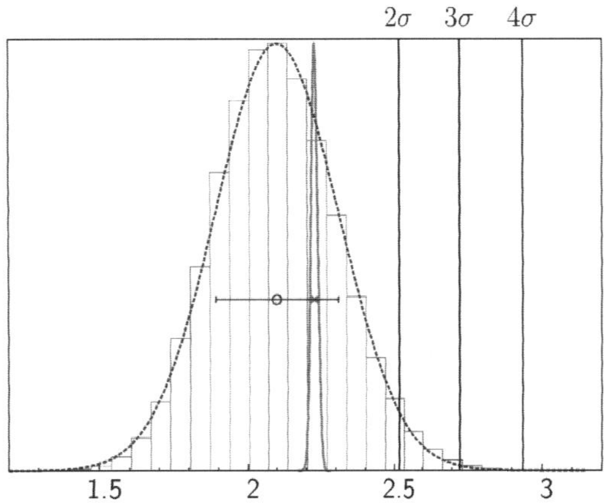

Figure 6: ε_K with inclusive V_{cb}.

Acknowledgments

We thank R. Van de Water for helpful discussion on V_{cb}. The research of W. Lee is supported by the Creative Research Initiatives Program (No. 20160004939) of the NRF grant funded by the Korean government (MEST). W. Lee would like to acknowledge the support from the KISTI supercomputing center through the strategic support program for the supercomputing application research (No. KSC-2014-G3-003). Part of computations were carried out on the DAVID GPU clusters at Seoul National University.

References

[1] Bailey J A, Jang Y C, Lee W and Park S (SWME) 2015 *Phys. Rev.* **D92** 034510 (*Preprint* 1503.05388)
[2] Bailey J A, Jang Y C, Lee W and Park S 2015 *PoS* **LATTICE2015** 348 (*Preprint* 1511.00969)
[3] Bailey J A, Bazavov A, Bernard C *et al.* 2014 *Phys.Rev.* **D89** 114504 (*Preprint* 1403.0635)
[4] DeTar C 2015 *Proceedings of the 27th International Symposium on Lepton Photon Interactions at High Energies* (*Preprint* 1511.06884)
[5] Gambino P, Healey K J and Turczyk S 2016 (*Preprint* 1606.06174)
[6] Bailey J A *et al.* (Fermilab Lattice, MILC) 2015 *Phys. Rev.* **D92** 014024 (*Preprint* 1503.07839)
[7] Flynn J M, Izubuchi T, Kawanai T, Lehner C, Soni A, Van de Water R S and Witzel O 2015 *Phys. Rev.* **D91** 074510 (*Preprint* 1501.05373)
[8] Amhis Y *et al.* (Heavy Flavor Averaging Group (HFAG)) 2014 (*Preprint* 1412.7515)
[9] Bailey J A *et al.* (MILC) 2015 *Phys. Rev.* **D92** 034506 (*Preprint* 1503.07237)
[10] Na H, Bouchard C M, Lepage G P, Monahan C and Shigemitsu J (HPQCD) 2015 *Phys. Rev.* **D92** 054510 [Erratum: Phys. Rev.D93,no.11,119906(2016)] (*Preprint* 1505.03925)
[11] Aubert B *et al.* (BaBar) 2009 *Phys. Rev.* **D79** 012002 (*Preprint* 0809.0828)
[12] Glattauer R 2015 *PoS* **EPS-HEP2015** 554
[13] Blum T *et al.* 2015 *Phys. Rev.* **D91** 074502 (*Preprint* 1502.00263)
[14] Bai Z *et al.* (RBC, UKQCD) 2015 *Phys. Rev. Lett.* **115** 212001 (*Preprint* 1505.07863)
[15] Christ N, Izubuchi T, Sachrajda C, Soni A and Yu J 2013 *Phys.Rev.* **D88** 014508 (*Preprint* 1212.5931)
[16] Garcia-Martin R, Kaminski R, Pelaez J R, Ruiz de Elvira J and Yndurain F J 2011 *Phys. Rev.* **D83** 074004 (*Preprint* 1102.2183)
[17] Colangelo G, Gasser J and Leutwyler H 2001 *Nucl. Phys.* **B603** 125–179 (*Preprint* hep-ph/0103088)
[18] https://indico.mitp.uni-mainz.de/event/48/contribution/5/material/slides/0.pdf
[19] Christ N, Izubuchi T, Sachrajda C T, Soni A and Yu J 2014 *PoS* **LATTICE2013** 397 (*Preprint* 1402.2577)
[20] https://conference.ippp.dur.ac.uk/event/470/session/15/contribution/407/material/slides/0.pdf
[21] Durr S, Fodor Z, Hoelbling C *et al.* 2011 *Phys.Lett.* **B705** 477–481 (*Preprint* 1106.3230)
[22] Laiho J and Van de Water R S 2011 *PoS* **LATTICE2011** 293 (*Preprint* 1112.4861)
[23] Blum T *et al.* (RBC, UKQCD) 2016 *Phys. Rev.* **D93** 074505 (*Preprint* 1411.7017)
[24] Choi B J *et al.* (SWME) 2016 *Phys. Rev.* **D93** 014511 (*Preprint* 1509.00592)
[25] Aoki S *et al.* 2016 (*Preprint* 1607.00299)
[26] Bae T *et al.* 2014 *Phys.Rev.* **D89** 074504 (*Preprint* 1402.0048)
[27] Charles J *et al.* 2015 *Phys. Rev.* **D91** 073007 (*Preprint* 1501.05013)
[28] http://www.utfit.org/UTfit/ResultsSummer2016SM
[29] Olive K *et al.* 2014 *Chin.Phys.* **C38** 090001
[30] http://www.utfit.org/UTfit/ResultsSummer2014PostMoriondSM
[31] Brod J and Gorbahn M 2012 *Phys.Rev.Lett.* **108** 121801 (*Preprint* 1108.2036)
[32] Buras A J and Girrbach J 2013 *Eur.Phys.J.* **C73** 2560 (*Preprint* 1304.6835)
[33] Buras A J and Guadagnoli D 2008 *Phys.Rev.* **D78** 033005 (*Preprint* 0805.3887)
[34] Brod J and Gorbahn M 2010 *Phys.Rev.* **D82** 094026 (*Preprint* 1007.0684)
[35] Chakraborty B, Davies C T H, Galloway B, Knecht P, Koponen J, Donald G C, Dowdall R J, Lepage G P and McNeile C 2015 *Phys. Rev.* **D91** 054508 (*Preprint* 1408.4169)
[36] Alekhin S, Djouadi A and Moch S 2012 *Phys.Lett.* **B716** 214–219 (*Preprint* 1207.0980)

Experimental studies of the kaon-nucleus interaction at low energy with x-ray spectroscopy of kaonic atoms

H Shi [1], M Bazzi [1], G Beer [2], G Bellotti [3,4], C Berucci [5,1], A M Bragadireanu [1,6], D Bosnar [7], M Cargnelli [5], C Curceanu [1], A D Butt [3,4], R Del Grande [1], L Fabietti [8], C Fiorini [3,4], F Ghio [9], C Guaraldo [1], R S Hayano [10], M Iliescu [1], T Ishiwatari [5], M Iwasaki [11], P Levi Sandri [1], J Marton [5], M Miliucci [1], S Okada [11], D Pietreanu [1,6], K Piscicchia [1,12], A Romero Vidal [13], A Scordo [1], D L Sirghi [1,6], F Sirghi [1,6], H Tatsuno [14,15], I Tucakovic [1], O Vazquez Doce [8], E Widmann [5], S Wycech [16] and J Zmeskal [5] (SIDDHARTA, SIDDHARTA-2 Collaboration)

[1]INFN, Laboratori Nazionali di Frascati, Via E. Fermi 40, I-00044 Frascati(Roma), Italy
[2]Department of Physics and Astronomy, University of Victoria, P.O. Box 1700 STN CNC, Victoria BC V8W 2Y2, Canada
[3]Politecnico di Milano, Dipartimento di Elettronica, Informazione e Bioingegneria, Milano, Italy
[4]INFN Sezione di Milano, Via Celoria 16 - 20133 Milano, Italy
[5]Stefan-Meyer-Institut für Subatomare Physik, Boltzmanngasse 3, 1090 Wien, Austria
[6]IFIN-HH, Institutul National pentru Fizica si Inginerie Nucleara Horia Hulubbei, Reactorului 30, Magurele, Romania
[7]Department of Physics, Facaulty of Science, University of Zagreb, Bijenicka 32, HR-10000 Zagreb, Croatia
[8]Excellence Cluster Universe, Technische Universität München, Boltzmannstraße 2, D-85748 Garching, Germany
[9]INFN Sezione di Roma I and Instituto Superiore di Sanita, I-00161 Roma, Italy
[10]Department of Physics, School of Science, The University of Tokyo, Bunkyo-ku Hongo 7-3-1, Tokyo, Japan
[11]RIKEN, Institute of Physical and Chemical Research, 2-1 Hirosawa, Wako, Saitama 251-0198, Japan
[12]Museo Storico della Fisica e Centro Studi e Ricerche "Enrico Fermi", Piazza del Viminale 1-00184Roma, Italy
[13]Universidade de Santiago de Compostela, Casas Reais 8, 15782 Santiago de Compostela, Spain
[14]National Institute of Standards and Technology (NIST), Boulder, CO, 80303, USA
[15]High Energy Accelerator Research Organization (KEK), Tsukuba, 305-0801, Japan
[16]National Center for Nuclear Studies, 00–681 Hoza 69, Warsaw, Poland

E-mail: `hexishi@lnf.infn.it`

Abstract. In the exotic atoms in which one electron is replaced by a negatively charged kaon, the kaon-nucleus hadronic interaction introduces an energy shift and broadening of the low-lying states of the kaonic atoms. The shift and width can be determined with high precision from the atomic x-ray spectroscopy, and this experimental method provides unique information to understand the low energy kaon-nucleus interaction at the production threshold.

Content from this work may be used under the terms of the Creative Commons Attribution 3.0 licence. Any further distribution of this work must maintain attribution to the author(s) and the title of the work, journal citation and DOI.
Published under licence by IOP Publishing Ltd

1. Introduction

There are two experimental approaches to probe the kaon-nucleus strong interaction. One is by studying the scattering and the reaction channels between the kaon and the nucleus, to directly search for bound states and extract the potential of the strong interaction. The other method of the precision x-ray measurement of the kaonic atoms, measures the shift and broadening of the energy levels of the atomic states, caused by the kaon-nucleus strong interaction. The method is of significant importance, since it is the unique laboratory that gives the information of the kaon-nucleus system at their production threshold.

In an exhaustive review in 1997 and the references therein, C. J. Batty and colleagues analyzed all the published results of kaonic atom x-ray measurements with nuclei from Li to U [1]. They fitted the full data set with an optical potential with an imaginary part that presents the absorption of the kaon by the nucleus, and a real part as the depth of the potential. The analysis concluded with an attractive and absorptive kaon-nucleus potential, however due to the limit in the precision of the experimental data, the real part of the potential can vary from about -40 MeV to -200 MeV, depending on the choice of phenomenological models. The indication of a very deep potential motivated the subsequent theoretical and experimental searches of deeply bound kaon-nucleus state, which remains be an open question today.

In this context, precision x-ray spectroscopy of the Z=1,2 kaonic atoms as the simplest kaon-nucleus systems is fundamental to understand the low-energy interaction between the kaon and the nucleus. Firstly, in case of kaonic hydrogen which is the simplest kaonic atom, the strong interaction induced shift and width of the $1s$ state can be deduced from the K-series x-rays, and they are related to the s-wave K^-p scattering length a_{K^-p} through the Deser-Trueman formula [2]. This scattering length consists of two isospin dependent components of a_0 (I=0) and a_1 (I=1), which are among the most fundamental parameters for the low-energy $\bar{K}N$ interaction, and which can only be derived experimentally from x-ray spectroscopy. To disentangle these two components, we need also the precise measurement of the kaonic deuterium $1s$ shift and width, further taking into account higher order contributions from the three-body interaction [3]. Experimentally this measurement is extremely difficult mainly due to the small x-ray yield and the broad width of the $1s$ level. Until now an unambiguous detection of the kaonic deuterium x-rays has not been achieved and the measurement remains to be the major challenge of upcoming experiments.

For the Z=2 kaonic atoms, there had been a long-standing discrepancy between the theoretical and the measured strong-interaction induced shift of the $2p$ level of kaonic helium-4. This "kaonic helium puzzle" was solved by the KEK E570 experiment which showed the $2p$ level shift was 2 ± 2(stat.) $\pm$ 2(syst.) eV [4], refuting the previous average shift of about - 40 eV from three early experiments [5][6][7]. More essentially, the precise measurement of the kaonic helium $2p$ level shift is a test ground for possible deeply-bound kaonic nuclear states predicted by Akaishi and Yamazaki [8][9]. A combination of electronvolt precision measurement of the shift and the width of the $2p$ level for both kaonic helium-3 and kaonic helium-4 atoms can effectively narrow down the allowed range of the nuclear potential of kaon in the search of deeply-bound states.

In the next section measurements performed in the SIDDHARTA experiment will be introduced in detail.

2. The SIDDHARTA experiment

2.1. *Experimental method*

The experiment made use of the K^+K^- pairs coming from the ϕ decays with a 49% branching ratio, at the DAΦNE e^-e^+ collider in Laboratori Nazionali di Frascati (LNF), Italy. As the beam energies are tuned to produce ϕ (1020) resonance at rest in the lab frame, the K^+K^- pair has a fairly good back-to-back configuration which could be identified by the kaon detectors made of two plastic scintillators placed above and below the interaction point as illustrated in Figure 1. The kaons are distinguished from the minimum ionizing particles using the time of flight information at the kaon detectors, and the coincidence of the two scintillators defines the kaon trigger, which marks the timing of the incident kaons.

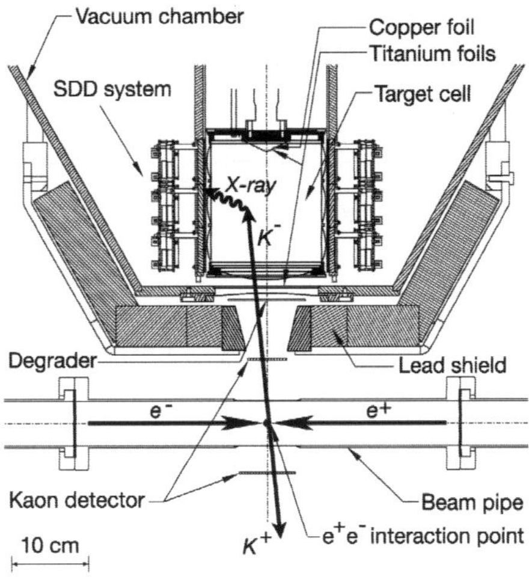

Figure 1: A schematic cutaway view of the SIDDHARTA setup at the DAΦNE interaction point. The charged kaon pairs are identified with two plastic scintillators, and the K^- induced x-rays detected by the SDDs are identified from the time correlation to the kaon pair events.

A fraction of the negatively charged kaons that made the kaon trigger will stop inside the volume of the gaseous target placed about 20 cm above the interaction point to form kaonic atoms. With the time correlation between the subsequent x-rays detected by the Silicon Drift Detectors (SDDs) and the kaon trigger, one can distinguish the kaon origin x-rays from the background coming from the processes that are asynchronous to the kaon pair production.

The SDDs had a total active area of 144 cm^2, covering about 10% of the solid angle around the target cell. The energy calibrations of the SDDs were done with an interval of few hours, using the K_α lines of Ti (4.5 keV) and Cu (8.0 keV) to determine the scale of the energy spectra. The energy resolution at 6 keV was stable at about 150 eV (FWHM) throughout our measurement. At the operating temperature of 140 K, the SDDs showed a timing resolution of 700 ns FWHM, with which we effectively rejected the asynchronous background by four orders of magnitude. More details about the configuration and the performance of the detectors can be found in previous publications of the SIDDHARTA experiment [10][11][12].

We accumulated data during the data taking in 2009 with gaseous targets of hydrogen (1.3 g/l), deuterium (2.50 g/l), helium-3 (0.96 g/l), and helium-4 (1.65 g/l and 2.15 g/l), and obtained physics results for all the measurements.

2.2. Kaonic hydrogen and deuterium

With the precise measurement of the kaonic hydrogen K-series x-rays as the main objective of the SIDDHARTA experiment, most of the beam time was dedicated to the hydrogen target measurement to a total integrated luminosity of 340 pb^{-1}. A total amount of 100 pb^{-1} integrated luminosity was dedicated to the first exploratory kaonic deuterium measurement.

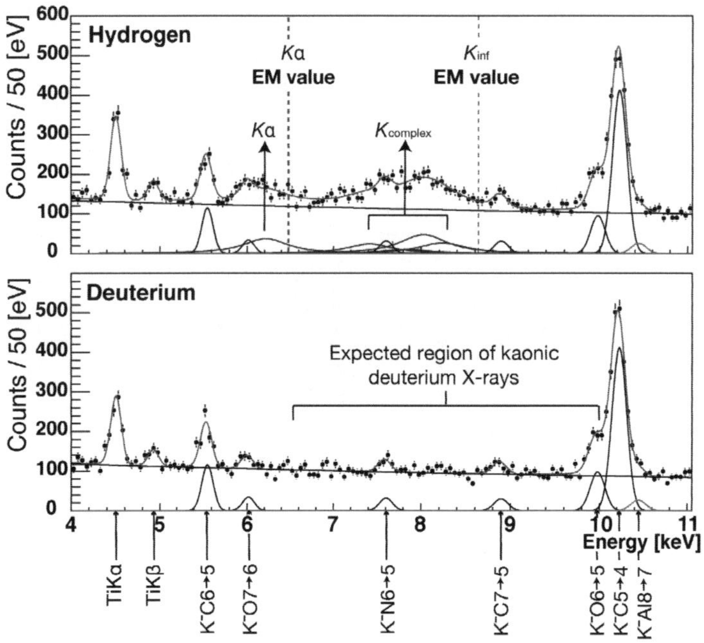

Figure 2: Result of the hydrogen and deuterium targets measurement in SIDDHARTA.

To derive the 1s shift ϵ_{1s} and width Γ_{1s}, as well as the x-ray yield of kaonic hydrogen, we performed a combined analysis of the hydrogen and deuterium targets measurement, by fitting simultaneously the kaonic x-ray spectra from the two data sets. Since the yield of the kaonic deuterium K-series x-rays is more than one order of magnitude smaller than that of the kaonic hydrogen, the spectrum from the deuterium target measurement helped determining the background x-rays from other kaonic atoms when the K^- stopped inside the target cell made of Kapton, as shown in Figure 2 [13]. This combined analysis contributed to reduce the systematic errors in the kaonic hydrogen results, which achieved the best precision up to date for the 1s shift ϵ_{1s} = -283 ± 36 (stat.) ± 6 (syst.) eV and for the width Γ_{1s} = 541 ± 89 (stat.) ± 22 (syst.) eV [12]. The results solved the discrepancy between two recent kaonic hydrogen x-ray experiments KpX in KEK [14][15] and DEAR at DAΦNE [16], in that the two results do not overlap with each other within their error bars. The improved precision presents more stringent constrains to the theoretical study of the low-energy QCD near the K^-p threshold.

Together with a full Monte Carlo simulation of the setup, we determined the absolute yields of kaonic x-rays for a negatively charged kaon stopped inside a hydrogen gaseous target at the density of 1.3 g/l to be $0.012^{+0.004}_{-0.003}$ for K_α and $0.043^{+0.012}_{-0.011}$ for all the K-series transitions K_{tot} [13]. This is the second data point of the kaonic hydrogen x-ray yield from an experiment, and is necessary to improve the kaonic atom cascade model which is verified by comparing the prediction of the absolute x-ray yields to the experimental values. For the deuterium target spectrum which showed no significant amount of signal of kaonic deuterium x-rays, we evaluated an upper limit for the yield of the kaonic deuterium K_α x-ray as $Y(K_\alpha)$ <0.0039 (C.L. 90%) [17]. It presents a solid reference to plan for the future precision measurement of kaonic deuterium x-rays.

Table 1: Results on the energy shifts (ΔE_{2p}) and widths (Γ_{2p}) of the kaonic helium-3 and kaonic helium-4 $2p$ states [10][11][18].

Target	ΔE_{2p} [eV]	Γ_{2p} [eV]
helium-4	+5 ± 3 (stat.) ± 4 (syst.)	14 ± 8 (stat.) ± 5 (syst.)
helium-3	-2 ± 2 (stat.) ± 4 (syst.)	6 ± 6 (stat.) ± 7 (syst.)

2.3. Kaonic helium-3 and helium-4

The strong interaction induced $2p$ level shift and width are measured for both kaonic helium-3 and helium-4 atoms [10][11][18], and it is the first time that kaonic helium-3 x-rays are measured. The results of the shift as listed in Table 1 are consistent with the results of E570 [4], thus a zero-compatible shift of the $2p$ level from experiment is established, which is in agreement with the theoretical estimations in [7][19]. We have not found abnormally large widths which can directly support the estimations made in conjunction with the prediction of a deeply-bound kaon states [8][9].

A decisive measurement that will give a conclusion to this topic is under preparation at J-PARC [20], in which the transition-edge sensor micro-calorimeter as the x-ray detector will be used to reach the precision requirement of sub-electronvolt.

3. Future perspectives

As a continuation of the SIDDHARTA experiment which successfully applied the timing capability in addition to the excellent energy resolution of the SDDs, the extended SIDDHARTA-2 collaboration is preparing a series of modifications and upgrades of the apparatus and experimental configurations, aiming to measure the kaonic deuterium x-rays with a precision compatible with that achieved in the hydrogen target measurement [21]. The main upgrades include the development of the new SDD chips with larger effective area to increase the acceptance and the application under temperature below 50 K to reach the best timing resolution to improve the signal to background ratio. The first precision measurement result of the kaonic deuterium x-ray from SIDDHARTA-2 will finally determine the isospin dependent components of the s-wave scattering lengths of the $\bar{K}N$ interaction, which are the missing building blocks for the Chiral SU(3) based effective field theories of the low-energy kaon-nucleus interaction.

A promising future perspective from the precision point of view comes from a recent publication on the measurement of exotic atom x-rays using the transition-edge sensor micro-calorimeters in a hadronic beam line [22]. The exploratory experiment done at the πM1 pion beam line at Paul Scherrer Institute successfully measured the x-ray transitions of pionic atoms, with an absolute energy uncertainty of 0.1 eV in the 6 keV region. This break-through demonstrated the feasibility of the application of the microcalorimeter for high precision spectroscopy at high energy particle beamlines. It is also a milestone towards the realization of the decisive high precision measurement of the kaonic helium-3 and helium-4 x-rays to be performed at J-PARC.

Acknowledgements

We thank C. Capoccia, G. Corradi, B. Dulach, and D. Tagnani from LNF-INFN; and H. Schneider, L. Stohwasser, and D. Stükler from Stefan-Meyer-Institut, for their fundamental contribution in designing and building the SIDDHARTA setup. We thank as well the DAΦNE staff for the excellent working conditions and permanent support. Part of this work was supported by the European Community-Research Infrastructure Integrating Activity

"Study of Strongly Interacting Matter" (HadronPhysics2, Grant Agreement No. 227431, and HadronPhysics3 (HP3) Contract No. 283286) under the Seventh Framework Programme of EU; HadronPhysics I3 FP6 European Community program, Contract No. RII3-CT-2004-506078; Austrian Science Fund (FWF) (P24756-N20); Austrian Federal Ministry of Science and Research BMBWK 650962/0001 VI/2/2009; Croatian Science Foundation under Project No. 1680; Romanian National Authority for Scientific Research, Contract No. 2-CeX 06-11-11/2006; and the Grant-in-Aid for Specially Promoted Research (20002003), MEXT, Japan.

References

[1] Batty C J, Friedmann E, Gal A 1997 *Phys. Rep.* **287** 385
[2] Deser S et al 1954 *Phys. Rev.* **96** 774
[3] Meissner U -G et al 2004 *Eur. Phys. J. C* **35** 349
[4] Okada S et al 2007 *Phys. Lett. B* **653** 387
[5] Wiegand C and Pehl R 1971 *Phys. Rev. Lett.* **27** 1410
[6] Batty C J et al 1979 *Nucl. Phys. A* **326** 455
[7] Baird S et al 1983 *Nucl. Phys. A* **392** 297
[8] Akaishi Y and Yamazaki T 2002 *Phys. Rev. C* **55** 044005
[9] Akaishi Y 2005 *Proc. Inter. Conf. on Exotic Atoms EXA05* p 45
[10] Bazzi M et al 2009 *Phys. Lett. B* **681** 310
[11] Bazzi M et al 2011 *Phys. Lett. B* **697** 199
[12] Bazzi M et al 2011 *Phys. Lett. B* **704** 113
[13] Bazzi M et al 2016 *Nucl. Phys. A* **954** 7
[14] Iwasaki M et al 1997 *Phys. Rev. Lett.* **78** 3067
[15] Ito T M et al 1998 *Phys. Rev. C* **58** 2366
[16] Beer G et al 2005 *Phys. Rev. Lett.* **94** 212302
[17] Bazzi M et al 2013 *Nucl. Phys. A* **907** 19
[18] Bazzi M et al 2012 *Phys. Lett. B* **714** 40
[19] Batty C J 1990 *Nucl. Phys. A* **508** 89
[20] Okada S 2015 *Proposal of E62 experiment at J-PARC PAC*
[21] SIDDHARTA-2 Collaboration 2010 *Proposal at Laboratori Nazionali di Frascati, INFN, "The upgrade of the SIDDHARTA apparatus for an enriched scientific case"*
[22] Okada S et al 2016 *Prog. Theor. Exp. Phys.* 091D01

LTH 1110

Towards NNLO accuracy for ε'/ε

M Cerdà-Sevilla[1], **M Gorbahn**[1], **S Jäger**[2] **and A Kokulu**[1]

[1] *Department of Mathematical Sciences, University of Liverpool, Liverpool L69 7ZL, UK*

[2] *Department of Physics and Astronomy, University of Sussex, Brighton BN1 9QH, UK*

E-mail: Maria.Cerda-Sevilla@liverpool.ac.uk, martin.gorbahn@liverpool.ac.uk, S.Jaeger@sussex.ac.uk, Ahmet.Kokulu@liverpool.ac.uk

Abstract. The quantity ε'/ε measures direct CP violation in Kaon decays. Recent SM predictions show a 2.9σ tension with data, with the theoretical uncertainty dominating. As rapid progress on the lattice is bringing nonperturbative long-distance effects under control, a more precise knowledge of short-distance contributions is needed. We describe the first NNLO results for ε'/ε and discuss future prospects, as well as issues of scheme dependence and the separation of perturbative and nonperturbative effects. Finally we also comment on the solution of the renormalisation-group evolution in one of the talks at this conference and present the correct solution.

1. Introduction

CP violation (CPV) is one of the most fascinating phenomena of high energy physics and is a natural place to search for physics beyond the Standard Model (SM). For instance, the SM is unable to account for the observed matter-antimatter asymmetry in the universe, which, in thermal baryogenesis scenarios, requires CPV. New sources of CPV, however, generally modify the SM predictions for CPV in flavour-violating decays. In the case of CPV in $K_L \to \pi\pi$, the SM predicts a particularly strong suppression due to mass and CKM hierarchies, which imply powerful Glashow-Iliopoulos-Maiani (GIM) cancellations. This mechanism need not apply to models of new physics. Therefore kaons are particularly promising for shedding light on physics at a more fundamental level. High-precision CP-violating kaon observables offer the exciting possibility of establishing the presence of new physics (NP).

While CPV in K_L decays was discovered 50 years ago, and direct CPV—quantified by the parameter ε'—has been firmly established for more than a decade, the theoretical treatment of the latter suffered from nonperturbative uncertainties which until very recently could not be computed in a controlled approximation. While short-distance contributions from scales $\sim m_c$ and above have been known to next-to-leading-order in perturbation theory since the mid-1990s [1–6], the long-distance corrections, usually represented as $K \to \pi\pi$ matrix elements of local four-quark operators, have been for the first time determined with controlled systematics only in 2015 [7,8]. This achievement, when combined with the short-distance calculations, opens the possibility of a precision theory prediction of ε'. A state-of-the-art analysis at NLO within the SM [9] gives

$$\left(\frac{\varepsilon'}{\varepsilon}\right)_{\rm SM} = (1.9 \pm 4.5) \times 10^{-4}. \tag{1}$$

Content from this work may be used under the terms of the Creative Commons Attribution 3.0 licence. Any further distribution of this work must maintain attribution to the author(s) and the title of the work, journal citation and DOI.
Published under licence by IOP Publishing Ltd

(The complex phase is very close to zero.) This is to be contrasted with the current experimental world average based on measurements at NA48 [10] and KTeV [11,12] of

$$\mathrm{Re}\left(\frac{\varepsilon'}{\varepsilon}\right)_{\mathrm{exp}} = (16.6 \pm 2.3) \times 10^{-4}. \tag{2}$$

The 2.9σ difference between experiment and theory could have one of several origins, the most exciting among them being a possible contribution from new particles beyond the SM. Clearly, a refined SM prediction is necessary. While at the moment the error budget [9] is dominated by the uncertainty on a single non-perturbative matrix element, this error is expected to shrink in the near future with further progress in lattice QCD. The next most important source of uncertainty at present is due to missing higher-order perturbative contributions, making NNLO accuracy imperative. In this proceedings, we preview (partial) results of an ongoing NNLO calculation [13], which roughly cuts the (total) perturbative error in half. We further describe a renormalisation-group-invariant formalism for separating perturbative and non-perturbative effects and briefly discuss future perspectives. Finally, we comment on and correct the expressions for the mixed QCD-QED renormalisation-group evolution [14] that were presented in another talk at this conference.

2. $\Delta S = 1$ Effective Hamiltonian

The key tool to disentangle the physics of the different scales ($\Lambda_{\mathrm{QCD}}, m_c, m_b, M_W$ and possibly M_{NP}) is the weak effective $\Delta S = 1$ Hamiltonian [1–6]. It takes the general form

$$\mathcal{H}_{\mathrm{eff}} = \sum C_i(\mu) Q_i(\mu), \tag{3}$$

where the process-independent Wilson coefficients $C_i(\mu)$ contain the physics of scales (virtualities) above μ (μ^2), while the degrees of freedom below μ are encoded in process-dependent hadronic matrix elements of local operators Q_i, in the present case $\langle \pi\pi | Q_i(\mu) | K \rangle$. These hadronic matrix elements are non-perturbative, and the lack of systematic predictions for them has precluded controlled predictions of ε' for a long time. This bottleneck has begun to be removed by recent developments in lattice QCD, in particular by the RBC and UKQCD collaborations [7,8].

Several choices of the scale μ are possible. So far, the state of the art [9] consisted of evaluating the Wilson coefficients in a three-flavour ($n_f = 3$) theory, obtained by integrating out the weak scale, as well as the bottom and charm quarks. Hadronic matrix elements with fully quantified systematic errors are available in the $n_f = 3$ theory from RBC and UKQCD [7,8] and agree well with earlier calculations in realistic models [15–17]. In the future, nonperturbative calculations with dynamical charm quark will likely become available, allowing to work based on a weak Hamiltonian in a theory with $n_f = 4$ quark flavours. It is convenient to split the Wilson coefficients as $C_i = \frac{4G_F}{\sqrt{2}} V_{ud} V_{us}^* (z_i + \tau y_i)$, where z_i and y_i are real and $\tau = -(V_{td}V_{ts}^*)/(V_{ud}V_{us}^*)$ is the phase-convention-independent CP-violating CKM combination governing CPV in K decays. For $n_f = 4$ (dynamical charm quark), we then have[1]

$$\mathcal{H}_{\mathrm{eff}} = \frac{4G_F}{\sqrt{2}} V_{ud} V_{us}^* \left(\sum_{i=1}^{2} z_i(\mu) [Q_i^u - Q_i^c] + \tau \left[\sum_{i=1}^{2} z_i(\mu) Q_i^c + \sum_{i=3..10,7\gamma,8g} y_i(\mu) Q_i \right] \right), \tag{4}$$

where we have made use of the fact that $z_{1,2}^c = y_{1,2}^c = -z_{1,2}$, while GIM cancellations ensure $z_{3...10} = 0$.

[1] Using the projection operators P_L and P_R instead of the traditional $V \pm A \times V \pm A$ introduces a factor of 4 in the overall normalisation of the Hamiltonian.

If the charm quark is integrated out ($n_f = 3$), we have instead

$$\mathcal{H}_{\text{eff}} = \frac{4G_F}{\sqrt{2}} V_{ud} V_{us}^* \left(\sum_{i=1}^{2} z_i(\mu) Q_i^u + \sum_{i=3...10,7\gamma,8g} z_i(\mu) Q_i + \sum_{i=3...10,7\gamma,8g} \tau y_i(\mu) Q_i \right). \tag{5}$$

The operators $Q_{1,2}^c$ involving the charm quark have disappeared. The hard GIM breaking in the $n_f = 3$ theory now entails nonzero values also for the "real parts" $z_{3...10,7\gamma,8g}$ of $C_{3...10,7\gamma,8g}$.

The Wilson coefficients $C_i(\mu)$, or equivalently $z_i(\mu)$ and $y_i(\mu)$, are determined using renormalisation-group-improved perturbation theory. Schematically,

$$C_i(\mu_3) = \left[U(\mu_3, \mu_c) M^c(\mu_c) U(\mu_c, \mu_b) M^b(\mu_b) U(\mu_b, \mu_W) \right]_{ij} C_j(\mu_W), \tag{6}$$

which is easily decomposed into expressions for z_i and y_i. The evolution operators $U(\mu_1, \mu_2)$ resum large logarithms $\ln \mu_2/\mu_1$, while the matching matrices $M(\mu)$ connect the Wilson coefficients across a flavour threshold, in particular:

$$C_i^{(n_f=3)}(\mu_c) = M_{ij}^c(\mu_c) C_j^{(n_f=4)}(\mu_c). \tag{7}$$

The operators can be classified as current-current ($Q_{1,2}^u$, $Q_{1,2}^c$), QCD penguin ($i = 3 \ldots 6, 8g$), and electroweak penguin ($i = 7 \ldots 10, 7\gamma$). At present, all ingredients in (6) are known to NLO accuracy (with the exception of the dipole operators Q_{8g} and $Q_{7\gamma}$), and some of them to NNLO accuracy. The work described here completes the QCD-penguin part to NNLO, which removes about half of the current perturbative uncertainty on ε'.

3. NNLO matching for the QCD penguins

To obtain the QCD-penguin Wilson coefficients to NNLO accuracy, we require the submatrices of M^c and M^b for $Q_{1...6,8g}$ and $Q_{1,2}^c$. We leave aside the dipole operator for the time being, which is formally suppressed by m_s/Λ_{QCD}. The new ingredients are then related to the removal of the operators Q_1^c and Q_2^c at a scale $\mu_c \sim m_c$, although we have recalculated the remaining, known [18] elements of the matching matrices M^c, M^b. The anomalous dimension matrices are known from [19].

To determine these threshold corrections, we match Green's functions with operator insertions in the n_f- and $n_f + 1$-flavour theories at the matching scale $\mu_i = \mathcal{O}(m_i)$ ($m_i = m_c, m_b$).

Novel effects at NNLO include current-current operator insertions getting contributions from virtual charm/bottom-quarks and the strong coupling constant being discontinuous at the threshold scale. In our calculation we use dimensional regularization and avoid the appearance of traces over γ_5 by employing the so-called "modern basis" [20,21]; however this requires extra steps to connect with the "traditional" basis employed in the existing literature on ε', and in the lattice-QCD calculations, which is nontrivial at NNLO. We expand in the external momenta (as appropriate for a matching onto dimension-six operators) and set the masses of the n_f light quarks to zero. After renormalisation, the n_f+1-flavour result still contains infrared divergences in the form of poles in $\epsilon = (4-d)/2$, which have to be reproduced in the n_f-flavour theory. As the Green's functions in the n_f-flavour theory contain only massless tadpole loop diagrams (after expansion in the external momenta), they are given entirely in terms of the ultraviolet counterterms in the n_f-flavour theory, which are related to known anomalous dimensions. The cancellation of divergences constitutes an important check of our calculation. In the end, the matching results in finite threshold corrections for the Wilson coefficients

$$C_i^{(n_f)}(\mu) = \left[M_{n_f, n_f+1}(\mu) \right]_{ij} C_j^{(n_f+1)}(\mu). \tag{8}$$

4. Scale and scheme independence

Physical observables cannot depend on the dimensional renormalisation scale μ. This means that the μ-dependence has to cancel between the Wilson coefficients $C_i(\mu)$ and matrix elements $\langle Q_i(\mu)\rangle$ order by order in perturbation theory. However, in the present case the matrix elements are found from non-perturbative calculations and their explicit μ-dependence is not known. Instead, we can verify explicitly that our Wilson coefficients $C_i(\mu)$ satisfy the NNLO renormalisation-group equations up to $\mathcal{O}(\alpha_s^3)$ remainder terms, which they do. However, this still does not allow to give an estimate of the impact of missing higher-order perturbative corrections, part of which are contained in the published lattice-QCD matrix elements and inseparable from the non-perturbative uncertainties from dynamics at scales $\sim \Lambda_{\rm QCD}$.

To allow for a full separation of scales, we note that the evolution operator $U(\mu_1, \mu_2)$ in fact factorizes,

$$U(\mu_1, \mu_2) = J(\mu_1) U^{(0)}(\mu_1) \left(U^{(0)}(\mu_2)\right)^{-1} J^{-1}(\mu_2), \tag{9}$$

where

$$U^{(0)}(\mu) = \exp\left(-\frac{\gamma^{0T}}{2\beta_0} \ln \alpha_s(\mu)\right), \tag{10}$$

$$J(\mu) = \hat{1} + \frac{\alpha_s(\mu)}{4\pi} J^{(1)} + \left(\frac{\alpha_s(\mu)}{4\pi}\right)^2 J^{(2)} + \dots . \tag{11}$$

$U^{(0)}(\mu)$ is just the leading-order evolution operator, with the usual ratio $\alpha_s(\mu_2)/\alpha_s(\mu_1)$ replaced by $1/\alpha_s(\mu)$.

We can then define the following object appearing in the intermediate stages in our calculation:

$$\hat{M}(\mu_i) = \left(U^{(0)}_{n_f}(\mu_i)\right)^{-1} \left(J_{n_f}(\mu_i)\right)^{-1} M_{n_f, n_f+1}(\mu_i) J_{n_f+1}(\mu_i) U^{(0)}_{n_f+1}(\mu_i). \tag{12}$$

Note that the matrix $\hat{M}(\mu_i)$ contains *all* the dependence on the scale μ_i in the expression (6) for the Wilson coefficient, and depends on no other scales. As a consequence, it must by itself be independent of the renormalisation scale. The way our calculation is set up, we obtain the matching matrix with dependence on both $\alpha_s^{(n_f)}(\mu)$ and on $\alpha_s^{(n_f+1)}(\mu)$. Expressing $\alpha_s^{(n_f+1)}$ in terms of $\alpha_s^{(n_f)}$ and taking into account both the explicit μ-dependence and the μ-dependence of the quark mass appearing in the matching, we observe that the scale dependence cancels, up to terms of order $\mathcal{O}(\alpha_s^3)$. The residual μ-dependence, which is of order α_s^3, gives a complete estimate of higher-order effects at the threshold μ_i. We find that these residual scale dependences are at the percent level and that the value of ε'/ε is only mildly shifted when going from NLO to NNLO. This points to an excellent behaviour of the perturbation series not only at the scale μ_b, but also at the scale μ_c. This means at least for the dominant contribution the perturbation theory seems to work.

Our results are best expressed in terms of renormalisation-group and scheme-independent Wilson coefficients and matrix elements,

$$\hat{C}_i^{(n_f)} = \left(U^{(0)}_{n_f}(\mu)\right)^{-1}_{ik} \left(J_{n_f}(\mu)\right)^{-1}_{kj} C_j^{(n_f)}(\mu), \tag{13}$$

$$\langle \pi\pi | \hat{Q}_i^{(n_f)} | K\rangle = \langle \pi\pi | Q_j^{(n_f)}(\mu) | K\rangle \left(J_{n_f}(\mu)\right)_{jk} \left(U^{(0)}_{n_f}(\mu)\right)_{ki} . \tag{14}$$

Note that we have not suppressed any μ-dependence: each hatted object is separately scale- and scheme-independent[2]. For this reason, relating the "modern" basis to the "traditional" basis is

[2] The explicit NNLO transformations of J, C and $\langle Q\rangle$ under a change of scheme and the resulting scheme cancellation can be found in Ref. [19].

trivial at the level of the hatted objects. Perturbative uncertainties are fully contained in the hatted coefficients, while non-perturbative uncertainties reside in the hatted matrix elements. The two matching steps together with RG evolution are then expressed as

$$\hat{C}_i^{(3)} = \hat{M}^c_{ik}\hat{M}^b_{kj}\hat{C}_j^{(5)}. \tag{15}$$

For dynamical charm, one simply removes the last matching step.

In fact, the hatted hadronic matrix elements could in principle be obtained nonperturbatively on the lattice, obviating or reducing the need to resort to technically demanding lattice-continuum matching. If this is not done, the resultant extra "perturbative" uncertainty may be considered as part of the lattice error budget.

5. ε'/ε within the Standard Model

Lattice computations of the relevant matrix elements for the analysis of ε'/ε are currently performed only in the isospin limit ($\alpha = 0$, degenerate masses), and the Standard-Model prediction is based on the following expression

$$\frac{\varepsilon'}{\varepsilon} = -i\frac{\omega_+}{\sqrt{2}|\varepsilon_K|}e^{i(\delta_2-\delta_0-\phi_{\varepsilon_K})}\left[\frac{\mathrm{Im}(\mathcal{A}_0)}{\mathrm{Re}(\mathcal{A}_0)}(1-\Omega_{\mathrm{eff}}) - \frac{1}{a}\frac{\mathrm{Im}(\mathcal{A}_2)}{\mathrm{Re}(\mathcal{A}_2)}\right], \tag{16}$$

which is accurate up to $\mathcal{O}(1/\omega_+)$ corrections. Here $\mathcal{A}_I \equiv \langle(\pi\pi)_I|\mathcal{H}_{\mathrm{eff}}|K\rangle$ $(I = 0, 2)$ are the amplitudes for the two isospin states with their strong phases $\delta_{0,2}$ removed. The latter, as well as the phase ϕ_{ε_K} and magnitude $|\varepsilon_K|$of ε_K and the isospin ratio ω_+ are all determined from experimental data. The coefficients a and Ω_{eff} are a (partial) parameterization of corrections to the isospin limit, with values computed in chiral perturbation theory in [22–24].

The evaluation of the ratios $\mathrm{Im}(\mathcal{A}_I)/\mathrm{Re}(\mathcal{A}_I)$ is the central ingredient to the theory prediction of ε'/ε and the formalism used for their determination is explained in reference [9]. The quantities

$$\mathrm{Re}(\mathcal{A}_0) \approx \frac{4G_F}{\sqrt{2}}V_{ud}V^*_{us}(z_+\langle Q_+\rangle_0 + z_-\langle Q_-\rangle_0), \quad \mathrm{Re}(\mathcal{A}_2) \approx \frac{4G_F}{\sqrt{2}}V_{ud}V^*_{us}z_+\langle Q_+\rangle_2 \tag{17}$$

are completely dominated by the two operators $Q^u_{1,2}$. ($Q_\pm = \frac{1}{2}(Q^u_2 \pm Q^u_1)$.) Operator identities among the $V-A\times V-A$ operators imply that the ratios

$$\left(\frac{\mathrm{Im}(\mathcal{A}_2)}{\mathrm{Re}(\mathcal{A}_2)}\right)_{V-A} = \mathrm{Im}(\tau)\frac{3(y_9+y_{10})}{2z_+} \tag{18}$$

$$\left(\frac{\mathrm{Im}(\mathcal{A}_0)}{\mathrm{Re}(\mathcal{A}_0)}\right)_{V-A} = \mathrm{Im}(\tau)\frac{2y_4}{(1+q)z_-} + \mathcal{O}(p_3), \tag{19}$$

where in the numerators only the terms involving the $V-A\times V-A$ operators are kept, are almost free from hadronic uncertainties. The factor q in Eq.(19) is given in terms of the current-current Wilson coefficients and operators, $q \equiv (z_+(\mu)\langle Q_+(\mu)\rangle_0)/(z_-(\mu)\langle Q_-(\mu)\rangle_0)$. The small ratio z_+/z_- implies that only a modest accuracy is needed on the hadronic matrix elements entering the isospin-0 ratio through q. The term $\mathcal{O}(p_3)$ also involves a small Wilson coefficient, which moreover multiplies a colour-suppressed hadronic matrix element.

Conversely, the contributions coming from the $(V-A)\times(V+A)$ operators are very sensitive to long-distance effects (hadronic matrix elements). To minimize the nonperturbative uncertainties, one notes that $\mathrm{Re}(\mathcal{A}_{0,2})$ govern CP-conserving $K\to\pi\pi$ decays and are well determined by data. One then has

$$\left(\frac{\mathrm{Im}(\mathcal{A}_0)}{\mathrm{Re}(\mathcal{A}_0)}\right)_{V+A} = -\frac{4G_F}{\sqrt{2}}V_{ud}V^*_{us}\,\mathrm{Im}(\tau)\,y_6\frac{\langle Q_6\rangle_0}{\mathrm{Re}(\mathcal{A}_0)} + \mathcal{O}(p_5) \tag{20}$$

$$\left(\frac{\mathrm{Im}(\mathcal{A}_2)}{\mathrm{Re}(\mathcal{A}_2)}\right)_{V+A} = -\frac{4G_F}{\sqrt{2}}V_{ud}V^*_{us}\,\mathrm{Im}(\tau)\,y_8^{\mathrm{eff}}\frac{\langle Q_8\rangle_2}{\mathrm{Re}(\mathcal{A}_2)}. \tag{21}$$

Again, the omitted terms are suppressed by small Wilson coefficients and/or colour-suppressed operator matrix elements. As a result, the prediction for ε' involves predominantly two hadronic matrix elements (often parameterised in terms of parameters $B_6^{(1/2)}$ and $B_8^{(3/2)}$), as well as perturbative Wilson coefficients $z_{1,2}$ and $y_{6,8}$. Our new calculation essentiall removes the perturbative uncertainty on y_6. The uncertainties on $z_{1,2}$ are already tiny, leaving y_8 and an improved treatment of isospin-breaking corrections as the main objectives for the future.

Note that using $\mathrm{Re}(\mathcal{A}_{0,2})$ from data also in the $V-A\times V-A$ operators introduces dependence on (mainly) the matrix element of the operator Q_4, and its Wilson coefficient, and should be avoided. This is the main reason why the prediction of [9] is more accurate than that of RBC and UKQCD, and leads to a more pronounced tension with the data, in spite of employing the same nonperturbative matrix elements and including an error estimate for isospin-breaking corrections.

6. Summary and outlook

The experimental data for ε'/ε is in mild tension with the SM. This conclusion is possible because of recent progress in lattice QCD, which has reduced the (still dominant) uncertainties on long-distance non-perturbative contributions. With further progress expected in the near future, currently subleading uncertainties, most importantly higher-order perturbative short-distance contributions, will start to dominate. We have described a calculation at NNLO accuracy of the short-distance contributions to the QCD penguin amplitudes, sufficient to achieve NNLO accuracy on the isospin-zero $K \to \pi\pi$ decay amplitude ratio. Our results leave the central value of the theoretical prediction nearly unchanged, while greatly reducing the perturbative uncertainty on this piece, leaving unknown NNLO corrections to the $I = 2$ amplitude ratio to dominate the perturbative uncertainty on ε', at about half the previous NLO error.

Several further steps will be needed before the theoretical uncertainty matches the experimental one, and perhaps a discrepancy with the SM can be confirmed. One of them is the computation of the $I = 2$ amplitude ratio to NNLO accuracy. Another is a better treatment of isospin-breaking corrections, and electromagnetic corrections in particular. The latter affect both the Wilson coefficients and the hadronic matrix elements, and require both perturbative and lattice calculations, as well as a consistent interfacing of the two.

Acknowledgments

MC, SJ and AK would like to thank the organisers of KAON 2016 conference for this wonderful experience. The work of MC, MG and AK has been supported by the STFC consolidated grant ST/L000431/1. SJ is supported by STFC consolidated grant ST/L000504/1, a Weizmann Institute "Weizmann-UK Making Connections" grant, and an IPPP Associateship. We would like to thank Teppei Kitahara for notifying us of a typo in Eq. (27) and sharing their revised version.

Comment on the solution of the renormalisation group evolution

At this conference one of the authors of Ref. [14] presented a solution to the renormalisation group equation

$$\mu \frac{d}{d\mu} C = \gamma^T C \tag{22}$$

for the Wilson coefficients C. Here

$$\gamma = \sum_{m+n\geq 1} \gamma^{(m,n)} \left(\frac{\alpha_s}{4\pi}\right)^m \left(\frac{\alpha_e}{4\pi}\right)^n \tag{23}$$

represent the anomalous dimensions relevant for ε'/ε. While the presented solution incorporates some corrections of $\mathcal{O}(\alpha_e^2/\alpha_s^2)$ not all effects of this order have been take into account in this work. In the following we will give the correct solution for the evolution matrix (9) in the case of combined QCD and QED corrections, where

$$J(\mu) = \left[1 + \frac{\alpha_e}{4\pi}\left(J_0^{(1,1)} + J_1^{(1,1)} \ln \alpha_s + J_2^{(1,1)} (\ln \alpha_s)^2\right)\right] \left[1 + \frac{\alpha_s}{4\pi}\left(J_0^{(1,0)} + J_1^{(1,0)} \ln \alpha_s\right)\right] \times \left[1 + \frac{\alpha_e}{\alpha_s}\left(J_0^{(0,1)} + J_1^{(0,1)} \ln \alpha_s\right) + \frac{\alpha_e^2}{\alpha_s^2}\left(J_0^{(0,2)} + J_1^{(0,2)} \ln \alpha_s\right)\right] \tag{24}$$

follows the form chosen in Ref. [14]. To determine a differential equation for $J(\mu)$ we use the scale invariance $\mu(d/d\mu)\hat{C} = 0$ of the Wilson coefficients defined in (13) and trade the scale dependence in the total derivative for an α_s dependence as

$$\mu\frac{d}{d\mu} = -2\alpha_s^2 \frac{\beta_s}{4\pi}\left(\frac{\partial}{\partial \alpha_s} - \frac{\alpha_e^2 \beta_e}{\alpha_s^2 \beta_s}\frac{\partial}{\partial \alpha_e}\right). \tag{25}$$

Here the implicit μ dependence in α_e is now understood as $\alpha_e(\mu) = \alpha_e(\alpha_s(\mu))$ and β_s and β_e are defined via the QCD $\mu(d/d\mu)\alpha_s = -2\alpha_s^2\beta_s/(4\pi)$ and QED $\mu(d/d\mu)\alpha_e = +2\alpha_e^2\beta_e/(4\pi)$ beta functions. The resulting equation for J then reads:

$$\frac{\partial J}{\partial \alpha_s} = \frac{\alpha_e^2 \beta_e}{\alpha_s^2 \beta_s}\frac{\partial J}{\partial \alpha_e} - \frac{4\pi\, \gamma^T J}{2\alpha_s^2 \beta_s} + \frac{J \gamma^{(1,0)^T}}{2\alpha_s \beta_s^{(0,0)}} \tag{26}$$

and which implies a set of simple algebraic equations for the matrices $J_l^{(i,j)}$ which read

$$\begin{aligned}
0 &= \frac{\left[J_1^{(1,0)}, \gamma^{(1,0)^T}\right]}{2\beta_s^{(0,0)}} - J_1^{(1,0)} \\
0 &= \frac{\left[J_0^{(1,0)}, \gamma^{(1,0)^T}\right]}{2\beta_s^{(0,0)}} + \frac{\beta_s^{(1,0)}\gamma^{(1,0)^T}}{2\beta_s^{(0,0)^2}} - \frac{\gamma^{(2,0)^T}}{2\beta_s^{(0,0)}} - J_0^{(1,0)} - J_1^{(1,0)} \\
0 &= \frac{\left[J_1^{(0,1)}, \gamma^{(1,0)^T}\right]}{2\beta_s^{(0,0)}} + J_1^{(0,1)} \\
0 &= \frac{\left[J_0^{(0,1)}, \gamma^{(1,0)^T}\right]}{2\beta_s^{(0,0)}} - \frac{\gamma^{(0,1)^T}}{2\beta_s^{(0,0)}} + J_0^{(0,1)} - J_1^{(0,1)} \\
0 &= \frac{\left[J_2^{(1,1)}, \gamma^{(1,0)^T}\right]}{2\beta_s^{(0,0)}} \\
0 &= \frac{\left[J_1^{(1,0)}, \gamma^{(0,1)^T}\right]}{2\beta_s^{(0,0)}} + \frac{\left[J_1^{(1,1)}, \gamma^{(1,0)^T}\right]}{2\beta_s^{(0,0)}} - 2J_2^{(1,1)} \\
0 &= \frac{\left[J_0^{(1,0)}, \gamma^{(0,1)^T}\right]}{2\beta_s^{(0,0)}} + \frac{\left[J_0^{(1,1)}, \gamma^{(1,0)^T}\right]}{2\beta_s^{(0,0)}} + \frac{\beta_s^{(1,0)}\gamma^{(0,1)^T}}{2\beta_s^{(0,0)^2}} + \frac{\beta_s^{(0,1)}\gamma^{(1,0)^T}}{2\beta_s^{(1,0)^2}} - \frac{\gamma^{(1,1)^T}}{2\beta_s^{(0,0)}} - J_1^{(1,1)} \\
0 &= \frac{\left[J_1^{(0,2)}, \gamma^{(1,0)^T}\right]}{2\beta_s^{(0,0)}} - \frac{\gamma^{(0,1)^T} J_1^{(0,1)}}{2\beta_s^{(0,0)}} + \frac{J_1^{(0,1)}\beta_e^{(0,0)}}{\beta_s^{(0,0)}} + 2J_1^{(0,2)} \\
0 &= \frac{\left[J_0^{(0,2)}, \gamma^{(1,0)^T}\right]}{2\beta_s^{(0,0)}} - \frac{\gamma^{(0,1)^T} J_0^{(0,1)}}{2\beta_s^{(0,0)}} + \frac{J_0^{(0,1)}\beta_e^{(0,0)}}{\beta_s^{(0,0)}} + 2J_0^{(0,2)} - J_1^{(0,2)}.
\end{aligned} \tag{27}$$

Here the expansion coefficients $\beta_s^{(n,m)}$ and $\beta_e^{(n,m)}$ of the beta functions are defined via

$$\beta_s = \sum_{n,m\geq 0} \beta_s^{(m,n)} \left(\frac{\alpha_s}{4\pi}\right)^m \left(\frac{\alpha_e}{4\pi}\right)^n \quad \text{and} \quad \beta_e = \sum_{n,m\geq 0} \beta_e^{(m,n)} \left(\frac{\alpha_e}{4\pi}\right)^m \left(\frac{\alpha_s}{4\pi}\right)^n . \tag{28}$$

In the original version of Ref. [14] terms proportional to $\beta_e^{(0,0)}$ and $\beta_s^{(0,1)}$ were absent. Our results agree with their revised version.

References

[1] A. J. Buras, M. Jamin, M. E. Lautenbacher and P. H. Weisz, *Effective Hamiltonians for $\Delta S = 1$ and $\Delta B = 1$ non-leptonic decays beyond the leading logarithmic approximation, Nucl. Phys.* **B370** (1992) 69-104 *Addendum: Nucl. Phys.* **B375** (1992) 501.

[2] A. J. Buras, M. Jamin, M. E. Lautenbacher and P. H. Weisz, *Two loop anomalous dimension matrix for $\Delta S = 1$ weak non-leptonic decays. 1. $\mathcal{O}(\alpha_s^2)$, Nucl. Phys.* **B400** (1993) 37-74, [hep-ph/9211304].

[3] A. J. Buras, M. Jamin and M. E. Lautenbacher, *Two loop anomalous dimension matrix for $\Delta S = 1$ weak non-leptonic decays. 2. $\mathcal{O}(\alpha\alpha_s^2)$, Nucl. Phys.* **B400** (1993) 75-102, [hep-ph/9211321].

[4] M. Ciuchini, E. Franco, G. Martinelli and L. Reina, *ε'/ε at the next-to-leading order in QCD and QED, Phys. Lett.* **B301** (1993) 263-271, [hep-ph/9212203].

[5] A. J. Buras, M. Jamin, M. E. Lautenbacher, *The anatomy of ε'/ε beyond leading logarithms with improved hadronic matrix elements, Nucl. Phys.* **B408** (1993) 209-285, [hep-ph/9303284].

[6] M. Ciuchini, E. Franco, G. Martinelli and L. Reina, *The $\Delta S = 1$ effective Hamiltonian including next-to-leading order QCD and QED corrections, Nucl. Phys.* **B415** (1994) 403-462, [hep-ph/9304257].

[7] T.Blum, P. Boyle, N.Christ, J.Frison, N. Garron, et al., *$K \to \pi\pi \Delta I = 3/2$ decay amplitude in the continuum limit, Phys. Rev. D* **91** (2015) no.7, 074502, [arXiv:1502.00263 [hep-lat]].

[8] Z. Bai,T.Blum, P. Boyle, N.Christ, J.Frison, et al., *Standard-model prediction for direct CP violation in $K \to \pi\pi$ decay, Phys. Rev. Lett.* **115** (2015) no.21, 212001, [arXiv:1505.07863 [hep-lat]].

[9] A. J. Buras, M. Gorbahn, S. Jäger, and M. Jamin, *Improved anatomy of ε'/ε in the Standard Model, JHEP* **1511** (2015) 202, [arXiv:1507.06345 [hep-ph]].

[10] NA48 Collaboration, J. Batley et al., *A Precision measurement of direct CP violation in the decay of neutral kaons into two pions, Phys. Lett.* **B544** (2002) 97-112, [hep-ex/0208009].

[11] KTeV Collaboration, A. Alavi-Harati et al., *Measurements of direct CP violation, CPT symmetry, and other parameters in the neutral kaon system, Phys. Rev.* **D67** (2003) 012005, [hep-ex/0208007].

[12] KTeV Collaboration, E. Worcester, *The Final Measurement of ε'/ε from KTeV,* [arXiv:0909.2555].

[13] M. Cerdà -Sevilla, M. Gorbahn, S. Jäger and A. Kokulu, in preparation.

[14] T. Kitahara, U. Nierste and P. Tremper, *Singularity-free Next-to-leading Order $\Delta S = 1$ Renormalization Group Evolution and $\varepsilon'_K/\varepsilon_K$ in the Standard Model and Beyond,* arXiv:1607.06727 [hep-ph].

[15] W. A. Bardeen, A. J. Buras, J.-M. Gérard, *The $K \to \pi\pi$ Decays in the Large-N Limit: Quark Evolution, Nucl. Phys.* **B293** (1987) 787.

[16] W. A. Bardeen, A. J. Buras and J. M. Gerard, *A Consistent Analysis of the $\Delta I = 1/2$ Rule for K Decays, Phys. Lett.* **B192** (1987) 138.

[17] A. J. Buras, J.-M. Gérard, W. A. Bardeen, *Large-N Approach to Kaon Decays and Mixing 28 years Later: $\Delta I = 1/2$ Rule, $\hat{B}_K$ and ΔM_K, Eur. Phys. J.* **C74** (2014), no. 5 2871, [arXiv:1401.1385].

[18] J. Brod and M. Gorbahn, *ϵ_K at Next-to-Next-to-Leading Order: The Charm-Top-Quark Contribution, Phys. Rev.* **D82** (2010) 094026, [arXiv: 1007.0684].

[19] M. Gorbahn and U. Haisch, *Effective Hamiltonian for non-leptonic $|\Delta F| = 1$ decays at NNLO in QCD, Nucl. Phys.* **B713** (2005) 291-332, [hep-ph/0411071].

[20] K.G. Chetyrkin, M. Misiak and M. Munz, *$|\Delta F| = 1$ nonleptonic effective Hamiltonian in a simpler scheme, Nucl. Phys.* **B520** (1998) 279, [aXiv:hep-ph/9711280].

[21] C. Bobeth, P. Gambino, M. Gorbahn and U. Haisch, *Complete NNLO QCD analysis of $\bar{B} \to X_s \ell^+ \ell^-$ and higher order electroweak effects, JHEP* **0404** (2004) 071, [hep-ph/0312090].

[22] V. Cirigliano, A. Pich, G. Ecker and H. Neufeld, *Isospin violation in ε' Phys. Rev. Lett.* **91** (2003) 162001, [hep-ph/0307030].

[23] V. Cirigliano, G. Ecker, H. Neufeld and A. Pich, *Isospin breaking in $K \to \pi\pi$, Eur. Phys. J.* **C33** (2004) 369, [hep-ph/0310351].

[24] V. Cirigliano, G. Ecker, H. Neufeld, A. Pich and J. Portoles, *Kaon Decays in the Standard Model, Rev. Mod. Phys.* **84** (2012) 399, [arXiv:1107.6001 [hep-ph]].

IOP Conf. Series: Journal of Physics: Conf. Series **800** (2017) 012009 doi:10.1088/1742-6596/800/1/012009

Updated limit for the decay $K^0_S \to \mu^+\mu^-$

Miguel Ramos Pernas, on behalf of the LHCb collaboration
University of Santiago de Compostela, Physics Faculty, 15782 Santiago de Compostela, Spain
E-mail: miguel.ramos.pernas@cern.ch

Abstract. A search for the decay $K^0_S \to \mu^+\mu^-$ is performed using the data sample of *pp* collisions at $\sqrt{s} = 8$ TeV collected by the LHCb experiment at the Large Hadron Collider in 2012, corresponding to an integrated luminosity of $2.0\,\text{fb}^{-1}$. In the Standard Model, this decay is expected to be suppressed at a level below the current experimental sensitivity, but could be enhanced by new sources of *CP* violation. The observed number of candidates is compatible with the expected background and allows to set an upper limit $\mathcal{B}(K^0_S \to \mu^+\mu^-) < 6.9(5.8) \times 10^{-9}$ at 95 (90)% confidence level. This limit improves the result published by LHCb using $1.0\,\text{fb}^{-1}$ of collisions collected in 2011 at $\sqrt{s} = 7$ TeV.

1. Introduction

In the Standard Model (SM), the still unobserved $K^0_S \to \mu^+\mu^-$ decay can proceed only through a Flavour-Changing Neutral-Current (FCNC) transition. This process cannot occur at tree level and is further suppressed by the small amount of *CP* violation in kaon decays, since the S-wave component of the decay is forbidden when *CP* is conserved. In the SM, the decay amplitude is expected to be dominated by long distance contributions, which can be precisely constrained using the observed decays $K^0_S \to \gamma\gamma$ and $K^0_L \to \pi^0\gamma\gamma$. This leads to the prediction $\mathcal{B}\left(K^0_S \to \mu^+\mu^-\right) = (5.0 \pm 1.5) \times 10^{-12}$ [1, 2]. The corresponding branching fraction for the K^0_L decay into two muons is $(6.85 \pm 0.32) \times 10^{-9}$ [3], in excellent agreement with the experimental world average $\mathcal{B}\left(K^0_L \to \mu^+\mu^-\right) = (6.84 \pm 0.11) \times 10^{-9}$ [4].

Due to its suppression in the SM, the $K^0_S \to \mu^+\mu^-$ decay amplitude can be enhanced by possible New Physics (NP) contributions, notably from light scalars with *CP*-violating Yukawa couplings [1]. Contributions up to one order of magnitude above the SM expectation naturally arise in many models and would be compatible with the present bounds from other FCNC processes. An upper limit on $\mathcal{B}(K^0_S \to \mu^+\mu^-)$ close to 10^{-11} could be translated into model-independent bounds on the *CP*-violating phase of the $s \to dl^+l^-$ amplitude. This would be very useful to discriminate among NP scenarios in case other modes, such as $K^+ \to \pi^+\nu\bar{\nu}$, indicated a non-standard enhancement of such transitions [2].

2. Kaons at LHCb

The LHCb [5] experiment is mainly devoted to the study of flavour physics and *CP* violation in the b and c hadron decays. However, it can also be considered a kaon factory, since around $10^{13}/\,\text{fb}^{-1}$ K^0_S decay inside the LHCb acceptance.

The LHCb detector is a single-arm forward spectrometer covering the pseudorapidity range $2 < \eta < 5$. It is equipped with a high precision tracking system, providing a very good resolution in the measurements of the Impact Parameter (IP), momenta and mass. This allows to

Content from this work may be used under the terms of the Creative Commons Attribution 3.0 licence. Any further distribution of this work must maintain attribution to the author(s) and the title of the work, journal citation and DOI.
Published under licence by IOP Publishing Ltd

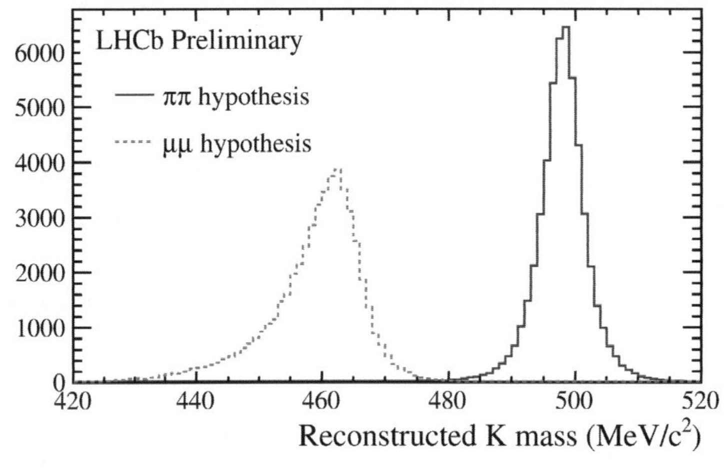

Figure 1. Invariant mass distributions for $K^0_S \to \pi^+\pi^-$ events, where the daughters are reconstructed as pions (solid blue), and as muons (dashed red).

reconstruct the invariant mass of $K^0_S \to \pi^+\pi^-$ events with a resolution of 4 MeV/c^2 (see Fig. 1). Another important feature of the LHCb detector is the very efficient particle identification. Most of the information used to do this task comes from two different Cherenkov detectors, complemented using other subdetectors. A new muon identification algorithm (μID) for low transverse momentum (p_T) tracks has been explicitly developed for this analysis, reaching a fraction of pions misidentified as muons of $\sim 0.49\%$, while maintaining an efficiency of 95% for the real muons. Most of the remaining misidentified pions come from $\pi \to \mu\nu$ decays inside the detector.

There are three different trigger systems to filter the large amount of data to be recorded by LHCb. The first level (L0), is implemented in hardware. Due to its p_T cuts, optimized for the study of b/c hadrons, it constitutes the biggest limiting factor in the study of kaons at LHCb. The two other trigger levels (Hlt1 and Hlt2) are implemented in software, being very flexible, so they can be optimized for any study. Although the L0 trigger could not be changed, the Hlt2 has been improved for kaon decays selection in the second part of the Run-I data-taking (2012). The trigger efficiency increased by a factor 2.5, leading to an overall trigger efficiency of $\sim 2.5\%$.

3. Analysis procedure

This document summarizes the preliminary results recently published by the LHCb collaboration [6]. The data used in this analysis corresponds to 2 fb^{-1} collected in 2012 by LHCb. The events from the data sample inside the signal region $m_{\mu\mu} \in [492, 504]$ MeV/c^2 are blinded, and only used in the final fit.

In order to exploit the maximum capability of the LHCb trigger, three different trigger selections have been used. Taking into account the previous LHCb result $\mathcal{B}\left(K^0_S \to \mu^+\mu^-\right) < 11 \times 10^{-9}$ [7], and considering the improvements coming from the optimization done for the 2012 data-taking period, the expected sensitivity of this analysis is $\mathcal{B}\left(K^0_S \to \mu^+\mu^-\right) \sim 10^{-9}$.

The $K^0_S \to \mu^+\mu^-$ candidates are obtained from tracks with opposite charge identified as muons, forming a vertex with invariant mass in the range $m_{\mu\mu} \in [400, 600]$ MeV/c^2. Due to the large branching fraction and abundance of $K^0_S \to \pi^+\pi^-$ events, this decay mode is used as normalization channel. In the case where the two pions are identified as muons, the invariant mass of the candidate is underestimated on average by ~ 40 MeV/c^2, which corresponds to ~ 10

times the resolution in that mass region.

In order to remove combinatorial background events, a Multivariate Analysis (MVA) has been done. The search for the algorithm with the best performance for each trigger category, lead to the utilization of a Boosted Decision Tree (BDT) with the same input variables for each one. The variables used correspond to kinematical and reconstruction quality variables, carefully chosen in order not to distinguish $\mu^+\mu^-$ and $\pi^+\pi^-$ modes, and to avoid the correlation with the mass of the candidate. The samples used to train the BDT correspond to data, trigger unbiased $K^0_S \to \pi^+\pi^-$ candidates for signal, and data $K^0_S \to \mu^+\mu^-$ candidates from the far right sideband ($m_{\mu^+\mu^-} \in [600, 950]$ MeV/c^2) for background. Each background sample consists of $K^0_S \to \mu^+\mu^-$ events passing the associated trigger conditions. Due to the presence of muon-ID cuts, and to avoid over-training effects, the different $K^0_S \to \pi^+\pi^-$ samples are built applying cuts to emulate the trigger effects.

The analysis is performed in bins of the BDT distribution, chosen after a loose cut to remove most of the combinatorial background events. A simultaneous maximum likelihood fit is done to all the trigger categories and BDT bins. The normalization factors are included so the branching fraction can be extracted directly from the fit. A constraint is applied on this magnitude using the previous LHCb result [7]. In order to extract the limit, the integral of the probability calculated from the $-\log\mathcal{L}$ is used.

3.1. Backgrounds

The combinatorial background is made from candidates reconstructed from secondary hadronic collisions in the detector material, which generate fake secondary vertices. Its structure is expected to exhibit a smooth mass distribution, with no peak in the di-muon mass. This contribution has been reduced using the aforementioned BDT algorithm.

The decay $K^0_L \to \mu^+\mu^-$ is probably the most dangerous background on the study of $K^0_S \to \mu^+\mu^-$. It becomes indistinguishable from $K^0_S \to \mu^+\mu^-$ at LHCb. However, due to the larger lifetime of K^0_L and after applying a lifetime acceptance correction, it has been proved that its effective branching fraction is $\mathcal{B}_{\text{eff.}}\left(K^0_L \to \mu^+\mu^-\right) \sim 10^{-11}$, thus negligible for this study.

A study on Monte-Carlo samples shows that the amount of $K^0_{S/L} \to \pi^+\mu^-\bar{\nu}$ events are also negligible. Contributions from $\omega \to \pi^0\mu^+\mu^-$ and $\eta \to \mu^+\mu^-\gamma$ are not expected to generate peaking structures in the di-muon spectrum either.

In conclusion, the main background remains $K^0_S \to \pi^+\pi^-$ decays, where the two pions are misidentified as muons. In Fig. 2 the probability of misidentifying a pion as a muon is shown as a function of its momentum. This performance is obtained after a cut based on the muon-ID algorithm, so there is an efficiency of 90% for real muons. A similar cut is applied in this analysis to reduce the large amount of misidentified pions, considerably reducing the background events in the signal region.

3.2. Normalization

The calculation of the normalization factors (α) has been done prior to the fit. Events from $K^0_S \to \pi^+\pi^-$ have been used to calculate these numbers, taking into account the trigger, selection, BDT and muon-ID algorithm efficiencies. The magnitude of interest is computed as

$$\mathcal{B}(K^0_S \to \mu^+\mu^-) = \mathcal{B}(K^0_S \to \pi^+\pi^-) \cdot \frac{\epsilon^{\pi\pi}}{\epsilon^{\mu\mu}} \cdot \frac{N^{\mu\mu}}{N^{\pi\pi}} \equiv \alpha N^{\mu\mu}, \tag{1}$$

where

$$\frac{\epsilon^{\pi\pi}}{\epsilon^{\mu\mu}} = \frac{\epsilon^{\pi\pi}_{sel}}{\epsilon^{\mu\mu}_{sel}} \times \frac{\epsilon^{\pi\pi}_{trig}}{\epsilon^{\mu\mu}_{trig}} \times \frac{1}{\epsilon^{\mu\mu}_{BDT}} \times \frac{1}{\epsilon_{\mu\text{ID}}}. \tag{2}$$

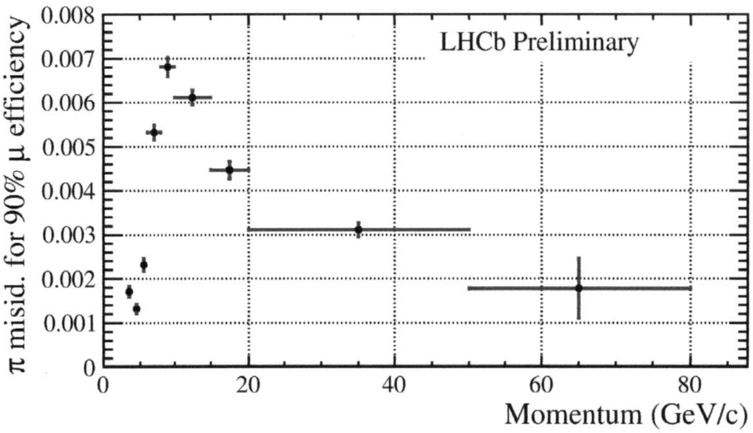

Figure 2. Probability to misidentify a pion from $K^0_S \to \pi^+\pi^-$ decays as a muon, as a function of its momentum, after applying a requirement based on the muon-ID having a 90% efficiency on real muons.

Source	TOS$_\mu$-TOS$_\mu$-TOS$_{\mu\mu}$	TOS$_\mu$-TOS$_{\mu\mu}$-TOS$_{\mu\mu}$	TIS-TIS-TOS$_{\mu\mu}$
	Uncertainties on normalisation factor		
Tracking	0.4%	0.4%	0.4%
Selection	3.3%	3.9%	1.1%
Trigger	8%	11%	-
K spectrum	3.3%	3.3%	3.3%
Muon ID	0.2%	0.3%	0.8%
	Uncertainties on signal yield		
Signal mass shape	0.8%	0.8%	0.8%
	Uncertainties on branching fraction		
Background shape	4×10^{-11}		

Table 1. Relevant systematic uncertainties considered in the analysis, split into relative uncertainties on the normalization factors. A relative uncertainty on the signal yield from the signal model used in the mass fit, and an uncertainty on the branching fraction, obtained combining the three categories, from the background model.

The equations (1) and (2) are applied to each of the trigger categories and BDT bins.

3.3. Systematic uncertainties

Several systematic effects contribute to the uncertainty on the normalization factors and the signal shape. In Tab. 1 the different uncertainties which have been taken into account in the fit are shown.

The biggest source of uncertainty comes from the trigger. This arises from the differences between data and Monte-Carlo samples in the transverse momentum of the kaon and its decay vertex radial position. For the category TIS-TIS-TOS$_{\mu\mu}$ no trigger systematic has been computed since the efficiency has been calculated directly from data.

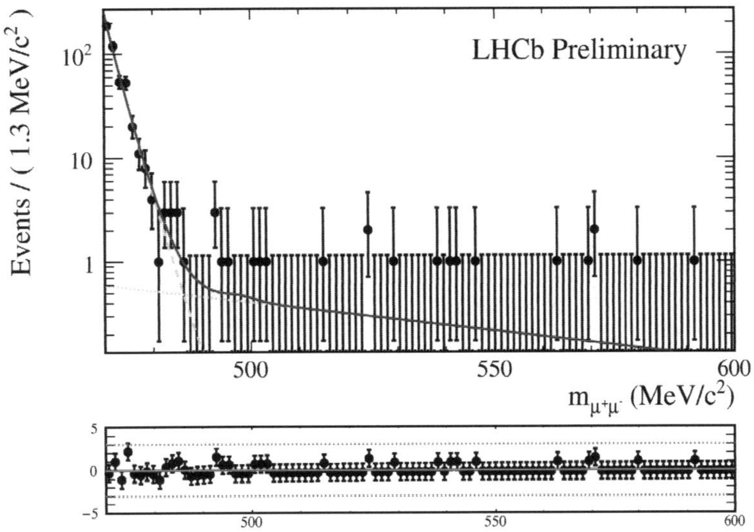

Figure 3. Fit to the reconstructed kaon mass distributions for the most sensitive category and BDT bin. The fitted model is shown in blue, and the background components are overdrawn for (orange) combinatorial background and (green) $K_S^0 \to \pi^+\pi^-$ double misidentification. The pull plot is shown below.

3.4. Fit and limit calculation

A simultaneous maximum likelihood fit has been done to all the trigger categories and BDT bins. The normalization factors are included in the fit so the branching fraction becomes directly observable. The previous LHCb result [7] has been taken into account including a constraint on the branching fraction. There are three different contributions in the di-muon spectra:

- Combinatorial background: parametrized by an exponential function.
- $K_S^0 \to \pi^+\pi^-$ events with two misidentificated pions: described by a power law function.
- $K_S^0 \to \mu^+\mu^-$ events: parametrized by an Hypatia function [8].

The parameters have been let free for the combinatorial background and the $K_S^0 \to \pi^+\pi^-$ shapes, while those describing the $K_S^0 \to \mu^+\mu^-$ peak have been fixed to values extracted from Monte-Carlo samples, leaving free only the yield. As can been seen in Fig. 3, no significant events have been observed in the signal region.

4. Results

Since the value obtained from the branching fraction was compatible with zero, a limit on the branching fraction is obtained. This quantity is computed integrating the posterior probability of the branching fraction, obtained from its $-\log \mathcal{L}$ profile (see Fig. 4).

The result obtained was

$$\mathcal{B}\left(K_S^0 \to \mu^+\mu^-\right) < 6.9(5.8) \times 10^{-9} \text{ at } 95(90)\% \text{ CL}, \tag{3}$$

improving by a factor 1.6 the previous LHCb result [7].

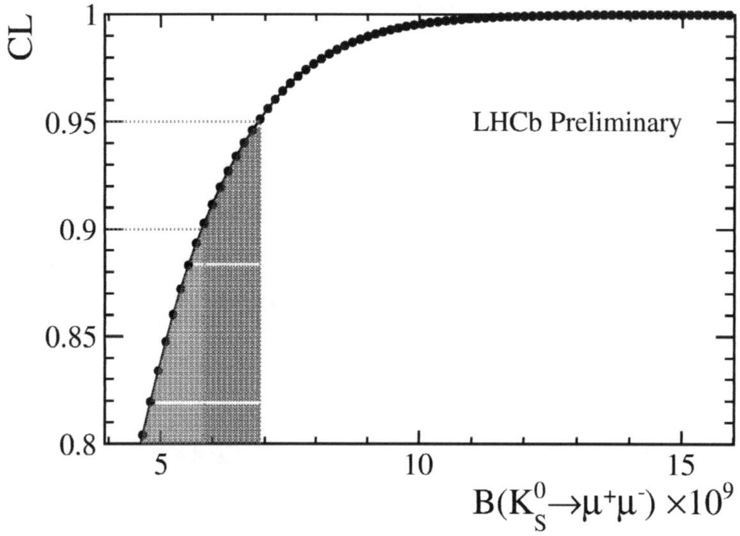

Figure 4. Exclusion confidence level (CL) for each value of the $K^0_S \to \mu^+\mu^-$ branching fraction. The regions corresponding to 90% and 95% CL are emphasised in magenta and blue, respectively.

5. Prospects for the Run-II and upgrade

Much effort is being put at LHCb to study strange decays. The new trigger lines for low p_T processes included in the Run-II of the LHC are a benchmark for the study of strange decays using this detector. However, during the new run, the L0 hardware trigger remains unchanged, thus the major source of inefficiency is still present ($\text{eff}_{\text{trigger}} \sim 3\%$). This translates into a limit on the sensitivities of 10^{-10} for $K^0_S \to \mu^+\mu^-$ during this period. After the LHCb upgrade, there will be no L0 trigger at all, thus efficiencies for strange decays up to $\sim 100\%$ can be reached. Therefore, the most interesting region for the studied decay $\mathcal{B}\left(K^0_S \to \mu^+\mu^-\right) \in \left[10^{-11}, 10^{-12}\right]$ will become accessible.

References

[1] G. Ecker and A. Pich, *The longitudinal muon polarization in* $K^0_L \to \mu^+\mu^-$, Nucl. Phys. **B366** (1991) 189.
[2] G. Isidori and R. Unterdorfer, *On the short distance constraints from* $K_{L,S} \to \mu + \mu-$, JHEP **01** (2004) 009, `arXiv:hep-ph/0311084`.
[3] G. D'Ambrosio, G. Ecker, G. Isidori, and H. Neufeld, *Radiative nonleptonic kaon decays*, in *2nd DAPHNE Physics Handbook:265-313*, pp. 265–313, 1994. `arXiv:hep-ph/9411439`.
[4] Particle Data Group, K. A. Olive *et al.*, *Review of particle physics*, Chin. Phys. **C38** (2014) 090001, and 2015 update.
[5] LHCb, A. A. Alves, Jr. *et al.*, *The LHCb Detector at the LHC*, JINST **3** (2008) S08005.
[6] LHCb Collaboration, *Updated limit for the decay* $K^0_S \to \mu^+\mu^-$, Tech. Rep. LHCb-CONF-2016-012. CERN-LHCb-CONF-2016-012, CERN, Geneva, Oct, 2016.
[7] LHCb, R. Aaij *et al.*, *Search for the rare decay* $K_S \to \mu^+\mu^-$, JHEP **01** (2013) 090, `arXiv:1209.4029`.
[8] D. Martínez Santos and F. Dupertuis, *Mass distributions marginalized over per-event errors*, Nucl. Instrum. Meth. **A764** (2014) 150, `arXiv:1312.5000`.

IOP Conf. Series: Journal of Physics: Conf. Series **800** (2017) 012010 doi:10.1088/1742-6596/800/1/012010

Study of rare B-decays with the CMS experiment

Sanjay Kumar Swain for the CMS collaboration
The School of Physical Sciences
National Institute of Science Education and Research (NISER)
Bhubaneswar, INDIA

1 Introduction

Heavy flavor physics is the study of high energy hadronic interactions among quark flavors. The production of heavy flavor plays an important role in testing Quantum Chromodynamics (QCD) calculations. Here we discuss the production and decay of B-mesons containing b quarks. The physics properties show up through different experimental observables. This enables us to search for physics phenomena beyond Standard Model (SM) by comparing the experimental result with theory prediction. Any significant deviation of the measured observable from the SM prediction would hint the presence of New Physics (NP). In this paper, the results from CMS [1] experiment using the data collected at 7/8/13 TeV center-of-mass energy are presented.

2 Dimuon spectrum with 13 TeV data

After long shut down of the LHC accelerator, the pp collisions started again around middle of 2015. The center-of-mass energy was 13 TeV. The CMS, ATLAS and LHCb experiments already produced many results using these 13 TeV data. CMS could reproduce the dimuon mass spectrum, which was seen earlier 7 TeV and 8 TeV data samples. Figure 1 shows the dimuon invariant mass with 13.1 fb^{-1} of data collected at 13 TeV center-of-mass energy. The colors correspond to dedicated dimuon triggers with low p_T thresholds, in specific mass windows, while the light gray continuous distribution represents events collected with a dimuon trigger with high p_T threshold. One can clearly see different resonances, such as ϕ, J/ψ, $\psi^{'}$, ω in the distribution.

3 B^+ production cross section with 13 TeV data

The measurements of b-hadron production cross sections provide essential information to understand QCD. Such studies have been carried out by several collider experiments. The recent studies of b-hadron production at the higher energies provided by the Run 2 of the LHC provide a new important test of theoretical calculations. Here we discuss the

Content from this work may be used under the terms of the Creative Commons Attribution 3.0 licence. Any further distribution of this work must maintain attribution to the author(s) and the title of the work, journal citation and DOI.
Published under licence by IOP Publishing Ltd

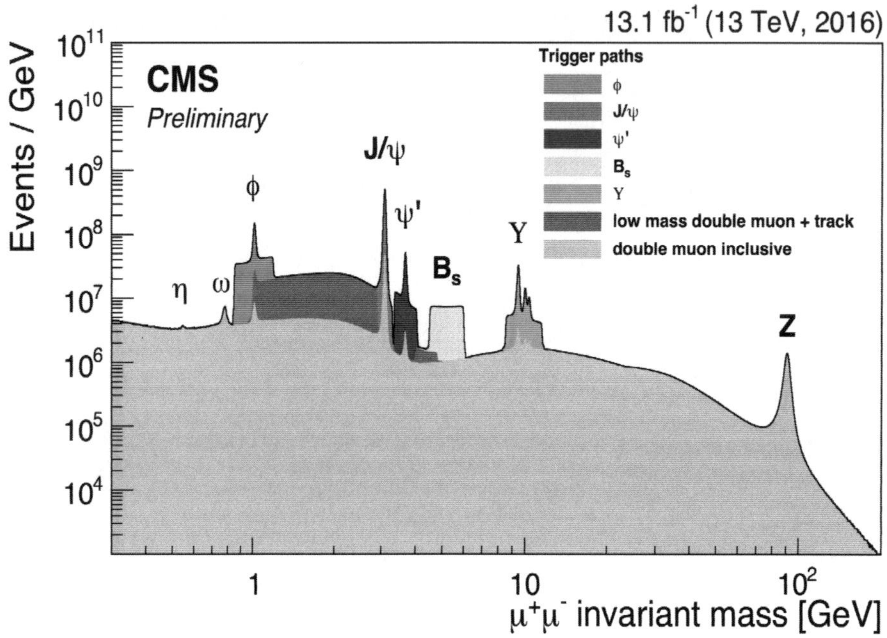

Figure 1: The dimuon mass distribution collected with various dimuon triggers, during the first 13 TeV data taking in 2015. Different resonances are clearly visible in the distribution.

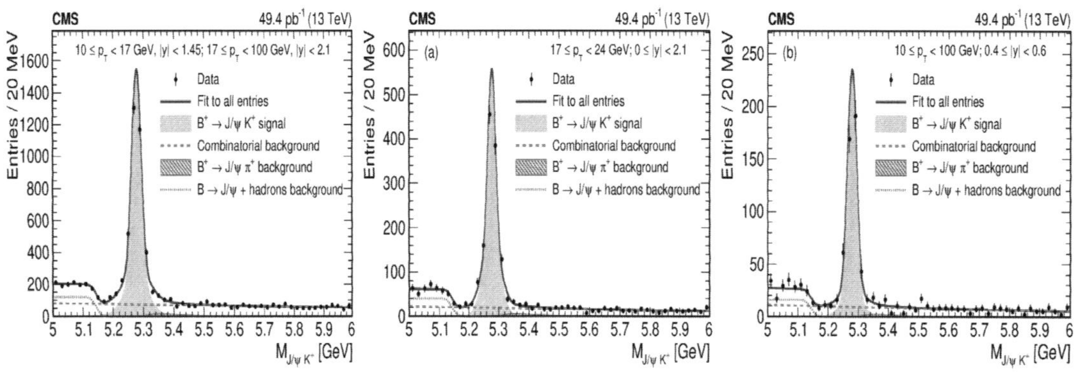

Figure 2: The plots show (left) the invariant mass distribution of the $B^+ \rightarrow J/\psi K^+$ candidates, integrated in the phase space region $10 < p_T^B < 17$ GeV with $|y^B| < 1.45$ and $17 < p_T^B < 100$ GeV with $|y^B| < 2.1$, (middle) in the region $17 < p_T^B < 24$ GeV with $|y^B| < 2.1$, and (right) $10 < p_T^B < 100$ GeV with $0.4 < |y^B| < 0.6$.

measurement of the B^+ differential cross section in pp collisions at 13 TeV, as function of transverse momentum (p_T^B) and rapidity (y^B). This result is based on a data sample collected by the CMS experiment, corresponding to an integrated luminosity of 49.4 pb^{-1}, and uses the channel pp$\rightarrow B^+X \rightarrow J/\psi K^+X$, selecting events where the J/ψ decays into a pair of muons. The muons are required to have at least one reconstructed segment in the muon stations that matches the extrapolated position of a track reconstructed in the silicon tracker, to satisfy $p_T > 4.2$ GeV and $|\eta| < 2.1$, and to have good quality in the fit to a track. The J/ψ candidates have an invariant mass within ± 150 MeV of the nominal J/ψ mass. Each J/ψ candidate must have $p_T > 8$ GeV and the χ^2 probability of the dimuon vertex fit is required to be larger than 10%. Both muons must be either within $|\eta| <1.6$ or one of the muons must have $p_T > 11$ GeV. Candidate B^+ mesons are reconstructed by combining a J/ψ candidate with a charged track of $p_T >1$ GeV. The track is assumed to be a kaon and the track-fit χ^2 must be less than five times the number of degrees of freedom. To reduce the peaking and non-peaking backgrounds, many kinematic and topological cuts have been applied. Finally, the signal yield is extracted with an extended unbinned maximum likelihood fit to the invariant mass distribution of the B^+ candidates, in each of the p_T^B and $|y^B|$ bins. Figure 2 shows the invariant mass distribution of all the B^+ candidates included in the analysis together with the corresponding fit result [2].

The differential cross sections as a function of p_T^B, integrated within $|y^B| < 1.45$ or for $|y^B| < 2.1$, and as a function of y^B, with $p_T^B > 10$ GeV or $p_T^B > 17$ GeV, are shown in Fig. 3 (left and right panels, respectively), where they are compared to FONLL (shaded boxes) and PYTHIA (dashed lines) calculations. The 7 TeV measurements are also displayed, for completeness. The bottom panels display the ratios of measured cross sections to those

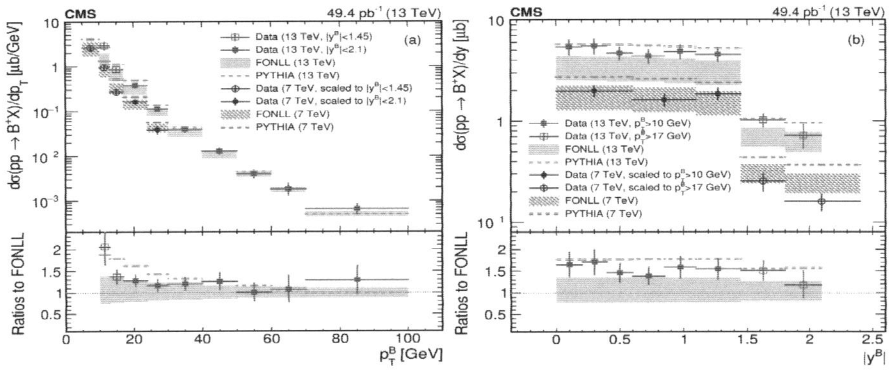

Figure 3: The differential cross section for $\frac{d\sigma}{dp_T^B}$ (left) and $\frac{d\sigma}{dy^B}$ (right). See the text for details.

predicted by FONLL; the PYTHIA/FONLL ratios are also shown, as dashed lines. The measured values show a reasonable agreement, both in terms of shape and of normalization, with FONNL calculations and with the results obtained with PYTHIA event generator.

4 Study of $B_s^0 \to \mu^+\mu^-$

In the standard model (SM) of particle physics, tree-level diagrams do not contribute to flavor-changing neutral-current (FCNC) decays. However, FCNC decays may proceed through higher-order loop diagrams, and this opens up the possibility for contributions from non-SM particles. In the SM, the rare FCNC decays $B_d \to \mu^+\mu^-$ have small branching fractions of $\mathcal{B}(B_s^0 \to \mu^+\mu^-) = (3.66 \pm 0.23) \times 10^{-9}$, corresponding to the decay-time integrated branching fraction, and $\mathcal{B}(B_d \to \mu^+\mu^-) = (1.06 \pm 0.09) \times 10^{-10}$. Charge conjugation is implied throughout this Letter. Several extensions of the SM, such as supersymmetric models with non-universal Higgs boson masses, specific models containing leptoquarks, and the minimal supersymmetric standard model with large $\tan\beta$, predict enhancements to the branching fractions for these rare decays. The decay rates can also be suppressed for specific choices of model parameters. Over the past 30 years, significant progress in sensitivity has been made, with exclusion limits on the branching fractions improving by five orders of magnitude.

The search for the $B \to \mu^+\mu^-$ signal, where B denotes B_s^0 or B^0, is performed by CMS in the dimuon invariant mass regions around the B_s^0 and B^0 masses. To avoid possible biases, the signal region $5.20 < m_{\mu\mu} < 5.45$ GeV was kept blind until all selection criteria were established. The $B \to \mu^+\mu^-$ candidates are constructed from two oppositely charged muons with $p_T > 4$ GeV and be consistent in direction and p_T with the muons that triggered the event. A boosted decision tree (BDT) constructed within the TMVA framework is trained to further separate genuine muons from those arising from misidentified charged hadrons. The variables used in the BDT can be divided into four classes: basic kinematic quantities, silicon-tracker fit information, combined silicon and muon track fit information, and muon detector information. The BDT is trained on MC simulation samples of B-meson decays to kaons and muons. Compared to the "tight" muons, the BDT working point used to select muons for this analysis reduces the hadron-to-muon mis-identification probability by 50% while retaining 90% of true muons. The probability to misidentify a charged hadron as a muon because of decay in flight or detector punch-through is measured in data from samples of well-identified pions, kaons, and protons. This probability ranges from $(0.5\text{–}1.3) \times 10^{-3}$, $(0.8\text{–}2.2) \times 10^{-3}$, and $(0.4\text{–}1.5) \times 10^{-3}$, for pions, kaons, and protons, respectively, depending on whether the particle is in the barrel or end cap, the running period, and the momentum. Each of these probabilities is ascribed an uncertainty of 50%, based on differences between data and MC simulation.

An unbinned maximum-likelihood fit to the $m_{\mu\mu}$ distribution is used to extract the signal and background yields. Events in the signal window can result from genuine signal, combinatorial background, background from semileptonic b-hadron decays, and the peaking background. The probability density functions (PDFs) for the signal, semileptonic, and peaking backgrounds are obtained from fits to MC simulation. The dimuon mass distributions for the four channels (barrel and endcap in 7 and 8 TeV data), further divided into categories corresponding to different bins in the BDT parameter, are fitted simultaneously. No significant excess is observed for $B_d \to \mu^+\mu^-$ and an upper limit of $\mathcal{B}(B_d \to \mu^+\mu^-) < 1.1 \times 10^{-9}$ is established at 95% CL. For $B_s^0 \to \mu^+\mu^-$, an excess of events with a significance of 4.3 standard deviations is observed, and a branching fraction of $\mathcal{B}(B_s^0 \to \mu^+\mu^-) = (3.0^{+1.0}_{-0.9}) \times 10^{-9}$ is determined, in agreement with the standard model expectations [3]. The CMS and LHCb reported the combined result for $B \to \mu^+\mu^-$ to exploit fully the statistical power of the data and to account for the main correlations between them [4]. The results are shown in Figure 4. The ATLAS experiment

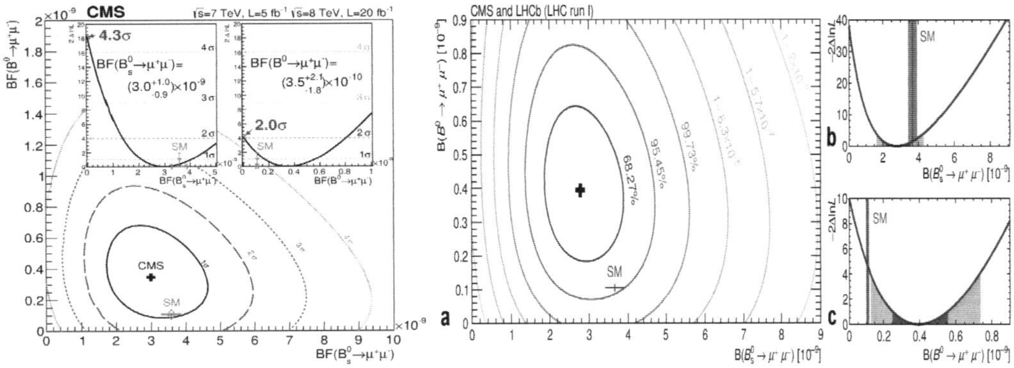

Figure 4: (Left plot) CMS result for scan of ratio of joint likelihood for $B_s^0 \to \mu^+\mu^-$ and $B_d \to \mu^+\mu^-$. As insets, the likelihood ratio scan for each of the branching fractions when the other is profiled together with other nuisance parameters; the significance at which the background-only hypothesis is rejected is also shown. (Right plot) CMS and LHCb combined result showing the likelihood contours in $\mathcal{B}(B_d \to \mu^+\mu^-)$ versus $\mathcal{B}(B_s^0 \to \mu^+\mu^-)$ plane. Also shown are the variations of the test statistics for $\mathcal{B}(B_d \to \mu^+\mu^-)$ and $\mathcal{B}(B_s^0 \to \mu^+\mu^-)$.

performed similar search for $B_s^0 \to \mu^+\mu^-$ and $B_d \to \mu^+\mu^-$ decays using pp collision data corresponding to an integrated luminosity of 25 fb^{-1}, collected at 7 and 8 TeV in the full LHC Run 1 data-taking period. This analysis supersedes the previous result based on 2011 data and exploits improved analysis techniques in addition to the larger dataset. For B^0 an upper limit $\mathcal{B}(B_d \to \mu^+\mu^-) < 4.2 \times 10^{-10}$ is placed at the 95% confidence level, based on the CL$_s$ method. The limit is compatible with the predictions based on the SM and with the combined result of the CMS and LHCb experiments. For B_s^0 the result is $\mathcal{B}(B_s^0 \to \mu^+\mu^-) = \left(0.9^{+1.1}_{-0.8}\right) \times 10^{-9}$, where the errors include both the statistical and systematic uncertainties [5]. An upper limit $\mathcal{B}(B_s^0 \to \mu^+\mu^-) < 3.0 \times 10^{-9}$ at 95% CL is placed, lower than the SM prediction, and in better agreement with the measurement of CMS and LHCb.

5 Angular analysis of the decay $B^0 \to K^{*0}\mu^+\mu^-$

The FCNC decay, $B^0 \to K^{*0}\mu^+\mu^-$, provides many opportunities to search for new phenomena beyond SM. In addition to the branching fraction, other properties of the decay can be measured, including the forward-backward asymmetry of the muons, $A_{\rm FB}$, and the longitudinal polarization fraction of the K^{*0}, $F_{\rm L}$. To better understand this decay, these quantities can be measured as a function of the dimuon invariant mass squared (q^2). The offline reconstruction requires two muons of opposite charge and two oppositely charged hadrons. The muons are required to match those that triggered the event readout, and also to pass general muon identification requirements. The hadron tracks are required to fail the muon identification criteria, have $p_T > 0.8$ GeV, and have an extrapolated distance of closest approach to the beamspot in the transverse plane greater than twice the sum in quadrature of the distance uncertainty and the beamspot transverse size. The two hadrons with invariant mass within 90 MeV of the nominal K^{*0} mass are selected for further consideration. The B^0 candidates are obtained by fitting the four charged tracks to a common vertex, and applying a vertex constraint to improve the resolution of the track parameters. Other offline selection requirement are also applied to select the B^0 signal candidates from data. For each q^2 bin, the observables of interest are extracted from an unbinned extended maximum-likelihood fit to three variables: the B^0 invariant mass and the two angular variables θ_K and θ_l. The results are consistent with standard model predictions and previous measurements [6]. They are shown in Figure 5.

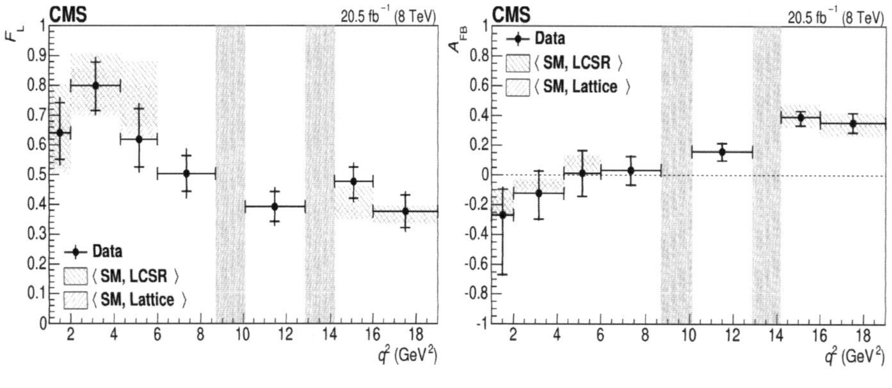

Figure 5: Measured values of $F_{\rm L}$ and $A_{\rm FB}$ versus q^2 for $B^0 \to K^{*0}\mu^+\mu^-$. The statistical uncertainty is shown by the inner vertical bars, while the outer vertical bars give the total uncertainty. The horizontal bars show the bin widths. The vertical shaded regions correspond to the J/ψ and ψ' resonances. The other shaded regions show the two SM predictions after rate averaging across the q^2 bins to provide a direct comparison to the data. Controlled theoretical predictions are not available near the J/ψ and ψ' resonances.

Acknowledgments

I would like to thank Department of Atomic Energy (DAE) and Department of Science and Technology(DST), Govt. of India for their support to attend this conference.

References

[1] S. Chatrchyan, et al., The CMS experiment at the CERN LHC, JINST 3 (2008) S08004.

[2] V. Khachatryan *et al.* [CMS Collaboration], arXiv:1609.00873 [hep-ex].

[3] S. Chatrchyan *et al.* [CMS Collaboration], Phys. Rev. Lett. **111**, 101804 (2013) doi:10.1103/PhysRevLett.111.101804 [arXiv:1307.5025 [hep-ex]].

[4] V. Khachatryan *et al.* [CMS and LHCb Collaborations], Nature **522**, 68 (2015) doi:10.1038/nature14474 [arXiv:1411.4413 [hep-ex]].

[5] M. Aaboud *et al.* [ATLAS Collaboration], arXiv:1604.04263 [hep-ex].

[6] V. Khachatryan *et al.* [CMS Collaboration], Phys. Lett. B **753**, 424 (2016) doi:10.1016/j.physletb.2015.12.020 [arXiv:1507.08126 [hep-ex]].

Latest discrete symmetries and Quantum Mechanics studies with KLOE-2

Daria Kamińska on behalf of the KLOE-2 Collaboration
Jagiellonian University, Cracow
E-mail: daria.kaminska@doctoral.uj.edu.pl

November 2016

Abstract. The KLOE-2 experiment at the INFN Laboratori Nazionali di Frascati (LNF) is currently taking data at the upgraded e^+e^- DAFNE collider, representing the continuation of KLOE with a new physics program including kaon interferometry and test of discrete symmetries among the main topics. In fact entangled neutral kaon pairs produced at DAFNE are a unique tool to probe discrete symmetries and/or QM basic principles at the utmost precision. The status of the latest ongoing analyses on KLOE data using the most refined analyses tools will be presented and discussed including prospects for KLOE-2: (i) search for decoherence and CPT violation effects in the $\phi \to K_S K_L \to \pi^+\pi^-\pi^+\pi^-$ decay, (ii) test of CP and CPT symmetries in K_S semileptonic decays, (iii) test of Time reversal and CPT in transitions in $\phi \to K_S K_L \to \pi e \nu, 3\pi^0(2\pi^0)$ decays, (iv) study of the $K_S \to \pi^+\pi^-\pi^0$ decay.

1. The KLOE experiment

DAΦNE is a e^+e^- collider localized in Frascati near Rome. Energy of the beams is adjusted to operate at the peak of ϕ meson resonance ($\sqrt{s} = m_\phi \approx 1019$ MeV). Due to a small transverse momenta of colliding beams, the created ϕ mesons are almost at rest ($\beta_\phi \approx 0.015$). The main decay channels of ϕ meson are pairs of neutral or charged kaons, with branching ratios of 34% and 49%, respectively. Due to conservation of quantum numbers in ϕ meson decay, the created K meson pairs are entangled, which allows one to obtain a pure beams of kaons (so called tagging technique) or perform a measurement based on quantum interferometry.

The above mentioned studies are performed on data collected by the KLOE detector, which surrounds e^+ and e^- beams interaction point. The cylindrically shaped KLOE detector consists of two main parts: a drift chamber and, surrounding it, a electromagnetic calorimeter, both inserted in a 0.52 T magnetic field. The detector collected in total about 2.5 fb^{-1} of data in two runs in the year 2001-2002 and 2004-2005, with loose trigger conditions which allows to a broad physics program.

Content from this work may be used under the terms of the Creative Commons Attribution 3.0 licence. Any further distribution of this work must maintain attribution to the author(s) and the title of the work, journal citation and DOI.
Published under licence by IOP Publishing Ltd

2. The Quantum Decoherence and CP, CPT and Lorentz Symmetries

The quantum interference between the two kaons initially in the entangled state and decaying into CP-violating channel $\phi \to K_S K_L \to \pi^+\pi^-\pi^+\pi^-$ is described by the following decay intensity distribution:

$$I(\pi^+\pi^-, \pi^+\pi^-; \Delta t) \propto e^{-\Gamma_L \Delta t} + e^{-\Gamma_S \Delta t} - 2(1-\zeta_{SL})e^{-\frac{\Gamma_S+\Gamma_L}{2}\Delta t}\cos(\Delta m \Delta t) \quad (1)$$

where Δt is the absolute value of the time differences of the two $\pi^+\pi^-$ decays. In the above expression the parameter ζ_{SL} is introduced to account for possible decoherence effect. Analogously, a $\zeta_{0\bar{0}}$ parameter can be used if Eq. 1 is expressed in the $\{K^0, \bar{K}^0\}$ basis. The decoherence parameter influence on $I(\pi^+\pi^-, \pi^+\pi^-; \Delta t)$ distribution is shown on Figure 1 (left). The most sensitive part of the spectra lies in the region of Δt close to 0. Using the standard quantum mechanics description, no decoherence should be observed ($\zeta = 0$), while the total decoherence corresponds to $\zeta = 1$, where kaons are no longer entangled. A analysis of a KLOE data sample corresponding to $L \approx 1.5$ fb^1

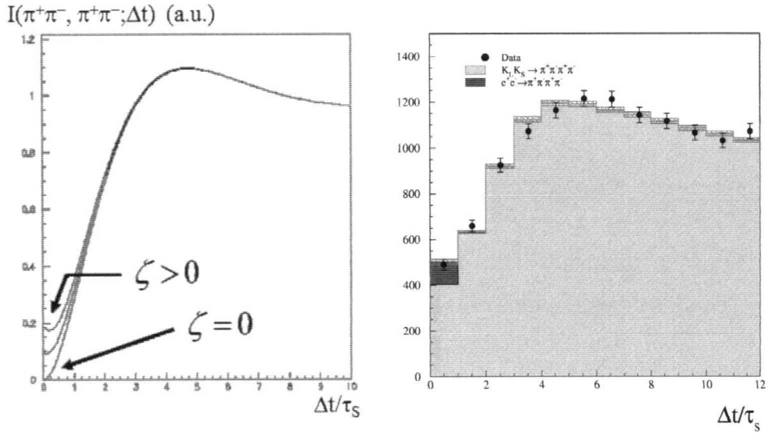

Figure 1. The $I(\pi^+\pi^-, \pi^+\pi^-; \Delta t)$ distribution with different values of the decoherence parameter $\zeta = \{0, 2 \cdot 10^{-6}, 4 \cdot 10^{-6}\}$ (left) compared with experimental one (right), where Monte Carlo simulation is fitted into data points. Figures are taken from [1, 2].

is based on the selection of two separate charged vertices, each formed by two opposite-curvature tracks inside the drift chamber, with an invariant mass and total momentum compatible with the two neutral kaon decays. The experimental points were fitted with Eq. 1 and taking into account detection resolution and efficiency as well as the background contribution from coherent and incoherent K_S regeneration on the beam-pipe wall and contamination from $e^+e^- \to \pi^+\pi^-\pi^+\pi^-$ channel. The result of the fit to the data is shown on Figure 1 (right) and is equal to [2]:

$$\zeta_{SL} = (0.3 \pm 1.8_{stat} \pm 0.6_{syst}) \cdot 10^{-2}, \quad (2)$$

$$\zeta_{0\bar{0}} = (1.4 \pm 9.5_{stat} \pm 3.8_{syst}) \cdot 10^{-7}, \quad (3)$$

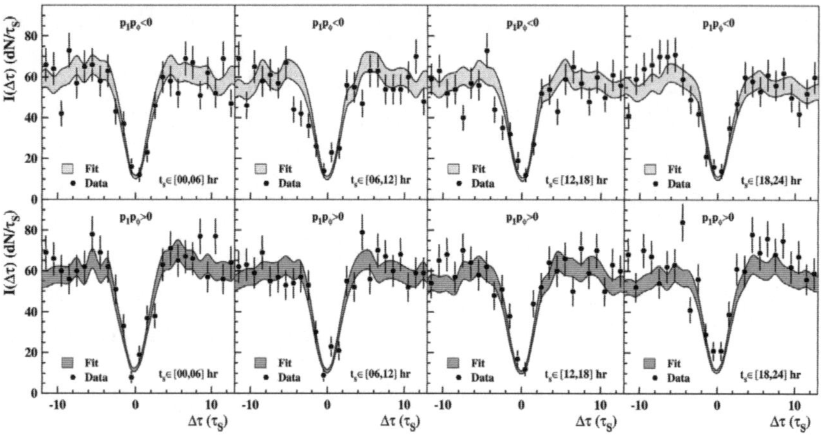

Figure 2. Results of the fit (color bands) to the double decay intensity distributions for data (black points) divided into 4 bins of sidereal time and two angular bins. Figure is adapted from [5]

showing no deviation from the quantum mechanics predictions. The bin width is optimized for the Δt resolution, which is approximately $1\tau_S$. This will be improved at KLOE-2 with the insertion of the Inner Tracker detector that allows to obtain better resolution on vertex reconstruction.

In addition, by using kaon interferometry with both kaons decaying in $\pi^+\pi^-$ pairs, the existence of CPT-violating and Lorentz-invariance violating effects predicted by the Standard Model Extension (SME) model was tested [3, 4]. Possible violation of the Lorentz invariance entails a dependence of the CPT-violating parameter δ on the sidereal time:

$$\delta \approx i \sin\phi_{SW} e^{i\phi_{SW}} \gamma_K (\Delta a_0 - \vec{\beta_K} \cdot \Delta\vec{a})/\Delta m, \tag{4}$$

where ϕ_{SW} stands for the superweak phase, γ_K and $\vec{\beta_K}$ are the kaon boost factor and velocity in the observer frame and a_μ are the four CPT- and Lorentz-violating coefficients. Figure 2 presents the KLOE-2 results on the kaon decay-time difference ratios, in units of the K_S lifetime τ_S for different orientations and sidereal times. The obtained results on Δa_μ parameters are consistent with no evidence of the CPT-violating effect and by reaching the sensitivity at the level $O(10^{-18})$ GeV are presently the most precise measurement in the quark sector of the SME.

Another component of KLOE-2 program concerning test of CPT symmetry violation is a test using asymmetries of the semileptonic decays:

$$A_{S,L} = \frac{\Gamma(K_{S,L} \to \pi^- e^+ \nu) - \Gamma(K_{S,L} \to \pi^+ e^- \bar{\nu})}{\Gamma(K_{S,L} \to \pi^- e^+ \nu) + \Gamma(K_{S,L} \to \pi^+ e^- \bar{\nu})} \tag{5}$$

$$= 2\left[Re\,(\epsilon_K) \pm Re\,(\delta_K) - Re\,(y) \pm Re\,(x_-)\right], \tag{6}$$

where ϵ_K and δ_K stand for CP and CPT violation in the mass matrix, respectively, while y parametrizes CPT violation assuming $\Delta S = \Delta Q$ rule and x_- is a small term that accounts for $\Delta S = \Delta Q$ rule violation. The difference:

$$A_S - A_L = 4[Re(\delta_K) - Re(x_-)] \tag{7}$$

is sensitive only to CPT violation effect originating from the mass matrix, while x_- value is experimentally limited to -0.002 ± 0.006 [8]. However, accuracy on A_L determination ($\sigma(A_L) = 0.7 \times 10^{-4}$ [10]) is more than two orders of magnitude bigger than this of the A_S and the uncertainty on A_S ($\sigma(A_S) = 0.01$[9]) is dominated by the data sample statistics. Therefore, further studies of $K_S \to \pi e \nu$ decay using sample four times bigger in statistics than the previous KLOE analysis can improve the precision of CPT test. More than 10^5 of $K_S \to \pi e \nu$ decays were reconstructed (see Figure 3, left) and will be used to determine the charge asymmetry and branching ratio for K_S semileptonic decays.

Another two-kaon observable sensitive to the real part of CPT symmetry violation term is defined by using ratios of probabilities of transitions constructed for $K_L \to 3\pi^0$ decay accompanied by $K_S \to \pi e \nu$ and its CPT-conjugated partner $K_L K_S \to \pi e \nu\ \pi^+\pi^-$ [6]:

$$R_{2,4}(\Delta t) = \frac{I(K_{e\mp 3}, 3\pi^0; \Delta t)}{I(\pi\pi, K_{e\mp 3}; \Delta t)} \frac{BR(K_L \to 3\pi^0)}{BR(K_S \to \pi\pi)} \frac{\Gamma_L}{\Gamma_S}. \tag{8}$$

In particular in asymptotic regime the discrepancy of $R_2(\Delta t)$ and $R_4(\Delta t)$ ratio from unity is related to the CPT symmetry violating parameter δ as:

$$R_{2,4}(\Delta t) \sim 1 \mp Re(\delta), \text{ for } \Delta t >> \tau_S. \tag{9}$$

In order to measure the double decay rates introduced in Eq. 8 the reconstruction of a pair of kaon decays separated by a time interval Δt, in asymptotic plateau region of $\Delta t >> \tau_S$ should be $O(1\tau_S)$ which translates to decay point spatial resolution $O(1cm)$. The preliminary spectrum for $\phi \to K_S K_L \to \pi^\pm l^\mp \nu 3\pi^0$ as a function of Δt is shown on right panel on Figure 3.

KLOE-2 has recently established an upper limit for the CP-violating rare decays $K_S \to 3\pi^0$ branching fraction $Br(K_S \to 3\pi^0) < 2.6 \times 10^{-8}$ and the amplitude ratio $\eta_{000} = A(K_S \to 3\pi^0)/A(K_L \to 3\pi^0) < 0.0088$ (90% c.l) [7]. The group is also looking for the $K_S \to \pi^+\pi^-\pi^0$ decay. An analysis based on $L = 1.7$ fb^{-1} will be the first measurement to date that makes use of the direct detection of this decay channel.

3. Summary

The neutral kaon system is an excellent laboratory for the study of CPT symmetry and the basic principles of Quantum Mechanics. The entangled pairs of $K_S K_L$ mesons allows to measure several parameters related to decoherence, the CP, CPT and Lorentz invariance. Using enhanced luminosity of DAΦNE and with new detectors, KLOE-2 is in the middle

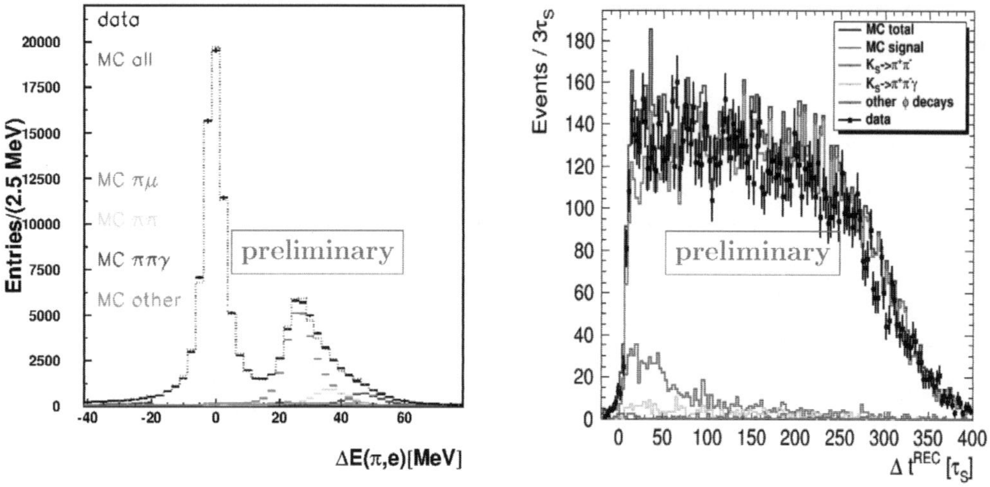

Figure 3. Left: distribution of $\Delta E(\pi, e) = E_{miss} - p_{miss}$ for all selected events after normalization procedure. Due to a non-registered neutrino in the decay, the $\Delta E(\pi, e)$ for $K_S \to \pi e \nu$ events takes value to zero. Right: spectrum of selected events in $\phi \to K_S K_L \to \pi^{\pm} l^{\mp} \nu 3\pi^0$ decay as a function of time interval Δt.

of data taking and is aiming to acquire more than 5 fb^{-1} of data. Significant progress is going to be made in all results concerning tests of CPT, CP and quantum mechanics.

Acknowledgments

We warmly thank our former KLOE colleagues for the access to the data collected during the KLOE data taking campaign. We thank the DAΦNE team for their efforts in maintaining low background running conditions and their collaboration during all data taking. We want to thank our technical staff: G.F. Fortugno and F. Sborzacchi for their dedication in ensuring efficient operation of the KLOE computing facilities; M. Anelli for his continuous attention to the gas system and detector safety; A. Balla, M. Gatta, G. Corradi and G. Papalino for electronics maintenance; M. Santoni, G. Paoluzzi and R. Rosellini for general detector support; C. Piscitelli for his help during major maintenance periods. This work was supported in part by the EU Integrated Infrastructure Initiative Hadron Physics Project under contract number RII3-CT- 2004-506078; by the European Commission under the 7th Framework Programme through the 'Research Infrastructures' action of the 'Capacities' Programme, Call: FP7-INFRASTRUCTURES-2008-1, Grant Agreement No. 227431; by the Polish National Science Centre through the Grants No. 2011/03/N/ST2/02652, 2013/08/M/ST2/00323, 2013/11/B/ST2/04245, 2014/14/E/ST2/00262, 2014/12/S/ST2/00459.

IOP Conf. Series: Journal of Physics: Conf. Series **800** (2017) 012011 doi:10.1088/1742-6596/800/1/012011

References

[1] A. Di Domenico, Talk at the Foundations of Probability and Physics-6 conference, Växjö, June 2011
[2] A. Di Domenico 2009 *J.Phys.Conf.Ser.* **171** 012008
[3] V. A. Kostelecky 2001 *Phys. Rev.* D **64** 076001
[4] O. W. Greenberg 2002 *Phys. Rev. Lett* **89** 231602
[5] D. Babusci *et al* (KLOE-2 Collaboration) 2014 *Phys. Lett.* B **730** 89
[6] J. Bernabeu, A. Di Domenico and P. Villanueva-Perez 2013 *Nucl.Phys.* B **868** 102
[7] D. Babusci *et al* (KLOE-2 Collaboration) 2013 *Phys. Lett.* B **723** 54
[8] K.A. Olive *et al.* (Particle Data Group) 2014 *Chin. Phys.* C **38** 090001
[9] F. Ambrosino *et al* (KLOE-2 Collaboration) 2006 *Phys. Lett.* B **636** 173
[10] A. Alavi-Harati *et al* (KTeV Collaboration) 2002 *Phys. Rev. Lett.* **88** 181601

Rare B Decays And Lepton Flavour Universality Tests at LHCb

Dr. Siim Tolk
University of Cambridge, LHCb, CERN
E-mail: siim.tolk@cern.ch

Abstract. In the lepton sector the Standard Model incorporates both lepton-flavour and lepton-number universality. In a generic New Physics scenario this is not necessarily the case and these symmetries need to be experimentally tested. Numerous searches have failed to find any signs of Lepton Flavour Violation (LFU) or Lepton Flavour Universality Violation (LFUV). New Physics could, however, generate LFUV in the heavy flavour sector where the existing experimental constraints are weaker. Recent searches at LHCb focus on precisely this unexplored territory by studying the B decays involving $b \to c\tau\nu$ and $b \to sl^+l^-$ transitions. Several theoretically and experimentally clean observables, such as $\mathcal{R}(D*)$ or $\mathcal{R}(K)$, diverge from the Standard Model predictions and could be the first signs of LFUV.

1. Introduction

The Standard Model (SM) is a Minimal Flavour Violating (MFV) model, where the flavour changing transitions in the quark sector are governed by the CKM matrix. The CKM matrix is non-diagonal and quarks from different generations are known to mix. Transitions between the lepton families (Lepton Flavour Violation, LFV) are not present in the SM, but known to exist in nature at least through neutrino mixing [1]. The expected LFV processes ($\mathcal{O}(10^{-40})$) are well below the current experimental reach. Many New Physics (NP) scenarios predict significantly larger and experimentally reachable LFV. LHCb has searched for LFV in many decay processes ($B_s \to e^{\pm}\mu^{\mp}$, $D^0 \to e^{\pm}\mu^{\mp}$). Thus far, no evidence of LFV has been seen.

Besides being lepton flavour conserving, the SM is also lepton favour universal (LFU) and the couplings of the different lepton families to the neutral and charged gauge bosons are identical. NP can violate LFU even if neutrino mixing is the only source of LFV.

The LFU assumption has been thoroughly tested at LEP, SLC, and many other colliders [2]. The most stringent limits on LFU Violation (LFUV) involve the first two generations. Thus, NP could easily evade the existing bounds from direct searches and still effect the third generation: many NP scenarios contain additional interactions and enhanced couplings only to the third generation. If NP is also MFV, the experimental limits on the kaon and pion decays need to be improved by an order of magnitude to be able to probe the same parameter space accessible to B decays [3]. Therefore, processes such as $b \to c\tau\nu$ and $b \to sl^+l^-$ transitions are well suited for LFU tests.

Content from this work may be used under the terms of the Creative Commons Attribution 3.0 licence. Any further distribution of this work must maintain attribution to the author(s) and the title of the work, journal citation and DOI.
Published under licence by IOP Publishing Ltd

2. $\mathcal{R}(D^*)$

According to the LFU, leptonic branching fractions only differ due to effects related to different lepton masses: Higgs couplings, the available phase space and the helicity suppression. Semi-leptonic B decays are in general well understood and described within the SM. The theoretical uncertainties arising from the hadronic form factors and common SM parameters can be reduced by constructing ratios of branching fractions, such as the relative ratio of the two $B^0 \to D^{*-}l^+\nu_\tau$ decays[1] [4]:

$$\mathcal{R}(D^*)_{SM} = \frac{\mathcal{B}(B^0 \to D^{*-}\tau^+\nu_\tau)}{\mathcal{B}(B^0 \to D^{*-}\mu^+\nu_\tau)} = 0.252(3). \tag{1}$$

Besides theoretical benefits the ratios can be also measured more precisely because the common detection uncertainties (D^* reconstruction, particle identification and tracking) cancel.

The $\mathcal{R}(D^*)$ ratio has been previously determined by BaBar and Belle at e^+e^- collisions [5, 6, 7]. The combined experimental result is in a significant tension with respect to the SM prediction at the 3.7σ level [8, 9]. The same quantity has been determined by the LHCb collaboration in a much more challenging hadronic environment which requires a different analysis strategy compared to what was used in Belle and BaBar.

In LHCb, the momentum of the signal B meson cannot be constrained by the "tagger" B from the other b-quark. Instead, the boost of the visible system is combined with the position information on the B and D decay vertices to reconstruct the B momentum. Only candidates where τ decays to a muon and two neutrinos are used to ensure the visible final state is the same for the nominator and denominator. The two decays are finally separated using mainly the three B rest frame variables: difference between the reconstructed B and the visible daughter candidate four-momenta (missing mass), squared four-momentum transfer to the lepton system (lepton pair recoil) and lepton energy.

The LHCb measurement on the full Run 1 dataset (integrated luminosity of 3 fb^{-1}) [10]

$$\mathcal{R}(D^*)_{LHCb} = 0.336 \pm 0.027(stat.) \pm 0.030(syst.), \tag{2}$$

compares well to the existing measurements in central value as well as in precision [5, 6, 7] - a remarkable achievement in a hadronic environment. It supports the tensions seen before and raises their significance to the level of 4σ (see Figure 1).

3. $\mathcal{R}(K)$

The $\mathcal{R}(D^*)$ is not the only observable where LHCb sees unexpected results. Another group of excellent LFU tests involves $b \to sl^+l^-$ type Flavour Changing Neutral Current (FCNC) transitions. FCNC's are highly suppressed in the SM and are therefore sensitive to a multitude of NP contributions (both at loop and tree level). The abundance of B decays to kaons and muons relative to the similar decays to kaons and electrons is precisely constrained to unity in the SM [11]:

$$\mathcal{R}(K)_{SM} = \frac{\mathcal{B}(B^+ \to K\mu^+\mu^+)}{\mathcal{B}(B^+ \to Ke^+e^+)} = 1 \pm \mathcal{O}(10^{-3}), \tag{3}$$

where small deviations arise from the lepton mass difference.

The $\mathcal{R}(K)$ ratio has been determined previously by the B factories, although with a meagre precision of 20-50% depending on the di-lepton mass squared (q^2) region [12, 13]. LHCb measures $\mathcal{R}(K)$ in the range $1 < q^2 < 6\ \texttt{GeV}^2/c^4$. This excludes the resonant J/ψ and $\Psi(2S)$ regions and allows for more precise theoretical estimates. The measurement is in practice performed as a double ratio between the relative signal mode yields and the respective resonant $B^+ \to J/\psi(\to l^+l^-)K^+$ modes to reduce the experimental uncertainties.

[1] Charge conjugated modes are included throughout this paper, unless specified otherwise.

IOP Conf. Series: Journal of Physics: Conf. Series **800** (2017) 012012 doi:10.1088/1742-6596/800/1/012012

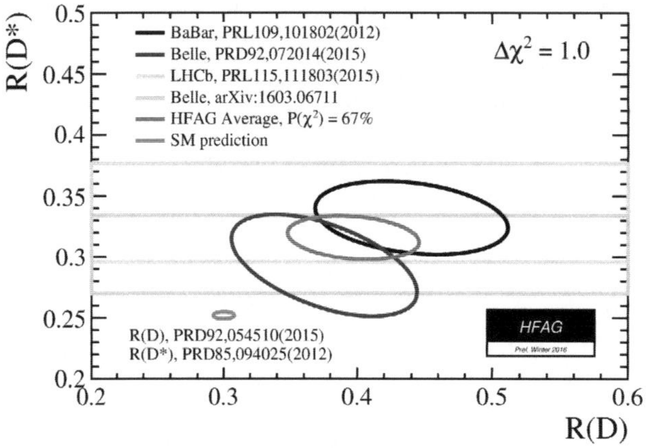

Figure 1. Experimental $\mathcal{R}(D*)$ and $\mathcal{R}(D)$ measurements from BaBar, Belle and LHCb, together with the latest SM predictions. The bands denote the confidence regions corresponding to one standard deviations around the central values. $\mathcal{R}(D*)$ and $\mathcal{R}(D)$ exceed the SM predictions by 1.9σ and 3.3σ respectively. Considering the correlation of -0.21 between the two, the resulting combined χ^2 is 19.73 for 2 degree of freedom, corresponding to a p-value of 5.2×10^{-5}. The difference with the SM predictions reported above is at 4.0σ level. Note, that the latest Belle $\mathcal{R}(D*)$ measurement [9] is not included. The effect of the new result on the combined significance is small.

The main uncertainties in this measurement are related to to the electron mode. The efficiency to detect an electron is around half of that of muons and the large radiative energy losses result in a broad mass distribution; the shape of which depends strongly on the number of associated photons in the detector and is thus difficult to simulate. The radiation also causes migration in the q^2 bins. All these effects, however, are well accounted for within the larger systematics of the electron mode.

The measured $\mathcal{R}(K)$ from LHCb's full Run 1 dataset [14]:

$$\mathcal{R}(K)_{LHCb} = 0.745^{+0.090}_{-0.074}(stat.) \pm 0.036(syst.), \tag{4}$$

is the most precise measurement of this quantity. It agrees well with the previous measurements in its di-lepton mass squared region and strengthens the claim for tensions in $b \to s$ transitions.

4. Individual leptonic B decays

The individual branching fractions measurements suggest that the tension in $\mathcal{R}(K)$ is caused by tensions in $B^+ \to K^+\mu^+\mu^-$ [15] rather than in $B^+ \to K^+e^+e^-$ [14]. Unfortunately the branching fraction of the electronic mode is not determined precisely enough to draw stronger conclusions.

Other LHCb branching fractions measurements of B decays that involve $b \to s$ FCNC transitions include $\mathcal{B}(B^0 \to K^*\mu^+\mu^-)$ [16, 17, 18], $\mathcal{B}(B^0 \to \phi\mu^+\mu^-)$ [19, 20, 21], and $\mathcal{B}(B_s \to \mu^+\mu^-)$ [22]. All these measurements are lower than their SM predicted values, especially in the low di-lepton mass squared region (in case of the semileptonic modes).

The physics involved in these transitions can additionally be studied with the angular observables of $B^0 \to K^*\mu^+\mu^-$ decays [23]. One of the theoretically cleanest observables (remapped to achieve a smaller form factor dependence) called P_5' differs from the SM predictions at (global) 3.4σ.

In the case of $B^0 \to K^*e^+e^-$ the analysis is again more difficult that that of the muon mode. Nevertheless, the latest results for the branching fraction and angular observables are in good agreement with the SM predictions [24]. All of these discrepancies could be explained by NP coupling differently to muons (and possibly tauons), but not electrons.

5. Summary

The tensions in the $\mathcal{R}(D^*)$, $\mathcal{R}(K)$ and individual $b \to s$ modes can not be ignored: the global $b \to sl^+l^-$ fits show global tensions with the SM at discovery $(4-5\sigma)$ level [25]. The situation is especially interesting because of the existing theoretical attempts able to explain many of the tensions simultaneously (e.g. leptoquarks, Z' models) [26, 27]. The case for the new analysis and better experimental precision in Run 2 is as strong as ever. In Run 2, LHCb is expected to record almost three times as many B decays as were recorded in Run 1. LHCb will update the present analysis as well as probe the tensions with additional measurements such as $\mathcal{R}(D^*)$ on hadronic τ decays, $\mathcal{R}(D)$, and extend the $\mathcal{R}$ measurements to other mesons and baryons $(D_s, \Lambda_c, \Lambda_c^*)$.

6. Acknowledgements

The author wishes to acknowledge the assistance and encouragement from the LHCb colleagues, financial support of Herschel Smith Foundation at Cambridge, and thank the KAON 2016 organisers for organising an inspiring conference.

References

[1] Fukuda Y *et al.* (Super-Kamiokande Collaboration) 1998 *Phys. Rev. Lett.* **81**(8) 1562–1567 URL `http://link.aps.org/doi/10.1103/PhysRevLett.81.1562`
[2] Patrignani C *et al.* (Particle Data Group) 2016 *Chin. Phys.* **C40** 100001
[3] Crivellin A, D'Ambrosio G, Hoferichter M and Tunstall L C 2016 *Phys. Rev.* **D93** 074038 (*Preprint* `1601.00970`)
[4] Fajfer S, Kamenik J F and Nisandzic I 2012 *Phys. Rev.* **D85** 094025 (*Preprint* `1203.2654`)
[5] Matyja A *et al.* (Belle) 2007 *Phys. Rev. Lett.* **99** 191807 (*Preprint* `0706.4429`)
[6] Aubert B *et al.* (BaBar) 2008 *Phys. Rev. Lett.* **100** 021801 (*Preprint* `0709.1698`)
[7] Bozek A *et al.* (Belle) 2010 *Phys. Rev.* **D82** 072005 (*Preprint* `1005.2302`)
[8] Lees J P *et al.* (BaBar) 2013 *Phys. Rev.* **D88** 072012 (*Preprint* `1303.0571`)
[9] Abdesselam A *et al.* 2016 (*Preprint* `1608.06391`)
[10] Aaij R *et al.* (LHCb) 2015 *Phys. Rev. Lett.* **115** 111803 [Addendum: Phys. Rev. Lett.115,no.15,159901(2015)] (*Preprint* `1506.08614`)
[11] Bobeth C, Hiller G and Piranishvili G 2007 *Journal of High Energy Physics* **2007** 040 URL `http://stacks.iop.org/1126-6708/2007/i=12/a=040`
[12] Lees J P *et al.* (BaBar) 2012 *Phys. Rev.* **D86** 032012 (*Preprint* `1204.3933`)
[13] Wei J T *et al.* (Belle) 2009 *Phys. Rev. Lett.* **103** 171801 (*Preprint* `0904.0770`)
[14] Aaij R *et al.* (LHCb) 2014 *Phys. Rev. Lett.* **113** 151601 (*Preprint* `1406.6482`)
[15] Aaij R *et al.* (LHCb) 2014 *JHEP* **06** 133 (*Preprint* `1403.8044`)
[16] Aaij R *et al.* (LHCb) 2016 (*Preprint* `1606.04731`)
[17] Bobeth C, Hiller G, van Dyk D and Wacker C 2012 *JHEP* **01** 107 (*Preprint* `1111.2558`)
[18] Bouchard C, Lepage G P, Monahan C, Na H and Shigemitsu J (HPQCD) 2013 *Phys. Rev. Lett.* **111** 162002 [Erratum: Phys. Rev. Lett.112,no.14,149902(2014)] (*Preprint* `1306.0434`)
[19] Aaij R *et al.* (LHCb) 2015 *JHEP* **09** 179 (*Preprint* `1506.08777`)
[20] Bharucha A, Straub D M and Zwicky R 2016 *JHEP* **08** 098 (*Preprint* `1503.05534`)
[21] Horgan R R, Liu Z, Meinel S and Wingate M 2014 *Phys. Rev.* **D89** 094501 (*Preprint* `1310.3722`)
[22] Khachatryan V *et al.* (LHCb, CMS) 2015 *Nature* **522** 68–72 (*Preprint* `1411.4413`)
[23] Aaij R *et al.* (LHCb) 2016 *JHEP* **02** 104 (*Preprint* `1512.04442`)
[24] Aaij R *et al.* (LHCb) 2015 *JHEP* **04** 064 (*Preprint* `1501.03038`)
[25] Descotes-Genon S, Hofer L, Matias J and Virto J 2016 *JHEP* **06** 092 (*Preprint* `1510.04239`)
[26] Bauer M and Neubert M 2016 *Phys. Rev. Lett.* **116**(14) 141802 URL `http://link.aps.org/doi/10.1103/PhysRevLett.116.141802`
[27] Greljo A, Isidori G and Marzocca D 2015 *JHEP* **07** 142 (*Preprint* `1506.01705`)

IOP Conf. Series: Journal of Physics: Conf. Series **800** (2017) 012013 doi:10.1088/1742-6596/800/1/012013

Supersymmetry and Kaon physics[1]

Kei Yamamoto

Theory Center, IPNS, KEK, Tsukuba, Ibaraki 305-0801, Japan

E-mail: `kei.yamamoto@kek.jp`

Abstract. Kaon physics has played an essential role in testing the Standard Model and in searching for new physics with measurements of CP violation and rare decays. Current progress of lattice calculations enables us to predict kaon observables accurately, especially for the direct CP violation, ϵ'/ϵ, and there is a discrepancy from the experimental data at the 2.9 σ level. On the experimental side, the rare kaon decays $K_L \to \pi^0\nu\bar{\nu}$ and $K^+ \to \pi^+\nu\bar{\nu}$ are ongoing to be measured at the SM accuracy by KOTO at J-PARC and NA62 at CERN. These kaon observables are good probes for new physics. We study supersymmetric effects; the chargino and gluino contributions to Z penguin, in kaon observables.

1. Introduction

Kaon physics is one of the most powerful probes of physics beyond the standard model (SM), and sensitive to high scale new physics (NP). The most exciting topic in kaon physics is the anomaly in the direct CP violation, ϵ'/ϵ. Recently, the hadron matrix elements of the $K \to \pi\pi$ decay have been determined with lattice QCD by the RBC-UKQCD collaborations [3], and the SM prediction of the direct CP violation is obtained as

$$\left(\epsilon'/\epsilon\right)_{\rm SM} = (1.38 \pm 6.90) \times 10^{-4}. \quad \text{[RBC-UKQCD]} \tag{1}$$

The hadronic uncertainties are reduced by the use of CP-conserving data as [4]

$$\left(\epsilon'/\epsilon\right)_{\rm SM} = (1.9 \pm 4.5) \times 10^{-4}, \quad \text{[Buras et al.]} \tag{2}$$

which has been confirmed [5]

$$\left(\epsilon'/\epsilon\right)_{\rm SM} = (0.96 \pm 4.96) \times 10^{-4}. \quad \text{[Kitahara et al.]} \tag{3}$$

All these results are below the experimental world average [6] from NA48 [7] and KTeV [8, 9] collaborations,

$$\left(\epsilon'/\epsilon\right)_{\rm exp} = (16.6 \pm 2.3) \times 10^{-4}. \tag{4}$$

In particular, Eqs. (2) and (3) disagree with the experimental data at the 2.9σ level. This may suggest a NP model providing enhancement of ϵ'/ϵ.

Rare kaon decays $K_L \to \pi^0\nu\bar{\nu}$ and $K^+ \to \pi^+\nu\bar{\nu}$ are also good probes for NP because they are theoretically clean modes and strongly suppressed in the SM. Especially, $K_L \to \pi^0\nu\bar{\nu}$ is a

[1] Based on the works with Motoi Endo, Satoshi Mishima, Daiki Ueda [1], and with Morimitsu Tanimoto [2].

Content from this work may be used under the terms of the Creative Commons Attribution 3.0 licence. Any further distribution of this work must maintain attribution to the author(s) and the title of the work, journal citation and DOI.
Published under licence by IOP Publishing Ltd

CP violating mode and provides the direct measurement of the CP violating phase in the CKM matrix. The SM predictions for these processes are given as [10]

$$\mathrm{BR}(K_L \to \pi^0 \nu\bar{\nu})_{\mathrm{SM}} = (3.4 \pm 0.6) \times 10^{-11}, \tag{5}$$

$$\mathrm{BR}(K^+ \to \pi^+ \nu\bar{\nu})_{\mathrm{SM}} = (8.4 \pm 1.0) \times 10^{-11}. \tag{6}$$

On the experimental side, KOTO and NA62 are ongoing to measure these modes at the SM accuracy. The data of these branching ratios are given as follows [11, 12]:

$$\mathrm{BR}(K_L \to \pi^0 \nu\bar{\nu})_{\mathrm{exp}} < 2.6 \times 10^{-8}\ (90\%\mathrm{C.L.}),\quad \mathrm{BR}(K^+ \to \pi^+ \nu\bar{\nu})_{\mathrm{exp}} = (1.73^{+1.15}_{-1.05}) \times 10^{-10}. \tag{7}$$

In this article, we discuss the effects of supersymmetry (SUSY) models on these kaon observables and the possibilities of solving the discrepancy in ϵ'/ϵ. We focus on chargino and gluino Z-penguin contributions in the scenario with large trilinear couplings, not proportional to the Yukawa couplings. They do not decouple even the SUSY breaking scale is high once relevant mass insertion (MI) parameters are fixed, and give large effects to kaon observables.

2. Z penguins in supersymmetry

In this article, we discuss chargino [1] and gluino [2] Z-penguin contributions. Let us consider chargino contributions to the Z-penguin diagrams (Fig. 1). They are described by the flavor-violating Z-boson vertex,

$$\mathcal{L}_{\mathrm{eff}} = \frac{-g^3}{8\pi^2 \cos\theta_W} Z_{ds}\, \bar{s}_L \gamma_\mu d_L Z^\mu + h.c.\,, \tag{8}$$

where the coupling Z_{ds} includes SM and SUSY contributions,

$$Z_{ds} = Z_{ds}^{(\mathrm{SM})} + Z_{ds}^{(\mathrm{SUSY})}. \tag{9}$$

This SUSY contribution $Z_{ds}^{(\mathrm{SUSY})}$ is written in terms of MI parameters $(\delta^q_{LR})_{ij}$,

$$(\delta^u_{LR})_{ij} = \frac{\frac{v_2}{\sqrt{2}}(T_U)^*_{ij}}{m^2_{\tilde{q}}}, \quad (\delta^u_{RL})_{ij} = \frac{\frac{v_2}{\sqrt{2}}(T_U)_{ji}}{m^2_{\tilde{q}}}, \tag{10}$$

where $(T_U)_{ij}$ is the trilinear scalar couplings. The coupling $Z_{ds}^{(\mathrm{SUSY})}$ is given by the up-type squark-chargino diagrams (Fig. 1). Focusing on the MI parameters, $(\delta^u_{LR})_{13}$ and $(\delta^u_{LR})_{23}$, one obtains

$$Z_{ds}^{(\mathrm{SUSY})} \simeq (\delta^u_{LR})^*_{13}(\delta^u_{LR})_{23} H_0(x_{\tilde{q}\tilde{W}}), \tag{11}$$

where $H_0(x_{\tilde{q}\tilde{W}})$ is the loop function. The coupling $Z_{ds}^{(\mathrm{SUSY})}$ does not decouple even if superparticles are heavy as long as the product of the mass insertion (MI) parameters $(\delta^u_{LR})^*_{13}(\delta^u_{LR})_{23}$ is fixed [13, 14]. These features are guaranteed by the $\mathrm{SU}(2)_L$ breaking, which is provided by $(\delta^u_{LR})_{13}$ and $(\delta^u_{LR})_{23}$ in Eq. (11). Although CP-violating FCNCs of Kaon are tightly constrained by the indirect CP violation of Kaon or electric dipole moments, SUSY contributions to them decouple in heavy SUSY scenarios. Thus, the discrepancy in ϵ'/ϵ may be explained by the Z-penguin contributions. Gluino contributions to the Z-penguin also have the same structures. In our analysis, we calculate in the mass eigenstates basis. We study two scenarios with different setup:

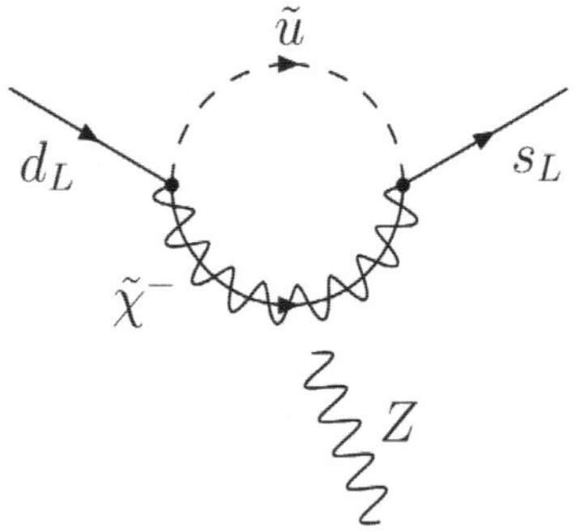

Figure 1. Chargino contributions to Z penguin are shown. SUSY particles are denoted in the mass eigenstate.

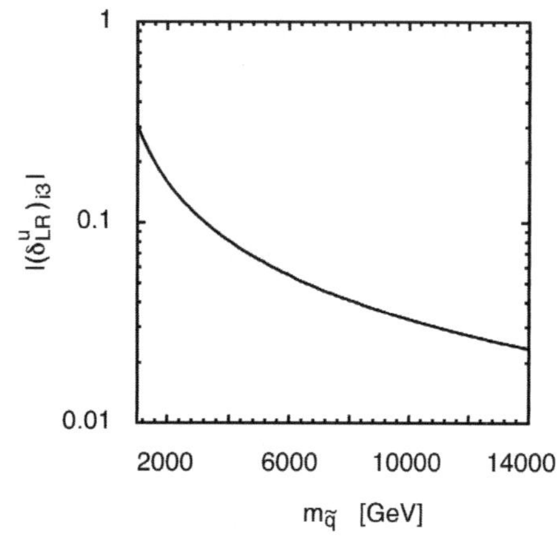

Figure 2. Vacuum stability constraint on $|(\delta^u_{LR})_{i3}|$ for $i = 1, 2$ as a function of $m_{\tilde{q}}$. Here, $m_{\tilde{q}} \equiv m_{\tilde{Q}_i} = m_{\tilde{U}_3}$ and $\tan\beta = 50$. It is assumed that the heavy Higgs bosons are decoupled and the stop left-right mixing is neglected.

- **Chargino contributions**
  We study the chargino contributions to the Z penguin especially with the vacuum stability constraint [1]. Since the MI parameters are proportional to scalar trilinear couplings, the chargino Z-penguin contributions are constrained by requiring the stability of the electroweak (EW) vacuum. This condition is not relaxed even if SUSY particles are heavy. In this work, charge-color breaking (CCB) vacua or potential directions unbounded from below (UFB) have been studied along with ϵ'/ϵ [13, 14]. However, their analyses follow the strategy of Ref. [15, 16], and the vacuum decay rate has not been examined. The vacuum decay is studied, and we discuss whether the current discrepancy of ϵ'/ϵ is explained by the chargino Z-penguin contributions.
- **Gluino contributions**
  The gluino contributions to the Z penguin are also studied [2]. The interesting point of the gluino contribution is that it gives the equal left-handed and right-handed Z couplings. In the scenario with the large right-handed Z couplings, we can enhance ϵ'/ϵ and the branching ratio of $K_L \to \pi^0\nu\bar{\nu}$ decay simultaneously [17]. We discuss whether the gluino contributions can enhance $\mathrm{BR}(K_L \to \pi^0\nu\bar{\nu})$ with satisfying the data of ϵ'/ϵ at the high-scale SUSY model.

3. Chargino contribution

We study the chargino contributions to Z penguin especially with considering the vacuum decay rate seriously. According to Eq. (11), large ϵ'/ϵ is achieved when $\tilde{u}_L$ and $\tilde{c}_L$ have a large mixing with $\tilde{t}_R$. The left-right mixing is proportional to the scalar trilinear coupling $(T_U)_{ij}$. Large flavor-violating trilinear couplings may generate instabilities of the EW vacuum [18]. Requiring that the lifetime of the EW vacuum is longer than the present age of the universe, the trilinear couplings, or equivalently $(\delta^u_{LR})_{13}$ and $(\delta^u_{LR})_{23}$, are constrained.

The vacuum decay rate per unit volume is expressed as

$$\Gamma/V = A \exp(-S_E), \tag{12}$$

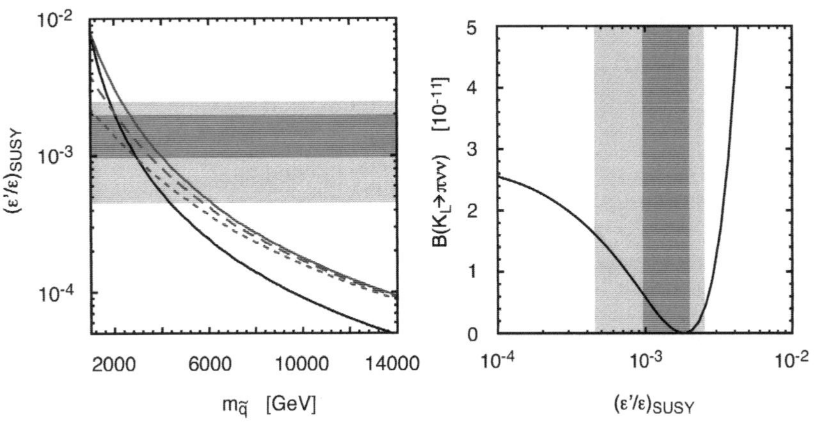

Figure 3. (*left*) $(\epsilon'/\epsilon)_{\rm SUSY}$ is shown as a function of $m_{\tilde{q}}$. The Wino mass $m_{\tilde{W}}$ is 1, 2, 3 TeV for the blue solid, dashed and dotted lines, respectively, while it is equal to $m_{\tilde{q}}$ on the black line. On the red (orange) region, $\Delta\,(\epsilon'/\epsilon)$ is saturated at the 1σ (2σ) level. The SM value follows Ref. [4]. (*right*) Correlation between $\mathrm{BR}(K_L \to \pi^0 \nu\bar{\nu})$ and $(\epsilon'/\epsilon)_{\rm SUSY}$ is shown.

where S_E is estimated at the semi-classical level, which is called the bounce action [19] and calculated by `CosmoTransition` 2.0a2 [20] in this work. The prefactor A is not determined at this level, and we adopt an order-of-estimation analysis : the typical energy scale $A \sim (100\,\mathrm{GeV})^4$ or $(10\,\mathrm{TeV})^4$. The lifetime of the EW vacuum is longer than the age of the universe if the bounce action satisfies

$$S_E \gtrsim 400. \tag{13}$$

We discuss whether the current discrepancy of ϵ'/ϵ is explained by the chargino Z-penguin contributions with satisfying the constraints especially from the vacuum stability condition. The vacuum decay rate is estimated to derive an upper bound on the size of $(T_U)_{i3}$ by requiring $S_E \gtrsim 400$. In fig 2, the bound of $(\delta^u_{LR})_{i3}$ is shown as a function of $m_{\tilde{q}} \equiv m_{\tilde{Q}_i} = m_{\tilde{U}_3}$. Due to the relation Eq. (10), the limit becomes severer as the SUSY scale increases. Therefore, the SUSY contributions to ϵ'/ϵ decrease according to Eq. (11).

In the left plot of Fig. 3, the SUSY contributions to ϵ'/ϵ are shown as a function of $m_{\tilde{q}}$. Here, $|(T_U)_{i3}|$ is set at $S_E = 400$, and $|(T_U)_{13}| = |(T_U)_{23}|$ is assumed. The CP-violating phase is taken to be maximal. There is a degree of freedom in choosing $m_{\tilde{W}}$, and it is set to be 1, 2, 3 TeV and $m_{\tilde{q}}$ as reference cases. It is found that the current discrepancy of ϵ'/ϵ can be explained; the SUSY scale can be as large as 4–6 TeV, depending on the choice of $m_{\tilde{W}}$. In the right plot of Fig. 3, correlation between $\mathrm{BR}(K_L \to \pi^0\nu\bar{\nu})$ and $(\epsilon'/\epsilon)_{\rm SUSY}$ is shown. $\mathrm{BR}(K_L \to \pi^0\nu\bar{\nu})$ decreases as ϵ'/ϵ increases unless ϵ'/ϵ is very large. The current discrepancy implies that $\mathrm{BR}(K_L \to \pi^0\nu\bar{\nu})$ is predicted to be less than 60% of the SM prediction.

4. Gluino contribution

The gluino contributions to the Z penguin is discussed in this section. We consider the high-scale SUSY model at $\mathcal{O}(10)$ TeV. As a simple set-up for it, we take the gluino, wino and bino masses $M_i (i = 3, 2, 1)$ with μ and $\tan\beta$, and the masses of sbottom $\tilde{b}_1$ and $\tilde{b}_2$ as:

$$M_3 = 10\ \mathrm{TeV}, \quad M_2 = 3.3\ \mathrm{TeV}, \quad M_1 = 1.6\ \mathrm{TeV}, \quad \mu = 10\ \mathrm{TeV}, \quad \tan\beta = 3, \qquad (14)$$
$$m_{\tilde{b}_1} = 10\ \mathrm{TeV}, \quad m_{\tilde{b}_2} = 15\ \mathrm{TeV}.$$

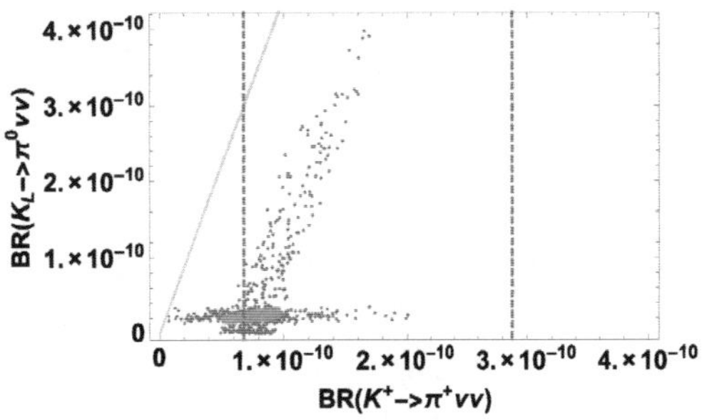

Figure 4. The predicted region for BR($K_L \to \pi^0\nu\bar{\nu}$) versus BR($K^+ \to \pi^+\nu\bar{\nu}$) without imposing ϵ_K. The green line corresponds to the Grossman-Nir bound [21]. The dashed red lines denote the 1σ experimental bounds for BR($K^+ \to \pi^+\nu\bar{\nu}$).

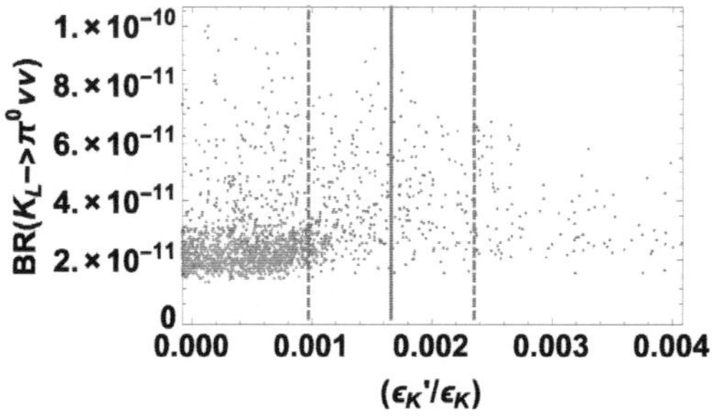

Figure 5. The predicted BR($K_L \to \pi^0\nu\bar{\nu}$) versus ϵ'/ϵ, where the Z coupling satisfies the condition of eq.(15). The vertical solid red line denotes the central value of the experimental data, and the dashed ones denote the experimental bounds with 3σ for ϵ'/ϵ. The pink denotes the SM prediction.

On the other hand, we take the masses of the first and second family up-type and down-type squarks around 15TeV within 5 − 15% relevantly. The flavor mixing parameters s^{dL}_{ij} are set as $s^{dL}_{i3}, s^{dR}_{d3} = 0 \sim 0.3$ $(i = 1, 2)$.

In Fig. 4, we show the predicted region for BR($K_L \to \pi^0\nu\bar{\nu}$) versus BR($K^+ \to \pi^+\nu\bar{\nu}$), with imposing ϵ_K. The left-right (LR) mixing angle θ^b_{LR}, which is correspond to the trilinear couplings, is fixed to 0.3. It is found that these two observables are enhanced significantly due to the large LR mixing angle. The direction of the enhancement of BR($K \to \pi\nu\bar{\nu}$) corresponds to the specific phase regions of the flavor mixing parameters, which are constrained by ϵ_K.

Let us discuss the correlation between BR($K_L \to \pi^0\nu\bar{\nu}$) and ϵ'/ϵ. The Z-penguin contributions to them are studied [17], and it is pointed out that a simultaneous enhancement of them is realized on the following condition for the flavor violationg Z couplings $\Delta^{sd}_{L,R}(Z)$:

$$|\text{Im}\Delta^{sd}_R(Z)| < \text{Im}\Delta^{sd}_L(Z) < 3.3|\text{Im}\Delta^{sd}_R(Z)| \ . \tag{15}$$

Since the gluino contributions give the equal left-handed and right-handed Z couplings, they can satisfy the above condition. In Fig. 5, we show the correlation between BR($K_L \to \pi^0\nu\bar{\nu}$) and ϵ'/ϵ, where Z coupling satisfies the above condition. The constraint from ϵ_K is also imposed. It is remarkable that the Z-penguin mediated by the gluino can enhance BR($K_L \to \pi^0\nu\bar{\nu}$) with satisfying the data of ϵ'/ϵ. While the estimated ϵ'/ϵ fits the observed value, the branching ratio of $K_L \to \pi^0\nu\bar{\nu}$ increases up to 1.0×10^{-10}. The correlations with $K_L \to \mu^+\mu^-$, and the B meson rare decays, $B^0 \to \mu^+\mu^-$ and $B_s \to \mu^+\mu^-$ are also discussed in [2].

5. Conclusion

In this article, we have discussed the SUSY effects on kaon observables ϵ'/ϵ and $K \to \pi\nu\bar{\nu}$. The recent analyses of the SM prediction of ϵ'/ϵ have reported a discrepancy from the experimental value, and we discuss the possibilities that the chargino and gluino Z-penguin contribution explain the discrepancy.

For the chargino contributions to the Z penguin, we have studied the vacuum stability constraint. The chargino contributions are constrained by the vacuum stability condition, and it is found that the SUSY contributions can bridge the current discrepancy of ϵ'/ϵ if the SUSY masses are smaller than 4–6 TeV. The discrepancy implies that BR($K_L \to \pi^0\nu\bar{\nu}$) is about less than 60% of the SM prediction.

The gluino contributions to the Z penguin have also been studied. Even the SUSY breaking scale is high, around 10 TeV, $\mathrm{BR}(K_L \to \pi^0 \nu\bar{\nu})$ can be enhanced with imposing the constraint from ϵ_K. It is remarkable that the gluino contributions can enhance $\mathrm{BR}(K_L \to \pi^0 \nu\bar{\nu})$ up to 1.0×10^{-10} with satisfying the data of ϵ'/ϵ.

The SM predictions of ϵ'/ϵ are expected to be improved in the near future. If the discrepancy would be confirmed, the chargino and gluino contributions could provide an attractive solution. Besides, the predicted correlations with the branching ratios of $K \to \pi\nu\bar{\nu}$ will be tested by KOTO and NA62 in future. KOTO is ongoing to measure $\mathrm{BR}(K_L \to \pi^0 \nu\bar{\nu})$ at the 10% level of the SM value [22], and NA62 will measure $\mathrm{BR}(K^+ \to \pi^+ \nu\bar{\nu})$ with ~10% precision by 2018 [23]. Kaon physics gives an exciting impact on NP in the coming years.

Acknowledgments
The author would like to thank Motoi Endo, Satoshi Mishima, Morimitsu Tanimoto, and Daiki Ueda for fruitful collaboration.

References

[1] M. Endo, S. Mishima, D. Ueda and K. Yamamoto, Phys. Lett. B **762** (2016) 493 [arXiv:1608.01444 [hep-ph]].
[2] M. Tanimoto and K. Yamamoto, PTEP **2016** (in press) arXiv:1603.07960 [hep-ph].
[3] T. Blum *et al.*, Phys. Rev. D **91**, no. 7, 074502 (2015) [arXiv:1502.00263 [hep-lat]].
[4] A. J. Buras, M. Gorbahn, S. Jäger and M. Jamin, JHEP **1511**, 202 (2015) [arXiv:1507.06345 [hep-ph]].
[5] T. Kitahara, U. Nierste and P. Tremper, arXiv:1607.06727 [hep-ph].
[6] K. A. Olive et al. (Particle Data Group), Chin. Phys. C, **38**, 090001 (2014) and 2015 update.
[7] J. R. Batley *et al.* [NA48 Collaboration], Phys. Lett. B **544**, 97 (2002) [hep-ex/0208009].
[8] A. Alavi-Harati *et al.* [KTeV Collaboration], Phys. Rev. D **67**, 012005 (2003) Erratum: [Phys. Rev. D **70**, 079904 (2004)] [hep-ex/0208007].
[9] E. T. Worcester [KTeV Collaboration], arXiv:0909.2555 [hep-ex].
[10] A. J. Buras, D. Buttazzo, J. Girrbach-Noe and R. Knegjens, JHEP **1511** (2015) 033 [arXiv:1503.02693 [hep-ph]].
[11] J. K. Ahn *et al.* [E391a Collaboration], Phys. Rev. D **81** (2010) 072004 [arXiv:0911.4789 [hep-ex]].
[12] A. V. Artamonov *et al.* [BNL-E949 Collaboration], Phys. Rev. D **79** (2009) 092004 [arXiv:0903.0030 [hep-ex]].
[13] G. Colangelo and G. Isidori, JHEP **9809**, 009 (1998) [hep-ph/9808487].
[14] A. J. Buras, G. Colangelo, G. Isidori, A. Romanino and L. Silvestrini, Nucl. Phys. B **566**, 3 (2000) [hep-ph/9908371].
[15] J. A. Casas and S. Dimopoulos, Phys. Lett. B **387**, 107 (1996) [hep-ph/9606237].
[16] J. A. Casas, Adv. Ser. Direct. High Energy Phys. **21**, 469 (2010) [hep-ph/9707475].
[17] A. J. Buras, JHEP **1604** (2016) 071 [arXiv:1601.00005 [hep-ph]].
[18] J. h. Park, Phys. Rev. D **83**, 055015 (2011) [arXiv:1011.4939 [hep-ph]].
[19] S. R. Coleman, Phys. Rev. D **15**, 2929 (1977) Erratum: [Phys. Rev. D **16**, 1248 (1977)].
[20] C. L. Wainwright, Comput. Phys. Commun. **183**, 2006 (2012) [arXiv:1109.4189 [hep-ph]].
[21] Y. Grossman and Y. Nir, Phys. Lett. B **398** (1997) 163 [hep-ph/9701313].
[22] J. K. Ahn *et al.*, arXiv:1609.03637 [hep-ex].
[23] M. Moulson [NA62 Collaboration], arXiv:1611.04979 [hep-ex].

Probing lepton flavour (universality) violation at NA62 and future kaon experiments

Lewis C. Tunstall,[0,1] **Andreas Crivellin,**[2]
Giancarlo D'Ambrosio[3] **and Martin Hoferichter**[4]

[1]Albert Einstein Centre for Fundamental Physics, Institute for Theoretical Physics, University of Bern, Sidlerstrasse 5, CH–3012 Bern, Switzerland
[2]Paul Scherrer Institut, CH–5232 Villigen PSI, Switzerland
[3]INFN-Sezione di Napoli, Via Cintia, I–80126 Napoli, Italy
[4]Institute for Nuclear Theory, University of Washington, Seattle, WA 98195-1550, USA

E-mail: tunstall@itp.unibe.ch

Abstract. Recent results from the LHC's first run have revealed intriguing departures from lepton flavour universality in the semi-leptonic decays of B-mesons. We discuss the complementary role that rare kaon decays can provide in testing new physics explanations of these flavour anomalies. In the framework of minimal flavour violation, we relate the chiral low-energy constants involved in $K \to \pi\ell\ell'$ and $K \to \ell\ell'$ ($\ell = \mu$ or e) with the new physics Wilson coefficients of the $b \to s$ effective Hamiltonian. We comment on the determination of these low-energy constants at NA62 and future kaon experiments, as well as the required improvements in sensitivity necessary to test the B-physics anomalies in the kaon sector.

1. Introduction

With the completion of Run 1 at the LHC, we find ourselves with several indirect hints that physics beyond the Standard Model (SM) may be lurking in the semi-leptonic decays of B-mesons. Among the recent experimental anomalies, two at LHCb have received considerable attention:

- a 2.6σ signal of lepton flavour universality violation (LFUV) in the measured [1] branching fractions of $B \to K\ell^+\ell^-$ decays ($\ell = \mu$ or e);
- the measured value [2, 3] of the angular observable P_5' [4] in the decay $B \to K^*\mu^+\mu^-$ deviates from the SM prediction at the 2–3σ level [5–7].

Taken separately, each deviation is at most a 3σ effect; however, when combined with other $b \to s$ transitions, global fits [8, 9] indicate that (a) new physics (NP) is preferred over the SM by 4–5σ, and (b) the effect is in the $\mu\mu$ modes only. Expressed in terms of the effective Hamiltonian for $b \to s$ transitions [9]

$$\mathcal{H}_{\text{eff}}^{|\Delta B|=1} = -\frac{4G_F}{\sqrt{2}} V_{tb}V_{ts}^* \sum_i C_i^B(\mu) Q_i^B(\mu) + \text{h.c.}\,, \tag{1}$$

[0] Speaker

Content from this work may be used under the terms of the Creative Commons Attribution 3.0 licence. Any further distribution of this work must maintain attribution to the author(s) and the title of the work, journal citation and DOI.
Published under licence by IOP Publishing Ltd

the potential NP can be interpreted as contributions to the Wilson coefficients $C^B_{9,10}$ of the semi-leptonic operators

$$Q^B_9 = \frac{e^2}{32\pi^2}\left[\bar{s}\gamma^\mu(1-\gamma_5)b\right]\sum_{\ell=e,\mu}\left[\bar{\ell}\gamma_\mu\ell\right] \quad \text{and} \quad Q^B_{10} = \frac{e^2}{32\pi^2}\left[\bar{s}\gamma^\mu(1-\gamma_5)b\right]\sum_{\ell=e,\mu}\left[\bar{\ell}\gamma_\mu\gamma_5\ell\right] . \qquad (2)$$

Here we discuss the complementary role that rare kaon decays can provide in testing NP explanations of the B-physics anomalies. Our analysis [10] is based on the observation that at low energy scales $\mu \ll m_{t,b,c}$, the strangeness-changing transitions are described in terms of the effective Lagrangian [11]

$$\mathcal{L}^{|\Delta S|=1}_{\text{eff}} = -\frac{G_F}{\sqrt{2}}V_{ud}V^*_{us}\sum_i C_i(\mu)Q_i(\mu) + \text{h.c.} , \qquad (3)$$

which contains semi-leptonic operators

$$Q_{7V} = \left[\bar{s}\gamma^\mu(1-\gamma_5)d\right]\sum_{\ell=e,\mu}\left[\bar{\ell}\gamma_\mu\ell\right] \quad \text{and} \quad Q_{7A} = \left[\bar{s}\gamma^\mu(1-\gamma_5)d\right]\sum_{\ell=e,\mu}\left[\bar{\ell}\gamma_\mu\gamma_5\ell\right] , \qquad (4)$$

that are the $s \to d$ analogues of the $b \to s$ operators $Q^B_{9,10}$ in (2). In the framework of minimal flavour violation (MFV), the Wilson coefficients of the two sectors *are correlated*, and we use this feature to convert knowledge of $C_{7V,7A}$ into bounds on $C^B_{9,10}$. The quality of the bounds is hampered by non-perturbative effects from QCD, which are parametrised in terms of the low-energy constants (LECs) arising in the 3-flavour chiral expansion. In Section 2 we focus on the experimental determination of the LECs involved in $K^\pm \to \pi^\pm\ell^+\ell^-$ and $K_L \to \ell^+\ell^-$, and comment on how measurements at NA62 and future kaon experiments may improve the resulting limits on $C^B_{9,10}$. A similar strategy is adopted in Section 3 to obtain bounds on lepton flavour violation (LFV) in the B-meson sector, while concluding remarks are given in Section 4. For the analysis of LFUV and LFV in other kaon decays not discussed in this article, we refer the reader to [10].

2. Kaon probes of lepton flavour universality violation

2.1. $K^\pm \to \pi^\pm\ell^+\ell^-$

At low energies, the dominant contribution to $K^+ \to \pi^+\ell^+\ell^-$ is due to single virtual-photon exchange. The resulting amplitude involves a vector form factor $V_+(z)$ which up to $O(p^6)$ in the chiral expansion, can be decomposed in the general form [11]

$$V_+(z) = a_+ + b_+z + V^{\pi\pi}_+(z) , \qquad z = q^2/m^2_K . \qquad (5)$$

Here the LECs a_+ and b_+ parametrise the polynomial part, while the rescattering contribution $V^{\pi\pi}_+$ can be determined from fits to $K \to \pi\pi$ and $K \to \pi\pi\pi$ data. Chiral symmetry alone does not constrain the values of the LECs,[1] so instead, we consider the differential decay rate $d\Gamma/dz \propto |V_+(z)|^2$ as a means to extract a_+ and b_+ from experiment. The resulting fit to the decay spectra from all available high-statistics experiments is given in Table 1.

Now for the crucial point: if lepton flavour universality applies, the coefficients a_+ and b_+ have to be equal for the ee and $\mu\mu$ channels, which within errors is indeed the case. Since the SM interactions are lepton flavour universal, deviations from zero in differences like $a^{\mu\mu}_+ - a^{ee}_+$ would then be a sign of NP, and the corresponding effect would be necessarily short-distance.

[1] Although there are recent attempts to determine a_+ and b_+ on the lattice [12] or via large-N_c methods [13].

Channel	a_+	b_+	Reference
ee	-0.587 ± 0.010	-0.655 ± 0.044	E865 [14]
ee	-0.578 ± 0.016	-0.779 ± 0.066	NA48/2 [15]
$\mu\mu$	-0.575 ± 0.039	-0.813 ± 0.145	NA48/2 [16]

Table 1. Fitted values of coefficients in the vector form factor (5).

To convert the allowed range on $a_+^{\rm NP}$ into a corresponding range in the Wilson coefficients $C_{7V}^{\ell\ell}$, we make use of the $O(p^2)$ chiral realization of the $SU(3)_L$ current

$$\bar{s}\gamma^\mu(1-\gamma_5)d \leftrightarrow iF_\pi^2(U\partial^\mu U^\dagger)_{23}\,, \qquad U = U(\pi, K, \eta)\,, \tag{6}$$

to obtain

$$a_+^{\rm NP} = \frac{2\pi\sqrt{2}}{\alpha} V_{ud}V_{us}^* C_{7V}^{\rm NP}\,. \tag{7}$$

Contributions due to NP in $K^+ \to \pi^+\ell^+\ell^-$ can then be probed by considering the *difference* between the two channels

$$C_{7V}^{\mu\mu} - C_{7V}^{ee} = \alpha\frac{a_+^{\mu\mu} - a_+^{ee}}{2\pi\sqrt{2}V_{ud}V_{us}^*}\,. \tag{8}$$

In the framework of MFV, this can be converted into a constraint on the NP contribution to C_9^B:

$$C_9^{B,\mu\mu} - C_9^{B,ee} = -\frac{a_+^{\mu\mu} - a_+^{ee}}{\sqrt{2}V_{td}V_{ts}^*} \approx -19 \pm 79\,, \tag{9}$$

where we have averaged over the two electron experiments listed in Table 1.

Evidently, the determination of $a_+^{\mu\mu} - a_+^{ee}$ needs to be improved by an $O(10)$ factor in order to probe the parameter space relevant for the B-anomalies, whose explanation involves Wilson coefficients $C_{9,10}^B = O(1)$ [9]. Improvements of this size may be possible at NA62, especially for the experimentally cleaner dimuon mode which currently has the larger uncertainty.

2.2. $K_L \to \ell^+\ell^-$

With an eye towards future kaon experiments involving K_L beams[2] (e.g. a side programme like KLEVER at NA62 [17]), we also consider $K_L \to \ell^+\ell^-$ decays as another potential probe of LFUV. These decays are complementary to $K^+ \to \pi^+\ell^+\ell^-$ since they provide the means to constrain NP effects due to axial-vector interactions.

The dominant long-distance contribution is due to $K_L \to \gamma^*\gamma^* \to \ell^+\ell^-$, where the dispersive component of the decay amplitude involves a counterterm χ that is decomposed in long- and short-distance parts $\chi(\mu) = \chi_{\gamma\gamma}(\mu) + \chi_{\rm SD}$. The SM prediction for $\chi_{\rm SD}$ is known [11], but $\chi_{\gamma\gamma}$ depends on two LECs whose values are not fixed by chiral symmetry. Nevertheless, we invoke the same argument applied to $K^+ \to \pi^+\ell^+\ell^-$, and observe that if lepton flavour universality holds, then the SM values of χ must be equal in both the ee and $\mu\mu$ channels. Then, using the chiral realization (6) of the $SU(3)_L$ current, one obtains an analogous relation to (7) for the NP Wilson coefficient:

$$C_{7A}^{\rm NP} = -\frac{\alpha}{F_K G_F V_{ud}V_{us}^*}\left(\frac{2\Gamma_{\gamma\gamma}}{\pi m_K^3}\right)^{1/2}\chi_{\rm NP}\,, \qquad \Gamma_{\gamma\gamma} = \Gamma(K_L \to \gamma\gamma)\,. \tag{10}$$

[2] LCT thanks Matthew Moulson for informative discussions on the challenges of measuring rare K_L decays with high precision.

Channel	χ (Solution 1)	χ (Solution 2)
ee	$5.1^{+15.4}_{-10.3}$	$-(57.5^{+15.4}_{-10.3})$
$\mu\mu$	3.75 ± 0.20	1.52 ± 0.20

Table 2. Values of the contact term $\chi(M_\rho)$ extracted from the measured $K_L \to e^+e^-$ and $K_L \to \mu^+\mu^-$ rates.

The final step is to observe that within the framework of MFV, the difference

$$C_{7A}^{\mu\mu} - C_{7A}^{ee} = -\frac{\alpha}{F_K G_F V_{ud} V_{us}^*}\left(\frac{2\Gamma_{\gamma\gamma}}{\pi m_K^3}\right)^{1/2}\left(\chi^{\mu\mu} - \chi^{ee}\right), \tag{11}$$

is directly related to the Wilson coefficients of the B-physics sector:

$$C_{10}^{B,\mu\mu} - C_{10}^{B,ee} = 2.6\left(\frac{3.5 \times 10^{-4}}{V_{td}V_{ts}^*}\right)\left(\chi^{\mu\mu} - \chi^{ee}\right). \tag{12}$$

Clearly, the quality of the bounds on $C_{10}^{B,\ell\ell}$ depends on the precision with which $\chi^{\ell\ell}$ can be determined. The present situation is as follows: χ can be determined (up to a two-fold ambiguity) from the measured $K_L \to \ell^+\ell^-$ rates, with the resulting values shown in Table 2.[3] Although Solution 2 for the ee channel is easily excluded, the current data are not sufficiently precise to distinguish among the remaining Solutions. Moreover, Solution 1 for the ee channel carries a large uncertainty which needs to be improved in order to test LFUV in the interesting regions of parameter space.

To gain an idea of the improvement in precision required, suppose the uncertainty in $\Gamma(K_L \to \ell^+\ell^-)$ could be reduced by a factor of 10 and that the central value remained unchanged. In this idealised scenario, Solution 2 for the $\mu\mu$ channel would be strongly disfavored, given that LFUV (if present at all) should manifest itself as a small effect. Substituting the resulting difference $\chi^{\mu\mu} - \chi^{ee} \sim 1.3 \pm 1.3$ into (12) would then yield the bound $C_{10}^{B,\mu\mu} - C_{10}^{B,ee} \approx 3.5 \pm 3.5$. We thus find that the improvement required to obtain competitive bounds on C_{10}^B is of similar magnitude to what we found in the analysis of C_9^B in $K^+ \to \pi^+\ell^+\ell^-$.

We note that, contrary to the K_L decay, short-distance contributions to the $K_S \to \ell^+\ell^-$ decay width do not interfere with any long-distance physics. In the expression for the decay width [11]

$$\Gamma(K_S \to \ell^+\ell^-) = \frac{m_K}{8\pi}\beta_\ell\left[\beta_\ell^2|B|^2 + |C|^2\right], \tag{13}$$

short-distance effects contribute to the second term [19], which in the notation of [11] reads

$$\mathrm{Im}\, C_{\rm SD} = -2\sqrt{2}F_K m_\ell\, \mathrm{Im}\left\{V_{ud}V_{us}^* C_{7A}\right\} = 2\sqrt{2}F_K m_\ell\, y_{7A}\, \mathrm{Im}\left\{V_{td}V_{ts}^*\right\}. \tag{14}$$

Recently, LHCb have strengthened their 2012 bound [21] on the $K_S \to \mu^+\mu^-$ rate by roughly a factor of two: $\mathrm{BR}(K_S \to \mu^+\mu^-) < 6.9(5.8) \times 10^{-9}$ at the 95% (90%) confidence level [22]. With improvements expected in the near future, LHCb will start probing the theoretically interesting region $\mathrm{BR}(K_S \to \mu^+\mu^-) \approx 10^{-11}$ [19] where meaningful bounds on NP contributions to the coefficient y_{7A} can be derived.

[3] One subtlety involved in the fit is that the equality $\chi^{\mu\mu} = \chi^{ee}$ is assumed to hold to all orders in the chiral expansion. In general [18, 19], higher-order corrections involve m_ℓ-dependent terms that produce a shift $\Delta\chi$ which must be subtracted to identify potential LFUV contributions. At the two-loop level, we have estimated [10] the size of the shift to be $\Delta\chi^{\mu\mu} - \Delta\chi^{ee} = -2.8$, in line with explicit calculations based on Canterbury approximants [20].

	$K_L \to \mu^\pm e^\mp$	$K^+ \to \pi^+\mu^\pm e^\mp$	$K^+ \to \pi^+\mu^\pm e^\mp$ (NA62 projection)
$\left(\lvert C^{\mu e}_{7V}\rvert^2 + \lvert C^{\mu e}_{7A}\rvert^2\right)^{1/2}$	$< 1.3 \times 10^{-6}$	$< 2.2 \times 10^{-5}$	$< 5.1 \times 10^{-6}$
$\left(\lvert C^{B,\mu e}_{9}\rvert^2 + \lvert C^{B,\mu e}_{10}\rvert^2\right)^{1/2}$	< 0.71	< 12	< 2.7

Table 3. Limits on LFV Wilson coefficients from kaon decays. The last line shows the corresponding limits in the B-system assuming MFV, while the rightmost column refers to the projected limit from NA62 [23].

3. Lepton flavour violating decays

NA62 is projected to improve existing limits on LFV in the kaon sector by an order of magnitude or more [23]. Here we adopt a similar approach to our analysis of LFUV in Section 2, and use MFV to convert limits on the LFV Wilson coefficients $C^{\mu e}_{7V,7A}$ of the kaon sector into bounds on the corresponding $b \to s$ transitions. The analysis is simplified by the fact that LFV is absent in the SM[4] so the decay amplitudes can be expressed directly in terms of $C^{\mu e}_{7V,7A}$ and quark operators based on the chiral realization (6).

In the context of LFV searches at NA62, the mode of interest to us is $K^+ \to \pi^+\mu^\pm e^\mp$, whose branching fraction takes the form

$$\mathrm{Br}\big[K^+ \to \pi^+\mu^\pm e^\mp\big] = 0.027\big\{\lvert C^{\mu e}_{7V}\rvert^2 + \lvert C^{\mu e}_{7A}\rvert^2\big\}\,. \tag{15}$$

As a point of comparison, the same combination of Wilson coefficients enters in the $K_L \to \mu^\pm e^\mp$ decay, with branching fraction

$$\mathrm{Br}\big[K_L \to \mu^\pm e^\mp\big] = 2.6\big\{\lvert C^{\mu e}_{7V}\rvert^2 + \lvert C^{\mu e}_{7A}\rvert^2\big\}\,. \tag{16}$$

Based on (15) and (16), the present experimental limits [23] can be used to constrain the term $(\lvert C^{\mu e}_{7V}\rvert^2 + \lvert C^{\mu e}_{7A}\rvert^2)^{1/2}$. The resulting limits are given in the first line of Table 3, where the limit from $K_L \to \mu^\pm e^\mp$ decays is an order of magnitude more stringent than the one from $K^+ \to \pi^+\mu^\pm e^\mp$. Although the projected limits from NA62 [23] will also fall short by a factor of 4, it would be interesting to examine whether removal of the GigaTracker could produce a sufficient increase in statistics to become competitive with the K_L limits.[5]

As in the case of LFUV, we use MFV to obtain limits on the B-physics coefficients. These are shown in the bottom line of Table 3, where in the case of the $K^+ \to \pi^+\mu^\pm e^\mp$ decay, the resulting constraints are slightly better than (9), but of similar magnitude. The strongest constraint is obtained from the limit on $K_L \to \mu e$.

4. Remarks and future prospects

Rare kaon decays offer a potential probe into NP explanations of the flavour anomalies observed by LHCb in semi-leptonic B-decays. In the framework of MFV, we have discussed how limits on LFUV and LFV at kaon experiments can be converted into bounds on the Wilson coefficients $C^B_{9,10}$ of the $b \to s$ effective Hamiltonian. In this respect, we have focused on the K^+ decay modes we consider to be of most relevance to the NA62 programme (with K_L modes included as a point of comparison):

- $K^+ \to \pi^+\ell^+\ell^-$ and $K_L \to \ell^+\ell^-$ as a means to constrain LFUV due to vector and axial-vector interactions. Although the LECs involved in these modes are poorly constrained

[4] Modulo negligible effects due to neutrino oscillations.

[5] The signal for $K^+ \to \pi^+\mu^\pm e^\mp$ is three charged tracks, so an accurate measurement of the pion and lepton momenta is not essential. LCT thanks A. Ceccucci for sharing this suggestion.

from theory, they can be determined via precise experimental measurements. In particular, bounds on the short-distance NP effects can be obtained by considering *differences* between the ee and $\mu\mu$ parameters. For this method to obtain meaningful bounds on $C^B_{9,10}$, we found that order-of-magnitude reductions on the LECs' current uncertainties are required. At NA62, this will require a measurement of the $K^+ \to \pi^+\ell^+\ell^-$ spectrum which significantly reduces the uncertainties of the NA48/2 data.

- the LFV modes $K^+ \to \pi^+\mu^\pm e^\mp$ and $K_L \to \mu^\pm e^\mp$. In these decays, the amplitude factorises into short- and long-distance components, so strong bounds can be obtained on the former. NA62's projected limits on $K^+ \to \pi^+\mu^\pm e^\mp$ will fall short of the existing E871 bound on $K_L \to \mu^\pm e^\mp$ [24], so it would be worthwhile considering if the experiment can be adapted to boost the sample of K^+ decays involving three charged particles.

We note that in order to convert bounds in the kaon sector into those in B-physics, we have worked within the framework of MFV. In general, the potential NP may not satisfy MFV, so from this point of view there are three logical possibilities which can be tested at NA62:

(i) if the NP explanations for the B-meson anomalies are consistent with MFV, then one should see a signal at the sensitivities discussed in Sections 2 and 3;

(ii) the experimental searches at a sensitivity expected from MFV turn out negative. In this case, one could immediately infer that any NP scenario explaining the B-anomalies would require departures from MFV;

(iii) a signal is observed at current or slightly improved sensitivity. As in point 2 above, we could then rule out NP explanations of the B-anomalies based on MFV.

Acknowledgements

Financial support by MIUR under the project number 2010YJ2NYW, the DOE (Grant No. DE-FG02-00ER41132), and the Swiss National Science Foundation is gratefully acknowledged.

References

[1] R. Aaij *et al.* (LHCb), Phys. Rev. Lett. **113** (2014) 151601 [arXiv:1406.6482].
[2] R. Aaij *et al.* (LHCb), Phys. Rev. Lett. **111** (2013) 191801 [arXiv:1308.1707].
[3] The LHCb Collaboration, LHCb-CONF-2015-002, CERN-LHCb-CONF-2015-002.
[4] S. Descotes-Genon, T. Hurth, J. Matias and J. Virto, JHEP **1305**, 137 (2013) [arXiv:1303.5794].
[5] S. Descotes-Genon, L. Hofer, J. Matias and J. Virto, JHEP **1412**, 125 (2014) [arXiv:1407.8526].
[6] W. Altmannshofer and D. M. Straub, Eur. Phys. J. C **75** (2015) 382 [arXiv:1411.3161].
[7] S. Jäger and J. Martin Camalich, Phys. Rev. D **93** (2016) 014028 [arXiv:1412.3183].
[8] W. Altmannshofer and D. M. Straub, arXiv:1503.06199.
[9] S. Descotes-Genon, L. Hofer, J. Matias and J. Virto, JHEP **1606**, 092 (2016) [arXiv:1510.04239].
[10] A. Crivellin, G. D'Ambrosio, M. Hoferichter and L. C. Tunstall, Phys. Rev. D **93** (2016) 074038 [arXiv:1601.00970].
[11] V. Cirigliano, G. Ecker, H. Neufeld, A. Pich and J. Portoles, Rev. Mod. Phys. **84** (2012) 399 [arXiv:1107.6001].
[12] N. H. Christ *et al.* [RBC and UKQCD Collaborations], Phys. Rev. D **92** (2015) 094512 [arXiv:1507.03094].
[13] E. Coluccio Leskow, G. D'Ambrosio, D. Greynat and A. Nath, Phys. Rev. D **93** (2016) 094031 [arXiv:1603.09721].
[14] R. Appel *et al.* [E865 Collaboration], Phys. Rev. Lett. **83** (1999) 4482 [arXiv:hep-ex/9907045].
[15] J. R. Batley *et al.* [NA48/2 Collaboration], Phys. Lett. B **677** (2009) 246 [arXiv:0903.3130].
[16] J. R. Batley *et al.* [NA48/2 Collaboration], Phys. Lett. B **697** (2011) 107 [arXiv:1011.4817].
[17] M. Moulson, these proceedings [URL: `http://tinyurl.com/na62klever`].
[18] G. D'Ambrosio, G. Isidori and J. Portoles, Phys. Lett. B **423** (1998) 385 [arXiv:hep-ph/9708326].
[19] G. Isidori and R. Unterdorfer, JHEP **0401**, 009 (2004) [arXiv:hep-ph/0311084].
[20] P. Masjuan and P. Sanchez-Puertas, JHEP **1608**, 108 (2016) [arXiv:1512.09292].
[21] R. Aaij *et al.* [LHCb Collaboration], JHEP **1301**, 090 (2013) [arXiv:1209.4029].
[22] M. Ramos Pernas, these proceedings [URL: `http://tinyurl.com/LHCblimit`].
[23] A. Ceccucci, in Proc. Sci., CD15 (2015) 006 [URL: `http://tinyurl.com/na62limits`].
[24] D. Ambrose *et al.* [BNL Collaboration], Phys. Rev. Lett. **81** (1998) 5734 [arXiv:hep-ex/9811038].

Dark sectors and their signatures in Kaon physics

Maxim Pospelov

Department of Physics and Astronomy, University of Victoria, Victoria, BC V8P 1A1, Canada
Perimeter Institute for Theoretical Physics, Waterloo, ON N2J 2W9, Canada

E-mail: pospelov@uvic.ca

Abstract. I review the set of theoretical ideas motivating experimental searches of light physics beyond the Standard Model. Specifically, I give examples of very weakly interacting candidates that can be probed in direct detection experiments. I then discuss a possibility of probing such states in the high-intensity kaon experiments by looking at the rare and radiate decay modes. The interesting modes to pursue include $K^{\pm} \to \pi^{\pm}$ + missing energy, $K^{\pm} \to \pi^{\pm}\pi^{0}e^{+}e^{-}$, as well as Dalitz and double-Dalitz decays of π^0.

1. Introduction

The discovery of the Higgs boson in the LHC experiments is the triumphant developement in particle physics. For a long time the existence of the Higgs boson was anticipated based on indirect electroweak precision tests, and from theoretical self-consistency of the Standard Model. Besides a discovery of the new particle, one can argue that the discovery of the Higgs boson - which for now is fully consistent with being an elementary boson - is also a manifestation of a new force in Nature - the Yukawa-type force. Given that all other forces in nature are of gauge origin, the Higgs discovery is really a "big deal" for our understanding of how the world works.

Another significant outcome of the LHC is the total absence of new physics that was widely anticipated to show its appearence around the 1 TeV energy scale. So far, there are no hints on supersymmetric partners, new gauge bosons with $\sim$ TeV scale masses, or additional missing energy signals. The list of new physics models that is "put in trouble" by the LHC, and/or pushed above the TeV scale goes on and on. Recent excitement about a "di-photon" resonance at 750 GeV turned out to be baseless, as the early excess was nothing else but a statistical fluctuation.

Given that we have such a resounding success of the Standard Model (SM) on one hand, and no new physics indications (so far) on the other hand, one may go back to the "drawing board" and re-assess the possibilities for new physics, as well as our ability to probe it in particle physics experiments.

I make an effort to do so in Figure 1, where the cartoon of the coupling constants and particle masses that mediate the interactions is given. All SM forces have their coupling constants around the $\sim 0.1 - 0.01$ range, and the masses of mediators vary from a photon (massless), to gluon (effectively getting a QCD scale mass via confinement), and W, Z bosons. Interaction strength similar to a Fermi constant, $\sqrt{2}G_F = g_w^2/(8M_W^2)$, is represented by a diagonal line, that divides this parameter space into the "SM backyard" and the "unknown frontier". The high-energy colliders are uniquely positioned to probe the upper-right corner in this picture, where the strength of the coupling constants is significant, and the masses of new hypothetical

Content from this work may be used under the terms of the Creative Commons Attribution 3.0 licence. Any further distribution of this work must maintain attribution to the author(s) and the title of the work, journal citation and DOI.
Published under licence by IOP Publishing Ltd

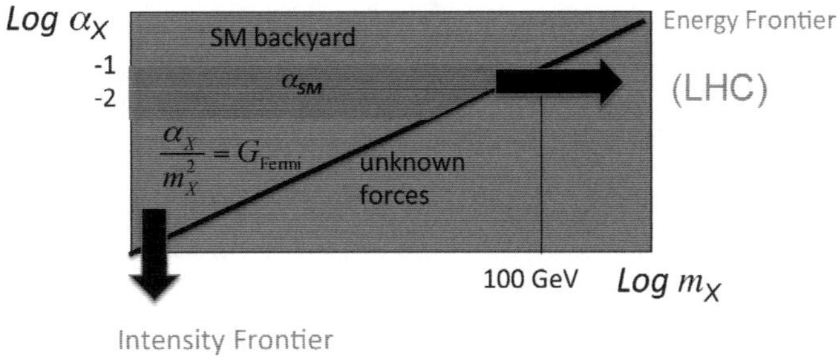

Figure 1. A cartoon representation of the coupling strength vs force mediator mass. The LHC is the way to explore high mass/not-so-small couplings for new forces. The intensity frontier direction (a downward pointing arrow) - very small couplings and masses much below the electroweak scale - is completely orthogonal, and can be explored with the new generation of electron beam facilities.

force carriers are large. Indeed, for a "sequential" Z' (i.e. coupled to the SM fermions the same way as the SM Z boson), the limits from CMS and ATLAS are $m_{Z'} > 2.8$TeV. This means that the analogue of the Fermi constant for the Z' compared to G_F is $\sim m_Z^2/m_{Z'}^2 < 10^{-3}$. Should a new force with the TeV scale mass mediator exist, its strength is already 1000 times smaller than the weak force! But what happens if there is new physics that is significanlty *lighter* than the TeV scale? If there are new gauge bosons with masses on the order of 100 MeV, and couplings $\alpha_X \sim 10^{-6}$ or so, one cannot use the LHC very efficiently to search for such particles. The reason for that is that the production cross section is not growing with energy, as it would be for a TeV scale bosons, but instead is falling rather steeply. Being weakly coupled, such new particles will be completely buried under the SM backgrounds. Therefore, the high-energy colliders have very limited ability to probe light and very weakly coupled sectors.

If the new light particles violate some exact or approximate symmetries of physics at the GeV-scale, such as parity or CP symmetry, one can still limit the corresponding analogue of Fermi constant rather efficiently using the precision symmetry tests at low energy (parity-violation experiments, electric dipole moments, and for certain types of new physics, flavor constraints). But certain types of new physics, such as "dark photon" to be discussed later, are completely "innocuous" as they preserve all discrete and flavor symmetries. In this case, the strength of such a force can be well above the SM G_F, and indeed such forces may be discovered in the SM backyard. To find them, or better to say, probe their existence, one needs a dedicated effort at the "intensity frontier" [1]. On the Figure 1 cartoon, intensity frontier activity is pictured by the arrow in the down-left corner.

To conclude the Introduction: the intensity frontier direction is in many ways an orthogonal direction to the energy frontier. Consequently, the absence of new physics at the LHC does not spell doom on [most of the] efforts at the intensity frontier. A sub-GeV scale is exactly the place where new high-intensity electron beams can have an unparallel sensitivity to light and very weakly coupled New Physics. In the rest of my talk I will go over some motivations for new physics at these low scales. The experiments with kaons, probing very minute standard model branching ratios, are uniquely positioned to probe the dark sectors as well.

2. Portals to light weakly coupled sectors

The portals to light neutral hidden sectors are an effective field theory classification of all lowest dimension operators that mediate interaction between SM and the hidden sector. Symbolically, we can write this type of contributions to the effective Lagrangian as

$$\Delta\mathcal{L} = \sum c_O O^{\rm SM} O^{\rm hidden}. \tag{1}$$

We assume that the hidden sector is not charged under the SM fields, and therefore both $O^{\rm SM}$ and $O^{\rm hidden}$ are SM singlet operators separately. Note, however, that they have to form a Lorentz singlet only in the product directly entering into $\Delta\mathcal{L}$, but separately they can transform under the Lorentz group. Given that, one can write down a set of terms of lowest dimension operators in $\Delta\mathcal{L}$:

$$B_{\mu\nu}F_{\mu\nu} \quad \text{kinetic mixing portal} \tag{2}$$
$$(H^\dagger H)(AS + \lambda S^2) \quad \text{Higgs portal} \tag{3}$$
$$(\bar{L}H)N \quad \text{neutrino portal} \tag{4}$$
$$(\bar{\psi}_{\rm SM}\gamma_\mu\psi_{\rm SM})A_\mu \quad \text{generic gauge portal} \tag{5}$$
$$(\bar{\psi}_{\rm SM}\gamma_\mu\gamma_5\psi_{\rm SM})\partial_\mu a/f_a \quad \text{axion − like portal} \tag{6}$$

In the expressions above, $B_{\mu\nu}$ is the SM hypercharge field strength, H is the SM Higgs doublet, L is a SM lepton doublet, and $\psi_{\rm SM}$ stands for a generic SM fermion. Particles from dark sector are denoted as A_μ, N and a. $F_{\mu\nu}$ stands for a usual field strength of A_μ.

Notice that the neutral singlet fermion N coupled to a composite SM effective operator of dimension $5/2$, $(\bar{L}H)$, naturally leads to masses and mixings in the neutrino sector. Therefore, it is quite possible that nature has "already used" one of these portals, and it makes sense to study possible phenomenological implications or all these structures. Sometimes, one can speculate that these simple structures may be behind certain long-standing anomalies and deviations in particle physics. One bright example is the case of the so-called dark photon, or a new particle coupled via the $\epsilon B_{\mu\nu}F_{\mu\nu}$ portal that can give a *positive* correction to the anomalous magnetic moments of charged leptons. For example, a few tens of MeV mass dark photon with $\epsilon \sim$ few$\times 10^{-3}$ could account for the long-standing theoretical deficit of about 3×10^{-9} in the muon anomalous magnetic moment [2].

Besides couplings to the SM, which is absolutely crucial for production of the dark sector particles, the dark sector states can interact among themselves. For example, interactions among dark particles, such as $(\bar{N}\gamma_\mu N)A_\mu$, which e.g. may lead to the decay of the dark photons into the dark fermionic states.

All of these hidden sector options have found vivid discussions in recent literature, with many potentially interesting applications to astrophysics and cosmology, collider physics and beam dump experiments [1]. In the rest of this talk, I will discuss the implications of the portal framework, and possible signatures of light particles in the kaon decay experiments.

3. Dark photons and dark scalars in the kaon decay experiments

To probe light weakly coupled states one needs high-luminosity yields of the SM particles, and rather precise detectors. Modern experiments with kaons generally fit this description. It is no surprise that some of the most exquisite limits come directly from studying kaon decays, or from the decays of the kaon decay products. Below, I go over the list of some decays and comment on their applications to the hidden sectors.

- $K^+ \to \pi^+$+missing energy, $K^+ \to \pi^+ e^+ e^-$. It is entirely possible that a light particle coupled either via the Higgs or kinetic mixing portals would be produced in an underlying $K^+ \to \pi^+ X$ decays. Subsequently, X can decay to a lepton pair. However, if the decay length is too long, as it would happen with a light scalar coupled via $\sim ASH^\dagger H$ portal and effective mixing to the SM Higgs below $\sim 10^{-3}$, the actual X decay may occur outside a detector. (Same signature will also occur if X decays invisibly, e.g. to a piar of light dark matter particles.) This may lead to a spectacular signatures of the underlying two-body decay for $K^+ \to \pi^+$+missing energy, and even a relatively modest excess of events may lead to a clear discovery. The details of constraints on light scalars mixed with the Higgs can be found in e.g. Ref. [3]. NA62 is poised to improve these bounds by a wide margin. If, on the other hand, the X particle is a vector, it is likely to decay quickly, in which case the invariant mass spectrum of e^+e^- may reveal a presence of a resonance via the bump hunt search.

- $K^+ \to \mu^+\nu e^+e^-$, $K^+ \to \mu^+\nu\mu^+\mu^-$. Rare radiative leptonic decays may be a signature of particles that couple predominantly to leptons [4]. In the above processes, kaons are a "controlled source of muons", and the underlying process $K^+ \to \mu^+\nu X$ may reveal itself in the spectrum of extra charged lepton pairs. Especially interesting is the regime of very small invariant masses, where such processes may be hidden behind substantial backgrounds from other kaon decay modes.
- $\pi^0 \to \gamma X \to \gamma e^+e^-$, $\pi^0 \to XX \to e^+e^-e^+e^-$. Dalitz and double-Dalitz decays of pions can be studied in the kaon beam experiments as well. In fact, a recent analysis of NA48/2 data [5] has contributed rather significantly into the dark photon story. It has shown that the Dalitz decays of π^0 show no unexpected peaking structures, setting very tight limits on the model, and excluding the remainder of the $g-2$ relevant parameter space in the minimal model. It is worth pointing out that a recent model [6] has been constructed to specifically circumvent these bounds. By "flipping" the dark charges of u and d quarks, one can eliminate the $\pi^0 \to X\gamma$ decays. However, $\pi^0 \to XX$ will remain, and can be analyzed for a "double peak" in the spectrum of four charged leptons.
- $K^+ \to \pi^+\pi^0e^+e^-$. This decay can also be sensitive to the $K^+ \to \pi^+\pi^0X$ models, and since the analysis of this decay mode is ongoing by the NA48 collaboration, it is also make sense to perform a search of peaking structures in the lepton pair invariant mass spectrum. This decay mode will also be senstivie to particles suggested in [6].

To conclude, kaon physics will remain vital for the searches of light weakly coupled particles. Future experiments, such as NA62, will make further significant progress in probing a variety of light particles coupled to the SM via well-motivated portals.

References

[1] J. Alexander *et al.*, arXiv:1608.08632 [hep-ph].
[2] M. Pospelov, Phys. Rev. D **80** (2009) 095002.
[3] M. Pospelov, A. Ritz and M. B. Voloshin, Phys. Lett. B **662** (2008) 53.
[4] B. Batell, N. Lange, D. McKeen, M. Pospelov and A. Ritz, arXiv:1606.04943 [hep-ph].
[5] J. R. Batley *et al.* [NA48/2 Collaboration], Phys. Lett. B **746** (2015) 178.
[6] J. L. Feng, B. Fornal, I. Galon, S. Gardner, J. Smolinsky, T. M. P. Tait and P. Tanedo, Phys. Rev. Lett. **117** (2016), 071803.

Results on kaon physics from OKA setup at U-70 .

V Obraztsov, on behalf of the "OKA" collaboration

Institute for High Energy Physics, Protvino, Moscow Region, RU

E-mail: Vladimir.Obraztsov@ihep.ru

Abstract. Some recent results from OKA setup are presented. First, the decay $K^+ \to \pi^0 e^+ \nu (K_{e3})$ is studied. About 3.15M events are selected for the analysis. The linear and quadratic slopes for the decay formfactor $f_+(t)$ are measured: $\lambda'_+ = (26.1 \pm 0.45 \pm 0.38) \times 10^{-3}$, $\lambda''_+ = (1.94 \pm 0.23 \pm 0.12) \times 10^{-3}$. For the exotic scalar and tensor interactions we get: $F_S/f_+(0) = (-0.44 \pm 0.7 \pm 0.24) \times 10^{-2}$; $F_T/f_+(0) = (0.16 \pm 2 \pm 1.3) \times 10^{-2}$. Several alternative parametrizations are tried: the Pole fit parameter is found to be $M_V = 890 \pm 3.7$ MeV ; the parameter of the Dispersive parametrization is measured to be $\Lambda_+ = (24.72 \pm 0.23) \times 10^{-3}$. Second, the results of a search for heavy neutrino in the $K_{\mu 2}$ decay are shown. The upper limits on the mixing parameter of the heavy neutrino with the muon neutrino $|U_{\mu H}|^2$ are obtained. Typically, $|U_{\mu H}|^2 \leq 10^{-6}$ for the region $225 \leq m_H \leq 375$ MeV. Third, a new study of the radiative $K_{\mu 3}$ decay are presented. The number of signal events is ~ 580 which is 4 times larger than in previous measurements. $R = Br(K^+_{\mu 3 \gamma}, 30 < E_\gamma < 60MeV)/Br(K_{\mu 3}) = (4.85 \pm 0.2(stat) \pm 0.5(syst)) \times 10^{-4}$ which should be compared with 4.7×10^{-4} from the theory. An estimate of the T-odd asymmetry gives $A_\xi = (-0.19 \pm 0.05 \pm 0.09)$. A space asymmetry over $cos(\theta^*_{\mu\gamma})$, where $\theta^*_{\mu\gamma}$ is the angle between photon and muon momenta in the kaon rest frame is measured to be $A(cos\theta^*) = 0.6 \pm 0.05 \pm 0.1$.

1. Introduction

The kaon decays provide unique information about the dynamics of the strong interactions. It has been a testing ground for such theories as current algebra, PCAC, Chiral Perturbation Theory (ChPT) etc. Another direction is a search for new physics: new light particles and new interactions. In this talk, we present a high-statistics study of K_{e3}; $K_{\mu 2}$ and $K_{\mu 3}\gamma$ decays from OKA detector at U-70 Proton Synchrotron.

2. OKA beam and detector

OKA is the abbreviation for 'Experiments with Kaons '. OKA beam is a RF-separated secondary beam of U-70 Proton Synchrotron of IHEP, Protvino. The beam is described elsewhere [1]. RF-separation with Panofsky scheme is realised. It uses two superconductive Karsruhe-CERN SC RF deflectors [2], donated by CERN. Sophisticated cryogenic system, built at IHEP [3] provides superfluid He for cavities cooling. The resulting beam has up to $\sim 20\%$ of kaons with an intensity of $\sim 10^6$ kaons per 3 sec U-70 spill. The OKA setup is a magnetic spectrometer, presented on Fig. 1. It includes:

(i) Beam spectrometer on the basis of 7 1mm pitch PC's ($BPC_{x,y}$) ~1500 channels in total, 4 2mm-thick scintillation counters and 2 threshold Cherenkov counters.

Content from this work may be used under the terms of the Creative Commons Attribution 3.0 licence. Any further distribution of this work must maintain attribution to the author(s) and the title of the work, journal citation and DOI.
Published under licence by IOP Publishing Ltd

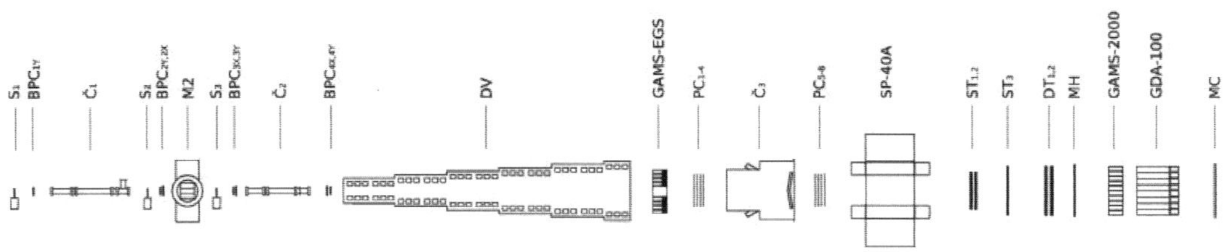

Figure 1. OKA setup

(ii) Decay volume(DV) with Veto system, 11m long, filled with He, veto system is composed of 670 Lead-Scintillator sandwiches $20\times$(5 mm Sc + 1.5 mm Pb) with WLS readout. The counters are grouped in 300 ADC channels.

(iii) Main magnetic spectrometer: 200×140 cm^2 aperture magnet wih $\int Bdl \sim 1$ Tm; 5K 2 mm pitch PC's; 1K 9 mm Straw's and 300 channels of 40 mm DT's.

(iv) Gamma detectors: GAMS-2000 (~ 2000 4×4 cm^2 lead glass blocks), large angle detector (EGS) (~ 1500 4×4 cm^2 lead glass blocks).

(v) Muon detector: GDA-100 Hadron Calorimeter (100 20×20 cm^2 iron-scintillator sandwiches with WLS plates readout); 4 1×1 m^2 Sc counters behind GDA-100.

3. Trigger and statistics

Very simple trigger, which is almost "minimum bias " one, has been used during data-taking: $Tr = S_1 \cdot S_2 \cdot S_3 \cdot \overline{\check{C}_1} \cdot \check{C}_2 \cdot \overline{S}_{bk} \cdot (\Sigma_{GAMS} > MIP)$. It is a combination of beam Sc counters, $\check{C}_{1,2}$ threshold Cerenkov counters ($\check{C}_1$ sees pions, $\check{C}_2$- pions and kaons), S_{bk} - a "beam-killer " counter located in the beam-hole of the GAMS gamma-detector. $\Sigma_{GAMS} > MIP$ is a requirement for the analog sum of amplitudes in the GAMS-2000 to be higher than a MIP signal.
The "OKA" is taking data since 2010, the total available statistics corresponds to $\sim 15MK_{e3}$ decays. In the present study we use part of the statistics taken in 2011 and 2012.

4. K_{e3} decay study.

The data processing starts with the beam particle reconstruction in $BPC_1 \div BPC_4$, then the secondary tracks are looked for in $PC_1 \div PC_8$; $ST_1 \div ST_3$; $DT_1 \div DT_2$ and events with one good positive track are selected. The decay vertex is searched for, and a cut is introduced on the matching of incoming and decay track. The next step is to look for showers in GAMS-2000 and EGS calorimeters. The electron identification is done using the ratio of the energy of the shower to the momentum of the associated track. The E/p distribution is shown in Fig. 2. The particles with 0.8 < E/p < 1.2 are accepted as electrons. The events with one charged track identified as electron and two additional showers in ECAL are selected for further processing. The mass spectrum of $\gamma\gamma$ shows a clean π^0 peak at $M_{\pi 0} = 134.9$ MeV with a resolution of ~ 8.5 MeV. To fight the main background from $K_{\pi 2}$ decay, the angle between the momentum of the beam kaon $\vec{p_K}$ and that of the $e\pi$-system i.e. $\vec{p_e} + \vec{p_\pi}$ is considered, see Fig. 2. The background is clearly seen as a peak at zero angles. The cut is $\alpha > 2$ mrad. Further selection is done by the requirement that the event passes 2C $K \rightarrow e\nu\pi^0$ fit. The event selection results in $\sim$3.15M events. The surviving background is estimated from MC to be less than 1%.

4.1. Analysis

The analysis is based on the fit of the distribution of the events over the Dalitz plot. The variables $y = 2E_e^*/M_K$ and $z = 2E_\pi^*/M_K$, where E_e^*, E_π^* are the energies of the electron and π^0 in the kaon

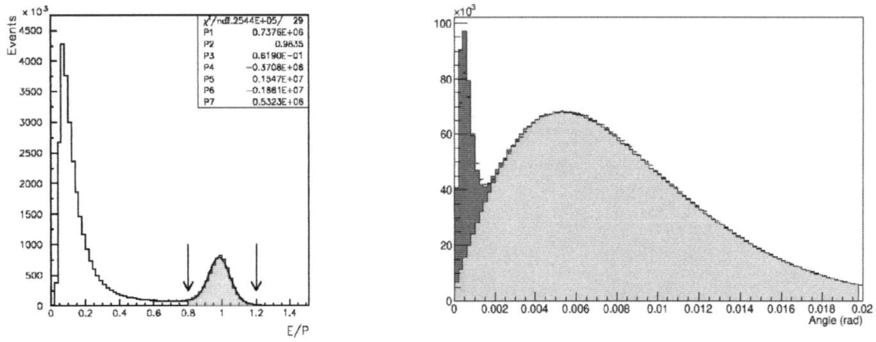

Figure 2. E/P plot - the ratio of the energy of the associated ECAL cluster to the momentum of the charged track (left); α - the angle between $\vec{p}_K$ and $\vec{p}_e + \vec{p}_\pi$ in the lab-system (right).

c.m.s are used. The background events, as MC shows, occupy the peripheral part of the plot. The most general Lorentz invariant form of the matrix element for the decay $K^+ \to l^+\nu\pi^0$ is [4]:
$M = \frac{-G_F V_{us}}{2}\bar{u}(p_\nu)(1+\gamma^5)[((P_K+P_\pi)_\alpha f_+ + (P_K-P_\pi)_\alpha f_-)\gamma^\alpha - 2m_K f_S - i\frac{2f_T}{m_K}\sigma_{\alpha\beta}P_K^\alpha P_\pi^\beta]v(p_l)$
It consists of scalar, vector and tensor terms. $f_\pm$ are the functions of $t = (P_K - P_\pi)^2$. In the Standard Model (SM) the W-boson exchange leads to the pure vector term. The term in the vector part, proportional to f_- is reduced(using the Dirac equation) to a scalar form-factor, proportional to $(m_l/2m_K)f_-$ and is negligible in the case of K_{e3}. Different parametrizations have been used for $f_+(t)$. First is just a Taylor series: $f_+(t) = f_+(0)(1 + \lambda'_+ t/m_\pi^2 + \frac{1}{2}\lambda''_+ t^2/m_\pi^4)$. It is usually used to compare with ChPT predictions. Alternative parametrization is the pole one: $f_+(t) = f_+(0)\frac{m_V^2}{m_V^2 - t}$. The last is a relatively new Dispersive parametrization [5]: $f_+(t) = f_+(0)exp(\frac{t}{m_\pi^2}(\Lambda_+ + H(t)))$. Here H(t) is a known function.
The procedure for the experimental extraction of the parameters λ_+, f_S, f_T, which was developed in [6] is used. This procedure allows avoiding systematic errors due to the "migration" of the events over the Dalitz plot because of the finite experimental resolution. The radiative corrections were taken into account by reweighting every MC event, according to [7].

4.2. Results and comparison with theory

The results of the fit are summarized in Table 1. The first line is just a linear fit, it gives average slope of the $f_+(t)$ formfactor. It could be compared to quite old $ChPT$ $O(p^4)$ result [8]: $\lambda_+^{ChPT} = (31.0 \pm 0.6) \times 10^{-3}$. The second line is the "standard" fit with two parameters - linear and quadratic slopes. The quadratic term is quite significant, there is a strong correlation between parameters as it is seen in Fig. 3. The $ChPT$ $O(p^6)$ prediction for the quadratic slope is [9]: $\lambda''_+(ChPT) = (1.1 \pm 0.1) \times 10^{-3}$. The quality of the fit is illustrated by the z projection of the Dalitz plot, shown on Fig. 3. The third and fourth lines of the Table 1. correspond to the fits, when on top of the standard V-A term the scalar or tensor terms are allowed. In the fourth line, the correlation between F_S and F_T is used, inspired by a LeptoQuark model. It is seen, that f_S and f_T are not significant. The main contribution to systematic is coming from the variation of the cut on -z- coordinate of the vertex and the cut on the angle α. The systematic errors are shown in the Table 1. as the second error. The result of the Pole fit is: $M_V = 890 \pm 3.7$ MeV. It can be compared to the PDG value for the mass of K^* [10]: $M_{K*} = 891.66 \pm 0.26$ MeV. The Dispersion fit gives $\Lambda_+ = (24.72 \pm 0.23) \times 10^{-3}$. An interpretation of limits on F_S and F_T is possible in the framework of the scalar LeptoQuark(LQ) model. Then a diagram with LQ

Table 1. The results of the fits of the $f_+(t), f_S, f_T$ formfactors.

$\lambda'_+ \times 10^3$	$\lambda''_+ \times 10^3$	$f_S/f_+(0) \times 10^2$	$f_T/f_+(0) \times 10^2$
29.56$\pm$ 0.28	0	0	0
26.1 $\pm$ 0.45 $\pm$ 0.38	1.94$\pm$ 0.23 $\pm$ 0.12	0	0
26.1 $\pm$ 0.45 $\pm$ 0.38	1.93 $\pm$ 0.24$\pm$ 0.12	-0.44 $\pm$ 0.7$\pm$ 0.24	0.16 $\pm$ 2$\pm$ 1.3
26.1 $\pm$ 0.45$\pm$ 0.38	1.93 $\pm$ 0.24$\pm$ 0.12	-0.41 $\pm$ 0.37$\pm$ 0.24	0

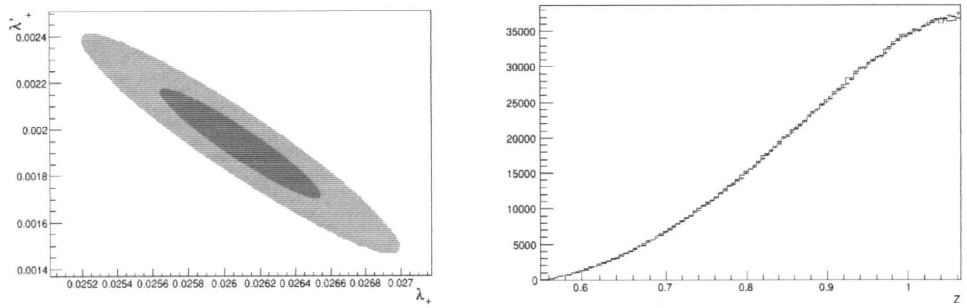

Figure 3. $\lambda' - \lambda''$ correlation plot (left); Projection of the Dalitz-plot on z (right) axis. Data is the points with errors; histogram is the fit, corresponding to the 2-nd line of the Table 1.

exchange should be added to the SM diagram with W. Applying Fiertz transformation to the LQ matrix element we get: $(\bar{s}\mu)(\bar{\nu}u) = -\frac{1}{2}(\bar{s}u)(\bar{\nu}\mu) - \frac{1}{8}(\bar{s}\sigma_{\alpha\beta}u)(\bar{\nu}\sigma^{\alpha\beta}\mu)$. The first term is the scalar, the second one - tensor. The relation between f_S, f_T and the Leptoquark scale Λ_{LQ} can be set out ([11]). As a result, a 95% lower limit for the LeptoQuark scale is $\Lambda_{LQ} > 3.5$ TeV.

5. Search for heavy neutrino in $K \to \mu\nu_H$ decay.

Two prescaled triggers are used for this study: $Tr_\mu = S_1 \cdot S_2 \cdot S_3 \cdot \overline{\check{C}_1} \cdot \check{C}_2 \cdot \overline{S}_{bk} \cdot S_\mu/4$ and $Tr_{Kdec} = S_1 \cdot S_2 \cdot S_3 \cdot \overline{\check{C}_1} \cdot \check{C}_2 \cdot \overline{S}_{bk}/10$, where S_μ is .OR. of 4 muon counters. Statistics of the November 2012 run is used for this analysis. One beam and one secondary track, identified as muon in GAMS, HCAL and muon counters are selected; decay vertex is required to be well inside the decay volume; total energy in the decay volume veto and in the large angle EGS e.m. calorimeter is required to be $E_{VETO} < 50$ MeV and $E_{EGS} < 100$ MeV. At this stage, $\sim 46M$ events are selected. Then several strong quality cuts are introduced for the beam and secondary track, which improve the missing mass resolution and suppress the background in the signal region. At this stage, we are left with $\sim 16M$ events. The distribution of the events over missing mass squared is shown in the left part of Fig. 4 together with the signal $K_{\mu2}$ MC distribution and the MC backgrounds $K_{\mu2\gamma}$; $K_{\mu3}$; $K_{\pi2}$. The results of the fit of the experimental distribution by the sum of the signal and the backgrounds is shown as well. The fit is good, as it is demonstrated by the residual distribution of the Fig. 4 (right). It is seen, that there is no significant extra signal from the heavy neutrino. To derive the upper limits, the residual spectrum is fitted by the gauss distributions with different masses and the MC-predicted width, which depends on the mass. From that, the one-sided 90% C.L. upper limits are constructed.

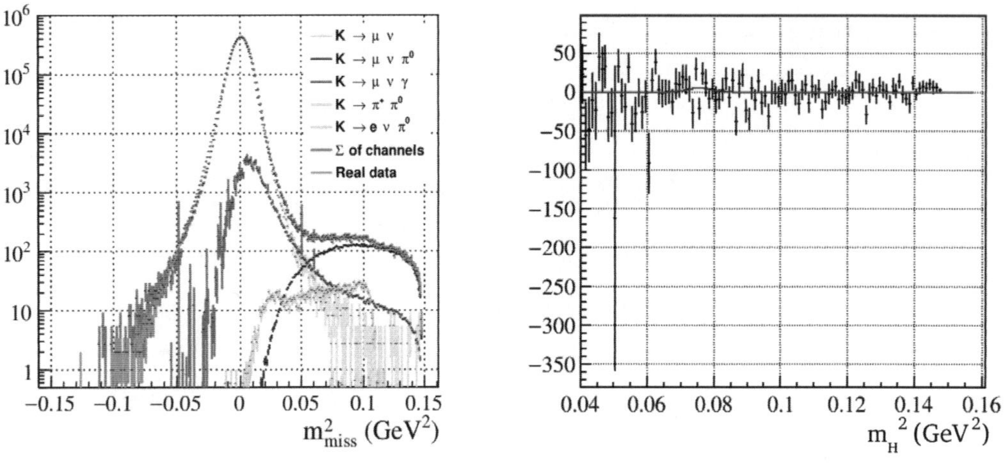

Figure 4. The missing mass squared for the selected events (left). The MC distributions for different components are also shown. Residuals distribution for the fit of the experimental data by the sum of $K_{\mu 2}$ and $K_{\mu 2\gamma}$; $K_{\mu 3}$; $K_{\pi 2}$ (right).

Using that, and normalizing to the $K \to \mu\nu$ it is straightforward to obtain the upper limit for the branching for $K^+ \to \mu^+ \nu_H$, which is shown in the left part of Fig. 5 and, finally, the upper limits for the mixing parameter $|U_{\mu H}|^2$, shown in the right part of Fig. 5. The last figure also shows the previous limits obtained by KEK E-010 experiment [12] and BNL-949 experiment [13]. It is seen, that our limits extend the excluded region up to ~ 375 MeV. The cosmological limit (BBN) is also shown [14] .

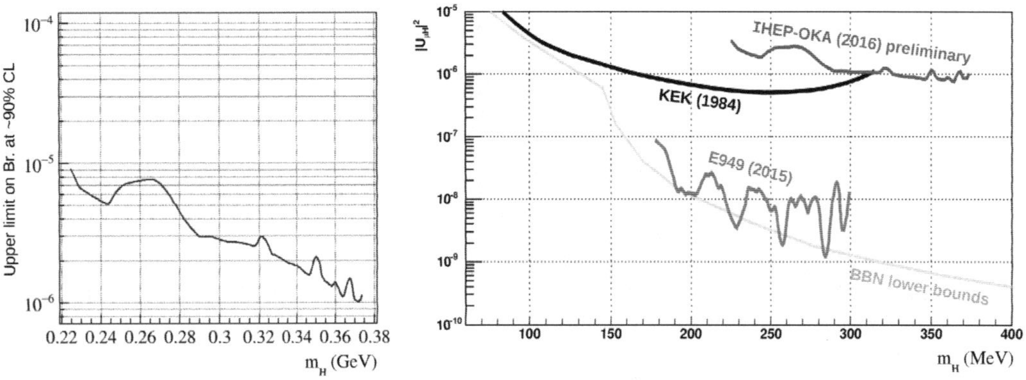

Figure 5. The 90% C.L. upper limits for the branching $K^+ \to \mu^+ \nu_H$ (left); 90% C.L. upper limits for the mixing parameter $|U_{\mu H}|^2$ (right). Existing limits from KEK E-010 and E949 BNL experiments together with the cosmological limit (BBN) are shown.

6. Study of the $K^+ \to \pi^0 \mu^+ \nu \gamma$ decay.

This decay was first seen by ISTRA+ [15] and KEK K470 [16] in 2006. The decay is interesting, in particular, because one can build up a non-zero T-odd correlation from the momenta of the particles $\xi = \vec{p}_\gamma \cdot (\vec{p}_\mu \times \vec{p}_\pi)/m_K^3$ and measure corresponding asymmetry $A_\xi = \frac{N(\xi>0)-N(\xi<0)}{N(\xi>0)+N(\xi<0)}$. There are calculations in χPT O(p4), in particular for the T-odd asymmetry [17]. For some extensions of SM $A_\xi \sim 3 \times 10^{-4}$ is expected. Statistics of 2012 and 2013 runs is used for the analysis. The selection requires one charged track identified as muon in GAMS, HCAL and μ-counters; then 3 extra showers are required, two of them from π^0 $0.11 < M_{\gamma\gamma} < 0.16$ GeV; missing momentum is required to point to active area of GAMS; total energy in veto is required to be $E_{VETO} < 50$ MeV and $E_{EGS} < 100$ MeV; to suppress $K^+ \to \pi^+\pi^0\pi^0(K_{\pi3})$ background two cuts are introduced $M^2_{miss} < 0.014$ or $M^2_{miss} > 0.0225$ GeV^2 ; finally the energy of the third photon in the kaon rest frame is required to be $E_\gamma > 30$MeV to suppress the $K_{\mu3}$ background. The resulting distribution over $\mu\pi^0\gamma\nu$ mass, calculated assuming $m_\nu = 0$ is presented in Fig. 6. The number of signal events from the fit is $\sim$ 580, which is about 4 times larger than in previous measurements. Using $K_{\mu3}$ events for the normalisation we get $R = Br(K^+_{\mu3\gamma}, 30 < E_\gamma < 60MeV)/Br(K_{\mu3}) = (4.85 \pm 0.2(stat) \pm 0.5(syst)) \times 10^{-4}$ which should be compared with 4.7×10^{-4} from the theory. An estimate of the T-odd asymmetry gives $A_\xi = (-0.19 \pm 0.05 \pm 0.09)$. A space asymmetry over $cos(\theta^*_{\mu\gamma})$, where $\theta^*_{\mu\gamma}$ is the angle between photon and muon momenta in the kaon rest frame is measured to be $A(cos\theta^*) = 0.6 \pm 0.05 \pm 0.1$.

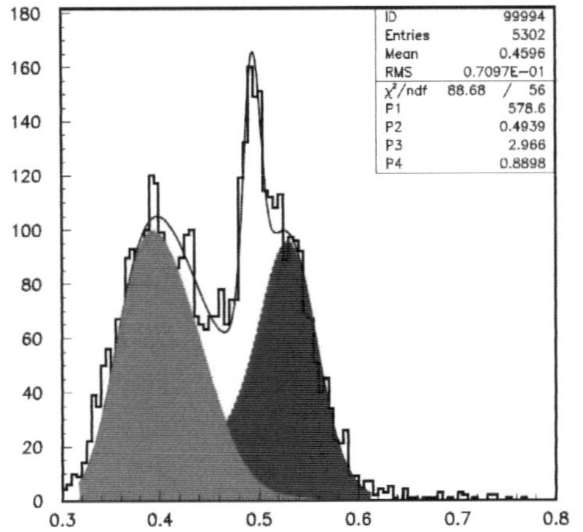

Figure 6. The invariant mass of the $\mu\pi^0\gamma\nu$ system calculated assuming $m_\nu = 0$. A fit with 3 components: the residual background from $K_{\mu3}$ (blue) ; $K_{\pi3}$ (red) and the signal $K_{\mu3\gamma}$ is shown.

References

[1] Garkusha V I et al. 2003 *Preprint* IHEP 2003-4.
[2] Citron A et al. 1979 *Nucl. Instr. and Meth.* **164** 31.
[3] Ageev A et al. 2008 *Proceedings of RUPAC* 282.
[4] Steiner H et al. 1971 *Phys. Lett.* B **36** 521.
[5] Bernard V et al. 2006 *Phys. Lett.* B **638** 480; Bernard V et al. 2009 *Phys.Rev.* D **80** 034034.
[6] Yushchenko O P et al. 2004 *Phys. Lett.* B **589** 111.
[7] Cirigliano V et al. 2002 *Eur. Phys. J.* C **23** 121.
[8] Gasser J and Leutwyler H 1985 *Nucl. Phys.* B **250** 517.
[9] Bijnens J,Talavera P 2003 *Nucl. Phys.* B **669** 341.
[10] Olive K A et al. 2014 *Chin. Phys.* C **38** 090001.
[11] Kiselev V V, Likhoded A K and Obraztsov V F 2002, *hep-ph/0204066/*.
[12] Asano Y et al. 1981 *Phys. Lett.* B **104**.
[13] Artamonov A V et al. 2015 *Phys. Rev.* D **91** 059903.
[14] Gorbunov D and Shaposhnikov M 2007 *JHEP* **0710** 015.
[15] Tchikilev O G et al. 2007 *Phys. Atom. Nucl.* **70** 29.
[16] Shimizu S et al. 2006 *Phys. Lett.* B **633** 190.
[17] Braguta V et al. 2003*Phys. Rev.* D **68**.

Precise measurement of the $K_{e2}/K_{\mu 2}$ branching ratio and search for new physics beyond the Standard Model

S Bianchin (on behalf of the TREK/E36 Collaboration)
TRIUMF, 4004 Wesbrook Mall, Vancouver, BC V6T 2A3, CANADA
E-mail: s.bianchin@triumf.ca

Abstract. The E36 experiment recently conducted at J-PARC by the TREK Collaboration will provide a precise mesurement of the decay ratio $R_K = \Gamma(K^+ \to e^+\nu_e)/\Gamma(K^+ \to \mu^+\nu_\mu)$ with the aim of testing lepton universality, a basic assumption of the Standard Model (SM). The SM prediction for the decay ratio R_K is highly precise ($\delta R_K/R_K = 4 \times 10^{-4}$) and any observed deviation from this prediction would clearly indicate the existence of new physics. The E36 experimental apparatus, designed to offer small systematic errors, was installed at the J-PARC K1.1BR kaon beam line in the fall of 2014, fully commissioned in the spring of 2015, and the production data were collected in the fall of 2015. A scintillating fiber target was used to stop a beam of $1.2 \times 10^6 K^+$ per spill. The K^+ decay products were then momentum analyzed by a 12-sector superconducting toroidal spectrometer and charged particles were tracked with high precision by Multi-Wire Proportional Chambers (MWPC) combined with a photon calorimeter with a large solid angle (75% of 4π) and 3 different particle identification systems. Details of the E36 experimental apparatus, as well as the status of the data analysis will be presented.

1. Introduction

High precision electroweak measurements have been used extensively to test the Standard Model (SM) predictions. The SM successfully describes the results of most particle and nuclear physics experiments and it is believed that any deviation from its predictions would indicate the existence of new physics. Lepton universality being a key characteristics of the SM, the $K^+ \to l^+\nu_l$ (K_{l2}) decay, which is the simplest leptonic decay among the K^+ decay channels, is therefore one of the best channels to perform such tests. Any violation of lepton universality clearly indicates the existence of new physics beyond the SM. By measuring the ratio of leptonic decay widths of the charged kaon, R_K, defined as

$$R_K = \frac{\Gamma(K^+ \to e^+\nu_e)}{\Gamma(K^+ \to \mu^+\nu_\mu)}, \tag{1}$$

the K_{l2} hadronic form factor cancels out and the sensitivity to new physics is greatly enhanced by the helicity suppression in K_{e2}. The SM prediction can be determined with very high precision ($\delta R_K/R_K = 0.04\%$) [1], giving a value of

$$R_K^{SM} = \frac{m_e^2}{m_\mu^2}\left(\frac{m_K^2 - m_e^2}{m_K^2 - m_\mu^2}\right)^2 (1 + \delta_r) = (2.477 \pm 0.001) \times 10^{-5}, \tag{2}$$

Content from this work may be used under the terms of the Creative Commons Attribution 3.0 licence. Any further distribution of this work must maintain attribution to the author(s) and the title of the work, journal citation and DOI.
Published under licence by IOP Publishing Ltd

where δ_r is a radiative correction due the internal bremsstrahlung process (IB) which, by definition, contributes to R_K (unlike the structure dependent process (SD)). A Minimal Super-Symmetric (SUSY) extension of the SM (MSSM) with R-parity conservation has recently been considered as a candidate for new physics to be tested by R_K [2–4]. In the case of K_{l2}, in addition to the $W^{\pm}$ exchange (Figure 1a), a charged Higgs-mediated SUSY lepton flavor violating (LFV) contribution can be enhanced if accompanied by the emission of a τ neutrino (Figure 1b). This effect can be described as

$$R_K^{LFV} = R_K^{SM}\left(1 + \frac{m_K^4}{M_{H+}^4} \cdot \frac{m_\tau^2}{m_e^2}\Delta_{13}^2 \tan^6\beta\right) \tag{3}$$

where M_{H+} is the mass of the charged Higgs and Δ_{13} ($\lesssim 10^{-3}$) is the term induced by the exchange of a Bino and a Slepton. Taking $\Delta_{13} = 5\times10^{-4}$, $\tan\beta = 40$ and $M_{H+} = 500$ GeV, would yield to a value of $R_K^{LFV} = 1.013 \times R_K^{SM}$. It is therefore possible to reach a contribution at the percent level due to possible LFV enhancements arising in SUSY models.

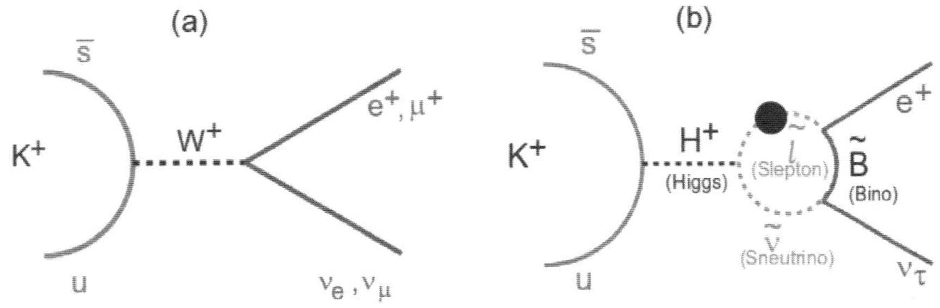

Figure 1. Contributions to R_K from (a) SM and (b) LFV SUSY. A charged Higgs-mediated LFV SUSY contribution can be strongly enhanced by the emission of a τ neutrino.

Recent in-flight decay experiments, such as KLOE [5] and NA62 [6], have measured the R_K ratio leading to a current world average of $R_K = (2.488 \pm 0.010) \times 10^{-5}$. The E36 experiment at the Japan Proton Accelerator Complex (J-PARC) by the TREK (Time Reversal Experiment with Kaons) collaboration aims to provide a competitive measurement of R_K with different systematics.

2. R_K measurements using stopped K^+ at J-PARC

The E36 experiment, which is part of the TREK program at J-PARC, was set up and fully commissioned at the K1.1BR kaon beam line between fall 2014 and spring 2015. The data collection was completed by the end of 2015 using an upgraded version of the KEK-PS E246 12-sector superconducting toroidal spectrometer [7] used in a previous T-violation experiment at KEK [8–10]. The incoming K^+ is tagged by the Fitch Cherenkov counter before stopping and decaying in the active target, which consists of 256 scintillating fibers ($3 \times 3 \times 200$ mm^3) oriented in the direction of the beam, providing a precise kaon stop location in the transverse plane. Wrapped around the fiber target is a Spiral Fiber Tracker (SFT), which consists of two pairs of scintillating fiber ribbons of opposite helicity bundled together around the target, providing the longitudinal coordinate of the outgoing decay particles [11]. This target+SFT system is surrounded by 12 time-of-flight counters (TOF1) and 12 aerogel [12] counters (AC) aligned with the 12 sectors of the toroidal spectrometer and forms the “Central Detector”.
A highy segmented (768 crystals) large acceptance CsI(Tl) photon calorimeter barrel covering

about 75% of the total solid angle is used to identify and correct for structure dependent (SD) background events. The calorimeter features 12 holes (known as "muon holes") aligned with the sectors of the spectrometer allowing charged particles, such as π^+, μ^+ and e^+, to be momentum analyzed by tracking with the Multi-Wire Proportional Chambers (MWPC) C2, C3 and C4. At the exit of each magnet sector, another set of fast scintillator detectors (TTC and TOF2) and lead glass counters (PGC) [13] provides both trigger signals and e^+/μ^+ particle identification. Figure 2 shows the end view (left panel) and the side view (right panel) of the TREK-E36 experimental apparatus.

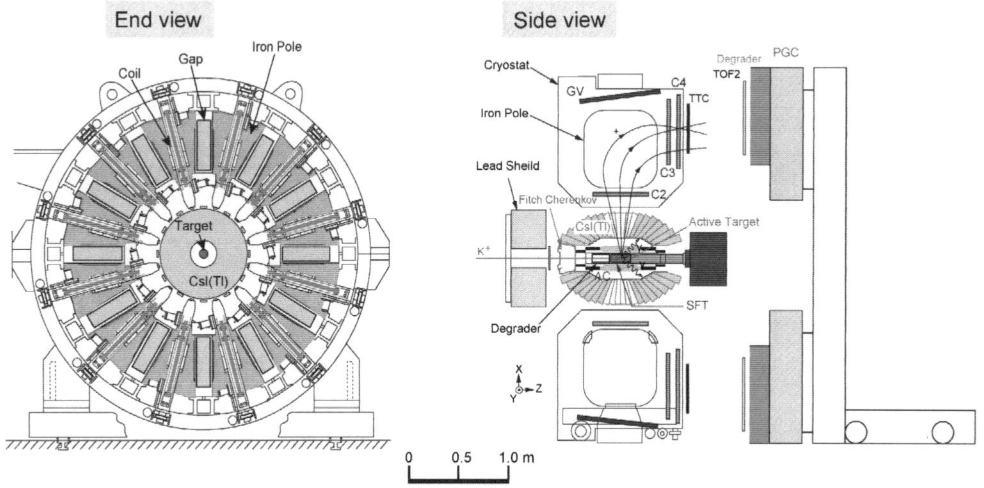

Figure 2. Schematic end and side views of the E36 detector system at J-PARC.

3. Status of analysis

Three particle identification systems allow to redundantly distinguish between positrons, muons and pions. The threshold aerogel Cherenkov (AC) counters sensitive to positrons surround the target bundle, the time-of-flight (TOF) is measured between scintillators near the target (TOF1) and in each gap (TOF2), and lead glass counters (PGC) are located at the end of each gap to identify positrons by their shower.

The e^+ detection efficiency is estimated to be greater than 98% for the aerogel (AC) and the lead glass (PGC) counters and the mis-identification probability to be about 3% and 4% respectively. The typical path length between TOF1 and TOF2 is about 250 cm and the time of flight is determined with a resolution of 200 ps; the time difference between e^+ and μ^+ at $p \sim 236 - 247$ MeV/c along this path being approximately 500 ps.

Figure 3 shows the particle momentum from tracking as a function of the ADC pulse height for AC (top left panel) and for PGC (top right panel). The bottom left panel shows the momentum versus the squared mass from the time-of-flight analysis. After applying threshold cuts in AC and PGC ADCs, the remaining events are shown in the bottom right panel where $K_{\pi 2}$ is suppressed, $K_{\mu 2}$ strongly reduced, and a clear separation between $K_{\mu 2}$ and K_{e3} events is observed. The expected K_{e2} region is circled.

It is important to note that the analysis of the E36 experiment data is at an early stage and that these results are very preliminary. Refinements of the particle identification method are being implemented and further improvements are expected.

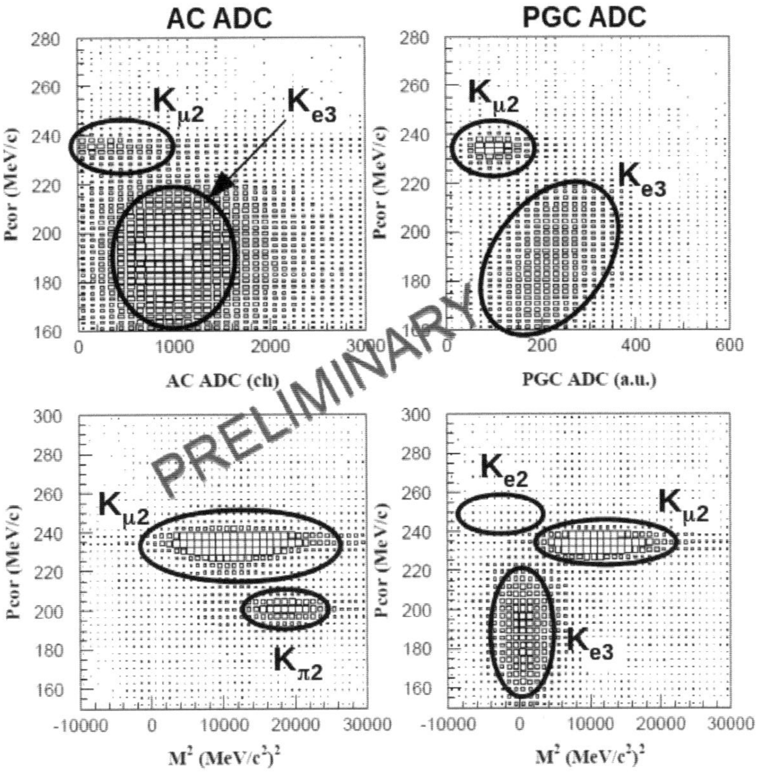

Figure 3. Particle identification analysis. Top left: aerogel counter (AC) analysis (momentum vs ADC pulse height); top right: lead glass (PGC) analysis; bottom left: time-of-flight (TOF) analysis; bottom right: time-of-flight (TOF) analysis with threshold cuts on AC and PGC ADCs.

Acknowledgments
This work is supported in part by NSERC and NRC in Canada, the KAKEN-HI Grants-in-Aid for Scientific Research in Japan, the DOE and NSF in the US, and in Russia by the Ministry of Science and Technology.

References

[1] Cirigliano V and Rosell I 2007 *Phys. Rev. Lett.* **99** 231801
[2] Masiero A, Paradisi P and Petronzio R 2006 *Phys. Rev.* D **74** 011701
[3] Masiero A, Paradisi P and Petronzio R 2008 JHEP **0811** 042
[4] Girrbach J, Nierste U 2012 [arXiv:1202.4906]
[5] Ambrosino F *et al.* (KLOE Collaboration) 2009 *Eur. Phys. J.* C **64** 627
[6] Lazzeroni C *et al.* (NA62 Collaboration) 2013 *Phys. Lett.* B **719** 326
[7] MacDonald J A *et al.* 2003 Nucl. Instr.. and Meth. A **506** 60
[8] Abe M *et al.* (E246 Collaboration) 1999 *Phys. Rev. Lett.* **83** 4253
[9] Abe M *et al.* (E246 Collaboration) 2004 *Phys. Rev. Lett.* **93** 131601
[10] Abe M *et al.* (E246 Collaboration) 2006 *Phys. Rev.* D **73** 072005
[11] Mineev O *et al.* (TREK Collaboration) The design and performance of a Spiral Fiber Tracker for the J-PARC E36 experiment [arXiv:1607.00219]
[12] Tabata M *et al.* (TREK Collaboration) 2015 *Nucl. Instr. and Meth.* A **795** 206
[13] Miyazaki Y *et al.* (TREK Collaboration) 2015 *Nucl. Instr. and Meth.* A **779** 206

IOP Conf. Series: Journal of Physics: Conf. Series **800** (2017) 012018 doi:10.1088/1742-6596/800/1/012018

Searches for Lepton Number Violation and resonances in $K^{\pm} \rightarrow \pi\mu\mu$ decays at NA48/2

M. Piccini[1]
INFN - sezione di Perugia - Italy
E-mail: Mauro.Piccini@pg.infn.it

Abstract. The NA48/2 experiment at CERN collected in 2003-2004 a large sample of charged kaon decays with multiple charged particles in the final state. A new upper limit on the rate of the lepton number violating decay $K^{\pm} \rightarrow \pi^{\mp}\mu^{\pm}\mu^{\pm}$ obtained from this sample is reported. Moreover searches for two-body resonances in the $K^{\pm} \rightarrow \pi\mu\mu$ decays (including heavy neutral leptons and inflatons) in the accessible range of masses and lifetimes are presented. During the 2016 run the NA62 experiment has collected data with a dedicated trigger for kaon decays with two muons in the final state. Preliminary results on the statistics collected during a period of two weeks for the decay $K^{+} \rightarrow \pi^{+}\mu^{+}\mu^{-}$ will be presented. The NA62 experiment has good potential for the studies of rare K^{+} decays in addition to its main goal, the measurement of the branching ratio of the decay $K^{+} \rightarrow \pi\nu\bar{\nu}$.

1. Introduction

The observation of neutrino oscillations[1] has unambiguously demonstrated the massive nature of neutrinos, thus right-handed neutrino states must be included in the Standard Model (SM). A natural extension of the SM involves the inclusion of sterile neutrinos which mix with ordinary neutrinos to explain several open questions. An example of such a theory is the Neutrino Minimal Standard Model (νMSM)[2]. In this model, three massive right-handed neutrinos are introduced to explain simultaneously neutrino oscillations, dark matter and baryon asymmetry of the Universe: the lightest has mass O(1 keV) and is a dark matter candidate; the other two, with masses ranging from 100 MeV/c to few GeV/c, are responsible for the masses of the

[1] On behalf of the NA48/2 Collaboration: G. Anzivino, R. Arcidiacono, W. Baldini, S. Balev, J.R. Batley, M. Behler, S. Bifani, C. Biino, A. Bizzeti, B. Bloch-Devaux, G. Bocquet, N. Cabibbo, M. Calvetti, N. Cartiglia, A. Ceccucci, P. Cenci, C. Cerri, C. Cheshkov, J.B. Ch'eze, M. Clemencic, G. Collazuol, F. Costantini, A. Cotta Ramusino, D. Coward, D. Cundy, A. Dabrowski, P. Dalpiaz, C. Damiani, M. De Beer, J. Derr e, H. Dibon, L. DiLella, N. Doble, K. Eppard, V. Falaleev, R. Fantechi, M. Fidecaro, L. Fiorini, M. Fiorini, T. Fonseca Martin, P.L. Frabetti, L. Gatignon, E. Gersabeck, A. Gianoli, S. Giudici, A. Gonidec, E. Goudzovski, S. Goy Lopez, M. Holder, P. Hristov, E. Iacopini, E. Imbergamo, M. Jeitler, G. Kalmus, V. Kekelidze, K. Kleinknecht, V. Kozhuharov, W. Kubischta, G. Lamanna, C. Lazzeroni, M. Lenti, L. Litov, D. Madigozhin, A. Maier, I. Mannelli, F. Marchetto, G. Marel, M. Markytan, P. Marouelli, M. Martini, L. Masetti, E. Mazzucato, A. Michetti, I. Mikulec, N. Molokanova, E. Monnier, U. Moosbrugger, C. Morales Morales, D.J. Munday, A. Nappi, G. Neuhofer, A. Norton, M. Patel, M. Pepe, A. Peters, F. Petrucci, M.C. Petrucci, B. Peyaud, M. Piccini, G. Pierazzini, I. Polenkevich, Yu. Potrebenikov, M. Raggi, B. Renk, P. Rubin, G. Ruggiero, M. Savrié, M. Scarpa, M. Shieh, M.W. Slater, M. Sozzi, S. Stoynev, E. Swallow, M. Szleper, M. Valdata-Nappi, B. Vallage, M. Velasco, M. Veltri, S. Venditti, M. Wache, H. Wahl, A. Walker, R. Wanke, L. Widhalm, A. Winhart, R. Winston, M.D. Wood, S.A. Wotton, A. Zinchenko, M. Ziolkowski.

Content from this work may be used under the terms of the Creative Commons Attribution 3.0 licence. Any further distribution of this work must maintain attribution to the author(s) and the title of the work, journal citation and DOI.
Published under licence by IOP Publishing Ltd

SM neutrinos (via the see-saw mechanism) and introduce extra CP-violating phases to account for baryon asymmetry. The νMSM can be further extended by adding a real scalar field, to incorporate inflation and provide a common source for electroweak symmetry breaking and for right-handed neutrino masses[3]. These SM extensions predict new particles, such as heavy neutrinos and inflatons, which could be produced in $K^{\pm} \to \pi\mu\mu$ decays. In this model the Lepton Number Violating (LNV) $K^{\pm} \to \pi^{\mp}\mu^{\pm}\mu^{\pm}$ decay, which is forbidden in the SM, could proceed via the production of on-shell Majorana neutrinos[4], while inflatons χ could be produced in $K^{\pm} \to \pi^{\pm}\chi$ decays, and promptly decaying to $\chi \to \mu^{+}\mu^{-}$[5, 6].

The NA48/2 experiments was taking data at CERN in 2003 and 2004. A detailed description of the NA48 detector is available elsewhere [7]. The main difference of the NA48/2 setup was the replacement of the neutral kaon beams with two high intensity secondary beams (K^{+} and K^{-}) with a momentum of 60 GeV/c. The two beams were simultaneous, generated from the same primary proton beam (400 Gev/c momentum) extracted from the SPS. The ration between the K^{+} and K^{-} fluxes was 1.8 and the rate of kaon decays was about 100 KHz. The main trigger used to collect the data analyzed in the study presented here was based on the selection of events with a decay vertex with three outgoing charged particles.

A neutral particle X produced in $K^{\pm} \to \mu^{\pm}X$ ($K^{\pm} \to \pi^{\pm}X$) will decay promptly to $\pi^{\pm}\mu^{\mp}$ ($\mu^{+}\mu^{-}$) and will produce a narrow spike in the invariant mass $M_{\pi\mu}$ ($M_{\mu\mu}$) spectrum. Therefore the invariant mass distributions of the collected $K^{\pm} \to \pi\mu\mu$ samples in NA48/2 have been scanned looking for such a signature. The results of these searches are presented, together with the prospects for the search for Lepton Number and Flavour Violation (LNFV) at the NA62 experiment, which aims to precisely measure the branching ratio of the $K^{+} \to \pi^{+}\nu\bar{\nu}$ decay.

2. Event reconstruction, selection and final data samples

The event selection is based on the reconstruction of a three-track vertex: given the resolution of the vertex longitudinal position ($\sigma_{vtx} \simeq 50$ cm), the decays $K^{\pm} \to \pi^{\mp}\mu^{\pm}\mu^{\pm}$ ($K^{LNV}_{\pi\mu\mu}$ from now on, kaon decays violating lepton number) and $K^{\pm} \to \pi^{\pm}\mu^{\pm}\mu^{\mp}$ ($K^{LNC}_{\pi\mu\mu}$ from now on, kaon decays conserving lepton number) mediated by a short-lived ($\tau \leq 10$ ps) resonant particle are indistinguishable from a genuine three-track decay. The size of the selected $K_{\pi\mu\mu}$ sample is normalised relative to the abundant $K^{\pm} \to \pi^{\pm}\pi^{+}\pi^{-}$ channel (denoted $K_{3\pi}$ below). The $K_{\pi\mu\mu}$ and $K_{3\pi}$ samples are collected concurrently using the same trigger logic. Since the μ and π masses are close ($m_{\mu}/m_{\pi} = 0.7$), the signal and the normalisation final states have similar topologies. This leads to first order cancellation of the systematic effects induced by imperfect kaon beam description, local detector inefficiencies, and trigger inefficiency. The selections for signal and normalization modes have a large common part; a vertex must satisfy the following criteria:

- The total electric charge of the three tracks is $Q = \pm 1$
- The vertex longitudinal position is within the 98 m long fiducial decay volume (i.e. downstream the final collimator).
- Each of the vertex tracks must have momentum p within the range (5;55) GeV/c, the total momentum of the three tracks must be consistent with the beam nominal range (55;65) GeV/c, and the total transverse momentum of the three tracks with respect to the beam axis direction is $p_T < 10^{-2}$ GeV/c. If several vertices fulfill the above conditions, the one with the lowest fit χ^2 is considered. The vertex tracks are required to be consistent in time and to be in the geometric acceptance of the main detectors. Track separations are required to exceed 2 cm in the first drift chamber (DCH1) plane to suppress photon conversions, and 20 cm in the Electromagnetic Calorimeter (LKr) and in the muon detector (MUV) front planes, to minimize particle misidentification due to shower overlaps and to multiple Coulomb scattering.

The following criteria are used to select the $K^{LNV}_{\pi\mu\mu}$ ($K^{LNC}_{\pi\mu\mu}$) candidates:

- The vertex must be composed of one $\pi^{\pm}$ candidate (with the ratio of energy E in the LKr calorimeter to momentum p measured in the spectrometer $E/p < 0.95$ to suppress electrons, and no in-time associated hits in the MUV), and a pair of identically (oppositely) charged $\mu^{\pm}$ candidates (with $E/p < 0.2$ and associated hits in the MUV). The $\pi^{\pm}$ candidate is required to have momentum above 15 GeV/c to ensure high muon rejection efficiency (above 99%).
- The invariant mass of the three tracks in the $\pi^{\mp}\mu^{\pm}\mu^{\pm}$ ($\pi^{\pm}\mu^{+}\mu^{-}$) hypothesis must satisfy the requirement $|M_{\pi\mu\mu} - M_K| < 5$ MeV/c2 ($|M_{\pi\mu\mu} - M_K| < 8$ MeV/c2), where M_K is the nominal $K^{\pm}$ mass. This range corresponds to 2 (3.2) times the resolution on $M_{\pi\mu\mu}$.

An additional requirement is applied to the $K_{\pi\mu\mu}$ samples, when searching for resonances: $|M_{ij} - M_X| < 2\sigma(M_{ij})$, where M_{ij} is the invariant mass of the ij pair ($ij = \pi^{\pm}\mu^{\mp}$, $\mu^{+}\mu^{-}$), M_X is the assumed resonance mass, and $\sigma(M_{ij})$ is the resolution on the invariant masses M_{ij}. Independently, the following criteria are applied to select the $K_{3\pi}$ sample:

- The pion identification criterion described above is applied to the odd-sign pion only, to symmetrize the selection of the signal and normalisation modes.
- The invariant mass of the three tracks in the 3π hypothesis is in the range $|M_{3\pi} - M_K| < 5$ MeV/c2. This interval corresponds approximately to 3 times the resolution on $M_{3\pi}$.

No restrictions are applied to additional energy depositions in the LKr calorimeter, nor to extra tracks not belonging to the three-track vertex, to decrease the sensitivity to accidental activity. The invariant mass distributions of data and MC events passing the $K^{LNV}_{\pi\mu\mu}$ ($K^{LNC}_{\pi\mu\mu}$) selections are shown in Fig. 1. The signal mass regions are indicated with vertical arrows. One event is observed in the signal region after applying the $K^{LNV}_{\pi\mu\mu}$ selection, while 3489 $K^{LNC}_{\pi\mu\mu}$ candidates are selected with corresponding selection. The number of expected background events in the sample $K^{LNV}_{\pi\mu\mu}$ is $N_{bkg} = 1.163 \pm 0.867_{stat} \pm 0.116_{syst} \pm 0.021_{ext}$, the main source of which is the $K_{3\pi}$ decay with two subsequent $\pi^{\pm} \to \mu^{\pm}\nu$ decays. The estimated $K_{3\pi}$ background contamination in the $K^{LNC}_{\pi\mu\mu}$ sample is $(0.36 \pm 0.10)\%$, based on MC simulations. Such a level of purity allows to consider the $K^{LNC}_{\pi\mu\mu}$ decay as the only background for the resonance searches over the collected $K^{LNC}_{\pi\mu\mu}$ sample. The number of $K^{\pm}$ decays in the 98 m long fiducial decay region is computed as $N_K = (N(K_{3\pi}) \cdot D)/(BR(K_{3\pi}) \cdot A(K_{3\pi})) = (1.64 \pm 0.01) \times 10^{11}$, where $N(K_{3\pi}) = 1.37 \times 10^{7}$ is the number of data candidates reconstructed within the $K_{3\pi}$ selection, $D = 100$ is the downscaling factor of the trigger used to collect the $K_{3\pi}$ sample, $BR(K_{3\pi})$ is the nominal branching ratio of the $K_{3\pi}$ decay mode and $A(K_{3\pi}) = (14.955 \pm 0.004)\%$ is the acceptance of the $K_{3\pi}$ selection evaluated with MC simulations.

3. Search for two-body resonances

A peak search assuming different mass hypotheses is performed over the distributions of the invariant masses M_{ij} ($ij = \pi^{\pm}\mu^{\mp}$, $\mu^{+}\mu^{-}$) of the selected $K_{\pi\mu\mu}$ samples. The precise evaluation of the acceptance for $K^{\pm} \to \mu^{\pm}X$ ($K^{\pm} \to \pi^{\pm}X$) decays with subsequent $X \to \pi^{\pm}\mu^{\mp}$ ($X \to \mu^{+}\mu^{-}$) decay as a function of the resonance mass and lifetime has been performed with dedicated MC simulations. The mass steps of the resonance searches and the width of the signal mass windows around the assumed mass are determined by the resolutions $\sigma(M_{ij})$ on the invariant masses M_{ij} ($ij = \pi^{\pm}\mu^{\mp}$, $\mu^{+}\mu^{-}$). The mass step is set to be equal to $\sigma(M_{ij})/2$, while the half-width of the signal mass window is $2\sigma(M_{ij})$. Therefore, the results obtained in the neighbouring mass hypotheses are highly correlated, as the signal mass window is about 8 times wider than the mass step of the resonance scan. In total, 284 (267) and 280 mass hypotheses are tested respectively for the search of resonances in the $M_{\pi\mu}$ distribution of the $K^{LNV}_{\pi\mu\mu}$ ($K^{LNC}_{\pi\mu\mu}$) candidates and in the $M_{\mu\mu}$

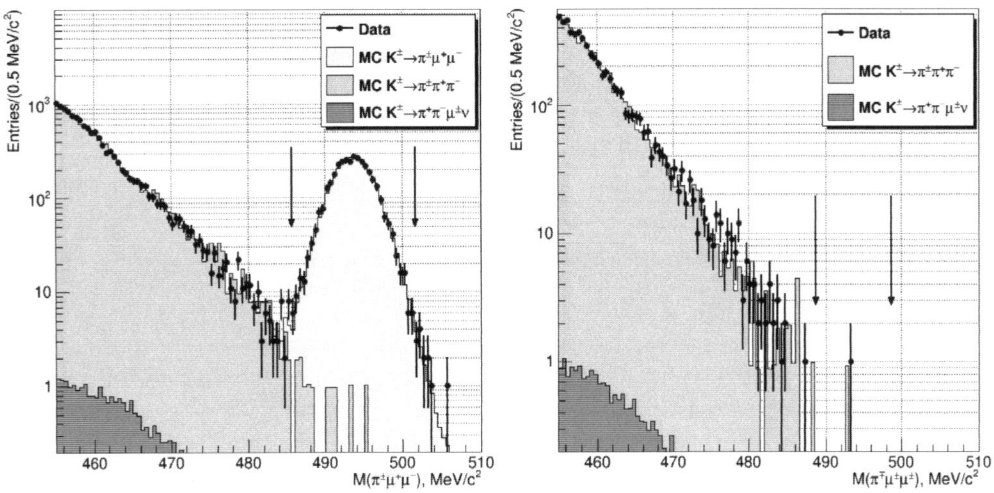

Figure 1. Invariant mass distributions of data and MC events passing the $K^{LNV}_{\pi\mu\mu}$ (left) and $K^{LNC}_{\pi\mu\mu}$ (right) selections. The boundaries of the signal mass regions are indicated with vertical arrows.

distribution of the $K^{LNC}_{\pi\mu\mu}$ candidates, covering the full kinematic ranges. The statistical analysis of the obtained results in each mass window is performed by using a quasi-Newton minimisation algorithm to find numerically the 90% confidence intervals for the case of a Poisson process in presence of unknown backgrounds, by applying an extension of the Rolke-Lopez method[8]. The upper limit at 90% CL on the number of signal events in the $K^{LNV}_{\pi\mu\mu}$ sample corresponding to the observation of one data event and a total number of expected background events N_{bkg} is obtained applying the Rolke-Lopez method to the total number of events in the KLNV sample: $N^{LNV}_{\pi\mu\mu} < 2.92$ at 90% CL. Using the values of the signal acceptance $A(K^{LNV}_{\pi\mu\mu}) = (20.62 \pm 0.01)\%$ estimated with MC simulations and the number N_K of kaon decays in the fiducial volume, the upper limit on the number of $K^{\pm} \to \pi^{\mp}\mu^{\pm}\mu^{\pm}$ signal events in the $K^{LNV}_{\pi\mu\mu}$ sample leads to a constraint on the signal branching ratio:

$$BR(K^{\pm} \to \pi^{\mp}\mu^{\pm}\mu^{\pm}) = \frac{N^{LNV}_{\pi\mu\mu}}{N_{3\pi} \cdot D} \cdot \frac{A(K_{3\pi})}{A(K^{LNV}_{\pi\mu\mu})} \cdot BR(K_{3\pi}) < 8.6 \times 10^{11} \text{ @ } 90\%\ CL \tag{1}$$

The total systematic uncertainty on the quoted upper limit is 1.5%. The largest source is the limited accuracy of the MC simulations (1.0%), followed by $BR(K^{\pm} \to \pi^{\pm}\mu^{+}\mu^{-})$ (0.8%), $BR(K^{\pm} \to \pi^{\pm}\pi^{+}\pi^{-})$ (0.73%), $BR(K^{\pm} \to \mu^{\pm}\mu^{+}\mu^{-}\nu)$ (0.24%) and $BR(K^{\pm} \to \mu^{\pm}\pi^{+}\pi^{-}\nu)$ (0.05%). This result improves the best previous limit[9] by more than one order of magnitude.

Concerning the search for two-body resonances, for each of the three resonance searches performed, the local significance $z = (N_{obs} - N_{exp})/\sqrt{\sigma N^2_{obs} + \sigma N^2_{exp}}$ of the signal has been evaluated for each mass hypothesis, where N_{obs} is the number of observed events, N_{exp} is the number of expected background events, and σN_{obs} (σN_{exp}) is the statistical uncertainty of N_{obs} (N_{exp}). No signal is observed, as the local significances never exceed 3 standard deviations. In absence of a signal, upper limits on the product $BR(K^{\pm} \to p_1 X)BR(X \to p_2 p_3)$ ($p_1p_2p_3 = \mu^{\pm}\pi^{\mp}\mu^{\pm}$, $\mu^{\pm}\pi^{\pm}\mu^{\mp}$, $\pi^{\pm}\mu^{+}\mu^{-}$) as a function of several resonance lifetime τ are obtained for each mass hypothesis m_i and shown in Fig. 2.

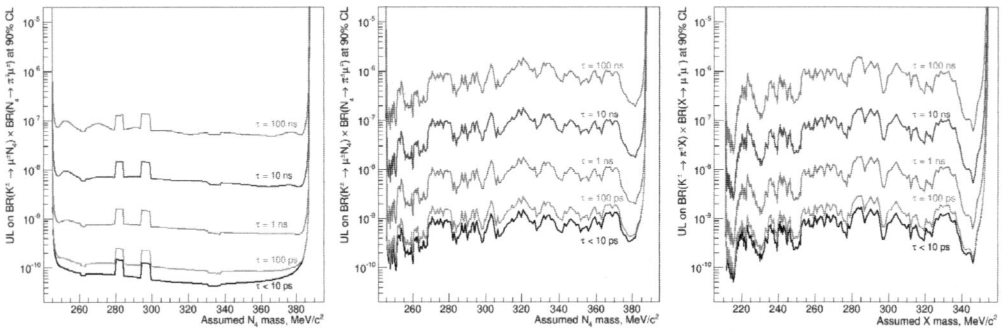

Figure 2. Obtained upper limits at 90% CL on the products of branching ratios as functions of the resonance mass and lifetime. $BR(K^{\pm} \to \mu^{\pm}N_4) \cdot BR(N_4 \to \pi^{\mp}\mu^{\pm})$ in the left; $BR(K^{\pm} \to \mu^{\pm}N_4) \cdot BR(N_4 \to \pi^{\pm}\mu^{\mp})$ in the center; $BR(K^{\pm} \to \mu^{\pm}\chi) \cdot BR(\chi \to \mu^{+}\mu^{-})$ in the right.

4. Prospects for the NA62 experiment

The NA62 experiment[10] has started to collect date in 2015 to perform the precise ($\sim$ 10%) measurement of the $K^+ \to \pi^+\nu\bar{\nu}$ branching ratio. A total of $\sim 1.5 \times 10^{13}$ K^+ decays is expected, with $\sim 3 \times 10^{12}$ π^0 decays from $K^+ \to \pi^+\pi^0$. Studies of the prospects for searches

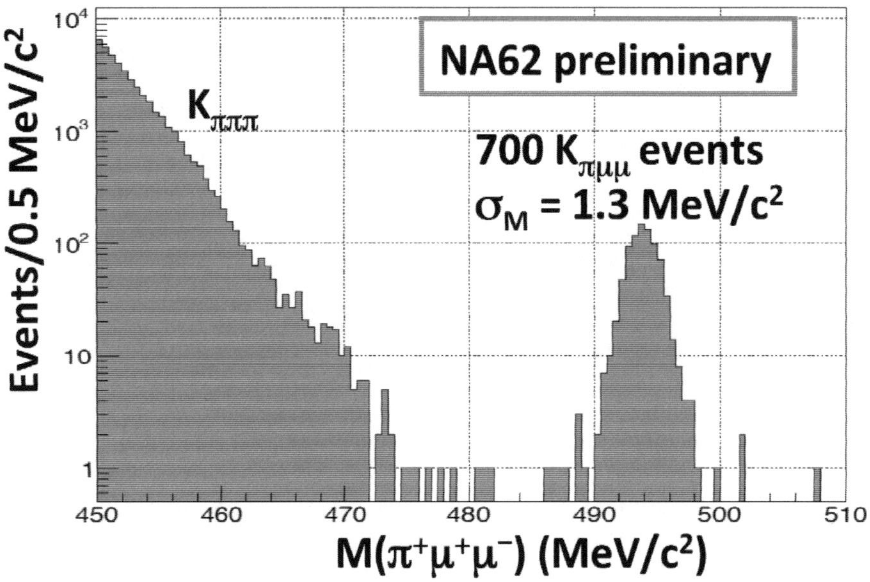

Figure 3. Sample of $K^{\pm} \to \pi^{\pm}\mu^{+}\mu^{-}$ collected by the NA62 experiment during 2 weeks in 2016.

for lepton-flavour (LF) or lepton-number (LN) violating and other forbidden decays with NA62 are underway. The expected acceptance for rare K^+ (π^0) decays is $\sim$ 10% (1%). Preliminary estimates of the single-event sensitivities (defined as the inverse of the number of accepted decays) give results at the level of 10^{-12} for K^+ decays to states such as $\pi^+\mu^{\pm}e^{\mp}$ (LFV),

$\pi^-\mu^+e^+$ (LFNV), and $\pi^-e^+e^+$ or $\pi^-\mu^+\mu^+$ (LNV) and at the level of 10^{-11} for π^0 decays to $\mu^\pm e^\mp$. During the 2016 run the experiment has collected data with a dedicated trigger for events with two muons in the final state. In two weeks of data taking, at 24% of the nominal beam intensity of NA62, ~700 $K^\pm \rightarrow \pi^\pm\mu^+\mu^-$ were collected as shown in Fig. 3. The resolution on the mass $M_{\pi\mu\mu}$ is 1.3 MeV/c^2, almost a factor two better than the one obtained in NA48/2. This will allow a better background rejection.

References

[1] K.A. Olive *et al.* (Particle Data Group) 2014 *Chin. Phys. C* **38**, 090001.
[2] T. Asaka and M. Shaposhnikov 2005 *Phys. Lett. B* **620**, 17.
[3] M. Shaposhnikov and I. Tkachev 2006 *Phys. Lett. B* **639**, 414.
[4] A. Atre, T. Han, S. Pascoli, and B. Zhang 2009 *JHEP* **0905**, 030.
[5] F. Bezrukov and D. Gorbunov 2010 *JHEP* **05**, 010.
[6] F. Bezrukov and D. Gorbunov 2014 *Phys. Lett. B* **736**, 494.
[7] V. Fanti *et al.* (NA48 collaboration) 2007 *Nucl. Instrum. Methods A* **574**, 433.
[8] W.A. Rolke and A.M. López 2001 *Nucl. Instrum. Methods A* **458**, 745.
[9] J.R. Batley *et al.* (NA48/2 collaboration) 2011 *Phys. Lett. B* **697**, 107.
[10] G. Ruggiero, this conference proceedings.

IOP Conf. Series: Journal of Physics: Conf. Series **800** (2017) 012019 doi:10.1088/1742-6596/800/1/012019

TTP16–048

Recent progress on CP violation in $K \to \pi\pi$ decays in the SM and a supersymmetric solution

T Kitahara[1,2,*], **U Nierste**[1], **and P Tremper**[1]

[1] Institute for Theoretical Particle Physics (TTP), Karlsruhe Institute of Technology, Engesserstraße 7, D-76128 Karlsruhe, Germany

[2] Institute for Nuclear Physics (IKP), Karlsruhe Institute of Technology, Hermann-von-Helmholtz-Platz 1, D-76344 Eggenstein-Leopoldshafen, Germany

E-mail: teppei.kitahara@kit.edu

Abstract. Using the recent first lattice results of the RBC-UKQCD collaboration for $K \to \pi\pi$ decays, we perform a phenomenological analysis of ϵ'_K/ϵ_K and find a discrepancy between SM prediction and experiments by $\sim 3\sigma$. We discuss an explanation by new physics. The well-understood value of ϵ_K, which quantifies indirect CP violation, however, typically prevents large new physics contributions to ϵ'_K/ϵ_K. In this talk, we show a solution of the ϵ'_K/ϵ_K discrepancy in the Minimal Supersymmetric Standard Model with squark masses above 3 TeV without fine-tuning of CP phases. In this solution, the *Trojan penguin* diagram gives large isospin-breaking contributions which enhance ϵ'_K, while the contribution to ϵ_K is suppressed thanks to the Majorana nature of gluinos.

KAON 2016 conference,
14-17 September 2016
University of Birmingham, United Kingdom

1. Introduction to ϵ'_K/ϵ_K discrepancy

In $K \to \pi\pi$ decays, one distinguishes between two types of charge-parity (CP) violation: direct and indirect CP violations which are parametrized by ϵ'_K and ϵ_K, respectively. Both types of CP violation have been quantified by many kaon experiments precisely. While ϵ_K is a per mille effect in the Standard Model (SM), ϵ'_K is smaller by another three orders of magnitude: $\epsilon'_K \sim \mathcal{O}(10^{-6})$. This strong suppression comes from the suppression of the isospin-3/2 amplitude w.r.t. the isospin-1/2 amplitude ($\Delta I = 1/2$ rule) and an accidental cancellation of leading contributions in the SM. In Fig. 1, the contributions of individual operators to ϵ'_K/ϵ_K are shown. Q_3–Q_6 are called QCD penguin operators, while Q_7–Q_{10} are called EW penguin operators. The leading contributions come from Q_6 and Q_8, having opposite sign, and thus a cancellation emerges. Remarkably, this figure also shows that even if one includes sub-leading contributions, the cancellation still exists with high precision.

The compilation of representative SM predictions and the experimental values for Re ϵ'_K/ϵ_K is given in Fig. 2. The SM predictions (colored bars) are taken from: Bertolini *et al.* (BEFL '97) [1], Pallante *et al.* (PPS '01) [2], Hambye *et al.* (HPR '03) [3], Buras and Gérard (BG '15) [4] with lattice result for $I = 2$ (BG '15+Lat.), RBC-UKQCD lattice result [5], Buras *et al.* (BGJJ '15) [6], and Kitahara *et al.* (KNT '16) [7]. The experimental values (black bars) are

[*] Speaker.

Content from this work may be used under the terms of the Creative Commons Attribution 3.0 licence. Any further distribution of this work must maintain attribution to the author(s) and the title of the work, journal citation and DOI.
Published under licence by IOP Publishing Ltd

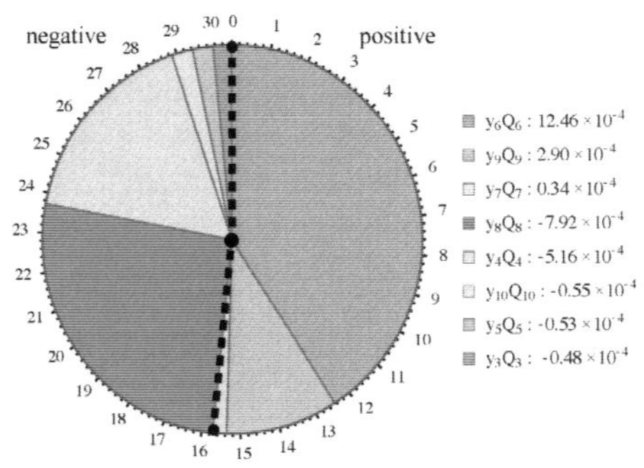

Figure 1. The composition of ϵ'_K/ϵ_K with respect to the operator basis. The right and left side of the dashed lines represent positive and negative contributions to ϵ'_K/ϵ_K, respectively. This figure is based on the result of Ref. [7].

taken from: E371 [8], NA31 [9], NA48 [10] and KTeV [11] collaborations, and the black thick one is the world average of the experimental values [12],

$$\mathrm{Re}\left(\frac{\epsilon'_K}{\epsilon_K}\right) = (16.6 \pm 2.3) \times 10^{-4} \text{ (PDG average)}. \tag{1}$$

In order to predict ϵ'_K in the SM, one has to calculate the hadronic matrix elements of four-quark operators with nonperturbative methods. The magenta bars in Fig. 2 have utilized analytic approaches to calculating them: chiral quark model (BEFL '97), chiral perturbation theory (PPS '01) with minimal hadronic approximation (HPR '03), and the dual QCD approach (BG '15). Note that the dual QCD approach predicts an upper bound on ϵ'_K/ϵ_K. On the other hand, a determination of all hadronic matrix elements from lattice QCD has been obtained only recently by the RBC-UKQCD collaboration [5], and the blue bars are based on the lattice result:

$$\frac{\epsilon'_K}{\epsilon_K} = \begin{cases} (1.9 \pm 4.5) \times 10^{-4} & \text{(BGJJ '15)}, \\ (1.06 \pm 5.07) \times 10^{-4} & \text{(KNT '16)}. \end{cases} \tag{2}$$

These results are obtained by next-to-leading order (NLO) calculations exploiting CP-conserving data to reduce hadronic uncertainties and include isospin-violating contributions [13] which are not included in the lattice result. Furthermore, the latter result includes an additional $\mathcal{O}(\alpha_{EM}^2/\alpha_s^2)$ correction, which appears only in this order, and also utilizes a new analytic solution of the renormalization group (RG) equation which avoids the problem of singularities in the NLO terms. The two numbers in Eq. (2) disagree with the experimental value in Eq. (1) by $2.9\,\sigma$ [6] and $2.8\,\sigma$ [7], respectively. The uncertainties are dominated by the lattice statistical and systematic uncertainties for the $I = 0$ amplitude. Therefore, in the near future, the increasing precision of lattice calculations will further sharpen the SM predictions in Eq. (2) and answer the question about new physics (NP) in ϵ'_K/ϵ_K.

The main difference between each result of analytic approaches and the lattice result is in the hadronic parameter $B_6^{(1/2)}$, which controls the largest positive contribution to ϵ'_K/ϵ_K, the y_6Q_6 component in the Fig. 1. In chiral perturbation theory, typically large values are obtained: $B_6^{(1/2)} \sim 1.6$ (BEFL '97), ~ 1.6 (PPS '01), and ~ 3 (HPR '03, [4]). On the other hand, the dual QCD approach predicts a smaller number, $B_6^{(1/2)} \leq B_8^{(3/2)} \sim 0.8$ (BG '15). The lattice result

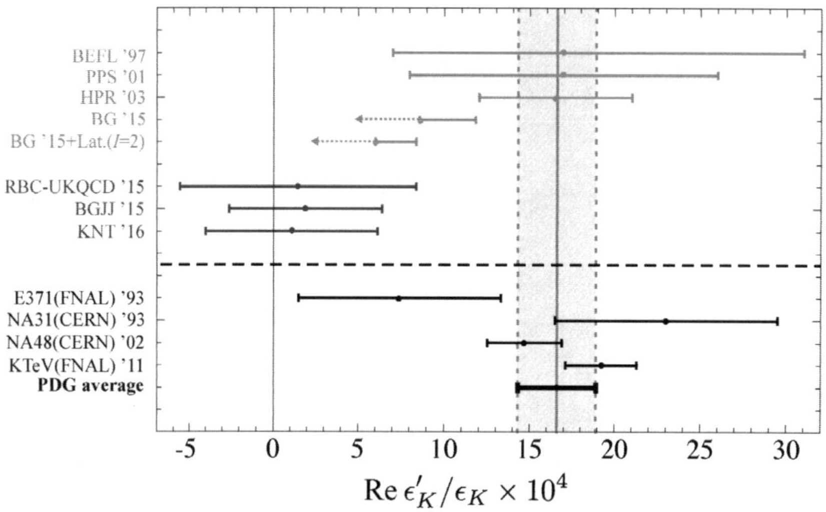

Figure 2. Compilation of representative SM predictions and the experimental values for Re ϵ'_K/ϵ_K. All error bars represent $1\,\sigma$ range. The SM predictions are taken from Bertolini *et al.* (BEFL '97) [1], Pallante *et al.* (PPS '01) [2], Hambye *et al.* (HPR '03) [3], Buras and Gérard (BG '15) [4], RBC-UKQCD lattice result [5], Buras *et al.* (BGJJ '15) [6], and Kitahara *et al.* (KNT '16) [7], where magenta bars are based on analytic approaches to hadronic matrix elements, while blue bars are based on lattice results. The black thick one is the world average of the experimental values [12].

is consistent with the latter result: $B_6^{(1/2)} = 0.56 \pm 0.20$ [5, 7]. Note that the lattice calculation *includes* final-state interaction of the two pions in accordance with Ref. [14].

We also should comment on the $\Delta I = 1/2$ rule, which denotes the largeness of the ratio of the CP-conserving amplitudes, $(\mathrm{Re}A_0/\mathrm{Re}A_2)_{\mathrm{exp.}} = 22.45 \pm 0.05$. Although none of the analytic approaches can explain such a large value, the first lattice calculation has found a consistent value within $1\,\sigma$, $(\mathrm{Re}A_0/\mathrm{Re}A_2)_{\mathrm{Lat.}} = 31.0 \pm 11.1$ [5, 15].

2. ϵ_K in the MSSM

An explanation of the puzzle between Eq. (1) and Eq. (2) by physics beyond the SM requires a NP contribution which is seemingly even larger than the SM contribution. However, it is known that once constraints from the corresponding $|\Delta S| = 2$ transition are taken into account, one expects that NP effects in a $|\Delta S| = 1$ four-quark process are highly suppressed. To explain the NP hierarchy in $|\Delta S| = 1$ vs $|\Delta S| = 2$ transitions, we specify to ϵ'_K and ϵ_K: The SM contributions are governed by the combination $\tau = -V_{td}V_{ts}^*/(V_{ud}V_{us}^*) \sim (1.5 - i0.6) \times 10^{-3}$ with $\epsilon_K'^{\mathrm{SM}} \propto \mathrm{Im}\,\tau/M_W^2$ and $\epsilon_K^{\mathrm{SM}} \propto \mathrm{Im}\,\tau^2/M_W^2$. If the NP contribution enters through the $\Delta S = 1$ parameter δ and is mediated by heavy particles of mass M, one obtains $\epsilon_K'^{\mathrm{NP}} \propto \mathrm{Im}\,\delta/M^2$, $\epsilon_K^{\mathrm{NP}} \propto \mathrm{Im}\,\delta^2/M^2$, and therefore the experimental constraint $|\epsilon_K^{\mathrm{NP}}| \leq |\epsilon_K^{\mathrm{SM}}|$ leads to

$$\left|\frac{\epsilon_K'^{\mathrm{NP}}}{\epsilon_K'^{\mathrm{SM}}}\right| \leq \frac{|\epsilon_K'^{\mathrm{NP}}/\epsilon_K'^{\mathrm{SM}}|}{|\epsilon_K^{\mathrm{NP}}/\epsilon_K^{\mathrm{SM}}|} = \mathcal{O}\left(\frac{\mathrm{Re}\,\tau}{\mathrm{Re}\,\delta}\right). \tag{3}$$

If NP enters through a loop with particles of mass $M \gtrsim 1$ TeV, the NP effects can be relevant only for $|\delta| \gg |\tau|$, and thus Eq. (3) seemingly forbids detectable NP contributions to ϵ'_K.

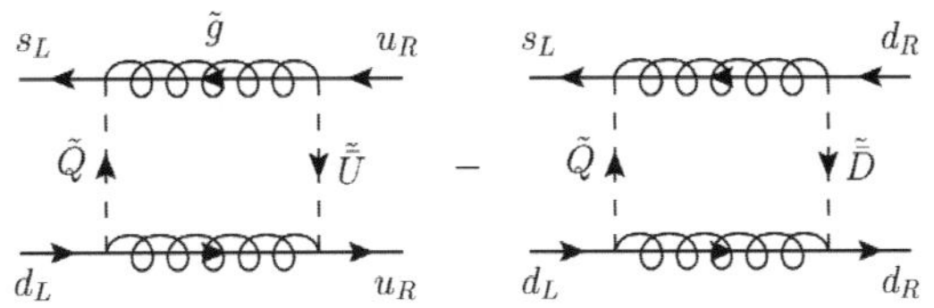

Figure 3. *Trojan penguin* contributions to $\mathrm{Im}\,A_2$ for $m_{\bar{U}} \neq m_{\bar{D}}$.

In the Minimal Supersymmetric Standard Model (MSSM), there is a bypass to Eq. (3). The Majorana nature of the gluino leads to a suppression of gluino-squark box contributions to ϵ_K. This is so, because there are two such diagrams (crossed and uncrossed boxes) with opposite signs. If the gluino mass $m_{\tilde{g}}$ equals roughly 1.5 times the average down squark mass M_S, both contributions to $\epsilon_K^{\rm SUSY}$ cancel [16]. For $m_{\tilde{g}} > 1.5M_S$, the gluino-box contribution approximately behaves as $[m_{\tilde{g}}^2 - (1.5M_S)^2]/m_{\tilde{g}}^4$, and the $1/m_{\tilde{g}}^2$ decoupling sets in. Note that this suppression appears only when a hierarchy $\Delta_{Q,12} \gg \Delta_{\bar{D},12}$ of $\Delta_{Q,12} \ll \Delta_{\bar{D},12}$ is satisfied, where the following notation is used for the squark mass matrices: $M^2_{X,ij} = m_X^2\left(\delta_{ij} + \Delta_{X,ij}\right)$, with $X = Q$, $\bar{U}$, or $\bar{D}$.

3. ϵ'_K/ϵ_K in the MSSM

The master equation for ϵ'_K is given by [6]

$$\frac{\epsilon'_K}{\epsilon_K} = \frac{\omega_+}{\sqrt{2}|\epsilon_K^{\rm exp}|{\rm Re}A_0^{\rm exp}}\left\{\frac{{\rm Im}A_2}{\omega_+} - \left(1-\hat{\Omega}_{\rm eff}\right){\rm Im}A_0\right\}, \tag{4}$$

with $\hat{\Omega}_{\rm eff} = (14.8 \pm 8.0) \times 10^{-2}$, the measured $|\epsilon_K^{\rm exp}|$, $\omega_+ = (4.53 \pm 0.02) \times 10^{-2}$, and the amplitudes $A_I = \langle(\pi\pi)_I|\mathcal{H}^{|\Delta S|=1}|K^0\rangle$ involving the effective $|\Delta S| = 1$ Hamiltonian $\mathcal{H}^{|\Delta S|}$. $I = 0, 2$ represents the strong isospin of the final two-pion state.

The MSSM contributions to ϵ'_K/ϵ_K have been widely studied in the past. However, the supersymmetry-breaking scale M_S was considered in the ballpark of the electroweak scale, so that the suppression mechanism inferred from Eq. (3) is avoided. The low-energy Hamiltonian in the case of small left-right squark mixing reads

$$\mathcal{H}^{|\Delta S|=1}_{\text{eff, SUSY}} = \frac{G_F}{\sqrt{2}}\sum_q\left[\sum_{i=1}^{2} c_i^q(\mu)Q_i^q(\mu) + \sum_{i=1}^{4}[c_i'^q(\mu)Q_i'^q(\mu) + \tilde{c}_i'^q(\mu)\tilde{Q}_i'^q(\mu)]\right] + \text{H.c.}, \tag{5}$$

where G_F is the Fermi constant and

$$\begin{aligned}
Q_1^q &= (\bar{s}_\alpha q_\beta)_{V-A}(\bar{q}_\beta d_\alpha)_{V-A}, \quad Q_2^q = (\bar{s}q)_{V-A}(\bar{q}d)_{V-A},\\
Q_1'^q &= (\bar{s}d)_{V-A}(\bar{q}q)_{V+A}, \quad Q_2'^q = (\bar{s}_\alpha d_\beta)_{V-A}(\bar{q}_\beta q_\alpha)_{V+A},\\
Q_3'^q &= (\bar{s}d)_{V-A}(\bar{q}q)_{V-A}, \quad Q_4'^q = (\bar{s}_\alpha d_\beta)_{V-A}(\bar{q}_\beta q_\alpha)_{V-A}.
\end{aligned} \tag{6}$$

Here $(\bar{s}d)_{V-A}(\bar{q}q)_{V\pm A} = [\bar{s}\gamma_\mu(1-\gamma_5)d][\bar{q}\gamma^\mu(1\pm\gamma_5)q]$, α and β represent color indices, and opposite-chirality operators $\tilde{Q}_i'^q$ are given by interchanging $V - A \leftrightarrow V + A$.

In our analysis [17] we found that the dominant supersymmetric contribution comes from a certain gluino box diagrams[#1], called a *Trojan penguin* in Ref. [19], which are shown in Fig. 3. It contributes to $\mathrm{Im}\,A_2$ when $m_{\bar{U}} \neq m_{\bar{D}}$. Because these contributions are governed by the strong

[#1] The other supersymmetric solution focusing the chargino Z-penguin contribution has been studied in Ref. [18].

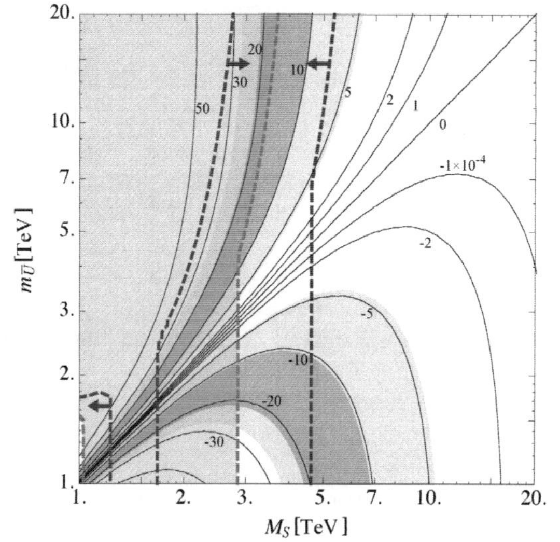

Figure 4. The black contour represents the supersymmetric contributions to ϵ'_K/ϵ_K in units of 10^{-4}. The ϵ'_K/ϵ_K discrepancy is resolved at $1\,\sigma\,(2\,\sigma)$ in the dark (light) green region. The red shaded region (region between the blue dashed lines) is excluded (preferred) by ϵ_K with inclusive (exclusive) $|V_{cb}|$ at 95 % C.L.

interaction and there is an enhancement factor $1/\omega_+ = 22.1$ for the ImA_2 term in (4), they easily become the largest contribution to ϵ'_K/ϵ_K [19]. In order to obtain the desired large effect in ϵ'_K, one needs a contribution to the operators $Q'_{1,2}$ with $(V-A)\times(V+A)$ Dirac structure, whose matrix elements are chirally enhanced by a factor $(m_K/m_s)^2$. Hence, the flavor mixing has to be in the left-handed squark mass matrix. The opposite situation with right-handed flavor mixing and $\tilde{u}_L$-$\tilde{d}_L$ mass splitting is not possible because of the SU$(2)_L$ invariance.

For the calculation of supersymmetric contributions to ϵ'_K/ϵ_K, one has to use the RG equations to evolve the Wilson coefficients calculated at the high scale M_S down to the hadronic scale $\mu_h = \mathcal{O}(1\text{ GeV})$ at which the hadronic matrix elements are calculated [5, 7]. To use the well-known NLO 10×10 anomalous dimensions for the SM four-fermion operator basis [20], we switch from Eq. (5) to

$$\mathcal{H}^{|\Delta S|=1}_{\text{eff, SUSY}} = \frac{G_F}{\sqrt{2}}\sum_{i=1}^{10}[C_i(\mu)Q_i(\mu) + \tilde{C}_i(\mu)\tilde{Q}_i(\mu)] + \text{H.c.}, \tag{7}$$

where $Q_{1,\dots,10}$ are given in Ref. [20] and

$$\begin{aligned} C_{1,2}(\mu) &= c^u_{1,2}(\mu), & \tilde{C}_{1,2}(\mu) &= 0,\\ C_{3,4,5,6}(\mu) &= \frac{1}{3}[c'^u_{3,4,1,2}(\mu) + 2c'^d_{3,4,1,2}(\mu)], & C_{7,8,9,10}(\mu) &= \frac{2}{3}[c'^u_{1,2,3,4}(\mu) - c'^d_{1,2,3,4}(\mu)], \end{aligned} \tag{8}$$

and the coefficients $\tilde{C}_{3,\dots 10}$ for the opposite-chirality operators can be obtained from $C_{3,\dots 10}$ by replacing $c'^q_i \to \tilde{c}'^q_i$. Note that the contribution of Fig. 3 is collected into the coefficients $C_{7,8}$. For the RG evolution of the coefficients, we use the new analytic solution of the RG equations discussed in Ref. [7].

Our main result is given in Fig. 4, where the portion of the squark mass plane which simultaneously explains ϵ'_K/ϵ_K and ϵ_K is shown. As input, we take the grand-unified theory relation for gaugino masses, $\alpha_s(M_Z) = 0.1185$, $m_{\tilde{g}}/M_S = 1.5$ for the suppressed ϵ_K, and $m_Q = m_{\bar{D}} = \mu_{\rm SUSY} = M_S$ with varying $m_{\bar{U}}$. Furthermore, the trilinear supersymmetry-breaking matrices A_q are set to zero, $\tan\beta = 10$, and the only nonzero off-diagonal elements of the squark mass matrices are $\Delta_{Q,12,13,23} = 0.1\exp(-i\pi/4)$ for the left-handed squark sectors. We have calculated all relevant one-loop contributions to the coefficients in Eq. (5) in the squark mass eigenbasis. The ϵ'_K/ϵ_K discrepancy can be resolved at $1\,\sigma\,(2\,\sigma)$ in the dark (light) green region. The red region is already excluded by the measurement of ϵ_K at 95 % C.L. in combination with the inclusive V_{cb}. On the other hand, the region between the blue dashed lines can explain the ϵ_K discrepancy at 95 % C.L. for the exclusive value of $|V_{cb}|$. The area in Fig. 4 labeled with negative values of ϵ'_K/ϵ_K becomes feasible by adding π to the phase of $\Delta_{Q,ij}$, which flips the sign of ϵ'_K while keeping ϵ_K unchanged.

4. Conclusions

In this talk, we have discussed the supersymmetric contributions to ϵ'_K, and it is found that the large contributions required to solve the discrepancy between Eq. (1) and Eq. (2) can be obtained in the multi-TeV squark mass range thanks to the Trojan penguin diagrams. Using a relatively heavy gluino, the severe constraint from ϵ_K can be fulfilled without fine-tuning.

Acknowledgments

The work of UN is supported by BMBF under grant no. 05H15VKKB1. PT acknowledges support from the DFG-funded doctoral school *KSETA*.

References

[1] Bertolini S, Eeg J O, Fabbrichesi M and Lashin E I 1998 Nucl. Phys. B **514**, 93 (*Preprint* hep-ph/9706260)
[2] Pallante E and Pich A 2001 Nucl. Phys. B **592**, 294 (*Preprint* hep-ph/0007208), Pallante E, Pich A and Scimemi I 2001 Nucl. Phys. B **617**, 441 (*Preprint* hep-ph/0105011)
[3] Hambye T, Peris S and de Rafael E 2003 JHEP **0305**, 027 (*Preprint* hep-ph/0305104)
[4] Buras A J and Gérard J M 2015 JHEP **1512**, 008 (*Preprint* 1507.06326 [hep-ph])
[5] Blum T *et al* 2012 Phys. Rev. Lett. **108**, 141601 (*Preprint* 1111.1699 [hep-lat]), Blum T *et al* 2012 Phys. Rev. D **86**, 074513 (*Preprint* 1206.5142 [hep-lat]), Blum T *et al* 2015 Phys. Rev. D **91**, no. 7, 074502 (*Preprint* 1502.00263 [hep-lat]), Bai Z *et al* [RBC and UKQCD Collaborations] 2015 Phys. Rev. Lett. **115**, no. 21, 212001 (*Preprint* 1505.07863 [hep-lat])
[6] Buras A J, Gorbahn M, Jäger S and Jamin M 2015 JHEP **1511**, 202 (*Preprint* 1507.06345 [hep-ph])
[7] Kitahara T, Nierste U and Tremper P (*Preprint* 1607.06727 [hep-ph])
[8] Gibbons L K *et al* 1993 Phys. Rev. Lett. **70**, 1203
[9] Barr G D *et al* [NA31 Collaboration] 1993 Phys. Lett. B **317**, 233
[10] Fanti V *et al* [NA48 Collaboration] 1999 Phys. Lett. B **465**, 335 (*Preprint* hep-ex/9909022), Batley J R *et al* [NA48 Collaboration] 2002 Phys. Lett. B **544**, 97 (*Preprint* hep-ex/0208009)
[11] Alavi-Harati A *et al* [KTeV Collaboration] 1999 Phys. Rev. Lett. **83**, 22 (*Preprint* hep-ex/9905060), Abouzaid E *et al* [KTeV Collaboration] 2011 Phys. Rev. D **83**, 092001 (*Preprint* 1011.0127 [hep-ex])
[12] Patrignani C 2016 Chin. Phys. C **40**, no. 10, 100001
[13] Cirigliano V, Pich A, Ecker G and Neufeld H 2003 Phys. Rev. Lett. **91**, 162001 (*Preprint* hep-ph/0307030)
[14] Lellouch L and Luscher M 2001 Commun. Math. Phys. **219**, 31 (*Preprint* hep-lat/0003023)
[15] Buras A J and Gérard J M (*Preprint* 1603.05686 [hep-ph])
[16] Crivellin A and Davidkov M 2010 Phys. Rev. D **81**, 095004 (*Preprint* 1002.2653 [hep-ph])
[17] Kitahara T, Nierste U and Tremper P 2016 Phys. Rev. Lett. **117**, no. 9, 091802 (*Preprint* 1604.07400 [hep-ph])
[18] Tanimoto M and Yamamoto K (*Preprint* 1603.07960 [hep-ph]), Endo M, Mishima S, Ueda D and Yamamoto K 2016 Phys. Lett. B **762**, 493 (*Preprint* 1608.01444 [hep-ph])
[19] Kagan A L and Neubert M 1999 Phys. Rev. Lett. **83**, 4929 (*Preprint* hep-ph/9908404), Grossman Y, Neubert M and Kagan A L 1999 JHEP **9910**, 029 (*Preprint* hep-ph/9909297)
[20] Buras A J, Jamin M, Lautenbacher M E and Weisz P H 1993 Nucl. Phys. B **400**, 37 (*Preprint* hep-ph/9211304), Buras A J, Jamin M and Lautenbacher M E 1993 Nucl. Phys. B **400**, 75 (*Preprint* hep-ph/9211321), Ciuchini M, Franco E, Martinelli G and Reina L 1994 Nucl. Phys. B **415**, 403 (*Preprint* hep-ph/9304257)

Search for the rare decays $K^+ \to \mu^+ X$ in E949 experiment

A. T. Shaikhiev

Institute for Nuclear Research RAS, 60 October Revolution Pr. 7a, 117312 Moscow, Russia

E-mail: shaykhiev@inr.ru

Abstract. The existence of a heavy neutrino, ν_H, in the $K^+ \to \mu^+\nu_H$ decays was tested using the E949 experimental data with an exposure of 1.70×10^{12} stopped kaons. The allowed heavy neutrino mass region for the analysis is from 175 MeV/c^2 to 300 MeV/c^2. With major background from the radiative $K^+ \to \mu^+\nu_\mu\gamma$ decay understood and suppressed, the preliminary new upper limits (90% C.L.) on the neutrino mixing matrix element between muon and heavy neutrino, $|U_{\mu H}|^2$, were set at the level of 10^{-8} to 10^{-9}. New limit on the $K^+ \to \mu^+\nu\bar{\nu}\nu$ branching ratio assuming Standard Model muon spectrum is set to be 2.4×10^{-6}.

1. Introduction

The Standard Model (SM) has been tested experimentally with high precision, but some phenomena cannot be explained in the SM framework: the neutrino oscillations experiments [1] confirm that neutrino have a mass and undergo mixing (while in the SM all neutrinos, ν_e, ν_μ, ν_τ, are massless). In other words, the weak eigenstates ν_e, ν_μ, ν_τ are linear superposition of the mass eigenstates ν_1, ν_2, ν_3. The SM also cannot explain baryon asymmetry of the Universe and dark matter. So far we know that there is a new physics beyond the SM, but we do not know exactly what is it. To search for new physics we are able either search for new particles (unpredicted by the SM) or search for rare processes and compare measured probabilities of these processes with the SM predicted ones.

An extension of the SM by three singlet fermions with masses smaller than the electroweak scale without adding any new physical principles (such as supersymmetry or extra dimensions) or new energy scales (like Grand Unified scale) allows to explain simultaneously the phenomena that cannot be fit to the SM. An example of such a theory is the renormalizable extension of the SM, the νMSM (neutrino Minimal Standard Model) [2, 3]. In this model, the lightest sterile neutrino, ν_{H1}, is likely to have a mass in the $\mathcal{O}(10)$ keV/c^2 region and is sufficiently stable to be a viable dark matter candidate. The masses of ν_{H2} and ν_{H3} should lie in the range from ~ 150 MeV/c^2 to ~ 100 GeV/c^2 and should be degenerate ($\Delta M_{2,3} \ll M_{2,3}$) to generate baryon asymmetry of the Universe.

Two strategies can be used for the experimental search of these particles. The first one is related to their production. Since they are massive, the kinematics of two body decays $K^\pm \to \mu^\pm\nu_\mu$ and $K^\pm \to \mu^\pm\nu_H$ is not the same. So the study of kinematics of rare meson decays can constrain the strength of the coupling of heavy leptons using the following expression [4]:

$$\Gamma(M^+ \to l^+\nu_H) = \rho\Gamma(M^+ \to l^+\nu_l)|U_{lH}|^2, \tag{1}$$

Content from this work may be used under the terms of the Creative Commons Attribution 3.0 licence. Any further distribution of this work must maintain attribution to the author(s) and the title of the work, journal citation and DOI.
Published under licence by IOP Publishing Ltd

where $M = \pi,\ K;\ l = e,\ \mu,\ \rho$ is a kinematical factor and lies in the range from 1 to 4 for $0 < m_{\nu_H} < 300$ MeV/c^2. The second strategy is to look for the decays of heavy leptons to hadrons and leptons.

The best constraints in the small mass region $m_{\nu_H} < 450$ MeV are coming from CERN PS191 experiment [5], giving roughly $|U_l H|^2 < 10^{-9}$ in the region 250 MeV/$c^2 < m_{\nu_H} <$ 450 MeV/c^2. The successful predictions of the Big Bang Nucleosynthesis (BBN) allow to establish a number of lower bounds on the couplings of neutral leptons [6, 7] which decrease considerably the admitted window for the couplings and masses of the neutral leptons.

The $K^+ \to \mu^+\nu\bar{\nu}\nu$ decay involves four fermions and cannot occur in first order in the SM. Therefore, its investigation provides information about higher-order weak effects. The most recent calculation of this process in the framework of the SM has been done by D. Gorbunov and A. Mitrofanov [8]. The SM predicts an extremely low total rate for this process ($\mathcal{O}(10^{-16})$) and any observed signal would be a clear evidence of new physics. This decay has been searched by one only experiment in 1973 [9], no evidence was found and upper limit $BR(K^+ \to \mu^+\nu\bar{\nu}\nu) < 6.0 \times 10^{-6}$ at 90% C.L. was set.

In this paper, we present result of a search for heavy neutrinos in $K^+ \to \mu^+\nu_H$ decays from the inclusive muon spectrum of $K^+ \to \mu^+ + nothing$ decays and search for rare decay $K^+ \to \mu^+\nu\bar{\nu}\nu$ using the kaon decay-at-rest data from the E949 experiment [10].

2. E949 experiment

The E949 K^+ beam was produced by a high-intensity proton beam from the Alternating Gradient Synchrotron (AGS) at Brookhaven National Laboratory (BNL). Protons were accelerated to a momentum of 21.5 GeV/c and hit a platinum production target.

The E949 detector had axial symmetry in beam direction (z-axis) and consisted of several components. Kaons were identified by the Čerenkov counter before the target. Incoming 710 MeV/c kaons stopped and decayed in scintillating fiber target. The momentum and trajectory of the outgoing charged particles were measured in drift chamber. These particles came to rest in a Range Stack (RS) of 19 layers of plastic scintillator. The primary functions of the RS were energy and range measurements of charged particles and their identification. The detection of any activities coincident with the charged track was very important for suppressing the backgrounds for $K^+ \to \mu^+\nu_H$ decay. Photons from $K_{\pi 2}$, $K_{\mu\nu\gamma}$, $K_{\mu 3}$ and other radiative decays were detected by hermetic photons detectors with $\simeq 4\pi$ solid angle coverage.

2.1. The trigger

The experimental signature of the $K^+ \to \mu^+\nu_H$ and $K^+ \to \mu^+\nu\bar{\nu}\nu$ decays is the same as for the $K^+ \to \pi^+\nu\bar{\nu}$ decay (one single charged track with no any detector activity). That's why we decided to use the main E949 trigger. It consisted of several requirements. First of all, a kaon must enter the target. To be sure that the kaon decayed at rest, the secondary charged particle must leave the target at least 1.5 ns later than the kaon hit in the Čerenkov detector. The 3-body K^+ decays were suppressed by the RS layer requirements. The charged particle must reach at least the sixth layer of the RS. The long tracks (in general, $K_{\mu 2}$ decay) were suppressed by the layer 19 veto requirement. There were also additional refined requirements of the charged track range taking into account the number of target fiber hits and the track's downstream position (z-coordinate) in RS layers 3, 11, 12, 13 as well as the deepest layer of penetration (refined range). The charged track must be within the fiducial region of all traversed RS layers.

The main E949 trigger included the online pion identification in the RS. It required a signature of $\pi^+ \to \mu^+$ decay in the online-selected stopping counter. The μ^+ from the $\pi^+ \to \mu^+\nu_\mu$ decay at rest had the kinetic energy of 4 MeV (few mm equivalent range in plastic scintillator) and rarely exited the stopping counter. So, pion pulses in the stopping counter recorded by the transient digitizers (TDs) had a double-pulse structure. Despite the online pion identification requirement, some muons remained in the final sample due to inefficiency.

Events were rejected if any activity in the photon detectors with energy above a threshold was detected. This condition removed events with photons. A similar requirement in the RS was also applied. The 24 sectors of the RS are conventionally grouped into six; a group of 4 sectors is called a "hextant". Only one hextant was allowed to have hits or two hextants if they were adjacent. This rejected events with multiple tracks and events with photon activity in the RS.

More detail description of the E949 experiment may be found in [10].

3. Data analysis

3.1. Total acceptance

In addition to the trigger requirements described above, we used several groups of offline cuts to select single muon track. The kinematic cuts were used to select events in the detector fiducial volume. Beam cuts were applied to identify incoming particle as a kaon and suppress extra beam particles at the track time. To suppress kaon decay-in-flight we applied delay coincidence cut. Numerous requirement were placed on the activity in the target to suppress background and ensure a reliable determination of the kinematic properties of the charged muon. Range-momentum cut was changed to select muons. In the main E949 analysis it was designed to check whether the range of a charged track is consistent with that for pions. To suppress photon activity in the detector we applied photon veto cut (loose for the background study and tight for the final result).

Acceptance for the $K^+ \to \mu^+\nu_H$ decay was measured using Monte-Carlo simulation and monitor triggers. The single event sensitivity (S.E.S.) for the heavy neutrino with mass $m_{\nu_H} = 250$ MeV/c^2 can be calculated as

$$S.E.S. = \frac{1}{Acc \times N_K} = 7.35 \times 10^{-10}, \tag{2}$$

where Acc is the total acceptance efficiency and N_K is the total number of stopped kaons. This sensitivity is roughly constant for the whole investigated region (Fig. 1).

To verify our acceptance measurement we calculated $K_{\mu 2}$ and $K_{\mu\nu\gamma}$ branching ratios. The $K_{\mu 2}$ branching ratio was measured to be 0.54 ± 0.15 and it is consistent with PDG value (0.6355 ± 0.0011) within the error [1]. The $K_{\mu\nu\gamma}$ branching ratio was measured to be $(1.3 \pm 0.4) \times 10^{-3}$ for $140 < p_\mu < 200$ and is also consistent with PDG value ($(1.4 \pm 0.2) \times 10^{-3}$) for the same muon momentum region [1].

3.2. Residual background

The search for $K^+ \to \mu^+\nu_H$ consists in a searching for additional peak below $K_{\mu 2}$ peak. So we should well understand all background sources that can fake or cover our signal. We simulated the main background sources, $K_{\mu\nu\gamma}$, $K_{\mu 3}$ and $K_{\pi 2\gamma}$ decays. After trigger requirements and offline selection criteria the $K_{\mu 3}$ contribution in the total number of background events is less than 1% of the $K_{\mu\nu\gamma}$ contribution due to two photons in the final state. The $K^+ \to \pi^+\pi^0\gamma$ decay can be ignored due to three photons in the final state and large range-momentum pion rejection. Therefore, the $K^+ \to \mu^+\nu_\mu\gamma$ is the dominant background source for the search of the $K^+ \to \mu^+\nu_H$ decay.

Given the agreement between the PDG values and our $K^+ \to \mu^+\nu_\mu$ and $K^+ \to \mu^+\nu_\mu\gamma$ branching ratio measurements, the experimental muon momentum spectra and the simulated $K_{\mu 2} + K_{\mu\nu\gamma}$ muon momentum spectra can be compared. To add $K_{\mu 2}$ and $K_{\mu\nu\gamma}$ decays together we take into account their branching ratios and the number of simulated events.

The dependence of acceptance after all cuts on the muon momentum and comparison between experimental (5% all data) and simulated muon momentum shape are shown in Fig. 1. There

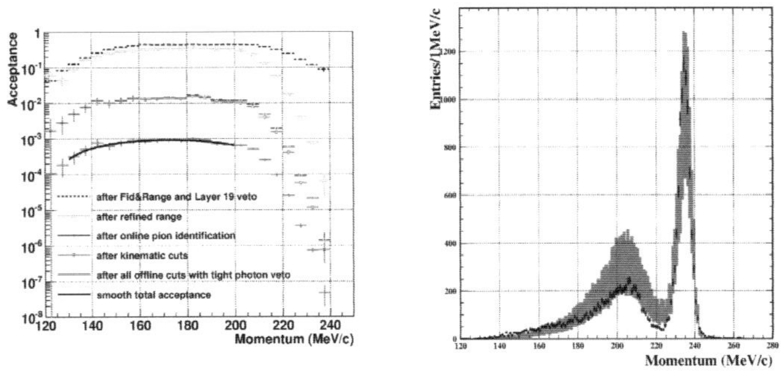

Figure 1. (Left) The dependence of acceptance on the muon momentum for the $K^+ \to \mu^+ \nu_H$ decay. (Right) Comparison between experimental (5% all data) and simulated muon momentum shape (in color).

are some discrepancies between data and MC in the muon momentum spectrum. Between 200 MeV/c and 220 MeV/c, the radiative gamma energy is low, the difference is caused by the difficulty in simulating detector activity or electronic noise of the low photon veto cut threshold. Beyond 220 MeV/c, it is caused by the uncertainty of layer 19 and refined range cuts.

Below 200 MeV/c, the simulated and experimental spectra are consistent, but MC is not the best fit of the data. Since the simulated shape does not show obvious bumps or valleys, we assume that the experimental background shape is also smooth.

3.3. Results

To search for additional peaks below the main $K_{\mu 2}$ peak we used asymptotic formula for the distribution of a test statistic, which was derived using the results of Wilks and Wald [11]. The method is a frequentist approach which is free of computationally expensive Monte Carlo calculations and is able to consider the shape of the signal. It thus avoids the ambiguity of selecting a signal region. Besides the mean value of the upper limit, an error band of the upper limit can be also calculated. The main feature of this approach is artificial data set (Asimov). An Asimov data set was used to evaluate the expected upper limit and its error band. For the heavy neutrino searching in this paper, the Asimov data set is the background-only expectation assuming no heavy neutrino signal in the region under test. The background shape was determined directly by fitting the momentum spectrum of data after all criteria. To avoid artificial peaks or valleys in the signal region, the range $\pm 6\sigma$ (the σ is the momentum resolution which is known from MC simulation) around the point of interest was chosen to fit for background with a second order polynomial function. With known background shape we are able to estimate expected number of background events in each bin in the testing region.

The muon momentum spectrum for the full data sample after all cuts and peak search result are shown in Fig. 2.

Since the observed upper limit is within the error band of the expected upper limit there is no evidence for a heavy neutrino signal.

The preliminary upper limit on mixing matrix element $|U_{\mu H}|^2$ set by this experiment is shown in Fig. 3. More details about heavy neutrino search in E949 experiment can be found in [13].

The final data set (Fig. 2 (left)) can be also used to constrain any decays $K^+ \to \mu^+ X$, where X is invisible (not detectable) set of neutral particles. Unfortunately, we cannot predict number

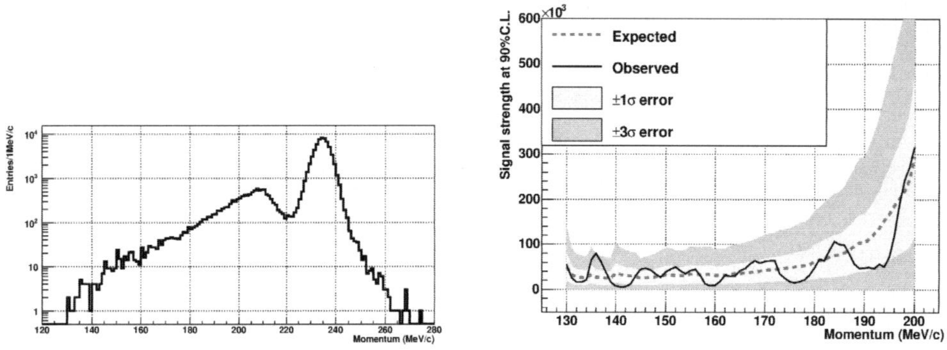

Figure 2. (Left) Muon momentum spectrum for the full data after all cuts applied. (Right) 90%C.L. expected upper limit with a $\pm 1\sigma$ error band and 99.8% C.L. error band. The black line is the observed upper limit result.

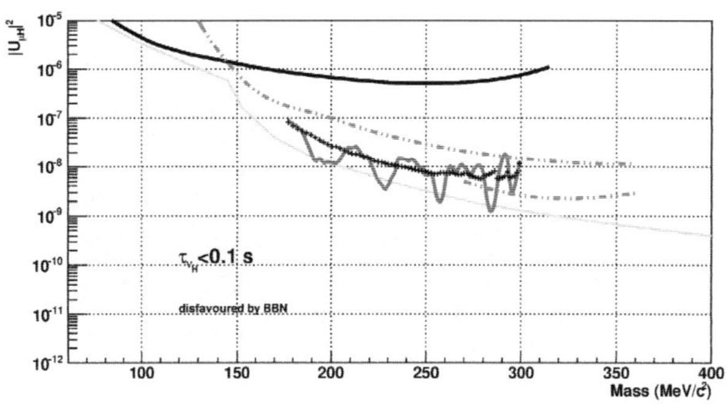

Figure 3. The preliminary upper limits on the mixing matrix element $|U_{\mu H}|^2$ set by this experiment (solid red curve, black crosses show expected upper limit) and others. Solid smooth black line shows the result of the peak search in kaon decays [12], dotted magenta lines show the results of the heavy neutrino decay experiment CERN PS191 [5] in two modes: top dotted line is derived from $K^+ \to \mu^+\nu_H \to \mu^+(\mu^- e^+ \nu_e) + c.c.$, bottom dotted line is derived from $K^+ \to \mu^+\nu_H \to \mu^+(\mu^-\pi^+) + c.c.$ The shaded region shows one of the possible BBN lower bounds [6, 7].

of background events with high precision, so we extract upper limit on the $K^+ \to \mu^+ X$ branching ratio assuming that all observed events is a signal (zero expected background). Acceptance for the single muon for different momentum is already known (Fig. 1 (left)). The lowest bound of the signal region, 130 MeV/c, was selected due to the acceptance drop off. The number of observed events is drastically increased with muon momentum. So, the upper bound of the signal region, 175 MeV/c, was selected to correspond to the previous experimental search for the decay $K^+ \to \mu^+\nu\bar{\nu}\nu$ ([9]). We get the following upper limit on the partial $K^+ \to \mu^+ X$ branching ratio:

$$BR(K^+ \to \mu^+ + X, 130 < p_\mu < 175 \text{ MeV/c}) < 7.5 \times 10^{-7} \quad (3)$$

For any assumed muon spectrum the total decay rate on the $K^+ \to \mu^+\nu\bar{\nu}\nu$ decay (or any other decay with single muon and set of invisible neutral particles) can be calculated using the following expression:

$$\frac{\Gamma(K^+ \to \mu^+\nu\bar{\nu}\nu)}{\Gamma(K^+ \to \text{all})} = 7.5 \times 10^{-7} \times \frac{\int_0^{p_\mu^{max}} (d\Gamma/dp_\mu)dp_\mu}{\int_{130}^{175} (d\Gamma/dp_\mu)dp_\mu} \tag{4}$$

where p_μ^{max} is the maximum muon momentum defined by kinematics ($p_\mu^{max} = (m_K^2 - m_\mu^2)/2m_K$). Using the SM spectrum [8] we get the following 90% C.L. upper limits on the total decay rate for $K^+ \to \mu^+\nu\bar{\nu}\nu$ decay:

$$\frac{\Gamma(K^+ \to \mu^+\nu\bar{\nu}\nu)}{\Gamma(K^+ \to \text{all})} < 2.4 \times 10^{-6} \tag{5}$$

More details about search for the $K^+ \to \mu^+\nu\bar{\nu}\nu$ decay in E949 can be found in [14].

4. Conclusion

The result of the search for heavy neutrinos in the $K^+ \to \mu^+\nu_H$ decay channel using the E949 data sample in an exposure of 1.70×10^{12} stopped kaons is presented. Since no evidence for extra peaks below the main $K^+ \to \mu^+\nu_\mu$ peak was found we set preliminary upper bounds on the mixing matrix element $|U_{\mu H}|^2$ in the mass region 175–300 MeV/c^2. In contrast to the CERN PS191 or BBN bounds our result is model-independent because we did not make any assumptions about heavy neutrino decay rates or couplings. Using the same data sample the upper limit on the $K^+ \to \mu^+\nu\bar{\nu}\nu$ branching ratio in the SM framework was set to be 2.4×10^{-6} which is about factor of 3 better than the current results from [9].

Acknowledgments

This research was supported in part by Grant #14-12-00560 of the Russian Science Foundation, the U.S. Department of Energy, the Ministry of Education, Culture, Sports, Science and Technology of Japan through the Japan-U.S. Cooperative Research Program in High Energy Physics and under Grant-in-Aids for Scientific Research, the Natural Sciences and Engineering Research Council (Grant no. 157985) and the National Research Council of Canada, National Natural Science Foundation of China, and the Tsinghua University Initiative Scientific Research Program.

References

[1] Olive K A *et al.* (Particle Data Group) 2014 *Chin. Phys.* **C38** 090001
[2] Asaka T and Shaposhnikov M 2005 *Phys. Lett.* **B620** 17–26
[3] Asaka T, Blanchet S and Shaposhnikov M 2005 *Phys. Lett.* **B631** 151–156
[4] Shrock R E 1981 *Phys. Rev.* **D24** 1232
[5] Bernardi G *et al.* 1988 *Phys. Lett.* **B203** 332
[6] Gorbunov D and Shaposhnikov M 2007 *JHEP* **0710** 015
[7] Boyarsky A, Ruchayskiy O and Shaposhnikov M 2009 *Ann. Rev. Nucl. Part. Sci.* **59** 191–214
[8] Gorbunov D and Mitrofanov A 2016 *e-print: arXiv:1605.08077 [hep-ph]*
[9] Pang C, Hildebrand R, Cable G and Stiening R 1973 *Phys. Rev.* **D8** 1989–2003
[10] Artamonov A V *et al.* (BNL-E949 Collaboration) 2009 *Phys. Rev.* **D79** 092004
[11] Cowan G, Cranner K, Gross E and Vitells O 2011 *Eur. Phys. J.* **C71** 1554
[12] Yamazaki T *et al.* 1984 *in Proceedings of Neutrino 84, Dortmund*
[13] Artamonov A V *et al.* (E949) 2015 *Phys. Rev.* **D91** 052001 [Erratum: Phys. Rev.D91,no.5,059903(2015)]
[14] Artamonov A V *et al.* (E949) 2016 *Phys. Rev.* **D94** 032012

Evidence for the rare decay $\Sigma^+ \to p\mu^+\mu^-$ at LHCb

Francesco Dettori

CERN European Organization for Nuclear Research, CH1211 Geneve 23, Switzerland

E-mail: francesco.dettori@cern.ch

Abstract. A search for the rare decay $\Sigma^+ \to p\mu^+\mu^-$ is performed using pp collision data recorded by the LHCb experiment at centre-of-mass energies $\sqrt{s} = 7$ and 8 TeV, corresponding to an integrated luminosity of $3\,\text{fb}^{-1}$. An excess of events is observed with respect to the background expectations with a signal significance of 4.0 standard deviations. No significant structure is observed in the dimuon invariant mass distribution.

1. Introduction

The $\Sigma^+ \to p\mu^+\mu^-$ decay[1] is a flavour changing neutral current process, allowed only at loop level in the standard model (SM). The process is dominated by long-distance contributions for a predicted branching fraction of $\mathcal{B}(\Sigma^+ \to p\mu^+\mu^-) \in [1.6, 9.0] \times 10^{-8}$ [1], while the short-distance SM contributions are suppressed at a branching fraction of about 10^{-12}. Evidence for this decay was seen by the HyperCP experiment [2] with a measured branching fraction $\mathcal{B}(\Sigma^+ \to p\mu^+\mu^-) = (8.6^{+6.6}_{-5.4} \pm 5.5) \times 10^{-8}$, which is compatible with the SM. Remarkably, the three observed decays have almost the same dimuon pair invariant mass of $m_{X^0} = 214.3 \pm 0.5\,\text{MeV}/c^2$, which lies close to the kinematic limit. Such a distribution, if confirmed, would point towards a decay with an intermediate particle coming from the Σ^+ baryon and decaying into two muons, *i.e.* a $\Sigma^+ \to pX^0(\to \mu^+\mu^-)$ decay, which would constitute evidence of physics beyond the SM (BSM). Various BSM explanations have been proposed to explain the HyperCP result. The intermediate X^0 particle could be for example a light pseudoscalar Higgs boson [3] [4] in a NMSSM or 2HDM model, or a sgoldstino [5] [6] in various supersymmetric models. Other interpretations and implications can be found in Ref. [7] [8] [9] [10] [11] [12] [13]; in general a pseudoscalar particle is favoured over a scalar one and a lifetime of order 10^{-14} s is estimated for the former case. This lifetime would correspond to a prompt X^0 signal, decaying in the same vertex as the Σ^+ baryon, in any present search for this particle. Attempts to confirm the presence of this X^0 particle have been made by different experiments in various initial and final states without finding any signal [14–21], and in LHCb through the final states $B^0_{(s)} \to \mu^+\mu^-\mu^+\mu^-$ [22] and $B^0 \to K^{*0}\mu^+\mu^-$ [23]. However, the search for the $\Sigma^+ \to p\mu^+\mu^-$ decay has not been repeated, mainly due to the absence of experiments with large hyperon production and due to experimental difficulties.

Hyperons are produced copiously in high energy proton-proton collisions at the Large Hadron Collider (LHC). A search for $\Sigma^+ \to p\mu^+\mu^-$ decays at the LHCb experiment could therefore confirm or disprove the HyperCP evidence, and measure its branching fraction. This search was also suggested in Ref. [24]. In this report, the search for the $\Sigma^+ \to p\mu^+\mu^-$ decay is presented as performed using pp collision data recorded by the LHCb experiment at centre-of-mass energies

[1] The inclusion of charge conjugated processes is implied throughout this note.

Content from this work may be used under the terms of the Creative Commons Attribution 3.0 licence. Any further distribution of this work must maintain attribution to the author(s) and the title of the work, journal citation and DOI.
Published under licence by IOP Publishing Ltd

$\sqrt{s} = 7$ and $8\,\text{TeV}$, corresponding to an integrated luminosity of $3\,\text{fb}^{-1}$. The measurement described here is detailed in Ref. [25].

2. Detector and data samples

The LHCb detector is a single-arm forward spectrometer covering the pseudorapidity range $2 < \eta < 5$, designed for the study of particles containing b or c quarks, and is detailed elsewhere [26] [27]. The online event selection is performed by a trigger, which consists of a hardware stage, based on information from the calorimeter and muon systems, followed by two software stages, the first performing a preliminary event reconstruction based on partial information while the second applying a full event reconstruction. Each of the three trigger stages is divided into several parallel trigger lines dedicated to different kinds of signals. Decay particles involved in this analysis are often not able to activate one or more trigger stages, owing mainly to their soft transverse momenta. Nevertheless, since Σ^+ baryons are copiously produced in the pp collisions recorded by LHCb, the present search can be performed on the events already recorded. In the offline processing, trigger decisions are associated with reconstructed candidates. A trigger decision on a particular line or on the full trigger can thus be ascribed to the reconstructed candidate, the rest of the event or both; events triggered as such are defined respectively as triggered on signal (TOS), independently of signal (TIS), or triggered on both. The estimation of the trigger efficiency when no specific path is selected is difficult, therefore a different strategy is adopted here: all the candidates passing the selection are used in the search for $\Sigma^+ \to p\mu^+\mu^-$ decays, while only the TIS candidates are used to convert the event yield into a branching fraction. This ensures a partial cancellation of the TIS trigger efficiency between signal and normalisation channels. Furthermore, control channels with large statistics can be exploited to estimate the trigger efficiency directly on data by measuring the overlap of events which are TIS and TOS simultaneously [28].

3. Analysis strategy and selection

After the trigger, a loose selection is applied based on geometric and kinematic variables. Afterwards candidates are selected by means of a multivariate selection based on a boosted decision trees algorithm (BDT) [29] [30]. The final search datasets are obtained by rejecting the background with a cut on the BDT output and on particle identification variables. The signal yield is obtained from a fit to the $p\mu^+\mu^-$ invariant mass and is converted into a branching fraction by normalising to the $\Sigma^+ \to p\pi^0$ control channel. To avoid experimenter bias, candidates in the signal regions were not examined until the analysis procedure had been finalised.

The analysis is designed in order to search for possible peaks in the dimuon invariant mass, pointing towards unknown intermediate particles. The resolution on the dimuon invariant mass as a function of the mass itself is shown in Figure 1(a). The selection is devised such that no fake structures are induced in the dimuon invariant mass distribution of signal candidates. The signal efficiency varies as a function of the dimuon invariant mass and is shown in Figure 1(b). The efficiency is larger at lower dimuon mass owing to the larger recoil momentum against the proton, which gives larger minimum transverse momentum to the two muons, ensuring a better tracking efficiency.

Together with the signal $\Sigma^+ \to p\mu^+\mu^-$ decay, the following channels are also selected in data: the $\Sigma^+ \to p\pi^0$ decay as normalisation channel, the $\Sigma^+ \to \overline{p}\mu^+\mu^+$ candidate decay as control channel for the combinatorial background, and the $K^+ \to \pi^+\pi^-\pi^+$ decay as a control channel for different parts of the efficiency evaluation. Candidate $\Sigma^+ \to p\mu^+\mu^-$ decays are selected by combining two good quality oppositely charged tracks with muon identification with a third track with proton identification. The three tracks are required to form a good quality secondary vertex (SV), displaced from any pp interaction vertex (PV) requiring a measured Σ^+ lifetime greater than 6 ps. Only Σ^+ candidates with transverse momentum $p_{\text{T}} > 0.5\,\text{GeV}/c$, $\chi^2_{\text{IP}} < 36$ and cosine of the angle between the flight direction and the reconstructed momentum larger than 0.9 are retained. Candidate $\Sigma^+ \to p\mu^+\mu^-$ decays are considered only if the invariant mass

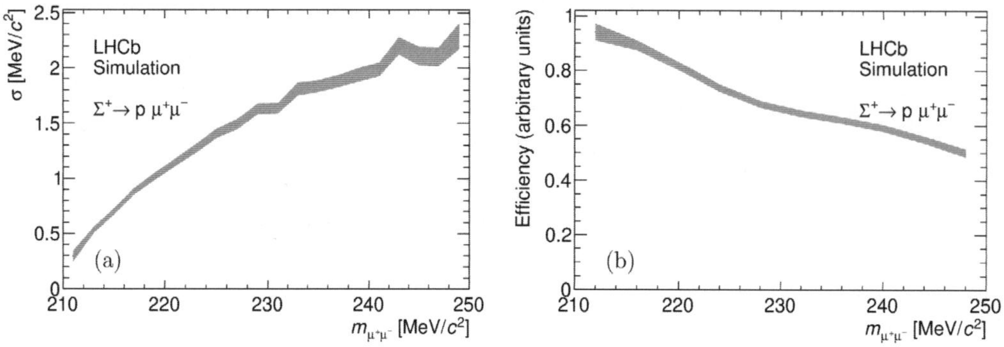

Figure 1: Resolution (a) on the dimuon invariant mass $m_{\mu^+\mu^-}$ as a function of the mass itself for simulated $\Sigma^+ \to p\mu^+\mu^-$ decays in LHCb. Efficiency (b) for simulated $\Sigma^+ \to p\mu^+\mu^-$ decays in LHCb as a function of the dimuon invariant mass.

satisfies $|m_{p\mu^+\mu^-} - m_{\Sigma^+}| < 500\,\text{MeV}/c^2$, where m_{Σ^+} is the known mass of the Σ^+ particle [31]. A large background component is present in data due to $\Lambda^0 \to p\pi^-$ decays, where the pion is misidentified as a muon and the two tracks are combined with a third one. This is vetoed discarding candidates having a $p\mu^-$ pair mass within 5 MeV/c^2 from the Λ^0 known mass when calculated with the $p\pi^-$ mass hypothesis. Possible backgrounds from exclusive decays peaking in the $p\mu^+\mu^-$ invariant mass have been examined, including $K^+ \to \pi^+\pi^-\pi^+$ and $K^+ \to \pi^+\mu^-\mu^+$ decays and various hyperon decays, and none has been found to contribute. In case of multiple candidates in a single event, all are retained in the selection. Multiple candidates are present in 5% of the events after the initial selection, while no multiple candidate is present in the final selection. The selection for the control channel $\Sigma^+ \to \overline{p}\mu^+\mu^+$ is identical to that of the signal but considering same-sign dimuon pairs. Candidate $\Sigma^+ \to p\pi^0$ decays are selected by combining one good quality track with proton identification with a π^0 reconstructed in the $\pi^0 \to \gamma\gamma$ mode from two clusters in the calorimeter. The selection of this decay is similar to that of the signal, but places a tighter requirement on the proton identification, and on the transverse momenta of the daughters in order to reduce the high combinatorial background. Finally, candidate $K^+ \to \pi^+\pi^-\pi^+$ decays are selected from three good quality tracks, one of which is oppositely charged with respect to the other two; the selection is similar but tighter than that of the signal to cope with the high level of combinatorial background. See Ref. [25] for further details on the selection criteria of the different channels.

The sample of $\Sigma^+ \to p\mu^+\mu^-$ candidates in data after the selection is dominated by combinatorial background, part of which is due to misidentified particles. This is rejected by cuts on the BDT and on multivariate particle identification variables [27] on the muons and on the proton. The BDT variable combines geometric and kinematic variables chosen so that the dependence on the $p\mu^+\mu^-$ invariant mass and on the dimuon invariant mass is linear and small to avoid biases. The BDT is optimised using simulated samples of $\Sigma^+ \to p\mu^+\mu^-$ events for the signal and candidates from the sidebands of the $\Sigma^+ \to \overline{p}\mu^+\mu^+$ selection in data for the background. The final cut values were chosen in order to optimise the sensitivity to a signal evidence for the smallest possible branching fraction [32]. No BDT selection is applied to the normalisation and control channels.

4. Normalisation

The number of signal candidates in the TIS sample is converted into a branching fraction with the formula

$$\mathcal{B}(\Sigma^+ \to p\mu^+\mu^-) = \frac{\varepsilon_{\Sigma^+\to p\pi^0}}{\varepsilon_{\Sigma^+\to p\mu^+\mu^-}} \frac{N_{\Sigma^+\to p\mu^+\mu^-}}{N_{\Sigma^+\to p\pi^0}} \mathcal{B}(\Sigma^+ \to p\pi^0) = \alpha \cdot N_{\Sigma^+\to p\mu^+\mu^-} \quad ,$$

where ε, N and $\mathcal{B}$ are the efficiency, candidate yield and branching fraction of the corresponding channel, respectively, and α is the single event sensitivity.

The ratio of signal and normalisation channel efficiencies, which includes the acceptance, the reconstruction efficiency of the final state particles and the selection efficiency, is computed with samples of simulated events corrected to take into account known differences between data and simulation. The reconstruction efficiency for the π^0 is calibrated using the ratio of $B^+ \to J/\psi K^{*+}(\to K^+\pi^0)$ and $B^+ \to J/\psi K^+$ decays reconstructed in data. The particle identification efficiency of protons and muons is calibrated exploiting control channels in data. Residual differences between data and simulation are treated as sources of systematic uncertainty. The signal and normalisation channels are required to be TIS at all trigger levels. The trigger efficiency ratio is thus expected to be unity. However, small differences in the average kinematic of the rest of the event in the two samples are present which cause the ratio to be different. The ratio of trigger efficiencies is thus evaluated with data-driven techniques [28] exploiting the large sample of $K^+ \to \pi^+\pi^-\pi^+$ decays. A systematic uncertainty is assigned for the applicability of this method to the relatively soft events of this analysis.

The invariant mass distribution of the $K^+ \to \pi^+\pi^-\pi^+$ control channel candidates in data is shown in Figure 2(a). A binned maximum likelihood extended fit is performed to the invariant mass distribution. The signal is described as an Hypatia function [33] while the background is described by a second-order polynomial. A total of $(966 \pm 2) \times 10^3$ $K^+ \to \pi^+\pi^-\pi^+$ candidates is measured.

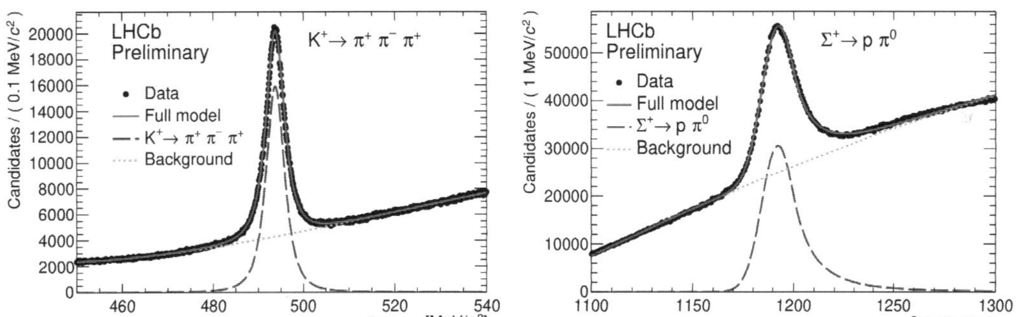

Figure 2: Invariant mass distribution of (a) $K^+ \to \pi^+\pi^-\pi^+$ candidates superimposed with the fit to data; (b) fit to the distribution of the corrected mass $m_\Sigma^{\rm corr}$ for $\Sigma^+ \to p\pi^0$ candidates superimposed with the fit to data.

The observed number of $\Sigma^+ \to p\pi^0$ candidates is $(1\,711 \pm 9) \times 10^3$ as obtained from a binned maximum likelihood extended fit to the corrected invariant mass distribution. The corrected invariant mass is defined as $m_\Sigma^{\rm corr} = m_{p\pi^0} - m_{\pi^0} + m_{\pi^0}^{\rm PDG}$ to correct for the π^0 mass reconstructed from the two photons. The $\Sigma^+ \to p\pi^0$ distribution is described as a Gaussian function with a power tail on the right side, while the background is described by a modified ARGUS function [34]. The invariant mass distribution is shown in Figure 2(b), superimposed with the fit. The single event sensitivity is $\alpha = (1.1 \pm 0.6) \times 10^{-8}$, where the uncertainty is dominated by the aforementioned systematic uncertainties, and corresponding to about 4×10^{11} Σ^+ particles produced in the LHCb acceptance in the full dataset in TIS events. This corresponds to 4.6 ± 4.2 $\Sigma^+ \to p\mu^+\mu^-$ candidates assuming a branching fraction of $(5 \pm 4) \times 10^{-8}$, to cover the SM predicted range.

The observed number of signal $\Sigma^+ \to p\mu^+\mu^-$ events is obtained with a fit to the $p\mu^+\mu^-$ invariant mass distribution in the range $1149.6 < m_{p\mu^+\mu^-} < 1409.6\,\text{MeV}/c^2$. The signal distribution is described by an Hypatia function [33]. The mass resolution and scale are calibrated using the control channel $K^+ \to \pi^+\pi^-\pi^+$ and comparing data and simulation distributions. No bias is seen in the peak position, while a 25% correction to the resolution has to be applied to match the one observed in data. A resolution of $4.28 \pm 0.19\,\text{MeV}/c^2$ is used as width of the signal $\Sigma^+ \to p\mu^+\mu^-$ distribution. The resolution is allowed to vary in the fit but constrained to the central value with a Gaussian constraint. The combinatorial background is described as a modified ARGUS function, with all parameters left free with exception of the threshold which is fixed to the kinematic limit.

While the normalisation is available for the TIS sample, further studies are needed to calculate the trigger efficiency for the full sample. Since the TIS dataset is a sub-sample of the full sample, the single event sensitivity of the latter will be equal or better than the one of the former.

5. Results and discussion

The invariant mass distribution of the $\Sigma^+ \to p\mu^+\mu^-$ candidates in data is shown in Figure 3(a) and (b) for the full and TIS datasets, respectively. A significant signal is present in the full dataset. The significance of the signal is of $4.0\,\sigma$, obtained from the comparison of the likelihood value of the full fit with that of the background-only fit, and includes the relevant systematic uncertainties as gaussian constrains to the likelihood. A total of $12.9^{+5.1}_{-4.2}$ signal candidates is observed. The signal in the TIS sample is found not to be significant. This is due to the observed signal candidates being needed for the related event to switch on at least one of the LHCb trigger stages. From the absence of a significant signal in the TIS dataset an upper limit on the $\Sigma^+ \to p\mu^+\mu^-$ branching fraction is derived, using the CLs method [35], of $\mathcal{B}(\Sigma^+ \to p\mu^+\mu^-) < 6.3 \times 10^{-8}$ at 95% confidence level (CL).

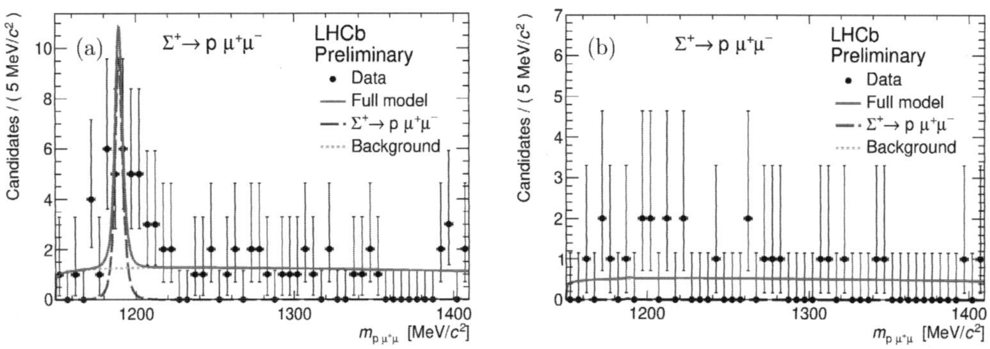

Figure 3: Invariant mass distribution for $\Sigma^+ \to p\mu^+\mu^-$ candidates: (a) all the candidates and (b) only the candidates satisfying the TIS requirement.

Considering candidates in the full selection, a scan for possible signals in the dimuon invariant mass is performed, restricted to within two times the resolution in the $\Sigma^+ \to p\mu^+\mu^-$ invariant mass around the known Σ^+ mass. The scan is performed considering a single Gaussian function for a putative X^0 resonance (signal) and a linear distribution for the remaining candidates (background). In Figure 4(a) the local p-value of the background-only hypothesis as a function of the dimuon mass is shown. No significant signal is found. In Figure 4(b) the fit corresponding to a mass of 214.3 MeV/c^2 is shown. The fitted number of events for this hypothesis is 1.6 ± 1.9, corresponding to a fraction 0.078 ± 0.092 of the considered candidates.

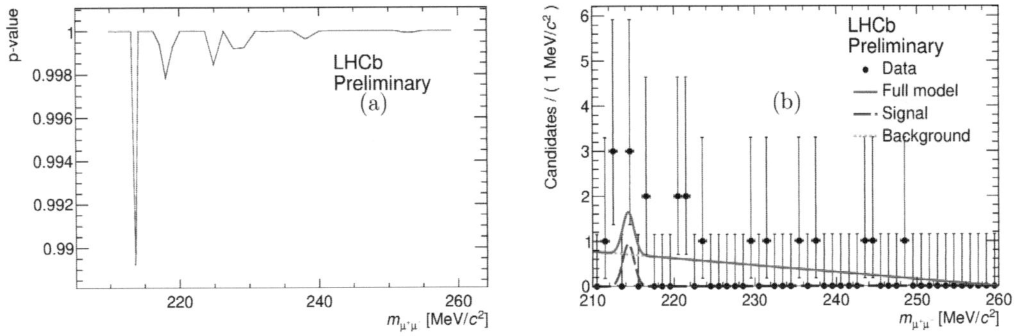

Figure 4: Plot of the local p-value as a function of the dimuon mass as obtained from the fits (see text) in (a) and (b) fit for a signal with mass $m_{X^0} = 214.3\,\text{MeV}/c^2$.

In summary, a search for the $\Sigma^+ \to p\mu^+\mu^-$ rare decay is performed by the LHCb experiment at centre-of-mass energies $\sqrt{s} = 7$ and 8 TeV, corresponding to an integrated luminosity of 3 fb^{-1}. Evidence for the $\Sigma^+ \to p\mu^+\mu^-$ decay is found in the full dataset, albeit not found when requiring the events to be triggered independently of the signal decay. The observed signal candidates show a dimuon invariant mass distribution consistent with phase space; no significant peak consistent with an intermediate particle is found in the dimuon invariant mass distribution. The upper limit on the branching fraction of the $\Sigma^+ \to p\mu^+\mu^-$ decay is 6.3×10^{-8} at 95% CL for a SM-like signal.

References

[1] He X G, Tandean J and Valencia G 2005 *Phys.Rev.* **D72** 074003
[2] Park H *et al.* (HyperCP collaboration) 2005 *Phys.Rev.Lett.* **94** 021801
[3] He X G, Tandean J and Valencia G 2007 *Phys.Rev.Lett.* **98** 081802
[4] He X G, Tandean J and Valencia G 2006 *Phys. Rev.* **D74** 115015
[5] Gorbunov D and Rubakov V 2006 *Phys.Rev.* **D73** 035002
[6] Demidov S V and Gorbunov D S 2007 *JETP Lett.* **84** 479–484
[7] He X G, Tandean J and Valencia G 2005 *Phys.Lett.* **B631** 100–108
[8] Geng C Q and Hsiao Y K 2006 *Phys. Lett.* **B632** 215–218
[9] Deshpande N, Eilam G and Jiang J 2006 *Phys.Lett.* **B632** 212–214
[10] Chen C H, Geng C Q and Kao C W 2008 *Phys.Lett.* **B663** 400–404
[11] Xiangdong G, Li C S, Li Z and Zhang H 2008 *Eur.Phys.J.* **C55** 317–324
[12] Mangano M L and Nason P 2007 *Mod.Phys.Lett.* **A22** 1373–1380
[13] Pospelov M 2009 *Phys. Rev.* **D80** 095002
[14] Love W *et al.* (CLEO collaboration) 2008 *Phys. Rev. Lett.* **101** 151802
[15] Tung Y C *et al.* (E391a collaboration) 2009 *Phys. Rev. Lett.* **102** 051802
[16] Abazov V M *et al.* (D0 collaboration) 2009 *Phys. Rev. Lett.* **103** 061801
[17] Aubert B *et al.* (BaBar collaboration) 2009 *Phys. Rev. Lett.* **103** 081803
[18] Hyun H J *et al.* (Belle collaboration) 2010 *Phys. Rev. Lett.* **105** 091801
[19] Abouzaid E *et al.* (KTeV collaboration) 2011 *Phys. Rev. Lett.* **107** 201803
[20] Ablikim M *et al.* (BESIII collaboration) 2012 *Phys. Rev.* **D85** 092012
[21] Lees J P *et al.* (BaBar collaboration) 2014 *Phys. Rev. Lett.* **113** 201801
[22] Aaij R *et al.* (LHCb collaboration) 2013 *Phys. Rev. Lett.* **110** 211801
[23] Aaij R *et al.* (LHCb collaboration) 2015 *Phys. Rev. Lett.* **115** 161802
[24] Park H K 2010 *JHEP* **10** 052
[25] Alves Jr A A *et al.* (LHCb Collaboration) 2016 Evidence for the rare decay $\Sigma^+ \to p\mu^+\mu^-$ LHCb-CONF-2016-013 URL https://cds.cern.ch/record/2224468

IOP Conf. Series: Journal of Physics: Conf. Series **800** (2017) 012021 doi:10.1088/1742-6596/800/1/012021

[26] Alves Jr A A *et al.* (LHCb collaboration) 2008 *JINST* **3** S08005
[27] Aaij R *et al.* (LHCb collaboration) 2015 *Int. J. Mod. Phys.* **A30** 1530022
[28] Aaij R *et al.* 2013 *JINST* **8** P04022
[29] Breiman L, Friedman J H, Olshen R A and Stone C J 1984 *Classification and regression trees* (Belmont, California, USA: Wadsworth international group)
[30] Schapire R E and Freund Y 1997 *J. Comput. Syst. Sci.* **55** 119
[31] Olive K A *et al.* (Particle Data Group) 2014 *Chin. Phys.* **C38** 090001 and 2015 update
[32] Punzi G 2003 *Statistical Problems in Particle Physics, Astrophysics, and Cosmology* ed Lyons L, Mount R and Reitmeyer R p 79 (*Preprint* physics/0308063)
[33] Martinez Santos D and Dupertuis F 2014 *Nucl. Instrum. Meth.* **A764** 150–155
[34] Albrecht H *et al.* (ARGUS collaboration) 1990 *Phys. Lett.* **B241** 278–282
[35] Read A L 2002 *J. Phys.* **G28** 2693–2704

Status of the J-PARC KOTO experiment

K Shiomi for the KOTO collaboration

Institute of Particle and Nuclear Studies, High Energy Accelerator Research Organization (KEK), Tsukuba, Ibaraki 305-0801, Japan

E-mail: shiomi@post.kek.jp

Abstract. The KOTO experiment is dedicated to observe the rare decay $K_L^0 \to \pi^0\nu\bar{\nu}$ at J-PARC. The rare decay $K_L^0 \to \pi^0\nu\bar{\nu}$ breaks the CP symmetry directly and is highly suppressed in the Standard Model. This mode is a powerful tool to measure Standard Model parameters and to search for new physics beyond the Standard Model because its theoretical uncertainty is small.

We performed the first physics run in May 2013 and collected 20 times more data than the first physics run since 2015. In this paper, results of the first physics run, detector upgrade work after the first physics run, and analysis status of 2015 data are reported.

1. Introduction

The rare decay $K_L^0 \to \pi^0\nu\bar{\nu}$ provides us one of the best probes for understanding CP violation. The decay breaks the CP symmetry directly. This decay is highly suppressed in the Standard Model because it occurs through a flavor-changing neutral current process. Owing to this, this decay can access indirectly very high energy scale which cannot be reached directly by accelerators [1, 2]. In addition, the contributions from new physics can be estimated precisely because the theoretical uncertainty on this mode is only a few percent. The branching ratio of this decay mode is predicted to be $(3.00 \pm 0.30) \times 10^{-11}$ in the Standard Model [3]. On the other hand, the current limit on this decay is 2.6×10^{-8} at the 90% confidence level, which was set by the KEK E391a experiment [4].

The KOTO experiment at J-PARC Main Ring accelerator is the successor of the E391a experiment and aims to observe the $K_L^0 \to \pi^0\nu\bar{\nu}$ decay. We had built a new beam line and constructed the new detector system and DAQ system to achieve a sensitivity of 10^{-11}.

We took the first physics run data in 2013, but we were able to take only 4 days of data in this time due to an incident at Hadron Experimental Facility. After that, the operation of the facility was stopped for two years to renovate safety of the facility. From April 2015, the facility restarted beam operation. We restarted physics data taking and collected 20 times more data than the first physics run, as shown in figure 1.

In this paper, results of the first physics run, detector upgrade work after the first physics run, and analysis status of 2015 data are reported.

2. Experiment

A proton beam of 30-GeV was extracted from the Main Ring accelerator to the hall of J-PARC Hadron Experimental Facility every 6 seconds with a 2-second duration, and was injected into the 66-mm-long gold target. The K_L^0's produced in the target were transported to the KOTO

Content from this work may be used under the terms of the Creative Commons Attribution 3.0 licence. Any further distribution of this work must maintain attribution to the author(s) and the title of the work, journal citation and DOI.
Published under licence by IOP Publishing Ltd

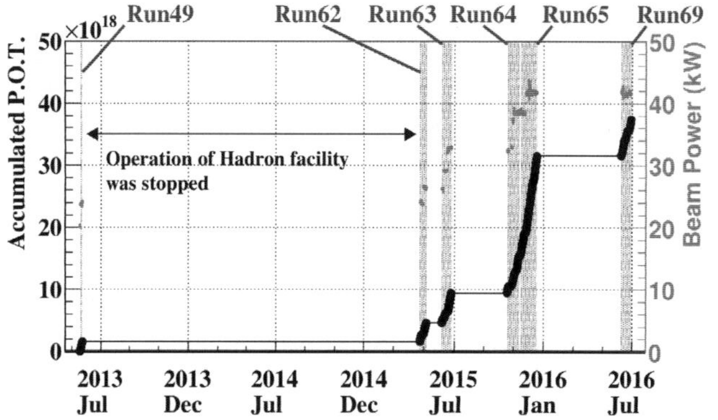

Figure 1. This plot shows the accumulated protons on the target (P.O.T.) as a function of time. Run49 is the first physics run. In this plot, beam power of the primary proton is also shown by the red points. The beam power increased gradually and the current beam power became two times higher than the beam power at the first physics run.

detector system through the "KL beam line" located along 16° line with respect to the primary proton beam line. The KL beam line consisted of two collimators, a sweeping magnet, and a lead photon absorber.

The schematic view of the KOTO detector system is shown in Fig. 2. Most of KOTO detectors were located inside a vacuum tank to avoid π^0 production through the interaction between beam neutrons and residual gas. The vacuum tank consisted of the front section, the middle section, and the end cap section. Large lead-scintillator sandwich counters called "Front Barrel (FB)" and "Main Barrel (MB)" were installed inside the front section and the middle section, respectively. A photon detector called "Neutron Collar Counter (NCC)" was located inside FB to separate the decay region from the upstream region. An electromagnetic calorimeter made of 2716 undoped CsI crystals was installed in the end cap section. The front of the calorimeter was covered with a set of plastic scintillators, named "Charged Veto (CV)". Behind the calorimeter, a photon veto detector called "Collar Counter04 (CC04)" was placed to detect particles that escape through the beam hole in the calorimeter. At the downstream of the vacuum tank, additional two "Colar Counters" and "Beam Hole" veto detectors were located to detect particles from K_L^0ś decays passing through the beam hole.

3. Results of the first physics run

At first, the analysis method is described briefly. The energies and positions of the two photons from the π^0 decay were measured with the CsI calorimeter. The decay position of the π^0 was reconstructed on the beam axis, assuming the invariant mass of the two photons to be the π^0 mass. The transverse momentum (P_T) of the π^0 was also calculated. We required a finite transverse momentum for signal candidates because neutrinos should have carried away a finite transverse momentum. A series of event selection criteria was imposed on the reconstructed π^0 kinematics and the hit information of the veto counters. Details on the selection criteria are described in Ref [5].

The signal search plane of the first physics run is shown in Fig. 3. Several events are clustered at z = 2000–2600 mm. In these events, neutrons in the beam halo (halo neutrons) interacted with NCC and produced π^0 . The two photons from the π^0 hit the CsI calorimeter. The decay vertexes of those were reconstructed at the NCC location that was outside the signal region. The

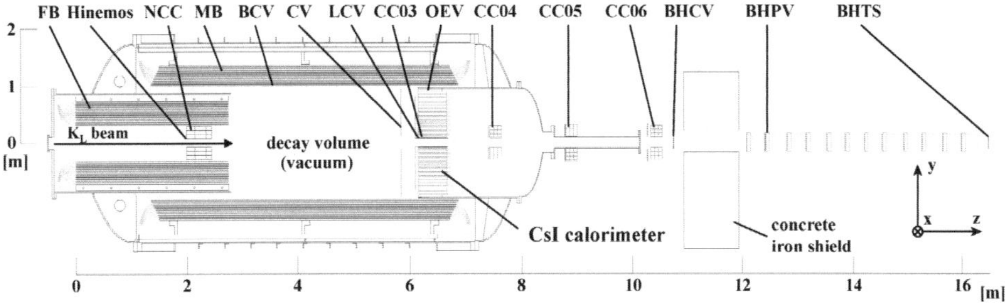

Figure 2. Cross-sectional side view of the KOTO detector system. The K_L^0 beam comes in from the left hand side. Also shown is the coordinate system used in this paper: x is the horizontal, y is the vertical, and z is the beam direction. The origin of the coordinate is set at the upstream end of FB.

events in the low P_T region came from $K_L^0 \to \pi^+\pi^-\pi^0$ decays. In those events, charged pions from the $K_L^0 \to \pi^+\pi^-\pi^0$ decay passed though the beam hole of the calorimeter and interacted with the beam pipe before they were detected by veto detectors located downstream. The event in Fig. 3 in the region above the signal box is likely to be caused by halo neutrons hitting the CsI calorimeter. We suspected that the single neutron could make two clusters through several hadronic interactions in the calorimeter.

We found one event in the signal box while 0.34 background (BG) events were predicted. We set an upper limit on the $K_L^0 \to \pi^0\nu\bar{\nu}$ decay to be 5.1×10^{-8} at the 90% confidence level [6]. We also set an upper limit on the $K_L^0 \to \pi^0 X^0$ decay to be 3.7×10^{-8} at the 90% confidence level, where X^0 is an invisible particle with the mass close to π^0; it was recently pointed out that the experimental search for a weakly-interacting light particle had a loophole around the π^0 mass region [7, 8].

The numbers of background events expected in the signal region are summarized in Table 1. The most serious background events in the first physics run were due to halo neutrons hitting the calorimeter.

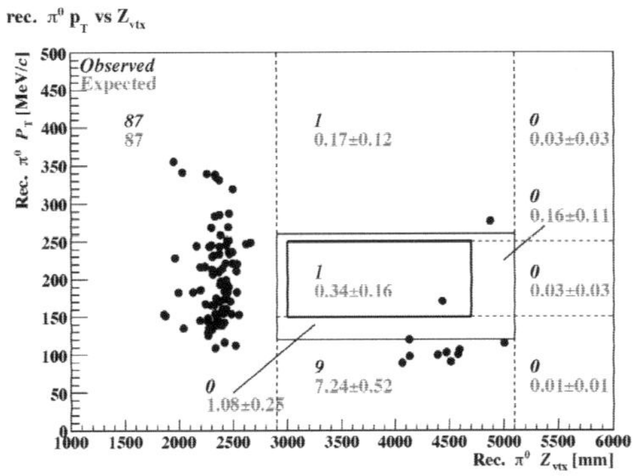

Figure 3. Scatter plot of reconstructed P_T vs z after imposing all selection criteria. The inner box is the signal region. The numbers in black show the numbers of observed events, and the numbers in red show the numbers of expected background events.

4. Detector Upgrade

After the first physics run, we upgraded several detectors to suppress BG events in the first physics run.

Background source	The number of background events
Halo neutrons hitting the CsI Calorimeter	0.18 ± 0.15
Kaon decay events	0.10 ± 0.04
Halo neutrons hitting the NCC	0.06 ± 0.06
Sum	0.34 ± 0.16

Table 1. Estimated numbers of background events in the signal region for the first physics run.

Downstream detectors were upgraded before the 2015 run. To suppress $K_L^0 \to \pi^+\pi^-\pi^0$ events, we replaced the vacuum pipe extended from the vacuum tank to the front of CC06 with thinner one, and we installed new scintillator counters outside of the vacuum pipe to detect charged particles generated by the interaction between charged pions and the vacuum pipe. To increase the detection efficiency for the photon escaping along the beam axis, we added four modules of "Beam Hole Photon Veto (BHPV)" (see Fig. 2.), which is made of lead and Cherenkov counter with aerogel [9] and to discriminate photons from neutrons. We also installed, in the downstream of BHPV, new photon counters made of lead and acrylic radiator to detect photons escaping detection by BHPV.

An additional barrel photon detector called "Inner Barrel(IB)" was installed to the inside of MB before the 2016 run to reduce $K_L^0 \to 2\pi^0$ BG events. The IB is a sandwich calorimeter which consists of 25 pairs of 5-mm-thick plastic-scintillator plates and 1-mm-thick lead plates.

5. Analysis status of 2015 data

We developed new cuts to suppress background events observed in the first physics run and checked the effect of new cuts for background reduction by using Run62 data (see Fig. 1). After that, we estimated the signal sensitivity and the number of background events in the signal box.

5.1. Study of background reduction

To suppress BG events due to halo neutrons, we tried to reduce the number of halo neutrons. There were two sources of halo neutrons. One was the surfaces of collimators located in the beam line. We re-aligned collimator positions according to the measurements of the beam profile. Another source was the vacuum window located in front of the KOTO detector system which separated the vacuum in the beam line from the vacuum in the decay region. The vacuum window was replaced with a thinner one. These modification reduced the number of background events due to halo neutrons hitting the calorimeter by 50%, as shown in Fig. 4. To get further reductions of background events due to halo neutrons, We studied hadron cluster events by using special data. In this data, a 10-mm-thick aluminum disk was inserted into the beam to scatter neutrons into the calorimeter. By using the special data, we developed new cuts to distinguish the clusters made by photons from the clusters made by neutrons. The new cuts improved the rejection of hadron cluster events by one order of magnitude while keeping signal efficiency almost the same as the previous analysis.

To suppress $K_L^0 \to \pi^+\pi^-\pi^0$ events, we installed new scintillator counters as mentioned in Section 4. We checked the veto effect of this new counters, as shown in Fig. 5. The reduction of events having small transverse momentum was consistent between data and MC within statistical uncertainties. The number of events expected in the signal region was 0.04 ± 0.01 after the veto cut.

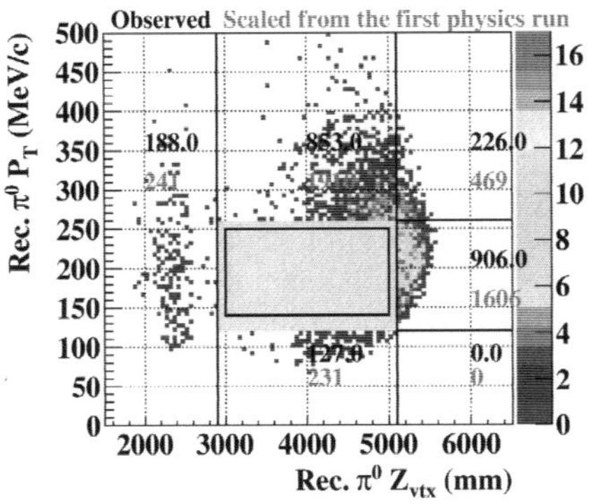

Figure 4. Scatter plot of reconstructed P_T vs z for samples enhanced neutron events. The gray box is the masked region in the analysis. The numbers in black show the number of events observed in Run62 and the numbers in red show the number of events scaled from the first physics run.

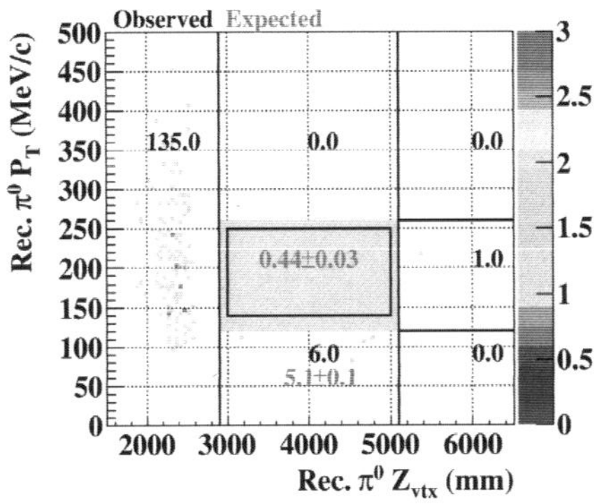

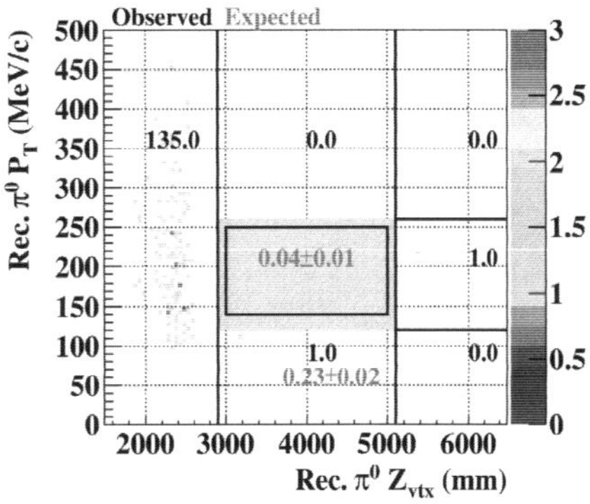

Figure 5. The left (right) plot shows scatter plot of reconstructed P_T vs z before (after) applying the veto cut of new scintillator counters. By applying the veto cut, the events in the low P_T region are removed effectively and the number of events expected in the signal box reduces by one order of magnitude.

5.2. Sensitivity and BG estimation

The total number of K_L^0ś in Run62 was estimated to be $(3.72 \pm 0.07) \times 10^{11}$ by using $K_L^0 \to 2\pi^0$ samples and it was 1.6 times larger than the first physics run. In Run62, background events due to $K_L^0 \to \pi^+\pi^-\pi^0$ and halo neutrons hitting the calorimeter were well suppressed by new cuts as mentioned in the previous subsection. We plan to use a larger signal region than that in the first physics run, which improved signal efficiency by 40%. The single event sensitivity (S.E.S) of the $K_L^0 \to \pi^0\nu\bar{\nu}$ decay for Run62 was estimated to be 5.9×10^{-9} preliminarily, which is two times better than the first physics run.

Figure 6 shows the reconstructed P_T vs z after imposing all selection criteria. The number of observed events in the downstream region (z>5000mm) was slightly larger than the number of expected background, but was still within a statistical uncertainty under this small data set. Table 2 summarizes the number of background events expected in the signal region.

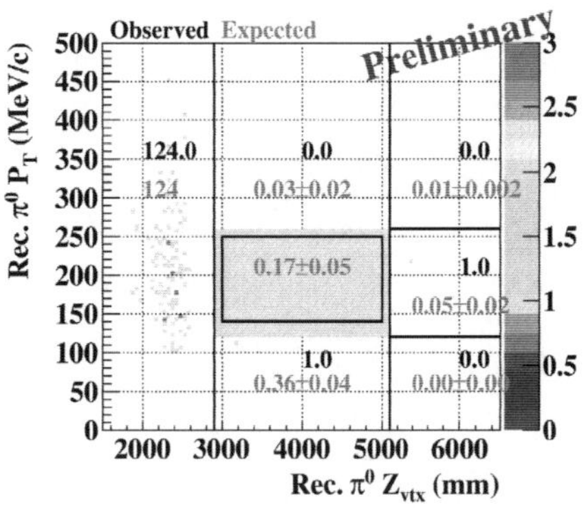

Figure 6. Scatter plot of reconstructed P_T vs z after imposing all selection criteria. The gray box is the masked region in the analysis. The inner box is the signal region. The numbers in black show the number of observed events, and the numbers in red show the number of expected background events.

Background source	The number of background events
$K_L^0 \to 2\pi^0$	0.04 ± 0.03
$K_L^0 \to \pi^+\pi^-\pi^0$	0.04 ± 0.01
Halo neutrons hitting the NCC	0.04 ± 0.04
Halo neutrons hitting the CsI calorimeter	0.05 ± 0.02
Sum	0.17 ± 0.05

Table 2. Estimated numbers of background events in the signal region for Run62 data.

6. Conclusions and Prospects

The KOTO experiment searches for the rare decay $K_L^0 \to \pi^0\nu\bar{\nu}$ at J-PARC. The first physics data was taken in 2013. One event was observed while 0.34 BG events were expected. We set a branching ratio on the $K_L^0 \to \pi^0\nu\bar{\nu}$ decay to be 5.1×10^{-8} at the 90% confidence level. We also set an upper limit on the $K_L^0 \to \pi^0 X^0$ decay to be 3.7×10^{-8} at the 90% confidence level, where X^0 is an invisible particle with the π^0 mass.

After the first physics run, we updated several detectors to reduce background events found in the first physics run and collected 20 times more data than the first physics run. We checked the reduction of background events by using small data set and confirmed that background events were well suppressed by new cuts.

Next, we will analyze more data to check background events with higher sensitivity and optimize cut conditions to achieve a higher sensitivity.

References

[1] M. Tanimoto and K. Yamamoto, *Prog. Theor. Exp. Phys.* **2015**, 053B07 (2015)
[2] A. J. Buras, D. Buttazzo, R. Knegjens, *J. High Energy Phys.* **1511**, 166 (2015)
[3] A. J. Buras, D. Buttazzo, J. Girrbach-Noe and R. Knegjens, *J. High Energy Phys.* **1511**, 033 (2015)
[4] J.K. Ahn *et al.*, *Phys. Rev.* D**81**, 072004 (2010)
[5] Y. Maeda, Ph.D. thesis, Kyoto University (2016) (Available at http://www-he.scphys.kyoto-u.ac.jp/theses, date last accessed September 5, 2016)
[6] J.K. Ahn *et al.*, arXiv:1609.03637 [hep-ex]
[7] K. Fuyuto, W.-S. Hos and M. Kohda, *Phys. Rev. Lett.* **114**, 171802 (2015)
[8] K. Fuyuto, W.-S. Hos and M. Kohda, *Phys. Rev.* D**93**, 054021 (2016)
[9] Y. Maeda *et al.*, *Theor. Exp. Phys.* **2015**, 063H01 (2015)

IOP Conf. Series: Journal of Physics: Conf. Series **800** (2017) 012023 doi:10.1088/1742-6596/800/1/012023

Status of the CERN NA62 Experiment

Giuseppe Ruggiero[1]
University of Liverpool, UK
E-mail: giuseppe.ruggiero@cern.ch

Abstract. The $K^+ \to \pi^+ \nu\bar{\nu}$ decay is one of the theoretically cleanest meson decays to look for indirect effects of new physics in a complementary way to direct searches at the LHC. The NA62 experiment at CERN SPS is designed to measure the branching ratio of this decay with 10% precision. NA62 is taking data since 2014, eventually reaching the final designed beam intensity. Using a sample of data taken in 2015, the experimental performances in view of the $K^+ \to \pi^+ \nu\bar{\nu}$ branching ratio measurement is reported.

[1] for the NA62 Collaboration: G. Aglieri Rinella, R. Aliberti, F. Ambrosino, R. Ammendola, B. Angelucci, A. Antonelli, G. Anzivino, R. Arcidiacono, I. Azhinenko, S. Balev, M. Barbanera, J. Bendotti, A. Biagioni, L. Bician, C. Biino, A. Bizzeti, T. Blazek, A. Blik, B. Bloch-Devaux, V. Bolotov, V. Bonaiuto, M. Boretto, M. Bragadireanu, D. Britton, G. Britvich, M.B. Brunetti, D. Bryman, F. Bucci, F. Butin, J. Calvo, E. Capitolo, C. Capoccia, T. Capussela, A. Cassese, A. Catinaccio, A. Cecchetti, A. Ceccucci, P. Cenci, V. Cerny, C. Cerri, B. Checcucci, O. Chikilev, S. Chiozzi, R. Ciaranfi, G. Collazuol, A. Conovaloff, P. Cooke, P. Cooper, G. Corradi, E. Cortina Gil, F. Costantini, F. Cotorobai, A. Cotta Ramusino, D. Coward, G. D'Agostini, J. Dainton, P. Dalpiaz, H. Danielsson, J. Degrange, N. De Simone, D. Di Filippo, L. Di Lella, S. Di Lorenzo, N. Dixon, N. Doble, B. Dobrich, V. Duk, V. Elsha, J. Engelfried, T. Enik, N. Estrada, V. Falaleev, R. Fantechi, V. Fascianelli, L. Federici, S. Fedotov, M. Fiorini, J. Fry, J. Fu, A. Fucci, L. Fulton, S. Gallorini, S. Galeotti, E. Gamberini, L. Gatignon, G. Georgiev, A. Gianoli, M. Giorgi, S. Giudici, L. Glonti, A. Goncalves Martins, F. Gonnella, E. Goudzovski, R. Guida, E. Gushchin, F. Hahn, B. Hallgren, H. Heath, F. Herman, T. Husek, O. Hutanu, D. Hutchcroft, L. Iacobuzio, E. Iacopini, E. Imbergamo, O. Jamet, P. Jarron, E. Jones, T. Jones K. Kampf, J. Kaplon, V. Kekelidze, S. Kholodenko, G. Khoriauli, A. Khotyantsev, A. Khudyakov, Yu. Kiryushin, A. Kleimenova, K. Kleinknecht, A. Kluge, M. Koval, V. Kozhuharov, M. Krivda, Z. Kucerova, Yu. Kudenko, J. Kunze, G. Lamanna, G. Latino, C. Lazzeroni, G. Lehmann-Miotto, R. Lenci, M. Lenti, E. Leonardi, P. Lichard, R. Lietava, V. Likhacheva, L. Litov, R. Lollini, D. Lomidze, A. Lonardo, M. Lupi, N. Lurkin, K. McCormick, D. Madigozhin, G. Maire, C. Mandeiro, I. Mannelli, G. Mannocchi, A. Mapelli, F. Marchetto, R. Marchevski, S. Martellotti, P. Massarotti, K. Massri, P. Matak, E. Maurice, M. Medvedeva, A. Mefodev, E. Menichetti, E. Minucci, M. Mirra, M. Misheva, N. Molokanova, J. Morant, M. Morel, M. Moulson, S. Movchan, D. Munday, M. Napolitano, I. Neri, F. Newson, A. Norton, M. Noy, G. Nuessle, T. Numao, V. Obraztsov, A. Ostankov, S. Padolski, R. Page, V. Palladino, G. Paoluzzi, C. Parkinson, E. Pedreschi, M. Pepe, F. Perez Gomez, M. Perrin-Terrin, L. Peruzzo, P. Petrov, F. Petrucci, R. Piandani, M. Piccini, D. Pietreanu, J. Pinzino, I. Polenkevich, L. Pontisso, Yu. Potrebenikov, D. Protopopescu, F. Raffaelli, M. Raggi, P. Riedler, A. Romano, P. Rubin, G. Ruggiero, V. Russo, V. Ryjov, A. Salamon, G. Salina, V. Samsonov, C. Santoni, G. Saracino, F. Sargeni, V. Semenov, A. Sergi, M. Serra, A. Shaikhiev, S. Shkarovskiy, I. Skillicorn, D. Soldi, A. Sotnikov, V. Sugonyaev, M. Sozzi, T. Spadaro, F. Spinella, R. Staley, A. Sturgess, P. Sutcliffe, N. Szilasi, D. Tagnani, S. Trilov, M. Valdata-Nappi, P. Valente, M. Vasile, T. Vassilieva, B. Velghe, M. Veltri, S. Venditti, P. Vicini, R. Volpe, M. Vormstein, H. Wahl, R. Wanke, P. Wertelaers, A. Winhart, R. Winston, B. Wrona, O. Yushchenko, M. Zamkovsky, A. Zinchenko.

Content from this work may be used under the terms of the Creative Commons Attribution 3.0 licence. Any further distribution of this work must maintain attribution to the author(s) and the title of the work, journal citation and DOI.
Published under licence by IOP Publishing Ltd

1. Introduction

The $K^+ \to \pi^+\nu\bar{\nu}$ and $K_L \to \pi^0\nu\bar{\nu}$ are flavour changing neutral current decays proceeding through box and electroweak penguin diagrams. A quadratic GIM mechanism and a strong Cabibbo suppression make these processes extremely rare. Using the value of tree-level elements of the Cabibbo-Kobayashi-Maskawa (CKM) triangle as external inputs, the Standard Model (SM) predicts [1, 2]:

$$\begin{aligned} BR(K^+ \to \pi^+\nu\bar{\nu}) &= (8.4 \pm 1.0) \times 10^{-11} \\ BR(K_L \to \pi^0\nu\bar{\nu}) &= (3.4 \pm 0.6) \times 10^{-11}. \end{aligned}$$

The knowledge of the external inputs dominate the uncertainties on the predictions. The theoretical accuracy, instead, is at percent level because: the short distance physics dominates thanks to the top quark exchange in the loop; the hadronic matrix elements cancel almost completely in the normalization of the $K \to \pi\nu\bar{\nu}$ branching ratios (BR) to the precisely measured BR of the $K^+ \to \pi^0e^+\nu$. The dependence on the CKM parameters partially cancels in the correlation between $K^+ \to \pi^+\nu\bar{\nu}$ and $K_L \to \pi^0\nu\bar{\nu}$. Therefore simultaneous measurements of the two BRs' would allow a theoretical clean exploitation of the CKM triangle using kaons only.

The $K \to \pi\nu\bar{\nu}$ decays are extremely sensitive to physics beyond the SM, probing the highest mass scales among the rare meson decays. The largest deviations from SM are expected in models with new sources of flavour violation, owing to weaker constraints from B physics [3, 4]. The experimental value of ε_K, the parameter measuring the indirect CP violation in neutral kaon decays, limits the range of variation expected for $K \to \pi\nu\bar{\nu}$ BRs' within models with currents of defined chirality, producing typical correlation patterns between charged and neutral modes [5]. Results from LHC direct searches strongly limit the range of variation mainly in supersymmetric models [6, 7]. Anyway thanks to the SM suppression and to the existing weak experimental constraints from K physics, significant variations of the $K \to \pi\nu\bar{\nu}$ BRs' from the SM predictions induced by new physics at mass scales up to 100 TeV are still possible providing a measurement with 10% precision at least.

Only the charged mode has been observed so far and the present experimental status is [8, 9, 10]:

$$\begin{aligned} BR(K^+ \to \pi^+\nu\bar{\nu})_{exp} &= \left(17.3^{+11.5}_{-10.5}\right) \times 10^{-11} \\ BR(K_L \to \pi^0\nu\bar{\nu})_{exp} &< 2.6 \times 10^{-8} \;\; 90\% \; CL. \end{aligned}$$

2. The NA62 apparatus for $K^+ \to \pi^+\nu\bar{\nu}$

A measurement of the BR of the $K^+ \to \pi^+\nu\bar{\nu}$ decay with 10% precision at least is the main goal of the NA62 experiment at CERN [11, 12]. The experiment plans to collect about 10^{13} kaon decays in few years using 400 GeV/c protons from SPS. As a consequence 10% signal acceptance and order of 10% signal to background ratio are required to attain the design precision [14]. The proton energy and the above requests both on kaon statistics and signal acceptance force the use of a kaon decay in flight technique. The 12 orders of magnitude of background rejection requires the use of almost independent experimental techniques to suppress unwanted final states.

Figure 1 shows the NA62 apparatus for $K^+ \to \pi^+\nu\bar{\nu}$ used to take data since 2014. Protons impinge on a Be target producing secondary charged particles, 6% of which are kaons. A 100 m long beam line selects, collimates, focuses and transports charged particles of 75 ± 0.8 GeV/c momentum. A Cerenkov counter (KTAG) filled with N_2 gas along the beam line identifies and timestamps kaons. Three silicon pixel stations (Gigatracker) of 6×3 cm^2 surface trace and timestamp all the beam particles. The Gigatracker faces the full 750 MHz beam rate, the KTAG the rate from kaons only. A guard ring detector (CHANTI) tags possible hadronic interactions of beam particles at the entrance of a 120 m long evacuated tank. The tank has a transverse

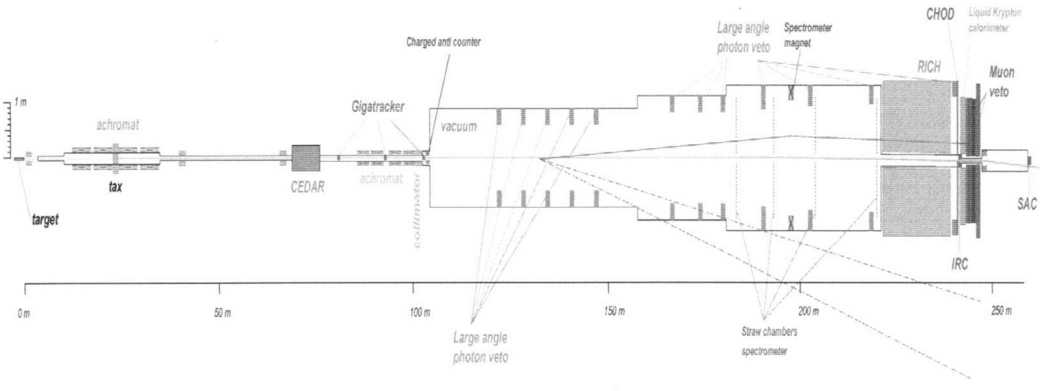

Figure 1. Scheme of the NA62 layout for $K^+ \to \pi^+ \nu\bar{\nu}$.

section with circular shape of radius varying from 1 to 1.4 m. About 10% of kaons decay in the first 80 m. The tank hosts 11 of 12 large angle annular electromagnetic calorimeters (LAV) made of lead glass blocks, surrounding the decay and downstream volumes to catch γs' emitted from kaon decays with angles up to 50 mrad with respect to the beam axis. A magnetic spectrometer made of straw tube arrays grouped in four chambers and located in vacuum along the last 40 m of the evacuated tank, traces charged particles. Holes of about 6.5 cm radius around the beam axis in the chambers let the undecayed beam particles passing through. An aluminum window closes off the tank just downstream of the last chamber of the spectrometer. The beam particles continue traveling in vacuum along the axis trough a 9cm radius beam pipe attached to the aluminum window and traversing the detectors downstream. A 17 m long RICH counter filled with Ne gas separates π^+, μ^+ and e^+ up to 40 GeV/c. The time of charged particles is measured both with RICH and with scintillator arrays (CHOD) placed downstream to the RICH. The liquid Krypton electromagnetic calorimeter (LKr) inherited from the NA48 CERN experiment [13] detects forward γs' and complements the RICH in particle identification. A shashlik small-angle annular calorimeter (IRC) in front of LKr detects γs' directed on the inner edges of the LKr hole around the beam axis. An hadronic calorimeter made of two modules of iron-scintillator sandwiches (MUV1 and MUV2) provides further π^+-μ^+ separation. A fast scintillator array (MUV3) identifies and triggers muons with sub-nanosecond time resolution. A shashlik calorimeter (SAC) placed on the beam axis downstream of a dipole magnet bending off-axis the rest of beam particles, detects γs' down to zero angle. The downstream detectors are illuminated by particles produced in kaon decays and upstream beam activity, corresponding to less than 1% of the beam rate. A multi-level trigger architecture is used. Timing information from CHOD, RICH, MUV3 and calorimetric variables from electromagnetic and hadronic calorimeters are build up on FPGAs' memories mounted on the readout TEL62 boards [15] to issue level zero trigger conditions. Software-based variables from KTAG, CHOD, LAV and magnetic spectrometer provide higher level trigger requirements.

NA62 has commissioned the beam line, detectors, trigger and data acquisition systems in 2014 and 2015. The Gigatracker operated in 2015 with a partial hardware configuration and has been fully commissioned n 2016. Low intensity data using a minimum bias trigger configuration have been collected in 2015 both for detector performance studies and physics analysis. NA62 has taken data in 2016 for physics at intensities up to 50% of the nominal one.

In the following sections, results from the low intensity 2015 sample to exploit the data quality in view of the $K^+ \to \pi^+ \nu\bar{\nu}$ measurement is described. Physics analysis on both 2015 and 2016 data are on-going.

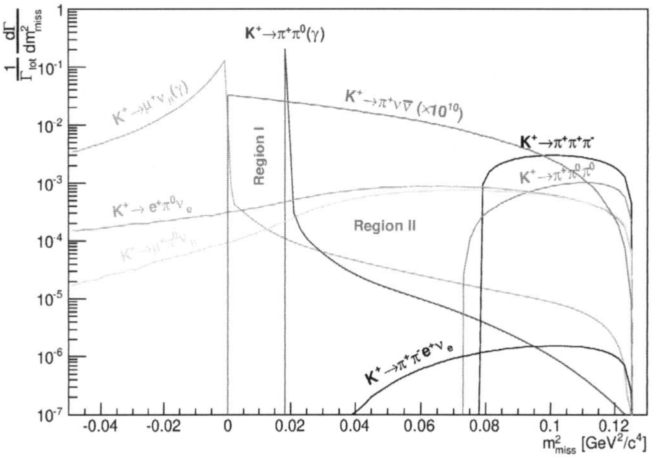

Figure 2. Theoretical m^2_{miss} distribution for signal and backgrounds from the main K^+ decay modes: the backgrounds are normalized according to their branching ratio; the signal is multiplied by a factor 10^{10}.

3. Principles of the Measurement

The signature of the signal is one track in Gigatracker compatible with a kaon hypothesis and one in the detector downstream. Kaon decays and beam–related activity are sources of background. Let P_K and P_{π^+} the 4-momenta of the kaon and the charged particles produced from kaon decay under the π^+ mass hypothesis, respectively. The squared missing mass distribution of the signal, $m^2_{miss} = (P_K - P_{\pi^+})^2$, has a three–body decay shape, while more than 90% of the charged kaon decays are mostly peaking, as shown in Figure 2. Signal is looked for in two signal regions around the $K^+ \to \pi^+\pi^0$ peak. Semileptonc decays, radiative processes, main kaon decay modes via reconstruction tails and beam induced tracks span across these regions. Therefore kinematic reconstruction, photon rejection, particle identification and sub-nanoseconds timing coincidences between subdetectors must be employed to get the final background rejection. A tight requirement on P_{π^+} between 15 and 35 GeV/c boosts the background suppression further, as will be shown in the next section. Monte Carlo studies performed in past years [14] have shown that NA62 can reach the goal, exploiting multiple and almost uncorrelated techniques to suppress the main background sources.

4. 2015 data quality analysis and prospects for the $K^+ \to \pi^+\nu\bar{\nu}$ measurement

The 2015 data offer the possibility to verify the detector performances in terms of timing, kinematic resolution, particle identification and γ rejection. To this purpose a single track selection has been set up as a preliminary step towards the $K^+ \to \pi^+\nu\bar{\nu}$ measurement.

Tracks reconstructed in the straw spectrometer matching in space energy depositions in calorimeters and signals in CHOD are selected. Matched CHOD signals define the time of the tracks with 200 ps resolution. A track not forming a common vertex within the decay region with any other in-time track defines a single track event. The last Gigatracker station and the first plane of the straw spectrometer bound the decay region. A vertex is defined as the average position of two tracks projected back in the decay region at the distance of closest (CDA) approach less than 1.5 cm. In order to select a single track event originated from kaon decays, a Gigatracker track is required to match the downstream track both in time and space,

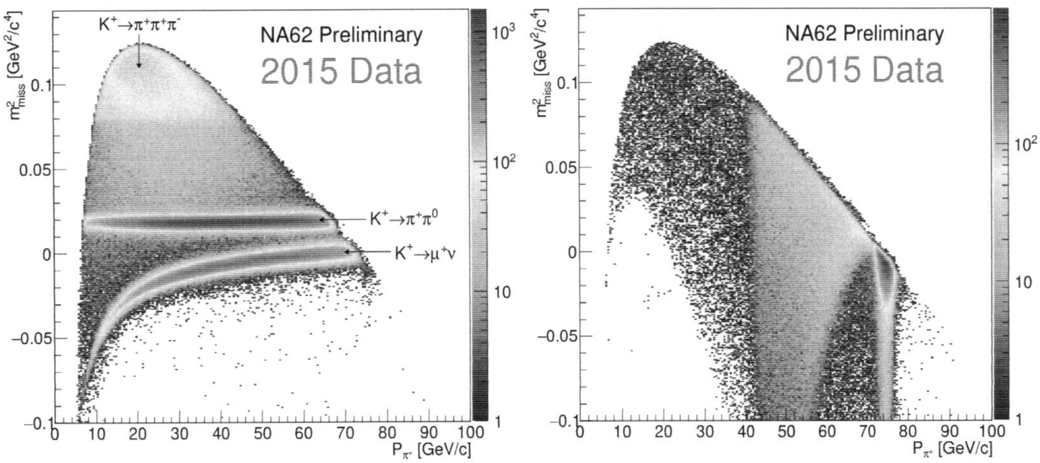

Figure 3. m^2_{miss} distribution under π^+ mass hypothesis as a function of the momentum of the track measured in the straw spectrometer after selection for single track from kaon decays (left). Same distribution as left-side picture, but asking for single track without a positive kaon tag in time in KTAG (right).

forming with it a vertex in the decay region, and to be in-time also with a kaon-like signal in KTAG. Figure 3 (left) shows the m^2_{miss} versus the spectrometer track momentum for 2015 data recorded at low intensity. The time resolutions of the KTAG and of a Gigatracker track has been measured in the range of 100 and 200 ps, respectively, matching the design values. The KTAG is also used in anti-coincidence with a Gigatracker track to select single track events not related to kaons (Figure 3 right). This technique shows that decay from beam π^+, elastic scattering of beam particles in the material along the beam line (KTAG and Gigatracker stations) and inelastic scatterings in the last Gigatracker station are the main sources of tracks downstream originated from beam−related activity. The sample of single track events from kaon decays selected above is used to study kinematic resolution, particle identification and γ rejection.

The resolution of the m^2_{miss} measured from the width of the $K^+ \to \pi^+\pi^0$ peak is in the range of 1.2×10^{-3} GeV$^2/c^4$, close to the 10^{-3} GeV$^2/c^4$ design value. The resolution is a factor 3 larger if the nominal kaon momentum is taken, instead of the event by event Gigatracker measured value. The tracking system of NA62 is also designed to provide a rejection factor in the range of $10^4 \div 10^5$ for $K^+ \to \pi^+\pi^0$ and $K^+ \to \mu^+\nu$ using m^2_{miss} to separate signal from backgrounds, respectively. The $K^+ \to \pi^+\pi^0$ kinematic suppression is measured using a sub-sample of single track events from kaon decays selected by requiring the additional presence of two γs' compatible with a π^0 in the LKr calorimeter. This constraint defines a sample of $K^+ \to \pi^+\pi^0$ with negligible background even in the signal m^2_{miss} regions, allowing the study of the far tails of the m^2_{miss}. The measured $K^+ \to \pi^+\pi^0$ suppression factor is of the order of 10^3. The partial hardware Gigatracker arrangement used in 2015 mainly limits the suppression because of m^2_{miss} tails due to beam track mis-reconstruction.

The particle identification of NA62 is designed to separate π^+ from μ^+ and e^+ in order to guarantee at least 7 order of magnitude suppression of $K^+ \to \mu^+\nu$ in addition to the kinematic rejection. RICH and calorimeters are employed together to this purpose. The sample of $K^+ \to \pi^+\pi^0$ used for kinematic studies and a sample of $K^+ \to \mu^+\nu$ selected among single track events from kaon decays by asking for the presence of signals in MUV3, are used to study the π^+-μ^+ separation in the RICH. About 10^2 muon suppression for 80% π^+ efficiency is measured in a momentum region between 15 and 35 GeV/c. Above 35 GeV/c the separation degrades quickly as expected from the Cerenkov threshold curves for π^+ and μ^+ in Neon. The RICH provides also an even better separation between π^+ and e^+. The same π^+ and μ^+ samples allow the calorimetric muon-pion separation to be investigated. Simple cut and count analysis provide a muon suppression factor within $10^4 \div 10^6$ for a π^+ efficiency in a $90\% \div 50\%$ range. Several analysis techniques are under study to optimise the separation.

NA62 is designed to suppress $K^+ \to \pi^+\pi^0$ by 8 orders of magnitude by detecting at least one γ from π^0 decay in one of the electromagnetic calorimeters (LAV, LKr, IRC, SAC) spanning an angular region up to 50 mrad. The π^0 suppression profits from the angle-energy correlation of the two π^0 γs' and the cut at 35 GeV/c maximum π^+ momentum at analysis level. This allows the above suppression to be reached even in presence of a single γ detection inefficiency of 10^{-5} for γ above 10 GeV and even higher for γ of lower energy. The suppression of π^0 from $K^+ \to \pi^+\pi^0$ is measured on data looking at the scaling of the m^2_{miss} π^0 peak after applying conditions for γ rejection to the single track events from kaon decays selected above. The measured π^0 veto inefficiency on the 2015 data is statistically limited at 10^{-6} (90% CL) as an upper limit. The corresponding signal efficiency is above 90%, the losses being mainly due to π^+ interactions in the RICH material producing extra clusters in LKr.

To conclude, the preliminary analysis of the low intensity 2015 data shows that NA62 is approaching the design sensitivity for measuring $K^+ \to \pi^+\nu\bar{\nu}$.

5. NA62 Physics Program beyond $K^+ \to \pi^+\nu\bar{\nu}$

The performances of the apparatus allow physics opportunities beyond the $K^+ \to \pi^+\nu\bar{\nu}$ to be addressed. NA62 can significantly improve the existing limits on lepton flavour and number violating decays like $K^+ \to \pi^+\mu^\pm e^\mp$ or $K^+ \to \pi^- l^+ l^+$. Experimentally π^0 physics can take advantage of the performances of the electromagnetic calorimeters and processes like $\pi^0 \to$ *invisible* or dark photon production can be investigated. Thanks to the quality of the kinematic reconstruction, searches for heavy neutrino produced in $K^+ \to l^+\nu$ decays can improve the present sensitivity. The longitudinal scale of the apparatus opens the possibility to search for long−living particles through their decays, like dark photon, heavy neutral leptons or axion-like particles produced at the target or in beam dump configurations. NA62 is already addressing part of the above physics program simultaneously with the $K^+ \to \pi^+\nu\bar{\nu}$ program. The full exploitation of this physics will constitute the core of the NA62 program beyond 2018.

References

[1] Brod J, Gorbahn M and Stamou E 2011 *Phys. Rev.* D **83** 034030.
[2] Buras AJ, Buttazzo D, Girrbach-Noe J and Knegjens R, 2015 *JHEP* **1511** 33.
[3] Blanke M, Buras AJ and Recksiegel S 2016 *Eur. Phys. J* C **76 no.4** 182.
[4] Blanke M, Buras AJ, Duiling B, Gemmler K and Gori S 2009 *JHEP* **903** 108.
[5] Buras AJ, Buttazzo D and Knegjens R 2015 *JHEP* **1511** 166.
[6] Isidori G, Mescia F, Paradisi P, Smith C and Trine S 2006 *JHEP* **0608** 64.
[7] Tanimoto M and Yamamoto K 2015 *PTEP* **2015 no.5** 053B07.
[8] Artamonov AV et al. (E949 Collaboration) 2008 *Phys. Rev. Lett.* **101** 191802.
[9] Artamonov AV et al. (E949 Collaboration) 2009 *Phys. Rev.* D **79** 092004.
[10] Ahn JK et al. (E391a Collaboration) 2010 *Phys. Rev.* D **81** 072004.
[11] Anelli A et al., CERN-SPSC-2005-013; SPSC-P-326.
[12] NA62 Technical Design Document NA62-10-07; https://cdsweb.cern.ch/record/14049857.
[13] Fanti V et al. (NA48 Collaboration) 2007 *Nucl. Instrum. Methods* A **574** 433.
[14] Ruggiero G (NA62 Collaboration) *PoS KAON* **13** 032 (2013).
[15] Angelucci B et al. 2012 *JINST* **7** C02046.

Loophole in $K \to \pi\nu\bar{\nu}$ Search & $K_L \to \pi^0\nu\bar{\nu}$ Beyond Grossman–Nir Bound

George W.S. Hou

Department of Physics, National Taiwan University, Taipei 10617, Taiwan

E-mail: `wshou@phys.ntu.edu.tw`

Abstract. The Grossman-Nir bound concept is robust, but the common perception of $K_L \to \pi^0\nu\bar{\nu} < 1.4 \times 10^{-9}$ may be circumvented. Because of the blinding $K\pi2(3)$ decay, K^+ decay experiments such as E787/E949 and the currently running NA62 kinematically exclude the "blinding zone". This offers an opportunity for the currently running KOTO experiment. With no kinematic control, if the KOTO experiment is able to veto all true background, it may discover $K_L \to \pi^0 + X^0$, where X^0 is an invisible "dark" boson with mass $m_{X^0} \sim m_{\pi^0}$. This could happen with 2015 data at hand, which supposedly should allow KOTO to reach nominal GN bound sensitivity. An explicit model of gauged $L_\mu - L_\tau$, linked with muon $g-2$ hence with very light Z', is given. Adding vector-like quarks that mix with usual quarks, W-boson loops induce $s \to dZ'$ and $b \to sZ'$ transitions. Besides realizing the scenario, the model provides further illustration of potential impact. Given that usual dark bosons linked with the photon are tightly constrained, KOTO has a unique opportunity to probe the muon-related dark sector.

1. The Kaon, as old as Particle Physics

If we count "V particles" as the start, kaon physics is 69 years and counting, coinciding also in time with Schwinger's calculation of $g - 2 = \alpha/\pi$, the first "penguin". The present day Standard Model (SM) was pretty much built up by learning from kaons, including the term "penguin diagram". But with NA62 at CERN and KOTO at J-PARC, Japan pursuing the measurements of $K^+ \to \pi^+\nu\bar{\nu}$ and $K_L \to \pi^0\nu\bar{\nu}$, respectively, kaons remain at the New Physics (NP) frontier.

I felt frustrated at the FPCP2016 conference held at Caltech in June, as the KOTO talk still referred to the Grossman-Nir (GN) bound [1] as if it were a firm ceiling. After relative quietness at ICHEP2016, I was gratified to learn, early in this conference, that the KOTO paper based on 2013 data ("100 hours") had appeared, with title

A new search for the $K_L \to \pi^0\nu\bar{\nu}$ and $K_L \to \pi^0 + X^0$ decays [2]

that acknowledges the possibility of $K_L \to \pi^0 + X^0$ decay [3], where X^0 is an invisible particle with $m_{X^0} \sim m_{\pi^0}$. The new 90% C.L. limit of $\mathcal{B}(K_L \to \pi^0\nu\bar{\nu}) < 5.1 \times 10^{-8}$ is worse than the existing bound from E391A [4] due to one observed event versus 0.34±0.16 expected background. The bound $\mathcal{B}(K_L \to \pi^0+X^0) < 3.7\times 10^{-8}$ fares better, indicating that KOTO finds the observed event less consistent with this two-body possibility.

The GN bound comes in two forms:

$$\mathcal{B}(K_L \to \pi^0\nu\bar{\nu}) < 4.3 \times \mathcal{B}(K^+ \to \pi^+\nu\bar{\nu}) \quad (1)$$

$$< 1.4 \times 10^{-9}. \quad (\text{"GN bound"}) \quad (2)$$

Content from this work may be used under the terms of the Creative Commons Attribution 3.0 licence. Any further distribution of this work must maintain attribution to the author(s) and the title of the work, journal citation and DOI.
Published under licence by IOP Publishing Ltd

The factor of 4.3 in Eq. (1) arises mostly from τ_{K_L}/τ_{K^+} [1] and isospin, and is basically sound. However, the commonly *perceived* "GN bound" is the numerical value given in Eq. (2), which follows by inserting the E787/E949 value [5] for $\mathcal{B}(K^+ \to \pi^+\nu\bar{\nu})$.[1] In this talk, we elucidate why the "GN bound" of Eq. (2) is not fool-proof. Since the 2015 data should [2] improve sensitivity by a factor of 20 or more, and should approach the "GN bound", contrary to common perception, there is discovery potential with data at hand.

2. Blinding $[K^+ \to \pi^+]\pi^0$: Blessed are the Blind

We, or at least I, have stared at the E949 plot so many times, that one becomes numb. This final plot of E787/E949 [5] shows two signal boxes, and in between the two, many red circles or blue triangles representing the blinding $K^+ \to \pi^+\pi^0$ decay. The $K\pi2$ branching ratio of 21% is 100 million times or more stronger than the target of $K^+ \to \pi^+ +$ *nothing*. When faced with blinding light, one would turn the head away, close one's eyes, and lift the arms to block it. The action corresponding to this figure of speech is *kinematic control*: use K^+ and π^+ momenta to block out the blinding zone. Though E787/E949 use stopped K^+ while NA62 exploits decay in-flight, kinematic exclusion leads to rather similar signal boxes.

E949 was not unaware of an issue at hand, and pursued [5] the measurement of $\mathcal{B}(K^+ \to \pi^+X)$ where X is an invisible particle. Nevertheless, this blinding spot lead to a bound that is worse than $\mathcal{B}(K^+ \to \pi^+\nu\bar{\nu})$ by a factor of 200 for $m_{X^0} \sim m_{\pi^0}$, corresponding to an earlier, relatively unimpressive bound [6] of $\mathcal{B}(\pi^0 \to \nu\bar{\nu}) < 2.7 \times 10^{-7}$. Thus, if Nature is so inclined to throw us a "dark" X^0 boson with π^0-like mass, then E787/E949 and NA62 would be caught unaware, because of the use of kinematic control. There is a second "window" due to $K\pi3$ decays. But somehow, through numerous proposal reviews, the notion of this "loophole" was not carried over to the K_L side, and the mindset was fixed on the pursuit of $K \to \pi\nu\nu$.

Our general observation [3] came as a surprise, as it is almost trivial: "Blessed are the Blind." It's been said that $K_L \to \pi^0\nu\bar{\nu}$ measurement is like searching for "Nothing to Nothing" (just $\gamma\gamma$). Not only the incoming K_L momentum is not known, one detects just two γs that are consistent with a π^0, but there is no ID to give the true π^0 energy-momentum. With such lack of kinematic control, the strategy is to veto everything. But clearly one cannot veto WILPs (Weakly Interacting Light Particles). Thus, if Nature throws in X^0s through the above two blinding spots, KOTO could luck out!

Put simply: the KOTO experiment at J-PARC could discover $K_L \to \pi^0X^0$ above the "GN bound" of Eq. (2). Given that 2015 data is supposedly able to bring sensitivity down to this level, this means discovery could start "now", at any value below the E391A limit [4].

So, what you learned at "school" was not quite right. But seeing this possibility came quite inadvertently for us, for our starting point was top changing neutral current (tCNC) search at the LHC, and this is a story of LHC-related work leading to interesting kaon physics. Let us therefore turn to a motivated model as an existence proof.

3. A Motivated Model (existence proof)

The explicit model is gauged $L_\mu - L_\tau$ related to muon $g-2$, plus extra vector-like quarks.

Gauging the difference between muon and tauon numbers is attractive in several ways. The difference renders it anomaly free [7] with SM particle content, while avoiding the electron (two other $L_e - L_i$ models) allows it to be less constrained, and the coupling with muon connects it potentially with muon $g-2$. The U(1) is broken by a singlet scalar $\langle\Phi\rangle = v_\phi/\sqrt{2}$, giving the Z' mass $m_{Z'} = g'v_\phi$, where g' is the gauge coupling. To link with quarks, one could add vector-like quarks Q, D, U that couple to the Z' boson, where Q stands for left- and right-handed doublets,

[1] Thus, if the upper bound for $\mathcal{B}(K^+ \to \pi^+\nu\bar{\nu})$ drops, so would the number in Eq. (2).

while D and U are left-right singlets. Q, D and U would have gauge invariant mass terms, but these exotic quarks mix with SM quarks through Yukawa couplings to the exotic ϕ field.

As stated, our original interest [3] was rare $t \to cZ'$ top decays, and we took interest in the model of Altmannshofer *et al.* [8] (AGPY). It is well known that a heavy Z' boson could account for the so-called P_5' anomaly uncovered by the LHCb experiment in angular analysis of $B \to K^*\mu^+\mu^-$ decays, where a 3.7σ deviation from SM expectation is seen for 1 fb^{-1} data. While the discrepancy persists for 3 fb^{-1} data, the significance unfortunately did not improve [9], hence this anomaly needs confirmation with more data at LHC Run 2. Nevertheless, invoking the gauged $L_\mu - L_\tau$ force on the muon side, AGPY constructed [8] the model as described above, where v_Φ that generates $m_{Z'}$ induces also the desired bsZ' couplings at tree level through Q and D quarks. We took note [3, 10] that the U quark induces $t_R \to c_R Z'$ transition (see the effective insertion inside the loop of Fig. 1) that is not linked with the P_5' anomaly, hence remain unconstrained. For the gauged $L_\mu - L_\tau$ interactions, $Z' \to \mu^+\mu^-$ decay typically occurs with 1/3 branching fraction, much higher than the Z boson, hence the ATLAS and CMS experiments should pursue $t \to cZ' \to c\mu^+\mu^-$ search.

Having invoked gauged $L_\mu - L_\tau$, it would be great if the model could explain P_5' and muon $g-2$ anomalies simultaneously. But checking constraints, (probably) to their surprise, AGPY discovered [11] that a not so well known process, called "neutrino trident" production, or $\nu_\mu N \to \nu_\mu \mu^+\mu^- N$, constrains the Z' to be rather light,

$$m_{Z'} < 400 \text{ MeV}. \tag{3}$$

This precludes the Z' to be simultaneously responsible for the muon $g-2$ and P_5' anomalies, as a heavy Z' is implied for the latter. But it is intriguing to think that *New Physics* behind the muon $g-2$ anomaly could arise from a *light particle*! As is well known from dark boson studies, this is possible only because Eq. (3) implies the gauge coupling $g' < 10^{-3}$, far weaker than the weak coupling (but the Z' is not Dark Matter, as it decays quite fast). However, being lighter than the kaon inspired us to study the impact of this Z' on kaon and B physics. Our original purpose was to see if this could rule out the model.

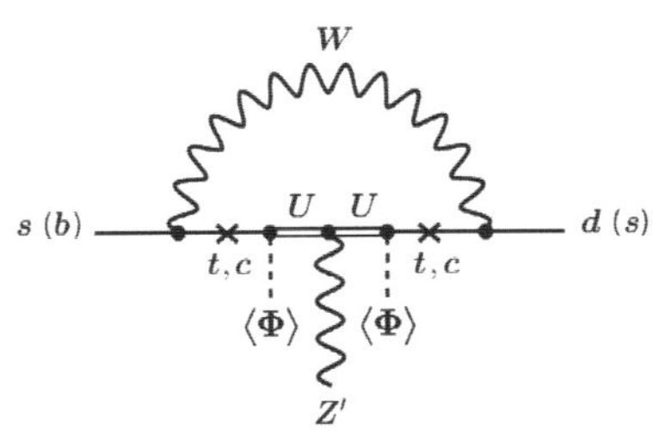

Figure 1. Effective dsZ' (sbZ') coupling, with Z' coupled to a vector-like U quark that mixes with c, t ("×" flips chirality) and connect with external d-type quarks via a W boson loop.

For the Z' related to muon $g-2$, its lightness and the very precise measurements in B and K sectors make it rather "precarious". To avoid fine-tuning, one could decouple the Q and D quarks by discrete Z_2 charge assignments. But even with just the U quark, as illustrated in Fig. 1, a SM W-boson loop around the effective ttZ' (t can be interchanged with c) coupling turns it into [3, 10] bsZ' and sdZ' couplings, with help of chirality flip from $m_{t(c)}$. The sdZ' effective coupling can precisely lead to $K \to \pi Z'$ decay, where the light Z' is a candidate for the X^0 that could lead to the surprise evasion of the GN bound of Eq. (2), if $m_{Z'} \sim m_{\pi^0}$. It was through this model and by comparing K^+ and K_L cases that we stumbled upon the observation outlined in previous Section. We note that chirality flip and CPV can both arise from SM, and straightforward formulas can be found in Ref. [10], including a more detailed discussion of $t \to cZ'$ for both heavy ("P_5'") and very light ("muon $g-2$") cases.

4. What Then? — an Illustration of Impact

As KOTO analyzes their 2015 data, what if they find $K_L \to \pi^0 +$ *nothing* above 1.4×10^{-9}? This is the nominal "GN bound" of Eq. (2), long perceived as impossible to breach, but could actually happen with 2015 data. We now go through possible hints and other possible rare B and K decays, using the model presented in Sec. 3 as an illustration of what might lie ahead. The discussion depends on the dimuon threshold. Below this threshold, one has $Z' \to \nu\bar{\nu}$ only, while above it one could pursue $\mu^+\mu^-$ bump search.

It is interesting that the BaBar experiment has in fact a mild hint for the analogous $B^+ \to K^+\nu\bar{\nu}$ decay. Note that $\mathcal{B}(B^+ \to K^+\pi^0) \ll \mathcal{B}(K^+ \to \pi^+\pi^0)$ is a rare decay itself at 10^{-5} level, therefore there is no analogue of a $B^+ \to K^+\pi^0$ induced "blinding spot". Besides, the detection environment is rather different. The BaBar experiment conducted [12] a binned $s_B \equiv m^2_{\nu\bar{\nu}}/m^2_B$ search for $B \to K^{(*)}\nu\bar{\nu}$, separating $q^2 \equiv m^2_{\nu\bar{\nu}}$, or missing mass-squared, into 10 bins. With 471M $B\bar{B}$ pairs, a *two-sided* 90% confidence interval was reported, with the lower bound driven by a mild excess in the lowest s_B bin.

For $m_{Z'} = 135$ MeV, i.e. in the π^0 mass window, hence $Z' \to \nu\bar{\nu}$ 100%, we plot in Fig. 2[left] the BaBar allowed region in the Y_{Ut}-Y_{Uc} plane, where Y_{Ui} (treated as real) are Yukawa couplings of Φ that mix the exotic U quark with quarks i in SM. We compare this with the E949 bound for $K^+ \to \pi^+ Z'$, the $K_L \to \pi^0\nu\bar{\nu}$ search bound from E391a, and the commonly perceived "GN bound" of Eq. (2). We see that the BaBar result could by itself be contentious with the latter, and pushing below the E391a bound could possibly lead to discovery.

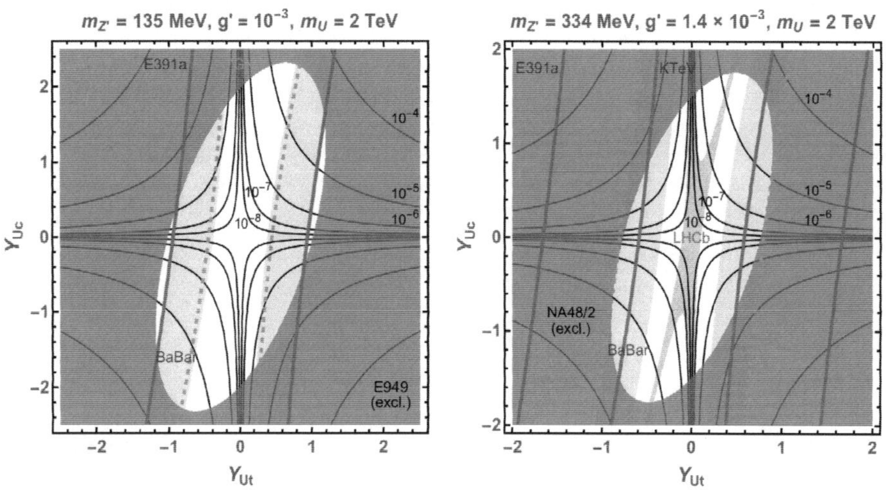

Figure 2. In Y_{Uc}-Y_{Ut} plane, for: [Left] $m_{Z'} = 135$ MeV ($Z' \to \nu\bar{\nu}$ at 100%), with $\mathcal{B}(K^+ \to \pi^+ Z') < 5.6 \times 10^{-8}$ exclusion (E949, dark grey region); [right] $m_{Z'} = 334$ MeV ($Z' \to \mu^+\mu^-$ at 1/3), with 95% C.L. allowed by $B^0 \to K^{*0}\chi(\to \mu^+\mu^-)$ (LHCb, pink region), 2σ exclusion by $K^+ \to \pi^+\mu^+\mu^-$ (NA48/2, dark grey region), and 90% C.L. exclusion by $K_L \to \pi^0\mu^+\mu$ (KTeV, solid purple line). Common to both are the 2σ allowed range for $\mathcal{B}(B^+ \to K^+Z')\mathcal{B}(Z' \to \nu\bar{\nu}) = (0.35^{+0.6}_{-0.15}) \times 10^{-5}$ (BaBar, light green region), $\mathcal{B}(K_L \to \pi^0 Z') < 2.6 \times 10^{-8}$ (E391a, blue solid line), and the usual "GN bound" of $\mathcal{B}(K_L \to \pi^0 Z')\mathcal{B}(Z' \to \nu\bar{\nu}) < 1.4 \times 10^{-9}$ (orange dashed line). We plot $\mathcal{B}(t \to cZ')$ contours in the backdrop.

So what about Belle? Belle pioneered the $B^+ \to K^+\nu\bar{\nu}$ search, but its followup [13] study just added 40% data without changing analysis strategy, where a cut on high p_{K^+} for sake of rejecting $B \to K^*\gamma$ background precisely cuts against the $B \to K^{(*)}Z'$ possibility. With existing full dataset, Belle should follow the path as done by BaBar, i.e. the binned $m^2_{\rm mis.}$ analysis, at least to crosscheck the BaBar result. If some excess is also found, then the B factories should attempt a combination. This should be followed up at Belle II, whether KOTO makes a discovery or not. Competition should be welcome.

There is a second kinematic exclusion window for $K^+ \to \pi^+\nu\bar{\nu}$ search, $m_{\rm mis.} > 260$ MeV, due to $K^+ \to \pi^+\pi\pi$ background. In Ref. [3], we had used the $m_{Z'} = 285$ MeV case to make estimates of what is allowed by a not so constraining NA48/2 study [14] of $K^+ \to \pi^+\mu^+\mu^-$, and we alluded to [3] an apparent mild excess at lowest q^2 above the mean over $1.0\ \text{GeV}^2 < q^2 < 6.0\ \text{GeV}^2$ in the $B^0 \to K^0\mu^+\mu^-$ result [15] of LHCb. Both cases have now been superseded, however, by the recent dark boson bump search in $B^0 \to K^{*0}\mu^+\mu^-$ decay by LHCb [16], and one is at best left with "pocket" physics, at the mercy of apparent statistical fluctuations. We give two examples, one near $m_{Z'} \sim 2m_\mu$ (below which efficiency vanishes), another at $m_{Z'} \sim 334$ MeV, as can be read off the supplementary material to Ref. [16]. Note that the Z' decays rather fast in the $L_\mu - L_\tau$ model, even for this light range where the associated coupling $g' \sim 10^{-3}$ is rather weak. For the three cases we mention, the lifetimes are at sub-fs level [10], hence one takes the $\tau = 0$ ps numbers from Ref. [16].

For $m_{Z'} = 219$ MeV, just above the dimuon threshold, though LHCb limit is weakest, the parameter space is still tightly constrained by E949 [5]. One finds $t \to cZ'$ must be less than 10^{-6} and typically much smaller, and the case is overall less interesting. Nevertheless, it should still be scrutinized by LHCb (and future Belle II) with more data, as one should keep in mind the HyperCP dimuon events [17].

Around $m_{Z'} = 334$ MeV, LHCb is again relatively tolerant, and it is outside the signal box for E949 (and NA62). We show the various constraints in Fig. 2[right]. The BaBar band (light green), now weaker than the "GN bound" of Eq. (2), is at some tension with the LHCb "allowed" region (pink), while NA48/2 [14] result on $K^+ \to \pi^+\mu^+\mu^-$, interpreted by us as a dimuon bump search, replaces E949 as a constraint. A search by KTeV for $K_L \to \pi^0\mu^+\mu^-$ [18] provides a different constraint. For this LHCb "pocket", $t \to cZ'$ could be enhanced beyond 10^{-5} in the funnel region, and would be rather interesting (if one believes in "pockets"). We have plotted the ($t \to cZ'$) decay contours in the backdrop of Fig. 2. More detailed discussion of $t \to cZ'$ decays can be found in Ref. [10], including the P_5'-motivated heavy Z' case. In general, the rather low branching ratios would probably need a 100 TeV pp collider to probe.

BaBar has recently searched [19] for muonic dark force, i.e. precisely the $L_\mu - L_\tau$ model, by looking for a Z' emitted from $e^+e^- \to \mu^+\mu^-$ which subsequently decays via $Z' \to \mu^+\mu^-$. They "exclude all but a sliver ... above dimuon threshold", i.e. some pockets are allowed below 400 MeV (and above as well). But the BaBar pockets do not seem to overlap with those of LHCb. However, note that this, as well as the tension between LHCb and the mild hint for $B \to K^{(*)}Z'(\to \nu\bar{\nu})$ in Fig. 2[right] might be reconciled by enriching Z' properties, such as coupling to dark sector that changes the branching ratios.

Our highlight, especially in light of null results for dimuon bump search from LHCb and BaBar, is the $m_{Z'} \sim m_{\pi^0}$ window (Fig. 2[left]), where opportunity lies for KOTO and Belle (II), both with existing data, and perhaps NA62, if the $\pi^0 \to$ *nothing* bound [6] could be improved. We remark that if KOTO reaches "GN bound" of Eq. (2) without discovery, our point [3] remains valid. That is, as NA62 (perhaps) improves the upper bound on $K^+ \to \pi^+\nu\bar{\nu}$, KOTO could still make discovery in the π^0 mass window, above the value implied by Eq. (1).

A very recent result by NA64 [20] claims strong exclusion of sub-GeV dark photons. We offer a few remarks. Dark "photon" invokes the idea of kinetic mixing with the photon, hence coupling to electrons. In this context, note that a small region remains near the π^0 mass window due to the "blinding spot" we discussed. But in the model we use to illustrate, viz. gauged $L_\mu - L_\tau$, the loop induced mixing of Z' with photon is much below NA64 sensitivity. This is because NA64 is an electron beam-dump experiment. A more challenging muon-beam experiment [21] needs to be explored to cover $L_\mu - L_\tau$ type of dark bosons.

5. Conclusion

$K_L \to \pi^0 +$ *nothing* can occur above "Grossman-Nir Bound" of 1.4×10^{-9}, and the KOTO experiment is in the position to make possible discovery with 2015 data at hand. If KOTO sees early events, of course they should dutifully try hard to reject them; but don't overkill — don't kill the baby. If one firmly finds events above the "Grossman-Nir Bound", then likely one has a dark object of π^0 mass, which could thereby slip through NA62 unnoticed. Even when KOTO reaches below 1.4×10^{-9}, the concept is still fertile, and one should not assume the new GN bound derived from $K^+ \to \pi^+\nu\bar{\nu}$ upper limits. This should keep KOTO and Belle (II) in the game for a while, and NA62 should try to improve the limit on $\pi^0 \to$ *nothing.*

"Blessed are the Blind, for they shall see the Heavens open." surely sounds like the 9th Beatitudes that Jesus taught. Well, this is not Scriptural, but all the same. The "blinding" bright light may be emanating from an Angel, or, as it is said "even the Devil may masquerades himself as an Angel of Light ...". But, no matter, for us mortals, we are just dying to see some new "light" from Heaven, and the Devil is nothing but a fallen Angel, which still offers a link.

So, please: Analyze, KOTO, Analyze!

Acknowledgments

We thank K. Fuyuto and M. Kohda for collaborative work, and P. Crivelli, Y.B. Hsiung, M. Kohda and M. Pospelov for discussions.

Reference

[1] Y. Grossman and Y. Nir, Phys. Lett. B **398**, 163 (1997).
[2] J.K. Ahn *et al.* [KOTO Collab.], arXiv:1609.03637 [hep-ex]; see also talk by K. Shiomi, this proceedings.
[3] K. Fuyuto, W.-S. Hou, M. Kohda, Phys. Rev. Lett. **114**, 171802 (2015).
[4] J.K. Ahn *et al.* [E391a Collab.], Phys. Rev. D **81**, 072004 (2010).
[5] A.V. Artamonov *et al.* [E949 Collab.], Phys. Rev. D **79**, 092004 (2009).
[6] A.V. Artamonov *et al.* [E949 Collab.], Phys. Rev. D **72**, 091102 (2005).
[7] X.-G. He, G.C. Joshi, H. Lew, R.R. Volkas, Phys. Rev. D **43**, 22 (1991).
[8] W. Altmannshofer, S. Gori, M. Pospelov, I. Yavin, Phys. Rev. D **89**, 095033 (2014).
[9] R. Aaij *et al.* [LHCb Collab.], JHEP **1602** (2016) 104.
[10] K. Fuyuto, W.-S. Hou and M. Kohda, Phys. Rev. D **93**, 054021 (2016).
[11] W. Altmannshofer, S. Gori, M. Pospelov, I. Yavin, Phys. Rev. Lett. **113**, 091801 (2014).
[12] J.P. Lees *et al.* [BaBar Collab.], Phys. Rev. D **87**, 112005 (2013).
[13] O. Lutz, S. Neubauer, M. Heck, T. Kuhr, A. Zupanc *et al.* [Belle Collab.], Phys. Rev. D **87**, 111103 (2013).
[14] J.R. Batley *et al.* [NA48/2 Collab.], Phys. Lett. B **697**, 107 (2011).
[15] R. Aaij *et al.* [LHCb Collab.], JHEP **1406**, 133 (2014).
[16] R. Aaij *et al.* [LHCb Collab.], Phys. Rev. Lett. **115**, 161802 (2015).
[17] H. Park *et al.* [HyperCP Collab.], Phys. Rev. Lett. **94**, 021801 (2005).
[18] A. Alavi-Harati *et al.* [KTEV Collab.], Phys. Rev. Lett. **84**, 5279 (2000).
[19] J.P. Lees *et al.* [BaBar Collab.], Phys. Rev. D **94**, 011102 (2016).
[20] D. Banerjee *et al.* [NA64 Collab.], arXiv:1610.02988 [hep-ex].
[21] S.N. Gninenko, N.V. Krasnikov and V.A. Matveev, Phys. Rev. D **91**, 095015 (2015).

IOP Conf. Series: Journal of Physics: Conf. Series **800** (2017) 012025 doi:10.1088/1742-6596/800/1/012025

Reduction of neutron-induced background in KOTO

Kota Nakagiri for the KOTO collaboration

Department of Physics, Faculty of Science, Kyoto University Kitashirakawa Oiwake-cho, Sakyo-ku, Kyoto, 606-8502, Japan

E-mail: n-kota@scphys.kyoto-u.ac.jp

Abstract.
The KOTO experiment aims to study the $K_L \to \pi^0\nu\overline{\nu}$ decay at J-PARC. In order to identify the signal, we measure two photons from a π^0 decay with an electromagnetic calorimeter consisting of undoped CsI crystals, and ensure that there are no other particles with hermetic veto counters. In the analysis of data taken in 2013, the neutron-induced background, which was caused by a beam-halo neutron hitting the calorimeter was dominant. The neutron makes a primary hadronic shower and a secondary neutron emitted from the shower makes a secondary shower after traveling inside the calorimeter. If these two shower clusters are observed in the calorimeter without any hits in veto detectors, it can mimic a signal event. We developed new methods to suppress this background, such as neutron-photon discrimination methods using cluster shape and pulse shape. We obtained $(3.8^{+2.4}_{-1.1}) \times 10^4$ reduction power for the neutron-induced background.

1. Introduction

1.1. $K_L \to \pi^0\nu\overline{\nu}$ decay

The $K_L \to \pi^0\nu\overline{\nu}$ decay is a CP-violating rare decay. This decay is a flavor changing neutral current process and is caused via loops. The branching ratio of this decay is predicted to be $(3.00 \pm 0.30) \times 10^{-11}$ in the Standard Model [1] with the theoretical uncertainty of 1–2%. It enables us to observe effects from new physics beyond the Standard Model which enhance the branching ratio.

1.2. KOTO experiment

The KOTO experiment aims to study the $K_L \to \pi^0\nu\overline{\nu}$ decay at the Japan Proton Accelerator Research Complex (J-PARC). In the KOTO experiment, signal candidates are identified by "two photons and nothing else." We measure two photons from a π^0 decay with an electromagnetic calorimeter consisting of undoped CsI crystals, and ensure that there are no other particles with hermetic veto counters (Figure 1). We also use π^0 kinematical information, which is reconstructed from energy and position information of two photon clusters on the CsI calorimeter for the identification. We set the signal region in a scatter plot of π^0 transverse momentum (P_t) versus π^0 decay Z vertex.

In 2013, we performed the first physics run and achieved single event sensitivity of 1.29×10^{-8} with only four days of data taking [2], which is equivalent to the sensitivity achieved by the KEK E391a experiment [3]. One event was observed, and it was consistent with the estimated number of background events. We found the mechanism of "neutron-induced background", which is explained in section 2, was a dominant background in the analysis.

Content from this work may be used under the terms of the Creative Commons Attribution 3.0 licence. Any further distribution of this work must maintain attribution to the author(s) and the title of the work, journal citation and DOI.
Published under licence by IOP Publishing Ltd

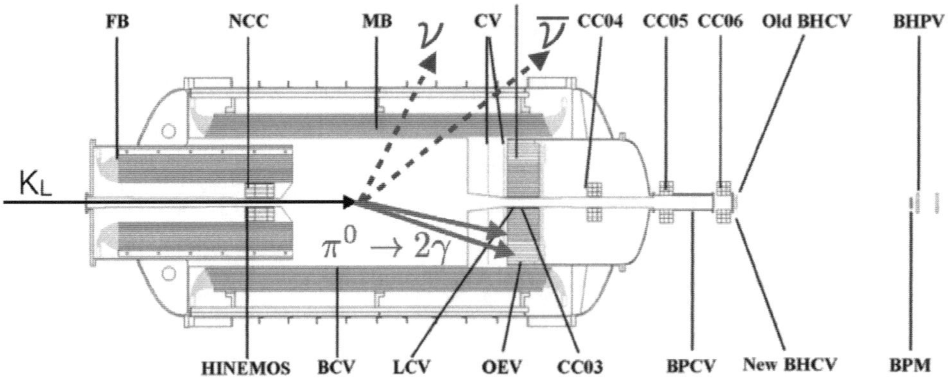

Figure 1. Schematic cross-sectional view of the KOTO detector.

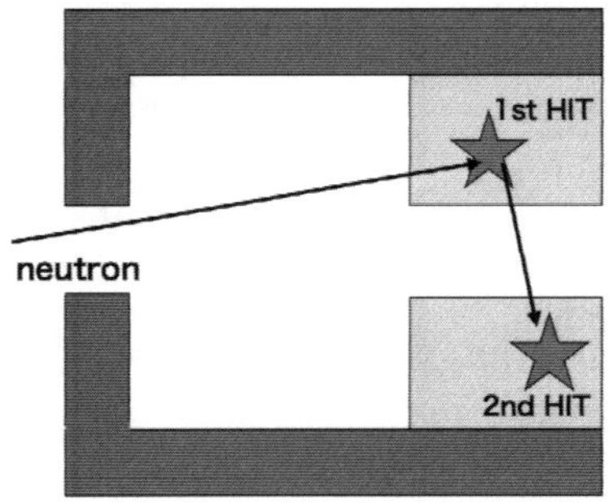

Figure 2. Mechanism of neutron-induced background. An incident primary neutron has hadronic interactions and makes a primary shower. An emitted secondary neutron from the shower makes another shower after traveling in the calorimeter.

We took data in 2015 and accumulated 20 times more data than that in the 2013 run. We also performed a special run to collect neutron samples, which is described in section 3.1.

2. Neutron-induced background

In the 2013 analysis, we found that beam-halo neutrons hitting the CsI calorimeter are the dominant source of our background. A beam-halo neutron hits the CsI calorimeter and forms a primary hadronic shower. In the interactions, a secondary neutron can be emitted and it produces a secondary shower after traveling inside the calorimeter (Figure 2). These two shower clusters can be observed without hits in any veto detectors. If such an event is reconstructed in the signal box, it will be a background. In 2013, the estimated number of this "neutron-induced background" was 0.18 out of the total expected background events of 0.34.

3. Methods to suppress neutron-induced background

We have developed several methods to suppress the neutron-induced background. The key of the background reduction is to distinguish a photon cluster from a neutron cluster using calorimeter hit information. The methods named "Shape χ^2," "Cluster Shape Discrimination," and "Pulse Shape Likelihood" that we developed will be explained in this paper. The first two use cluster shape information, and the last one uses waveform information from the hits in the calorimeter. The Shape χ^2 method has been used in the 2013 analysis, and the Cluster Shape Discrimination and the Pulse Shape Likelihood are newly developed after that.

3.1. Special run to collect neutron samples : Al target run

Evaluation of the neutron-induced background by means of simulation is difficult due to uncertainties in incident halo-neutron flux, energy, and their interactions. An "Al target run" was performed during the beam time to collect neutron-rich data as control samples. These data were used to evaluate the reduction power of neutron-induced background in each cut.

In the Al target run, a 10-mm-thick aluminum plate was inserted into the beam core at the upstream of the KOTO detector to scatter beam-core neutrons. Scattered neutrons hit the CsI calorimeter and made the events similar to the neutron-induced background.

3.2. Shape χ^2 cut

The dimensions of the cross section of our CsI crystals are 25 mm × 25 mm for this inner 1.2 × 1.2 m^2 region and 50 mm × 50 mm for the outer region, respectively. This small size of crystal gives us the information of two-dimensional shower shape. The Shape χ^2 method compares the observed energy deposit in each crystal in a cluster with its expected energy derived from a simulation. We qualify the validity of shower shape with the χ^2 defined as follows with an assumption that the cluster is made by a photon :

$$\chi^2 = \frac{1}{N} \sum_{i}^{27\times 27 \text{ region}} \left(\frac{e_i/E_{inc} - \mu_i}{\sigma_i} \right)^2, \tag{1}$$

where E_{inc} is the measured photon energy, e_i is the measured deposit energy in i-th crystal, μ_i and σ_i are the expected mean and standard deviation of e_i/E_{inc} respectively, obtained from a simulation with photon. The summation is taken over 27×27 crystals around the cluster center. We reject the event if a cluster has larger Shape χ^2 value than a pre-defined threshold. The background reduction of 300 is obtained with 80% signal acceptance, which is consistent with the 2013 analysis.

3.3. Cluster Shape Discrimination cut

Cluster Shape Discrimination also uses the cluster shape information, and uses a neural net based on several variables from a cluster shape. The main difference from the Shape χ^2 is that it uses the time difference among the crystals in a cluster. Figure 3 shows the energy and time difference distributions derived from a simulation. We can obtain expected time distribution in the cluster and compare it with data. Figure 4 shows distributions of the neural net output. The black histogram shows photon cluster samples and the red histogram shows neutron cluster samples. The background reduction of 1500 is obtained with 90% signal acceptance.

3.4. Pulse Shape Likelihood cut

Pulse Shape Likelihood method uses pulse shape information. The waveforms of all channels of the calorimeter were recorded by using 125 MHz ADC modules with a Gaussian filter in the KOTO experiment. This method extracts the information of the waveform by fitting with an asymmetric gaussian as follows.

$$A(t) = |A| exp\left(-\frac{(t-t_0)^2}{2\sigma(t)^2} \right), \qquad \sigma(t) = \sigma_0 + a(t - t_0). \tag{2}$$

In this function, the σ depends on the distance from the mean (t_0). Due to the difference of dE/dx of electromagnetic showers and hadronic showers, scintillating mechanism may be different and appears in the waveform difference. In the neutron case, a hadronic shower may have a larger tail component. It makes the "σ_0" and "a" parameters in the fitting function larger. By using such information, the likelihood ratio is calculated to determine which is more

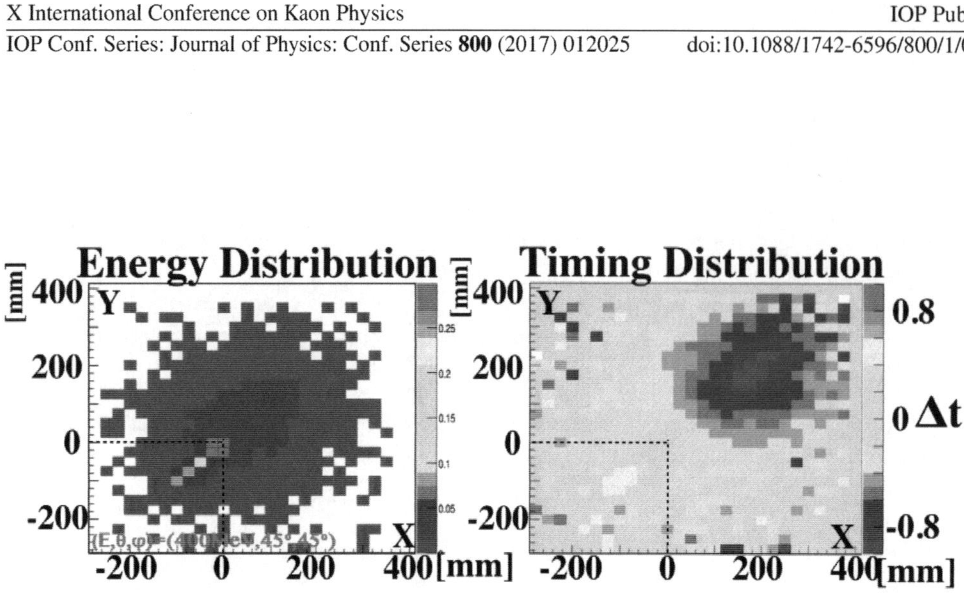

Figure 3. Energy (left) and time difference (right) distributions in a cluster derived from simulation. The X and Y axes show the positions relative to the position with highest energy, and color means energy and time difference.

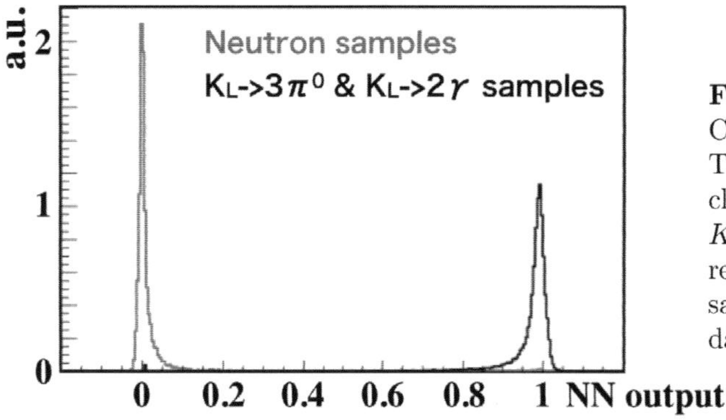

Figure 4. Neural net output of Cluster Shape Discrimination cut. The black histogram shows photon cluster samples obtained by using the $K_L \to 3\pi^0$ and $K_L \to 2\gamma$ modes, and red histogram shows neutron cluster samples obtained using Al target run data.

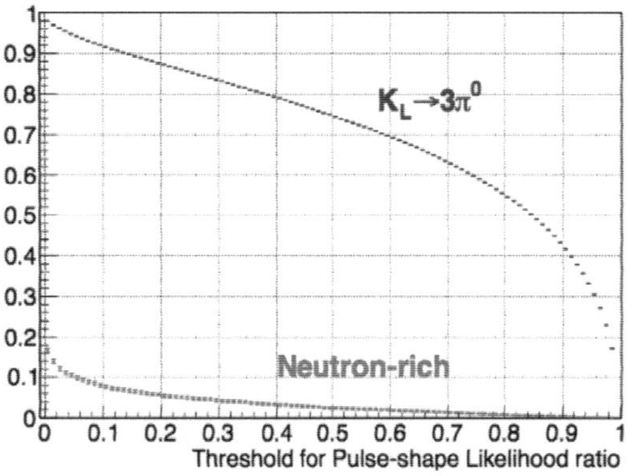

Figure 5. Acceptance of two-photon and neutron-induced events of Pulse Shape Likelihood cut as a function of the likelihood ratio threshold. Blue dots are for the two-photon events derived from $K_L \to 3\pi^0$ samples, and red dots are for the neutron-induced events derived from neutron samples collected in the Al target run.

likely to be a neutron-induced event or a two-photon event. Figure 5 shows the acceptance of two-photon and neutron-induced events as a function of the likelihood ratio threshold. The blue dots are for two-photon events and the red dots are for neutron-induced events. The reduction factor of 10 was obtained with 90% signal acceptance.

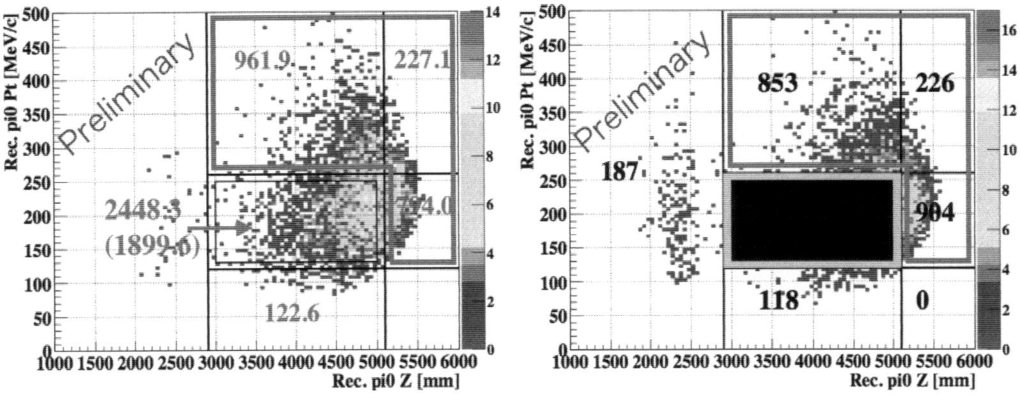

Figure 6. The π^0 P_t v.s. Z scatter plots of the Al target run (left) and the physics run (right) after applying all the cuts except for the neutron cuts. The signal box in the physics plot is blinded. The number "2448.3" in the left figure is the normalized number of events in the box drawn in gray and "1899.6" in parenthesis is the number in the signal box drawn in black.

4. Integration of reduction power of the methods

Both the Shape χ^2 method and the Cluster Shape Discrimination method use the cluster shape information and they are expected to have a strong correlation. The reduction power of these two cuts together was $3.4^{+2.1}_{-0.9} \times 10^3$. The reduction of the Pulse Shape Likelihood was evaluated independently, which was $11.1^{+0.4}_{-0.9}$. The reduction of the Pulse Shape Likelihood with waveform information of individual channels is independent of the reduction from the cluster shape discriminations. [1] The total reduction was obtained as a product of these reduction power , which result in $3.8^{+2.4}_{-1.1} \times 10^4$.

5. Estimation of number of neutron-induced background in physics analysis

In this section, we estimate the number of neutron-induced background events in physics data. The data set used here is 10% of the data taken in 2015. Figure 6 shows π^0 P_t v.s. Z plots for the Al target run and the physics run after applying all the cuts except for the neutron cuts. To estimate the number of neutron-induced background in the physics data, we used the Al target run data. The numbers in black show the number of observed events in each region. To compare the physics and Al target data, the number of events in the Al target run was normalized to the number of events in the physics run data. The normalization factor was calculated by taking the ratio of the number of events in the masked regions to avoid contamination from other background sources. The numbers in red in the Al target run are normalized. After the normalization, we multiply the reduction factor of neutron cuts to the numbers in red.

Figure 7 shows the π^0 P_t versus Z after applying all the neutron cuts. The number of neutron-induced background events in the signal box was estimated to be 0.049.

6. Future prospects : Both-end readout of the calorimeter

To reach the better sensitivity with larger statistics in the future, more reduction power would be needed. We are planning the upgrade of the CsI calorimeter readout.

[1] The Cluster Shape Discrimination uses the timing difference between crystals as an input variable ,which also does not make any correlation to the Pulse Shape Likelihood.

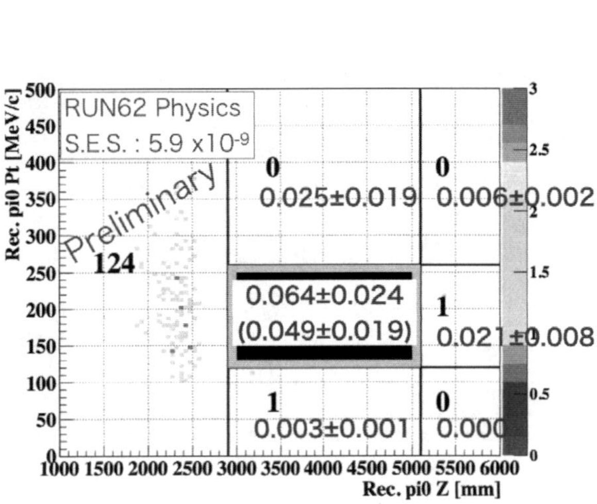

Figure 7. The π^0 P_t versus Z scatter plots of physics run after applying all the cuts including the neutron cuts. The numbers in black show the number of observed events. The numbers in blue show the expected number of neutron-induced background events derived from Al target run data. A number in parenthesis is for the signal box. Only neutron-induced background is considered and contributions from the other background sources are not considered in the estimation.

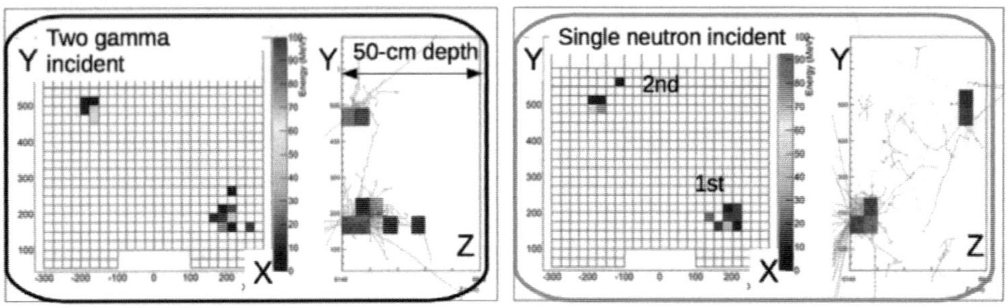

Figure 8. Event displays of a two-photon event (left) and a neutron-induced background event (right) from simulation.

The radiation length of CsI for photons is shorter than the interaction length for neutrons. Thus, if we know the depth of the interaction point, we can distinguish photons from neutrons. Figure 8 shows event displays of a two-photon event and a neutron-induced background event. For photon events, interaction points are in the upstream. Whereas for neutron events, they would distribute broadly. Currently our CsI calorimeter uses PMTs attached downstream-end of the CsI crystals. We plan to add a new photosensor on the upstream-end of the crystals. By using the time difference between the upstream and downstream ends, we would get the Z information of the cluster. A factor of 10 more background reduction is expected by this upgrade. We plan to upgrade the calorimeter in 2018.

7. Conclusion

We have developed several methods to suppress neutron-induced background which is a dominant background in the KOTO experiment. We obtained $(3.8^{+2.4}_{-1.1}) \times 10^4$ reduction power for this background and number of background events was estimated to be 0.049 for the data corresponding to the single event sensitivity of 5.9×10^{-9}.

We also have a plan to upgrade our CsI calorimeter readout. The new both-end readout system will give us a factor of 10 more background reduction. This upgrade is planned to be made in 2018.

IOP Conf. Series: Journal of Physics: Conf. Series **800** (2017) 012025 doi:10.1088/1742-6596/800/1/012025

References

[1] A. J. Buras, D. Buttazzo, J. Girrbach-Noe and R. Knegjens, *J. High Energy Phys.* **1511**, 033 (2015).
[2] J.K. Ahn *et al.*, arXiv:1609.03637 [hep-ex].
[3] J. K. Ahn *et al.*, *Phys. Rev.* **D** 81, 072004 (2010).

A dispersive treatment of $K_{\ell 4}$ decays

Peter Stoffer

Helmholtz-Institut für Strahlen- und Kernphysik (Theory) and
Bethe Center for Theoretical Physics, University of Bonn, 53115 Bonn, Germany

E-mail: stoffer@hiskp.uni-bonn.de

Gilberto Colangelo

Albert Einstein Center for Fundamental Physics, Institute for Theoretical Physics,
University of Bern, Sidlerstrasse 5, 3012 Bern, Switzerland

Emilie Passemar

Department of Physics, Indiana University, Bloomington, IN 47405, USA
Center for Exploration of Energy and Matter, Indiana University,
Bloomington, IN 47403, USA
Theory Center, Thomas Jefferson National Accelerator Facility,
Newport News, VA 23606, USA

Abstract. $K_{\ell 4}$ decays have several features of interest: they allow an accurate measurement of $\pi\pi$-scattering lengths; the decay is the best source for the determination of some low-energy constants of chiral perturbation theory (χPT); one form factor of the decay is connected to the chiral anomaly.

We present the results of our dispersive analysis of $K_{\ell 4}$ decays, which provides a resummation of $\pi\pi$- and $K\pi$-rescattering effects. The free parameters of the dispersion relation are fitted to the data of the high-statistics experiments E865 and NA48/2. By matching to χPT at NLO and NNLO, we determine the low-energy constants L_1^r, L_2^r and L_3^r. In contrast to a pure chiral treatment, the dispersion relation describes the observed curvature of one of the $K_{\ell 4}$ form factors, which we understand as an effect of rescattering beyond NNLO.

1. Motivation

$K_{\ell 4}$, the semileptonic decay of a kaon into two pions and a lepton-neutrino pair, plays a crucial role in the context of low-energy hadron physics, because it provides almost unique information about some of the $SU(3)$ $\mathcal{O}(p^4)$ low-energy constants (LECs) of chiral perturbation theory, the low-energy effective theory of QCD [1, 2, 3]. Although the $K_{\ell 4}$ decay offers similar information as $K\pi$ scattering, it happens at lower energies, where the chiral expansion is more reliable. Besides, as the hadronic final state contains two pions, $K_{\ell 4}$ is also one of the best sources of information on $\pi\pi$ interaction [4, 5, 6].

On the experimental side, we are confronted with impressive precision from the high-statistics measurements of the E865 experiment at BNL [7, 8] and the NA48/2 experiment at CERN [6, 9]. The statistical errors of the S-wave of one form factor reach in both experiments the sub-percent level. Matching this precision requires a theoretical treatment beyond one-loop order in the chiral

Content from this work may be used under the terms of the Creative Commons Attribution 3.0 licence. Any further distribution of this work must maintain attribution to the author(s) and the title of the work, journal citation and DOI.
Published under licence by IOP Publishing Ltd

expansion. A first treatment beyond one loop, based on dispersion relations, was already done twenty years ago [10]. The full two-loop calculation became available in 2000 [11]. However, as we will show below, even at two loops χPT is not able to predict the curvature of one of the form factors.

Here, we present the results of a new dispersive treatment of $K_{\ell 4}$ decays [12, 13]. We do not solve an exact dispersion relation for this process, but an approximate form, which follows if the contribution of D- and higher waves to the discontinuities are neglected.[1] This approximation is violated only at $\mathcal{O}(p^8)$ in the chiral counting. The effects due to $\pi\pi$ and $K\pi$ rescattering in S- and P-wave are resummed to all orders. We expect this to capture the most important contributions beyond $\mathcal{O}(p^6)$. Indeed it turns out that the dispersive description is able to reproduce the curvature of the form factor.

Our final analysis of $K_{\ell 4}$ decays represents an extension and a major improvement of our previous dispersive framework [14, 15, 16]. Instead of a single linear combination of form factors, now we describe the two form factors F and G simultaneously, including more experimental data in the fits. The new framework is valid also for non-vanishing invariant energies of the lepton pair. We apply corrections for isospin-breaking effects in the fitted data that have not been taken into account in the experimental analyses [17]. Besides a matching to one-loop χPT, we also study the matching at two-loop level.

2. Dispersion relation for $K_{\ell 4}$

In this section, we provide only a very short summary of our dispersive treatment of the $K_{\ell 4}$ form factors. For more details, we kindly refer the reader to [12, 13].

2.1. Matrix element and form factors

We consider the charged decay mode

$$K^+(p) \to \pi^+(p_1)\pi^-(p_2)\ell^+(p_\ell)\nu_\ell(p_\nu), \tag{1}$$

where $\ell \in \{e, \mu\}$ is either an electron or a muon. So far, experimental data is only available on the electron mode.

After integrating out the W boson, we end up with a Fermi type current-current interaction and the matrix element splits up into a leptonic times a hadronic part. The leptonic matrix element can be treated in a standard way. The hadronic matrix element exhibits the usual $V-A$ structure of weak interaction. Its Lorentz structure allows us to write the two contributions as

$$\langle \pi^+(p_1)\pi^-(p_2)|V_\mu(0)|K^+(p)\rangle = -\frac{H}{M_K^3}\epsilon_{\mu\nu\rho\sigma}L^\nu P^\rho Q^\sigma, \tag{2}$$

$$\langle \pi^+(p_1)\pi^-(p_2)|A_\mu(0)|K^+(p)\rangle = -i\frac{1}{M_K}\left(P_\mu F + Q_\mu G + L_\mu R\right), \tag{3}$$

where $P = p_1 + p_2$, $Q = p_1 - p_2$, $L = p - p_1 - p_2$. The form factors F, G, H and R are dimensionless scalar functions of the usual Mandelstam variables s, t and u. In experiments on the electron mode K_{e4}, R is not accessible. H gets a first contribution only at $\mathcal{O}(p^4)$ due to the chiral anomaly. Here we focus on the form factors F and G.

2.2. Reconstruction theorem and integral equations

Let us for the moment regard the di-lepton invariant squared energy $s_\ell = L^2$ as a fixed parameter. Based on fixed-$s/t/u$ dispersion relations, one can derive a decomposition of the form factors

[1] This means that D- and higher partial waves have no right-hand unitarity cut. However, they are still non-zero and have left-hand cuts due to unitarity in the crossed channels.

into functions of only one Mandelstam variable, known as 'reconstruction theorem' [18, 19]. In [20] the reconstruction theorem is used for a similar dispersive description of the $K_{\ell 4}$ form factors in order to study isospin-breaking effects in the phases at two loops. The derivation neglects the imaginary parts of D- and higher partial waves, an $\mathcal{O}(p^8)$ effect:

$$\begin{aligned} F(s,t,u) &= M_0(s) + \frac{u-t}{M_K^2} M_1(s) + (\text{terms involving functions of } t \text{ or } u) + \mathcal{O}(p^8), \\ G(s,t,u) &= \tilde{M}_1(s) + (\text{terms involving functions of } t \text{ or } u) + \mathcal{O}(p^8), \end{aligned} \tag{4}$$

where the functions of one variable M_0, ... are defined to contain only the right-hand cut of the partial waves of the form factors F and G in the three channels. E.g. the function M_0 contains the right-hand cut of the s-channel S-wave f_0 of the form factor F:

$$M_0(s) = P(s) + \frac{s^2}{\pi} \int_{4M_\pi^2}^{\infty} ds' \frac{\mathrm{Im} f_0(s)}{(s'-s-i\epsilon)s'^2}, \tag{5}$$

where $P(s)$ is a subtraction polynomial. Eight more functions M_1, ... take care of the right-hand cuts of S- and P-waves in all channels, such that all the discontinuities are divided up into functions of a single variable. They satisfy inhomogeneous Omnès equations with the solution

$$M_0(s) = \Omega_0^0(s) \left\{ \tilde{P}(s) + \frac{s^3}{\pi} \int_{4M_\pi^2}^{\Lambda^2} ds' \frac{\hat{M}_0(s') \sin \delta_0^0(s')}{|\Omega_0^0(s')|(s'-s-i\epsilon)s'^3} \right\}, \tag{6}$$

where $\tilde{P}(s)$ is a new subtraction polynomial and the Omnès function is given by

$$\Omega_0^0(s) = \exp\left\{ \frac{s}{\pi} \int_{4M_\pi^2}^{\infty} ds' \frac{\delta_0^0(s')}{(s'-s-i\epsilon)s'} \right\}. \tag{7}$$

In total, 9 subtraction constants appear. We need the following elastic $\pi\pi$ and $K\pi$ phase shifts as input, which we assume to reach a multiple of π at the cut-off Λ^2:

- δ_0^0, δ_1^1: elastic $\pi\pi$-scattering phase shifts [21, 22],
- $\delta_0^{1/2}$, $\delta_1^{1/2}$, $\delta_0^{3/2}$, $\delta_1^{3/2}$: elastic $K\pi$-scattering phase shifts [23, 24].

The inhomogeneities in the Omnès problem are given by the differences of the functions M_0, ... and the corresponding partial wave, e.g. $\hat{M}_0(s) = f_0(s) - M_0(s)$. These 'hat functions' contain the left-hand cut of the partial wave and we compute them by projecting out the partial wave of the decomposed form factor (4). The inhomogeneities $\hat{M}_0$, ... are then given as angular averages of all the functions M_0, ... Hence, we face a set of coupled integral equations: the functions M_0, ... are defined by dispersion integrals involving the inhomogeneities $\hat{M}_0$, ..., which are again defined as angular integrals of the functions M_0, ... This system can be solved by iteration.

2.3. Numerical solution of the dispersion relation

We note that the integral equations are linear in the subtraction constants. Therefore, for each subtraction constant we construct a basis solution, which we obtain by solving numerically the integral equations in an iterative procedure. The final result is a linear combination of these basis solutions.

We determine the subtraction constants using three sources of information: first, we fit the experimental data on the form factors F and G from the high-statistics experiments NA48/2 [6, 9] and E865 [7, 8]. Secondly, we use as an additional constraint the well-known soft-pion theorem [25, 26, 27], which establishes the following relations between F, G and f_+, the $K_{\ell 3}$ vector form factor:

$$\begin{aligned} F(M_\pi^2, M_K^2, M_\pi^2 + s_\ell) - G(M_\pi^2, M_K^2, M_\pi^2 + s_\ell) &= \mathcal{O}(M_\pi^2), \\ F(M_\pi^2, M_\pi^2 + s_\ell, M_K^2) + G(M_\pi^2, M_\pi^2 + s_\ell, M_K^2) &= \frac{\sqrt{2}M_K}{F_\pi} f_+(M_\pi^2 + s_\ell) + \mathcal{O}(M_\pi^2). \end{aligned} \tag{8}$$

Finally, we fix the subtraction constants that are not well determined by the data with chiral input.

3. Results

3.1. Fits to data

We perform a fit of the dispersion relation to both, the E865 [7, 8] and NA48/2 data sets [6, 9], corrected for additional isospin-breaking effects that were not taken into account in the experimental analyses [17]. Recently, a two-dimensional data set on the S-wave of F has become available (addendum to [9]): in this set, not only a single bin but up to 10 bins are used in s_ℓ-direction. If we allow for varying values of the di-lepton invariant squared energy s_ℓ, the subtraction constants become functions of this parameter and the functions $M_0, \ldots$ depend on two variables, e.g. $M_0(s, s_\ell)$. We perform our fits in the two-dimensional (s, s_ℓ)-plane using the full available data sets on the S- and P-waves of the form factors, given by

$$\begin{aligned} F_s(s, s_\ell) &= \left(M_0(s, s_\ell) + \hat{M}_0(s, s_\ell)\right) e^{-i\delta_0^0(s)}, \\ \tilde{F}_p(s, s_\ell) &= \left(M_1(s, s_\ell) + \hat{M}_1(s, s_\ell)\right) e^{-i\delta_1^1(s)}, \\ G_p(s, s_\ell) &= \left(\tilde{M}_1(s, s_\ell) + \hat{\tilde{M}}_1(s, s_\ell)\right) e^{-i\delta_1^1(s)}. \end{aligned} \tag{9}$$

Figure 1 shows the fit results for the S-wave of the form factor F. The two-dimensional phase space is projected on the s-axis and only the data sets with a single bin in s_ℓ-direction are plotted. The dispersive description reproduces beautifully the observed curvature of the form factor F_s. Note that χPT alone is not able to describe this curvature, which can be understood as a higher-order effect of $\pi\pi$ rescattering, fully taken into account in the dispersive Omnès representation of the form factor.

3.2. Matching to χPT

We perform the matching to χPT directly on the level of the subtraction constants. This means that we decompose the chiral expression at NLO or NNLO according to the reconstruction theorem and write the functions $M_0, \ldots$ in a chirally expanded Omnès form. This allows us to directly identify the subtraction polynomials $\tilde{P}(s)$ in (6) with chiral expressions. Such a procedure separates the resummation of rescattering effects from the chiral matching. Note also that we subtract all the functions $M_0, \ldots$ at zero energy.

By matching the dispersion relation to χPT, we are able to determine the LECs L_1^r, L_2^r and L_3^r. Using the information on the s_ℓ-dependence of the form factors, also L_9^r can be extracted, though the present experimental data does not allow a precise determination.

If we perform the matching at two-loop level, many NNLO LECs C_i^r enter the matching equations. We compare different input values for the C_i^r [28, 29, 30] and assign a 50% uncertainty to the contribution of the C_i^r to the subtraction constants. This C_i^r contribution is then fitted as well, using constraints on the chiral convergence of the subtraction constants. We find that the C_i^r input values of the BE14 global fit [30] lead to the best chiral convergence and a good χ^2 of the whole fit.

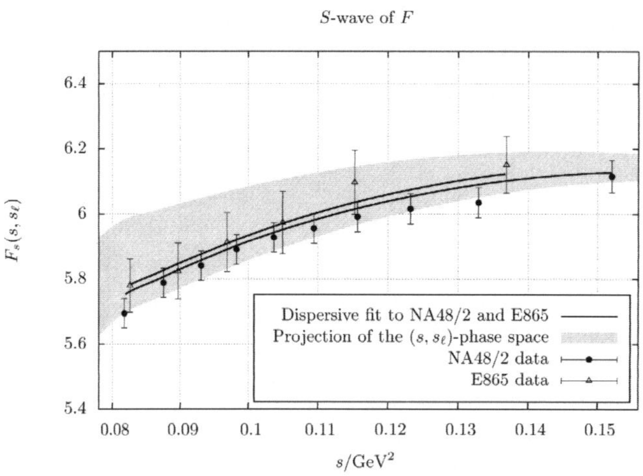

Figure 1. Fit result for the S-wave of the form factor F. The dispersive description reproduces beautifully the curvature of the form factor. The (s, s_ℓ)-phase space is projected on the s-axis, the plotted lines correspond to splines through the (s, s_ℓ)-values of the data sets with a single bin in s_ℓ-direction.

In table 1, we show the results of the matching at NLO and NNLO for the low-energy constants L_1^r, L_2^r and L_3^r. For the NNLO matching, the BE14 C_i^r input values were used. For comparison, we also quote the values of the BE14 global fit [30].

Table 1. Results for the LECs ($\mu = 770$ MeV).

	$10^3 \cdot L_1^r$	$10^3 \cdot L_2^r$	$10^3 \cdot L_3^r$	χ^2/dof
Dispersive treatment, NLO matching	0.51(6)	0.89(9)	−2.82(12)	$141/116 = 1.2$
Dispersive treatment, NNLO matching	0.69(18)	0.63(13)	−2.63(46)	$122/122 = 1.0$
BE14 global fit [30]	0.53(6)	0.81(4)	−3.07(20)	

At NLO, a large contribution to the uncertainties comes from the high-energy behaviour of the phase shifts, either from the $\pi\pi$ phases in the case of L_1^r and L_2^r or the $K\pi$ phases in the case of L_3^r. At NNLO, the largest uncertainty is due to the fitted contribution of the C_i^r.

4. Conclusions

We have presented a dispersive representation of $K_{\ell 4}$ decays that provides a model independent parametrisation valid up to and including $\mathcal{O}(p^6)$. The dispersion relation is based on unitarity, analyticity and crossing. It includes a full resummation of $\pi\pi$- and $K\pi$-rescattering effects. The dispersion relation is parametrised by subtraction constants that we determine by fitting experimental data and by using the soft-pion theorem as well as chiral input.

In contrast to a pure chiral description, the dispersion relation describes perfectly the experimentally observed curvature of the S-wave of the form factor F, which we interpret as a result of significant $\pi\pi$-rescattering effects. This is yet another case in which high-precision data clearly call for effects which go even beyond NNLO in χPT.

By using the matching equations to χPT we have extracted the values of the low-energy constants L_1^r, L_2^r and L_3^r. The correction from NLO to NNLO, when matching the chiral and dispersive representations and fitting the latter to the data are smaller than the corrections from NLO to NNLO observed in direct χPT fits. Constraints on the chiral convergence of the subtraction constants allow us to reduce the dependence on the input values for the C_i^r. Still, the poorly known values of the C_i^r are responsible for the larger uncertainties in the matching at NNLO.

The two-dimensional NA48/2 data set for the S-wave of F, which shows both the s- as well as the s_ℓ-dependence, has allowed us to extract a value for L_9^r [13], which is roughly compatible with previous determinations. In accuracy, however, it cannot compete yet, as it reflects the low precision in the measurement of the s_ℓ-dependence of F. A new high-statistics measurement of K_{e4} could improve on this and provide a new determination of L_9^r. Regarding prospects for forthcoming experiments, we refer to [31].

Acknowledgements

The speaker (PS) thanks the organisers for the invitation and support and for the perfect organisation of the KAON2016 conference. We are very grateful to J. Bijnens, B. Bloch-Devaux, G. Ecker, J. Gasser, I. Jemos, B. Kubis, S. Lanz, H. Leutwyler, S. Pislak, P. Truöl and A. van der Schaaf for valuable discussions and diverse support.

Financial support by the Swiss National Science Foundation, the DFG (CRC 16, "Subnuclear Structure of Matter", CRC 110 "Symmetries and the Emergence of Structure in QCD") and the U.S. Department of Energy (contract DE-AC05-06OR23177) is gratefully acknowledged.

References

[1] Weinberg S 1968 *Phys. Rev.* **166** 1568
[2] Gasser J and Leutwyler H 1984 *Annals Phys.* **158** 142
[3] Gasser J and Leutwyler H 1985 *Nucl. Phys.* **B250** 465
[4] Shabalin E P 1963 *J. Exp. Theor. Phys. (USSR)* **44** 765 [1963 *Sov. Phys. JETP* **17** 517]
[5] Cabibbo N and Maksymowicz A 1965 *Phys. Rev.* **137** B438–B443 [Erratum-ibid. 1968 **168** 1926]
[6] Batley J *et al.* (NA48-2 Collaboration) 2010 *Eur. Phys. J.* **C70** 635–657
[7] Pislak S *et al.* (BNL-E865 Collaboration) 2001 *Phys. Rev. Lett.* **87** 221801 [Erratum-ibid. 2010 **105** 019901]
[8] Pislak S *et al.* (BNL-E865 Collaboration) 2003 *Phys. Rev.* **D67** 072004 [Erratum-ibid. 2010 **D81** 119903]
[9] Batley J *et al.* (NA48/2 Collaboration) 2012 *Phys. Lett.* **B715** 105–115 [Addendum-ibid. 2015 **B740** 364]
[10] Bijnens J, Colangelo G and Gasser J 1994 *Nucl. Phys.* **B427**
[11] Amoros G, Bijnens J and Talavera P 2000 *Nucl. Phys.* **B585** 293–352
[12] Stoffer P 2014 *Dispersive Treatments of $K_{\ell 4}$ Decays and Hadronic Light-by-Light Scattering* Ph.D. thesis University of Bern (*Preprint* `arXiv:1412.5171[hep-ph]`)
[13] Colangelo G, Passemar E and Stoffer P 2015 *Eur. Phys. J.* **C75** 172
[14] Stoffer P 2010 *A Dispersive Treatment of $K_{\ell 4}$ Decays* Master's thesis University of Bern
[15] Colangelo G, Passemar E and Stoffer P 2012 *EPJ Web Conf.* **37** 05006
[16] Stoffer P, Colangelo G and Passemar E 2013 *PoS* **CD12** 058
[17] Stoffer P 2014 *Eur. Phys. J.* **C74** 2749
[18] Stern J, Sazdjian H and Fuchs N H 1993 *Phys. Rev.* **D47** 3814–3838
[19] Ananthanarayan B and Buettiker P 2001 *Eur. Phys. J.* **C19** 517–528
[20] Bernard V, Descotes-Genon S and Knecht M 2013 *Eur. Phys. J.* **C73** 2478
[21] Ananthanarayan B, Colangelo G, Gasser J and Leutwyler H 2001 *Phys. Rept.* **353** 207–279
[22] Caprini I, Colangelo G and Leutwyler H 2012 *Eur. Phys. J.* **C72** 1860
[23] Buettiker P, Descotes-Genon S and Moussallam B 2004 *Eur. Phys. J.* **C33** 409–432
[24] Boito D, Escribano R and Jamin M 2010 *JHEP* **1009** 031
[25] Callan C and Treiman S 1966 *Phys. Rev. Lett.* **16** 153–157
[26] Weinberg S 1966 *Phys. Rev. Lett.* **17** 336–340
[27] Treiman S B, Witten E, Jackiw R and Zumino B 1985 *Current Algebra and Anomalies* (World Scientific)
[28] Bijnens J and Jemos I 2012 *Nucl. Phys.* **B854** 631–665
[29] Jiang S Z, Zhang Y, Li C and Wang Q 2010 *Phys. Rev.* **D81** 014001
[30] Bijnens J and Ecker G 2014 *Ann. Rev. Nucl. Part. Sci.* **64** 149–174
[31] Colangelo G, Passemar E and Stoffer P *A dispersive treatment of $K_{\ell 4}$ decays* (to appear in the NA62 Physics Handbook)

IOP Conf. Series: Journal of Physics: Conf. Series **800** (2017) 012027 doi:10.1088/1742-6596/800/1/012027

Radiative corrections in kaon decays

M Knecht

Centre de Physique Théorique (CPT), UMR 7332 CNRS/Aix-Marseille Univ./Univ. du Sud Toulon-Var, Marseille, France

E-mail: knecht@cpt.univ-mrs.fr

Abstract. This account provides a brief overview of various non-perturbative methods that have been developed in order to address the issue of radiative corrections to kaon decays: chiral lagrangians, non-relativistic effective field theory and dispersive constructions. Only a small number of applications is mentioned, mainly for illustrative purposes.

1. Introduction

The experimental study of the kaon decay modes has made considerable progress during the last decade or so. This progress results from an increasing number of high-precision measurements, performed at several kaon facilities, like Fermilab with KTeV, DaΦne with KLOE, IHEP with ISTRA+, or the CERN SPS with NA48, NA48/2, and now NA62. In many "classical" (i.e. usually not considered as rare) decay channels, the number of collected events has multiplied the size of the existing statistical samples by several factors, sometimes even by one or several orders of magnitude. Many limits on lepton flavour violation, on violation of lepton flavour universality, on the violation of CKM unitarity, or on admixtures of non-standard components (e.g. right-handed or tensor currents) have already been obtained using these recent high-precision data on kaon decays (see the contributions of A. Buras, L. Tunstall, M. González-Alonso, E. Passemar to this conference for a survey of some of these aspects from a theoretical perspective).

Besides offering a window into possible new degrees of freedom, the study of kaon decays also provides information on low-energy constants, which can then be used in order to make predictions for other observables, or, more generally, to test our understanding of the properties of QCD in the non-perturbative, low-energy regime. This has been the original motivation for studying the most frequent decay modes, like the semi-leptonic $K_{\ell n}$ channels, with $n = 2, 3, 4$, and even 5. A rather complete description of these theoretical studies can be found in [1].

This very positive evolution allows today to look seriously for far more rare decay channels, which, due to both their extreme suppression and quite solid theoretical control in the standard model, offer excellent possibilities to unravel degrees of freedom that describe physics beyond the standard model. This is the motivation, for instance, for setting up experiments like NA62 at CERN or KOTO at J-PARC, in order to measure the branching fractions of the $K^\pm \to \pi^\pm \nu\bar{\nu}$ and $K_L \to \pi^0 \nu\bar{\nu}$ decay modes, respectively. See the contributions of K. Shiomi (KOTO) and of G. Ruggiero (NA62) to this conference.

But the increase of the experimental precision has also made it necessary to handle and to include effects that had so far been neglected. This is the case, for instance, of isospin-breaking corrections. Isospin-breaking effects have two origins: the difference $m_u - m_d$ of the masses of the

Content from this work may be used under the terms of the Creative Commons Attribution 3.0 licence. Any further distribution of this work must maintain attribution to the author(s) and the title of the work, journal citation and DOI.
Published under licence by IOP Publishing Ltd

light *up* and *down* quarks, and the electromagnetic interaction, $\alpha \neq 0$. Theoretical predictions are often made in a world where $\alpha = 0$ and $m_u = m_d$, so that one needs to connect this "theoretician's paradise" [2] to the real world, where experiments are made, and where isospin-breaking effects are definitely present. In particular, the inclusion of isospin-breaking corrections due to the mass difference between the charged and neutral pions (mainly an electromagnetic effect!), often turns out to be crucial in order to obtain agreement between theoretical prediction and experimental measurement [2].

Concerning radiative corrections to total decay rates Γ or to branching ratios, they usually turn out to be rather small, of the order of a few percents at most,

$$\Gamma = \Gamma_0 \left[1 + \frac{\Delta\Gamma}{\Gamma_0}\right] \qquad \left|\frac{\Delta\Gamma}{\Gamma_0}\right| \lesssim 1 - 2\%, \tag{1}$$

as expected from the size of the fine-structure constants α. On the other hand, radiative corrections to *differential* decay rates can become quite larger in certain regions of phase space,

$$\frac{d^2\Gamma(x,y)}{dxdy} = \frac{d^2\Gamma_0(x,y)}{dxdy}\left[1 + \delta(x,y)\right] \qquad |\delta(x,y)| \lesssim 10\%. \tag{2}$$

One then recovers the expected size of the correction $\Delta\Gamma/\Gamma_0$ to the total decay rate only after the integration over *all* phase space has produced a certain amount of cancellation between the values taken by $\delta(x,y)$ in different regions. For the same reason, when the decay rate is measured with experimental cuts on some phase space variables, radiative corrections can also be more important than mentioned above. It is thus even more important to have an evaluation, as accurate and as model-independent as possible, of isospin-breaking corrections if one wants to extract information from, for instance, the dependence of form factors with respect to some kinematic variable(s).

The remainder of this overview is devoted to the description of the main theoretical tools that have been developed in order to address the issue of radiative correction, or, more generally, of isospin-breaking effects, in kaon decays, and to illustrate them with a few applications. Section 2 is devoted to the low-energy effective framework using chiral lagrangians. Section 3 addresses the use of non-relativistic effective theories for the treatment of the cusp effect in $K \to \pi\pi\pi$ decays, and Section 4 describes the use of more dispersive-oriented approaches. The focus will definitely be on the effective theory techniques. Another way to deal with the issue of radiative corrections would be to rely on calculations from lattice QCD. This aspect is covered by the contribution of G. Martinelli to this conference.

2. Computing radiative corrections to kaon decay modes with chiral lagrangians

At energies well below the electroweak scale, the $\Delta S = 1$ non-leptonic transitions are described by an effective lagrangian involving a set of four-fermion operators $Q_i(\mu)$, modulated by Wilson coefficients $C_i(\mu)$. The latter sum up perturbative QCD corrections from M_W down to a scale $\mu \lesssim m_c$. In the absence of radiative corrections ($\alpha = 0$), the description of the semi-leptonic transitions involves a similar structure, but since there are no short-distance QCD corrections, it actually takes a simple factorized form, (leptonic current)×(hadronic current).

2.1. Structures of the chiral lagrangians

The description of semi-leptonic decays thus amounts to the evaluation of form factors of the relevant hadronic currents. This evaluation can be addressed in the following way. At the energy scale $\mu << \Lambda_{\rm had} \sim 1{\rm GeV}$, where kaon physics takes place, the relevant degrees of freedom are no longer quarks, but the lightest pseudo-scalar mesons that become the Goldstone bosons of the spontaneous breaking of chiral symmetry in the limit $m_{u,d,s} \to 0$ of massless light quarks. It is possible to construct an effective lagrangian that describes the interactions among these pseudo-scalar mesons in a systematic low-energy expansion. The structure of the effective lagrangian involved in the computation of the hadronic currents is fixed by the symmetry properties of the QCD lagrangian in the chiral limit, and is given by [3, 4, 5, 6]

$$\mathcal{L}_{\text{eff}}^{\text{str}} = \mathcal{L}_2^{\text{str}}(2+0) + \mathcal{L}_4^{\text{str}}(10+0) + \mathcal{L}_6^{\text{str}}(90+23) + \cdots . \tag{3}$$

It describes the coupling of the pseudo-scalar mesons to the hadronic currents, as well as the strong interaction effects among the mesons themselves. The low-energy expansion starts at order $\mathcal{O}(E^2)$ and proceeds systematically to higher orders [6, 7, 8, 9, 10, 11, 12, 13, 14]. At each order of the expansion, a certain number of low-energy constants, which are not fixed through symmetry considerations, and which contain the true information on the dynamical aspects of the strong interactions at long distances, need to be introduced. The number of these low-energy constants is indicated between parentheses, the first (second) number being the number of low-energy constants describing the sector of even (odd, from NLO onwards) intrinsic parity. The effective description of the $\Delta S = 1$ non-leptonic transitions of the kaons takes a similar form [15, 16, 17, 18]:

$$\mathcal{L}_{\text{eff}}^{\Delta S=1} = \mathcal{L}_2^{\Delta S=1}(1+1) + \mathcal{L}_4^{\Delta S=1}(22+28) + \cdots . \tag{4}$$

The numbers between parentheses now indicate, at each order, the number of independent low-energy constants with flavour quantum numbers corresponding to the octet or the 27-plet representations. At lowest order, there are just two such constants, G_8 and G_{27}. At order $\mathcal{O}(E^4)$, the number of counterterms corresponds to the sector of even intrinsic parity only. The contribution from the sector of odd intrinsic parity to $\mathcal{L}_4^{\Delta S=1}$ is discussed in Refs. [19, 20, 21].

The description of strong interactions and non-leptonic transitions among the light pseudo-scalar mesons given so far does not account for radiative corrections. Adding electromagnetic interactions requires to also include the photon as a low-energy degree of freedoms in the effective lagrangian [22, 23]. But this is not yet sufficient in order to provide an adequate description of semi-leptonic transitions. Since virtual photons can connect the quark and lepton currents, the factorized structure mentioned above is in fact lost. The light charged leptons, $e^\pm$ and $\mu^\pm$, and the corresponding neutrinos must then be included as well among the light degrees of freedom [24]. At the level of the Fermi interaction between quarks and leptons, these modifications have been investigated and described in Refs. [25, 26, 27]. At the level of the low-energy effective description in terms of pseudo-scalar mesons, the low-energy effective lagrangian for semi-leptonic processes becomes more involved,

$$\mathcal{L}_{\text{eff}}^{\text{str}} \longrightarrow \mathcal{L}_{\text{eff}}^{\text{str}} + \mathcal{L}_{\text{eff}}^{\text{str+EM}} + \mathcal{L}_{\text{eff}}^{\text{lept}} . \tag{5}$$

Loops involving virtual photons and/or leptons will then produce their own divergences, which in turn require, at each order in the low-energy expansion, an additional set of low-energy constants. For the electromagnetic interactions of the pseudo-scalar mesons only, the structure at leading and at next-to-leading orders reads

$$\mathcal{L}_{\text{eff}}^{\text{str+EM}} = \mathcal{L}_2^{\text{str+EM}}(1) + \mathcal{L}_4^{\text{str+EM}}(13+0) + \cdots \tag{6}$$

where [28, 22, 23]

$$\mathcal{L}_2^{\text{str+EM}} = e^2 C \langle QU^\dagger QU \rangle \qquad \mathcal{L}_4^{\text{str+EM}}(13+0) = \sum_{i=1}^{13} K_i \mathcal{O}_i^{\text{str;EM}} . \tag{7}$$

Finally, adding the light leptons requires some additional structures [24]:

$$\mathcal{L}_{\text{eff}}^{\text{lept}} = \mathcal{L}_2^{\text{lept}}(0) + \mathcal{L}_4^{\text{lept}}(5) + \cdots \qquad \mathcal{L}_4^{\text{lept}} = \sum_{i=1}^{5} X_i \mathcal{O}_i^{\text{lept}} . \tag{8}$$

The same thing happens for the description of electromagnetic corrections to the $\Delta S = 1$ non-leptonic transitions,

$$\mathcal{L}_{\text{eff}}^{\Delta S=1} \longrightarrow \mathcal{L}_{\text{eff}}^{\Delta S=1} + \mathcal{L}_{\text{eff}}^{\Delta S=1+EM}, \tag{9}$$

with [29, 30]

$$\mathcal{L}_{\text{eff}}^{\Delta S=1+EM} = \mathcal{L}_{2}^{\Delta S=1+EM}(1) + \mathcal{L}_{4}^{\Delta S=1;EM}(14+?) + \cdots \tag{10}$$

$$\mathcal{L}_{2}^{\Delta S=1+EM} = e^2 G_8 F_0^6 g_{\text{weak}} \langle \lambda_{23} U^\dagger Q U \rangle \qquad \mathcal{L}_{4}^{\Delta S=1+EM} = e^2 G_8 F_0^4 \sum_{i=1}^{14} Z_i \mathcal{O}_i^{\Delta S=1;EM}. \tag{11}$$

2.2. Determination of the low-energy constants

As we have just seen, including virtual photons and leptons in the low-energy description of semi-leptonic and non-leptonic transitions involves new low energy constants. We briefly discuss how these new low-energy constants have been determined. Those already present in $\mathcal{L}_{\text{eff}}^{\text{str}}$ and in $\mathcal{L}_{\text{eff}}^{\Delta S=1}$, i.e. in the absence of electromagnetic interactions, will not be addressed here. A recent update of the determination of the low-energy constants in the mesonic sector, for instance, can be found in Ref. [31].

The low-energy constant C appearing in $\mathcal{L}_2^{\text{str;EM}}(1)$ can be determined by a sum rule involving the spectral densities of the vector-vector and axial-axial two-point correlators [32],

$$C = -\frac{1}{16\pi^2}\frac{3}{2\pi}\int_0^\infty ds\, s\, \ln\frac{s}{\mu^2}\left[\rho_V(s) - \rho_A(s)\right]. \tag{12}$$

Saturating this sum rule with narrow-width vector and axial resonances, and using the Weinberg sum rules [33], gives $C = 3/(32\pi^2) M_V^2 M_A^2/(M_A^2 - M_V^2)\ln(M_A^2/M_V^2)$. The low-energy constants K_i appearing in $\mathcal{L}_4^{\text{str;EM}}(13+0)$ have been determined along similar lines in Refs. [34, 35]. One first needs to express the K_i's in terms of suitable QCD correlators in the chiral limit (in this case one needs also to consider three- and four-point functions), convoluted with the free photon propagator. From the study of the short-distance properties of these correlators, one can then establish spectral sum rules similar to the one given above for C, and saturate them with the lowest-lying resonances in the various relevant channels.

The low-energy constants X_i appearing in $\mathcal{L}_4^{\text{lept}}$ have been determined in Ref. [27] using a two-step matching procedure. One first computes the radiative corrections to the process $\bar{q}q' \to \ell\nu$ in the SM and in the four-fermion theory. Then one matches the radiatively corrected four-fermion theory to the chiral lagrangian, by identifying the QCD correlators (convoluted with the free photon propagator) that describe the X_i's. Finally, the resulting spectral sum rules can be saturated with the lowest-lying resonance states.

Finally, the low-energy constants g_{weak} and Z_i have been estimated in the large-N_c limit in Ref. [36]. In this approximation, one finds, for instance,

$$(g_8 e^2 g_{\text{weak}})^\infty = -\left(\frac{\langle\bar{\psi}\psi\rangle}{F_0^3}\right)^2\left[3C_8(\mu) + \frac{16}{3}e^2 C_6(\mu)(K_9 - 2K_{10})\right]. \tag{13}$$

The dependence on the short-distance scale μ vanishes at leading-order in the large-N_c limit. A scale dependence remains at subleading order in $1/N_c$, but the subleading contributions to g_{weak} and Z_i cannot be computed in general. In the case of g_{weak}, the subleading contribution induced by Q_7 has however been obtained in Ref. [37],

$$(g_8 e^2 g_{\text{weak}})^{1/N_c;Q_7} = -\frac{9}{8\pi^2}C_7(\mu)\frac{M_\rho^2}{F_0^2}\left[\ln\frac{\mu^2}{M_\rho^2} + \frac{1}{3} - 2\ln 2\right], \tag{14}$$

but this does not completely remove the residual scale dependence.

One may worry about the uncertainties on the values of the low-energy constants K_i, X_i and Z_i that come from saturation by lowest-lying narrow-width resonances, or from using the large-N_c limit (the former being itself an approximation to the latter). Typically, one should ascribe an uncertainty of about 30% to the values obatined as described above. This might not look as very precise, and it would certainly not be precise enough in order to make sufficiently accurate predictions for the dominant contributions, involving the low-energy constants of, say, $\mathcal{L}_4^{\rm str}$ or even $\mathcal{L}_6^{\rm str}$. However, radiative corrections represent only small effects, for which a precision of 30% in the corresponding low-energy constants is most of the time sufficient.

Radiative and isospin breaking corrections have therefore been computed within the low-energy expansion described above for many semi-leptonic and non-leptonic processes, for instance: $\pi \to \ell\nu_\ell(\gamma)$ and $K \to \ell\nu_\ell(\gamma)$ [24, 38, 39, 40], $K \to \pi\ell\nu_\ell(\gamma)$ [41, 42, 43, 44], $\pi^+ \to \pi^0 e\nu_e$ [45], $K^+ \to \pi^+\pi^-\ell\nu_\ell$ [46, 47, 48, 49], $K \to \pi\pi$ [50, 36, 51], $K \to \pi\pi\pi$ [52, 53, 54]. This list is certainly not extensive, and a more complete account can be found in Ref. [55].

3. Non-relativistic effective field theory

The high-precision experimental study of the decay mode $K \to \pi\pi^0\pi^0$ has revealed an important experimental feature, namely a cusp at $M_{00} = 2M_\pi$ (M_π denotes the mass of the charged pion) in the invariant mass distribution M_{00} of the two neutral pions. This cusp was first observed by NA48/2 [56] in a sample of $2.3 \cdot 10^7$ $K^\pm \to \pi^\pm\pi^0\pi^0$ events. It was promptly interpreted [57] as a final-state rescattering effect [58] $\pi^+\pi^- \to \pi^0\pi^0$ ($M_\pi \neq M_{\pi^0}$) corresponding to the combination $a_0 - a_2$ of S-wave scattering lengths. But simple phenomenological parameterizations [59, 60], or one-loop calculations in the low-energy expansion including isospin breaking [52, 53, 54], either do not reproduce the correct analyticity properties, or do not give a sufficiently accurate description of the cusp. A better description can be obtained by combining a non relativistic EFT framework, where $|\mathbf{p}|/M_\pi \sim \mathcal{O}(\epsilon)$, and a systematic expansion in powers of the scattering lengths, treated as free parameters, including orders ϵ^2, $a\epsilon^3$ and $a^2\epsilon^2$ [61, 62]. Radiative corrections were also treated within this non-relativistic framework [63]. This theoretical work, when applied to the data, allows for a very accurate determination of the relevant combination of scattering lengths, $a_0 - a_2 = 0.2571 \pm 0.0056$ [65]. This theoretical study was also extended to the $K_L \to \pi^0\pi^0\pi^0$ mode [64], for which the KTeV collaboration had collected a sample of $6.8 \cdot 10^7$ $K_L \to \pi^0\pi^0\pi^0$ events. But the rescattering effect is quite smaller than in the $K^\pm \to \pi^\pm\pi^0\pi^0$ mode, so that the combination of scattering lengths can only be determined with significantly less accuracy, $a_0 - a_2 = 0.215 \pm 0.031$ [66].

4. Dispersive construction of form factors and amplitudes

The systematic use of general properties like relativistic invariance, unitarity, crossing and analyticity, combined with the low-energy counting allows to construct, in a dispersive way, representations of scattering amplitudes or of form factors that include up to next-to-next-to leading order effects in the low-energy expansion. This "reconstruction theorem" was first proven in Ref. [67] for the $\pi\pi$ scattering amplitude in the isospin limit, and was subsequently implemented in order to obtain an explicit two-loop representation of the $\pi\pi$ amplitude [68]. But it actually is of a more general validity and, in particular, it is also applicable when isospin symmetry is explicitly broken, as has been discussed, for instance, in Ref. [69]. This "reconstruction theorem" also furnishes a convenient representation of amplitudes and form factors, which in turn provides a convenient starting point for studies that go beyond the strict framework of the low-energy expansion, when it becomes necessary to account for important final-state rescattering effects in $\pi\pi$ final states. For an illustration of the later aspect in the case of the $K_{\ell 4}$ form factors in the isospin limit, see the contribution of P. Stoffer to this conference, as well as [70].

The example that will be used as an illustration here concerns $M_\pi \neq M_{\pi^0}$ effects in the phases of form factors describing the amplitude for the $K_{\ell 4}$ decay mode $K^\pm \to \pi^+\pi^-e^\pm\nu_e$. Standard angular analysis of these form factors [71, 72] provides information on low-energy $\pi\pi$ scattering (thanks to Watson's theorem) through the difference $[\delta_S(s) - \delta_P(s)]_{\rm exp}$ of the phases of the $\pi\pi$ amplitude in the S and P waves, which is measurable in the interference of the F^{+-} and G^{+-} form factors (for the notation, see [1]). Comparison with solutions of the Roy equations [73]

$$[\delta_S(s) - \delta_P(s)]_{\rm exp} = f_{\rm Roy}(s; a_0^0, a_0^2), \tag{15}$$

allows to extract the values of the $\pi\pi$ S-wave scattering lengths in the isospin channels $I = 0, 2$. The Roy equations themselves follow from dispersion relations (analyticity, unitarity, crossing, the Froissard bound), $\pi\pi$ data at energies $\sqrt{s} \geq 1$ GeV, *and* isospin symmetry. Solutions of these equations can been constructed [74, 75] as long as the scattering lengths a_0^0 and a_0^2 lie within some range called the Universal Band.

Once standard radiative corrections have been taken care of, it is still important to take isospin-breaking corrections due to $M_\pi \neq M_{\pi^0}$ into account before analyzing data [2]. These effects were evaluated to order one loop in the low-energy expansion in Ref. [76], and they amount to replace the equation above by

$$[\delta_S(s) - \delta_P(s)]_{\rm exp} = f_{\rm Roy}(s; a_0^0, a_0^2) + \delta f_{\rm IB}(s; (a_0^0)^{\rm LO}_{\rm ChPT}, (a_0^2)^{\rm LO}_{\rm ChPT}). \tag{16}$$

Since the correction factor results from a one-loop calculation, it depends on the scattering lengths only through their lowest-order expressions, whereas they occur as free parameters in the solutions to the Roy equations. This could possibly constitute a drawback: if the radiative corrections depend stronly on the actual values of the scattering lengths, analyzing the data with the above determination of the correction factor could introduce a bias. This drawback is shared by other studies devoted to isospin breaking in $K_{\ell 4}$ decays [46, 47, 48, 49]. The issue whether it is actually possible to compute the correction factor in a way that it also involves the scattering lengths as free parameters,

$$[\delta_S(s) - \delta_P(s)]_{\rm exp} = f_{\rm Roy}(s; a_0^0, a_0^2) + \delta f_{\rm IB}(s; a_0^0, a_0^2), \tag{17}$$

was successfully addressed in Ref. [77]. A reanalysis of the NA48/2 data [78, 79] using the correction factor obtained in Ref. [77] leads to the following determination of the scattering lengths:

$$a_0^0 = 0.221 \pm 0.018 \qquad a_0^2 = -0.0453 \pm 0.0106. \tag{18}$$

Apart from slightly larger error bars, these values show no significant difference from those obtained [78, 79] using the correction factor with scattering lengths fixed at their lowest-order values [76],

$$a_0^0 = 0.2220(128)_{\rm stat}(50)_{\rm syst}\,(37)_{\rm th} \qquad a_0^2 = -0.0432(86)_{\rm stat}(34)_{\rm syst}(28)_{\rm th}. \tag{19}$$

5. Conclusions

The high precision reached by the data concerning non-leptonic and semi-leptonic decay modes of the kaons has made the treatment of isospin-breaking effects ($m_u \neq m_d$ and $\alpha \neq 0$) an unavoidable issue. A lot of activity has been going on in this field, extending the scope of the low-energy effective field theory in order to meet this necessity (inclusion of photons, leptons). Only a tiny fraction of the many applications has been mentioned in this overview. The good news is that the issue of additional low-energy constants that this extension brings with it has been dealt with in a rather satisfactory manner, although some progress on estimates of the Z_i's would be welcome. The effects due to $M_\pi \neq M_{\pi^0}$ were found to be important, especially for the description of the unitarity cusp in the $K \to \pi\pi\pi$ mode, or for the extraction of the $\pi - \pi$ scattering lengths from K_{e4}. The low-energy expansion at next-to-leading order is then not always sufficient, and more elaborate or better adapted approaches, like non-relativistic effective field theory or dispersive representations, need to be implemented in order to achieve the required precision. Finally, although it has been checked not to be the case for the determination of the phase difference $\delta_S(s) - \delta_P(s)$ from K_{e4} data, one should in general watch out for possible biases if the radiative corrections to form factors and/or decay distributions are computed for fixed values of the parameters one actually wants to extract from data.

Acknowledgments
I would like to thank the organizers for the kind invitation to participate to the KAON 2016 conference, and my colleagues V. Bernard and S. Descotes-Genon for a pleasant and fruitful collaboration on some of the topics reviewed here.

References

[1] Bijnens J, Colangelo G, Ecker G and Gasser J 1995 *The Second DAPHNE Physics Handbook* Vol 1, ed Maiani L, Pancheri G and Paver N (Frascati: INFN) p 315
[2] Gasser J 2008 *PoS KAON* 033
[3] Weinberg S 1979 *Physica* A **96** 327
[4] J. Gasser and H. Leutwyler 1983 *Phys. Lett.* B **125** 321
[5] J. Gasser and H. Leutwyler 1984 *Annals Phys.* **158** 142
[6] J. Gasser and H. Leutwyler 1985 *Nucl. Phys.* B **250** 465
[7] Wess J and Zumino B 1971 *Phys. Lett.* B **37** 95
[8] Witten E 1983 *Nucl. Phys.* B **223** 422
[9] Issler D 1990 Nonrenormalization of the chiral anomaly in chiral perturbation theory *Preprints* SLAC-PUB-4943-REV- and SLAC-PUB-4943.
[10] Akhoury R and Alfakih A 1991 *Annals Phys.* **210** 81
[11] Fearing H W and Scherer S 1996 *Phys. Rev.* D **53** 315
[12] Bijnens J, Colangelo G and Ecker G 1999 *JHEP* **9902** 020
[13] Ebertshauser T, Fearing H W and Scherer S 2002 *Phys. Rev.* D **65** 054033
[14] Bijnens J, Girlanda L and Talavera P 2002 *Eur. Phys. J.* C **23** 539
[15] Cronin J A 1967 Phys. Rev. **161** 1483
[16] Kambor J, Missimer J M and Wyler D 1990 *Nucl. Phys.* B **346** 17
[17] Esposito-Farese G 1991 *Z. Phys.* C **50** 255
[18] Ecker G, Kambor J and Wyler D 1993 *Nucl. Phys.* B **394** 101
[19] Ecker G, Neufeld H and Pich A 1992 *Phys. Lett.* B **278** 337
[20] Bijnens J, Ecker G and Pich A, *Phys. Lett.* B **286** 341
[21] Ecker G, Neufeld H and Pich A 1994 *Nucl. Phys.* B **413** 321
[22] Urech R 1995 *Nucl. Phys.* B **433** 234
[23] Neufeld H and Rupertsberger H 1996 *Z. Phys.* C **71** 131
[24] Knecht M, Neufeld H, Rupertsberger H and Talavera P 2000 *Eur. Phys. J.* C **12** 469
[25] Braaten E and Li C S 1990 *Phys. Rev.* D **42** 3888
[26] Braaten E and Li C S 1993 *J. Phys.* G **19** 1825
[27] Descotes-Genon S and Moussallam B 2005 *Eur. Phys.* J. C **42** 403
[28] Ecker G, Gasser J, Pich A and de Rafael E 1989 *Nucl. Phys.* B **321** 311
[29] Bijnens J and Wise M B 1984 *Phys. Lett.* **137B** 245
[30] Ecker G, Isidori G, Müller G, Neufeld H and Pich A 2000 *Nucl. Phys.* B **591** 419
[31] Bijnens J and Ecker G 2014 *Ann. Rev. Nucl. Part. Sci.* **64** 149
[32] Das T, Guralnik G S, Mathur V S, Low F E and Young J E 1967 *Phys. Rev. Lett.* 18 759
[33] Weinberg S 1967 *Phys. Rev. Lett.* **18** 507
[34] Moussallam B 1997 *Nucl. Phys.* B **504** 381
[35] Ananthanarayan B and Moussallam B 2004 *JHEP* **0406** 047
[36] Cirigliano V, Ecker G, Neufeld H and Pich A 2004 *Eur. Phys. J.* C **33** 369
[37] Knecht M, Peris S and de Rafael E 1999 *Phys. Lett.* B **457** 227
[38] Cirigliano V and Rosell I 2007 *JHEP* **0710** 005

[39] Gasser J and Zarnauskas G R S 2010 *Phys. Lett.* B **693** 122
[40] Cirigliano V and Neufeld H 2011 *Phys. Lett.* B **700** 7
[41] Cirigliano V, Knecht M, Neufeld H, Rupertsberger H and Talavera P 2002 *Eur. Phys. J.* C **23** 121
[42] Kastner A and Neufeld H 2008 *Eur. Phys. J.* C **57** 541
[43] Cirigliano V, Giannotti M and Neufeld H 2008 *JHEP* **0811** 006
[44] Gasser J, Kubis B, Paver N and Verbeni M 2005 *Eur. Phys. J.* C **40** 205
[45] Cirigliano V, Knecht M, Neufeld H and Pichl H 2003 *Eur. Phys. J.* C **27** 255
[46] Cuplov V 2004 Brisure d'Isospin et Corrections Radiatives au Processus $K_{\ell 4}$ *PhD thesis* Université de la Méditerranée - Aix-Marseille II
[47] Cuplov V and Nehme A 2003 Isospin breaking in K(l4) decays of the charged kaon *Preprint* hep-ph/0311274
[48] Nehme A 2004 *Phys. Rev.* D **69** 094012
[49] Stoffer P 2014 *Eur. Phys. J.* C **74** 2749
[50] Cirigliano V, Pich A, Ecker G and Neufeld H 2003 *Phys. Rev. Lett.* **91** 162001
[51] Cirigliano V, Ecker G and Pich A 2009 *Phys. Lett.* B **679** 445
[52] Bijnens J and Borg F 2004 *Nucl. Phys.* B **697** 319
[53] Bijnens J and Borg F 2005 *Eur. Phys. J.* C **39** 347
[54] Bijnens J and Borg F 2005 *Eur. Phys. J.* C **40** 383
[55] Cirigliano V, Ecker G, Neufeld H, Pich A and Portoles J 2012 *Rev. Mod. Phys.* **84** 399
[56] Batley J R *et al* [NA48/2 Collaboration] 2006 *Phys. Lett.* B **633** 173
[57] Cabibbo N 2004 *Phys. Rev. Lett.* **93** 121801
[58] Budini P, Fonda L 1961 *Phys. Rev. Lett.* **6** 419
[59] Cabibbo N and Isidori G 2005 *JHEP* **0503** 021
[60] Gamiz E, Prades J and Scimemi I 2007 *Eur. Phys. J.* C **50** 405
[61] Colangelo G, Gasser J, Kubis B and Rusetsky A 2006 *Phys. Lett.* B **638** 187
[62] Gasser J, Kubis B and Rusetsky A 2011 *Nucl. Phys.* B **850** 96
[63] Bissegger M, Fuhrer A, Gasser J, Kubis B and Rusetsky A 2009 *Nucl. Phys.* B **806** 178
[64] Bissegger M, Fuhrer A, Gasser J, Kubis B and Rusetsky A 2008 *Phys. Lett.* B **659** 576
[65] Batley J R *et al* 2009 *Eur. Phys. J.* C **64** 589
[66] Abouzaid E *et al* [KTeV Collaboration] 2008 *Phys. Rev.* D **78** 032009
[67] Stern J, Sazdjian H and Fuchs N H 1993 *Phys. Rev.* D **47** 3814
[68] Knecht M, Moussallam B, Stern J and Fuchs N H 1995 *Nucl. Phys.* B **457** 513
[69] Descotes-Genon S and Knecht M 2012 *Eur. Phys. J.* C **72** 1962
[70] Colangelo G, Passemar E and Stoffer P 2015 *Eur. Phys. J.* C **75** 172
[71] Cabibbo N and Maksymowicz A 1968 *Phys. Rev.* **137** B438; Erratum-ibid. **168** 1926
[72] Berends F A, Donnachie A and Oades G C 1968 *Phys. Rev.* **171** 1457
[73] Roy S M 1971 *Phys. Lett.* B **36** 353
[74] Ananthanarayan B, Colangelo G, Gasser J and Leutwyler H 2001 *Phys. Rept.* **353** 207
[75] Descotes-Genon S, Fuchs N H, Girlanda L and Stern J 2002 *Eur. Phys. J.* C **24** 469
[76] Colangelo G, Gasser J and Rusetsky A 2009 *Eur. Phys. J.* C **59** 777
[77] Bernard V, Descotes-Genon S and Knecht M 2013 *Eur. Phys. J.* C **73** 2478
[78] Batley J R *et al* [NA48/2 Collaboration] 2008 *Eur. Phys. J.* C **54** 411
[79] Batley J R *et al* [NA48-2 Collaboration] 2010 *Eur. Phys. J.* C **70** 635

Rare and radiative kaon decays

Giancarlo D'Ambrosio
INFN-Sezione di Napoli, Via Cintia, 80126 Napoli, Italia
E-mail: gdambros@na.infn.it

Abstract. We discuss theoretical issues in radiative rare kaon decays. The interest is twofold: to extract useful short-distance information and understand the underlying dynamics. We emphasize channels where either we can understand non-perturbative aspects of QCD or there is a chance to test the Standard Model. An interesting channel, $K^+ \to \pi^+\pi^0 e^+e^-$, is studied also in connection with the recent experimental NA48 results. Motivated by LHCB results on $K_S \to \mu^+\mu^-$ we discuss other channels like $K_{S,L} \to l^+l^-l^+l^-$. Motivated by recent theoretical work by Buras and collaborators we study also the $K^\pm \to \pi^\pm l^+l^-$ form factor.

1. Introduction

Kaon decays are an important place to study non-perturbative aspects of QCD and test the Standard Model. Indeed some channels are completely dominated by long-distance dynamics, such as the CP-conserving amplitude for $K \to \pi\pi$ and others, like $K \to \pi\nu\overline{\nu}$ [1], which are described in terms of pure short-distance physics. In this review we will be mostly concerned with kaon decays involving electromagnetic interactions and thus long-distance phenomena are not negligible. However, as we shall see, it is still possible in these channels to extract the short-distance component with an accurate analysis. Indeed there are plenty of motivations to look for new physics (NP) in these kaon decays [2].

The channels which will be considered here are $K \to \pi\pi\gamma$, $K \to \pi\pi ee$, $K_{S,L} \to l^+l^-$, $K_{S,L} \to l^+l^-l^+l^-$ and $K_S \to \pi^+\pi^-l^+l^-$. Recently, Buras and collaborators, motivated by their previous work where weak hadronic matrix elements are evaluated in their model (BBG model) and also by recent lattice results, are investigating New Physics to explain the experimental value of ϵ' [2, 3, 4, 5], Motivated by this interest we study also the $K^\pm \to \pi^\pm l^+l^-$ form factor in the BBG model [6].

2. $K \to \pi\pi\gamma$ and $K \to \pi\pi ee$-decays

2.1. $K \to \pi\pi\gamma$

Let's discuss $K(p) \to \pi(p_1)\pi(p_2)\gamma(q)$ decays: according to gauge and Lorentz invariance we decompose $K(p) \to \pi(p_1)\pi(p_2)\gamma(q)$ decays, in electric (E) and magnetic (M) amplitudes [7]. Particularly interesting are the recent interesting NA48/2 data regarding $K^+ \to \pi^+\pi^0\gamma$ decays [8].

Due to the $\Delta I = 3/2$ suppression of the bremsstrahlung, interference between E_B and the electric dipole (E_{DE}) and magnetic transitions (M_{DE}) can be measured. Defining $z_i = p_i \cdot q/m_K^2 \quad z_3 = p_K \cdot q/m_K^2$ and $z_3 z_+ = \frac{m^2_{\pi^+}}{m_K^2} W^2$, (W^2 is defined in this equation) we can study

Content from this work may be used under the terms of the Creative Commons Attribution 3.0 licence. Any further distribution of this work must maintain attribution to the author(s) and the title of the work, journal citation and DOI.
Published under licence by IOP Publishing Ltd

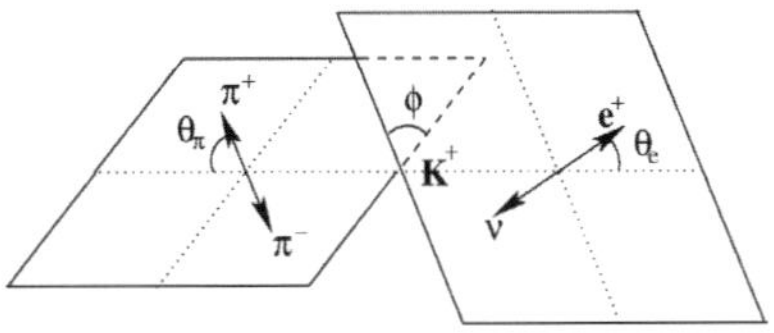

Figure 1. $K^+ \to \pi^+\pi^0 e^+e^-$ kinematical planes: N. Cabibbo and A. Maksymowicz definition of the angles [9]

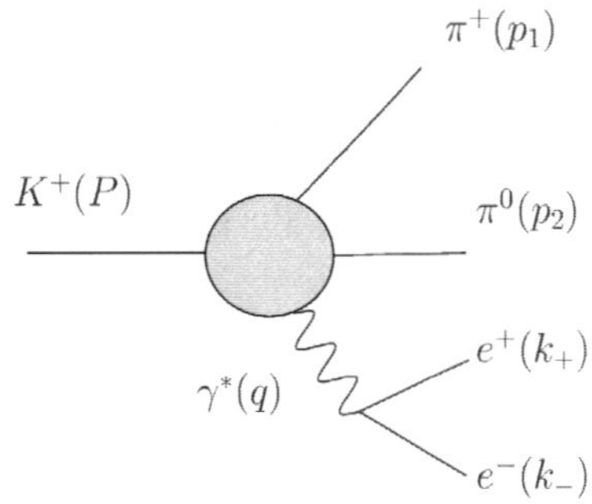

Figure 2. $K^\pm \to \pi^\pm\pi^0 e^+e^-$: long distance contributions dominated by one photon exchange.

the deviation from bremsstrahlung

$$\frac{\partial^2\Gamma}{\partial T_c^*\partial W^2} = \frac{\partial^2\Gamma_{IB}}{\partial T_c^*\partial W^2}\left[1+\frac{m_{\pi^+}^2}{m_K}2Re\left(\frac{E_{DE}}{eA}\right)W^2+\frac{m_{\pi^+}^4}{m_K^2}\left(\left|\frac{E_{DE}}{eA}\right|^2+\left|\frac{M_{DE}}{eA}\right|^2\right)W^4\right], \quad (1)$$

where $A = A(K^+ \to \pi^+\pi^0)$. Study of the Dalitz plot relative to Eq. (1) has lead to measure separately the linear and the quadratic term in W^2: indeed integrating over the Dalitz plot and dividing for the IB Branching ratio the right-hand side of Eq. (1) leads these two terms (linear and quadratic) respectively to the NA48/2 measurement of the two fractions, $Frac(INT)$ and $Frac(DE)$, (see Ref. [8])

NA48/2 $K^+ \to \pi^+\pi^0\gamma$ results

NA48/2		$T_c^* \in [0, 80]$ MeV
$Frac(DE)$	=	$(3.32\pm0.15 \pm 0.14)\times10^{-2}$
$Frac(INT)$	=	$(-2.35\pm0.35 \pm 0.39)\times10^{-2}$

Also the interesting CP bound was obtained [8]:

$$\frac{\Gamma(K^+\to\pi^+\pi^0\gamma)-\Gamma(K^-\to\pi^-\pi^0\gamma)}{\Gamma(K^+\to\pi^+\pi^0\gamma)+\Gamma(K^-\to\pi^-\pi^0\gamma)} < 1.5\cdot10^{-3} \quad \text{at} \quad 90\% \quad \text{CL}. \quad (2)$$

2.2. $K \to \pi\pi ee$-decays

Historically kaon four body semileptonic decays, K_{e4} have been studied as a tool to tackle final state rescattering effects in $K \to \pi\pi$-decays: crucial to this goal has been finding an appropriate set of kinematical variables which would allow i) to treat the system as two body decay in dipion mass $M_{\pi\pi}$ and dilepton mass $M_{l^+l^-}$ and ii) to identify appropriate kinematical asymmetries to extract observables crucially dependent on final state interaction [9].

In Fig. 1 we show the traditional kinematical variables for the four body kaon semileptonic decay which allow to write the four body phase space Φ in terms of the two two-body phase spaces: Φ_π and Φ_ℓ from [9, 10]. We extend the K_{e4} kinematical and dynamical description of the amplitude to describe the $K \to \pi\pi e^+e^-$ decays; in particular we write the long distance contribution in Fig. 2 as

$$\mathcal{M}_{LD} = \frac{e}{q^2}[\overline{u}(k_-)\gamma^\mu v(k_+)]H_\mu(p_1,p_2,q), \quad (3)$$

where H_μ is the electroweak hadronic vector, which can be written in terms of three form factors $F_{1,2,3}$:

$$H^\mu(p_1, p_2, q) = F_1 p_1^\mu + F_2 p_2^\mu + F_3 \varepsilon^{\mu\nu\alpha\beta} p_{1\nu} p_{2\alpha} q_\beta. \tag{4}$$

Then

$$\begin{aligned}\frac{d^5\Gamma}{dE_\gamma^* dT_c^* dq^2 d\cos\theta_\ell d\phi} &= \mathcal{A}_1 + \mathcal{A}_2 \sin^2\theta_\ell + \mathcal{A}_3 \sin^2\theta_\ell \cos^2\phi \\ &+\mathcal{A}_4 \sin 2\theta_\ell \cos\phi + \mathcal{A}_5 \sin\theta_\ell \cos\phi + \mathcal{A}_6 \cos\theta_\ell \\ &+\mathcal{A}_7 \sin\theta_\ell \sin\phi + \mathcal{A}_8 \sin 2\theta_\ell \sin\phi + \mathcal{A}_9 \sin^2\theta_\ell \sin 2\phi,\end{aligned} \tag{5}$$

where θ_ℓ and ϕ are two variables for K_{l4} decays and $\mathcal{A}_i$ are dynamical functions that can be parameterized in terms of the three form factors. The amplitudes $\mathcal{A}_{8,9}$, odd in θ_ℓ, are also linearly dependent on the final state, establishing a clear way to determine them; while $\mathcal{A}_{5,6,7}$ are generated by interference with the axial leptonic current.

One can easily show that the Bremsstrahlung, direct emission and electric interference terms contribute to $\mathcal{A}_{1-4}$. In contrast, $\mathcal{A}_{8,9}$ receive contributions from the electric-magnetic interference terms (BM and EM) and therefore capture long-distance induced P-violating terms. $\mathcal{A}_{5,6,7}$ are also P-violating terms but generated through the interference of Q_{7A} with long distances.

Short distance physics can be studied in $\mathcal{A}_{5,6,7}$ [11] and in the diplane angular asymmetry proportional to $\mathcal{A}_{8,9}$. This last observable is large and has been measured by KTeV and NA48 [12, 13]. However this observable is proportional to electric (bremsstrahlung) and magnetic interference, both contributions known already from $K_L \to \pi^+\pi^-\gamma$. In fact it was known that these contributions were large and they may obscure smaller but more interesting short distance physics effects.

We have performed a similar analysis for the decay $K^+ \to \pi^+\pi^0 e^+ e^-$ trying to focus on i) short distance physics and ii) all possible Dalitz plot analyses to disentangle all possible interesting long and short distance effects [10]. This decay has not been observed yet, and the interesting physics is hidden by bremsstrahlung [10, 14, 15]

$$\begin{aligned}\mathcal{B}(K^+ \to \pi^+\pi^0 e^+ e^-)_B &\sim (330 \pm 15)\cdot 10^{-8} \\ \mathcal{B}(K^+ \to \pi^+\pi^0 e^+ e^-)_M &\sim (6.14 \pm 1.30)\cdot 10^{-8},\end{aligned} \tag{6}$$

and so Dalitz plot analysis is necessary in order to capture the more interesting direct emission contributions. The $K^+ \to \pi^+\pi^0 e^+ e^-$-amplitude is then written as in Eq. 3.

We have studied the decay $K^\pm \to \pi^\pm\pi^0 e^+ e^-$, dominated by long distance through one photon exchange in Fig. 2. We computed, the large long distance contributions and the relatively small short distance ones in Ref. [10, 15]. We have succeeded in describing $K^\pm \to \pi^\pm\pi^0 e^+ e^-$ optimizing the theoretical and experimental knowledge of $K(p) \to \pi(p_1)\pi(p_2)\gamma(q)$ decays.

Relatively to the $K^\pm \to \pi^\pm\pi^0\gamma$ the possibility to go kinematically at large q^2 opens the possibility to beat the bremsstrahlung: at large dilepton invariant mass the bremsstrahlung can be even 100 times smaller than the magnetic contribution, at the price, however, of decreasing the statistics. Indeed compared to $K^\pm \to \pi^\pm\pi^0\gamma$ differential decay rates we have the possibility to measure also the q^2-differential rates

$$\frac{\partial^2\Gamma_{IB}}{\partial T_c^* \partial W^2 \partial q^2} \tag{7}$$

$$d\Phi = \frac{1}{4m_K^2}(2\pi)^5 \int ds_\pi \int ds_\ell \lambda^{1/2}(m_K^2, p_\pi^2, q^2)\Phi_\pi \Phi_\ell. \tag{8}$$

Then defining $q^2 = M_{ee}^2$ and the $\pi\pi$ invariant mass $p_\pi^2 = M_{\pi\pi}^2$, we can write

$$d^5\Phi = \frac{1}{2^{14}\pi^6 m_K^2}\frac{1}{s_\pi}\sqrt{1-\frac{4m_\ell^2}{q^2}}\lambda^{1/2}(m_K^2, p_\pi^2, q^2)\lambda^{1/2}(p_\pi^2, m_{\pi^+}^2, m_{\pi^0}^2)dp_\pi^2 dq^2 d\cos\theta_\pi d\cos\theta_\ell d\phi, \quad (9)$$

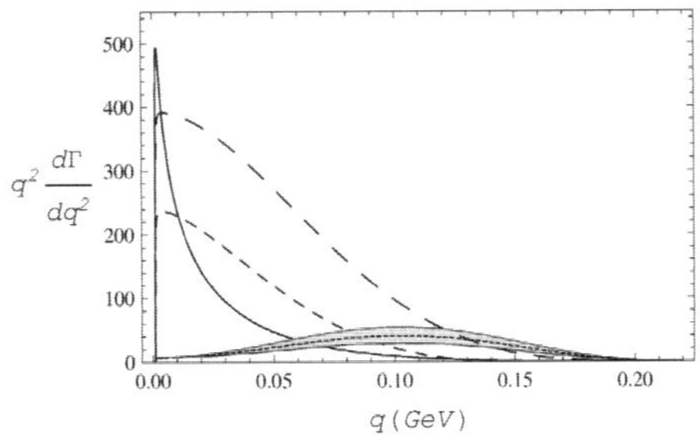

Figure 3. q-dependence of Bremsstrahlung (solid line), the dashed lines (from bigger to smaller) are 100M, 100BE and 300E. Error bars are omitted. The band corresponds to changing counterterms .

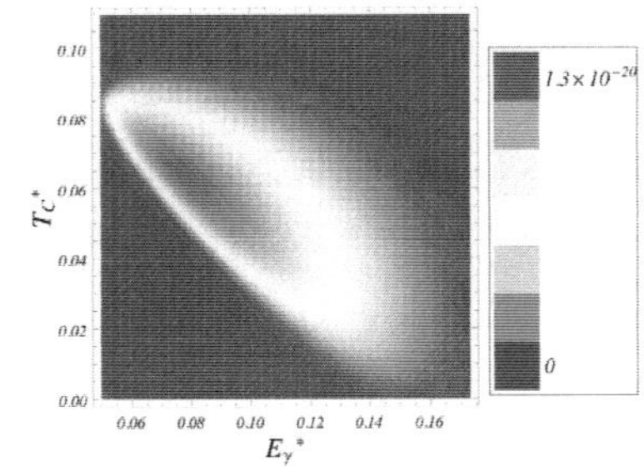

Figure 4. Dalitz plot BM contribution: two-dimensional density projection

The further advantage is that at each q^2 we can measure the different Dalitz plot, as we can see in (4), obtained at $\sqrt{q^2} = 50$ MeV. Also several short distance observables can be measured by appropriate kinematical analyses [10, 15]. In Table 1 we show the ratios of Bremsstrahlung (B) to magnetic (M) and to various interferences. Our effort has been to write the $K^\pm \to \pi^\pm\pi^0 e^+e^-$ amplitude in terms of experimental known Bremsstrahlung (B) and DE, electric (E) and magnetic (M), transitions [10]

Table 1. Branching ratios for the Bremsstrahlung and the relative weight of the rest of the contributions for different cuts in q, starting at q_{min} (first row) and ending at 180 MeV. In the last column we have also included the parity-odd magnetic-electric interference term.

q_c (MeV)	B $[10^{-8}]$	B/M	B/E	B/BE	B/BM
$2m_l$	418.27	71	4405	128	208
55	5.62	12	118	38	44
100	0.67	8	30	71	36
180	0.003	12	5	-19	44

The big news is that NA48/2 has reported a final measurement of this branching: the experiment selects 3 reconstructed tracks coming from one decay vertex, then Particle ID for $e^\pm/\pi^\pm$ separation; then two reconstructed γ clusters are compatible with π^0 mass [16] The number of $K^\pm$ decays (kaon flux) is measured by using the reference channel $K^\pm \to \pi^\pm\pi^0(\gamma)$. 5076 events have been selected with 289 background; the error is dominated by external error on $\mathcal{B}(\pi^0 \to e^+e^-\gamma)$

$$(4.06 \pm 0.06_{stat.} \pm 0.04_{syst.} \pm 0.13_{ext.}) \times 10^{-6}$$

to be compared to our prediction dominated from Bremsstrahlung [10, 15]

$$\mathcal{B}(IB) = (4.19) \times 10^{-6} \qquad \text{no Isospin breaking} \qquad \mathcal{B}(TOT) = (4.29) \times 10^{-6}$$

$$\mathcal{B}(IB) = (4.10) \times 10^{-6} \qquad \text{with Isospin breaking} \qquad \mathcal{B}(TOT) = (4.19) \times 10^{-6}$$

The amplitudes $\mathcal{A}_{8,9}$, odd in θ_ℓ, are also linearly dependent on the final state, establishing a clear way to determine them; while $\mathcal{A}_{5,6,7}$ are generated by interference with the axial leptonic current.

CP violation has been also studied in the $K \to \pi\pi\gamma$ and $K \to \pi\pi ee$ decays.

3. Bardeen Buras Gerard approach and $K^\pm \to \pi^\pm l^+ l^-$ form factor

In Refs. [3, 4], the authors compute loop calculations employing a cut-off regulator instead of the dimensional regularization method. Consequently, their results exhibit a quadratic dependence on the physical cut-off M which is lost in the usual chiral perturbative calculations. This quadratic dependence is a crucial ingredient in the matching of the meson and quark pictures. They argue that one can obtain a parametrization of non-perturbative QCD effects by matching a low-energy Lagrangian valid up to the scale M, to the logarithmic behavior of relevant Wilson coefficients at high-energy. In this work we refer to this computational method as the *Bardeen-Buras-Gérard framework* (BBG). Recentely Buras has advocated this determination to claim New Physics to explain the experimental value of ϵ' [5]. BBG evaluate $K \to \pi\pi$ chiral loop with a dimensional cut-off (M) and match the quadratic divergences of the $Q_i(M)$ weak matrix elements with the log of the short distance Wilson coefficient as described by the equation

$$H_{\text{eff}} = \sum_i c_i(\mu)\, Q_i(\mu) \tag{10}$$

We have studied in the BBG contest the $K^\pm \to \pi^\pm l^+ l^-$ form factor described in CHPT by a loop function, $W_+^{\pi\pi}(z)$ and a polynomial in z with experimentally determined coefficients a_+ and b_+ [17]. These parameterize the intermediate region between low and high energy regimes. Our goal is to predict the values of these two coefficients using BBG framework [6]. Indeed in Fig. 3 we show: i) (left plot) the a_+ as a function of M without the vectors contributions and ii) (right plot) adding the vectors contributions particularly important are the novel weak vector couplings.

4. $K_{S,L} \to l^+l^-$, $K_{S,L} \to l^+l^-l^+l^-$ and $K_S \to \pi^+\pi^-l^+l^-$

The recent LHCB limit on $K_S \to \mu\overline{\mu}$ [13] in Table 4 is close to test interesting New Physics (NP) models [18]. A high precision measurement can test the short distance (SD) SM but it requires to improve the long distance (LD) prediction [18, 19] with auxiliary channels [20]. $K_L \to \mu\mu$: the small ratio SD/LD $\sim \frac{1}{30}$ may obscure an experimental improvement on the rate [18]. The situation would be a bit ameliorated if the still unknown sign of $A(K_L \to \gamma\gamma)$ would be either theoretically or experimentally determined. Help to this ambiguity could come from the experimental study of $K_{S,L} \to l^+l^-l^+l^-$ [20] As shown in Table 4 these channels are at reach in a high intensity machine and they may also give LD distance info needed for a better control of $K_L \to \mu\mu$. These four body decays have also a peculiar feature, similarly to $K_{S,L} \to \pi^+\pi^-e^+e^-$, the two different helicity amplitudes interfere; then one can measure the sign $K_L \to\to \gamma^*\gamma^* \to l^+l^-l^+l^-$ by studying the time interference $K_S\ K_L$ which it has a decay length $2\Gamma_S$ [20].

$K_S \to \pi^+\pi^-l^+l^-$

Actually only $K^+ \to \pi^+\pi^0 e^+e^-$ has been studied so far [10] while $K_S \to \pi^+\pi^-\mu^+\mu^-$, more feasible experimentally, is in progress; however generic features can be already extracted from

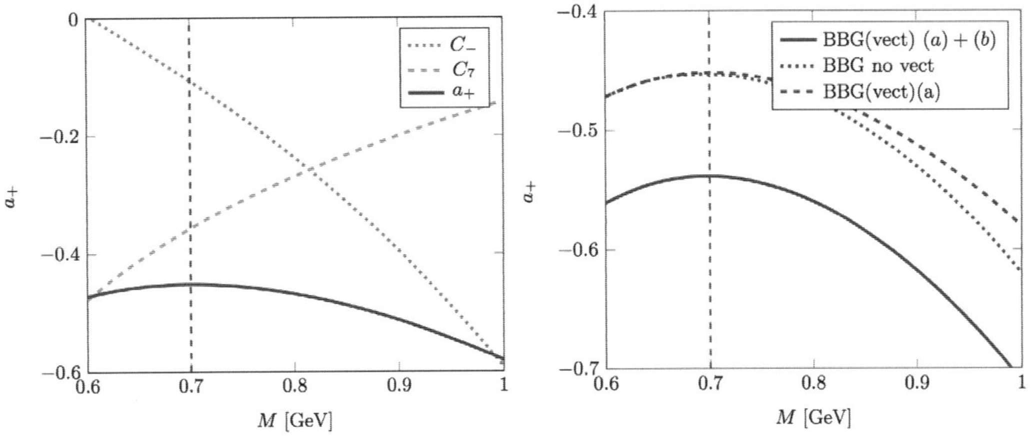

Figure 5. Left: In blue, the variation of a_+ as a function of M in GeV. The dotted green curve represents the contribution proportional to $C_-(M^2)$ and the dashed orange curve the one proportional to $C_7(M^2)$. The vertical dashed line stands for the matching scale Right: a_+ as a function of M in the three different frameworks: 'BBG no vect.' where vectors are not included, 'BBG(vect)(a)' represents the contribution coming only from diagrams with vectors coupled strongly and 'BBG(vect) $(a)+(b)$' is the case where vectors with weak couplings were included. The vertical line indicates the value $M = 0.7$ GeV.

	PDG	Prospects
$K_S \to \mu\mu$	$< 9 \times 10^{-9}$ at 90% CL	(LD)$(5.0 \pm 1.5) \cdot 10^{-12}$ NP $< 10^{-11}$
$K_L \to \mu\mu$	$(6.84 \pm 0.11) \times 10^{-9}$	difficult : SD $<<$ LD
$K_S \to \mu\mu\mu\mu$	–	SM LD $\sim 2 \times 10^{-14}$
$K_S \to ee\mu\mu$	–	$\sim 10^{-11}$
$K_S \to eeee$	–	$\sim 10^{-10}$
$K_S \to \pi^+\pi^-\mu^+\mu^-$	–	SM LD $\sim 10^{-14}$

Table 2. Interesting channels: PDG values vs theoretical estimates

Table I in Ref. [10]: we have less than 10 MeV phase space which can be extracted from the last lines of the Table I in Ref. [10] telling us that i) the Branching is expected $\mathcal{O}(10^{-14})$ and the novel purely electric and magnetic contribution are relatively enhanced with respect to the less interesting bremsstrahlung.

5. Conclusions

In these proceedings we have studied theoretical issues in radiative rare kaon decays. An interesting channel, $K^+ \to \pi^+\pi^0 e^+e^-$, is studied also in connection with the recent experimental NA48 results. Motivated by LHCB results on $K_S \to \mu^+\mu^-$ we discuss other channels like $K_{S,L} \to l^+l^-l^+l^-$. Motivated by recent theoretical work by Buras and collaborators we study also the $K^\pm \to \pi^\pm l^+l^-$ form factor.

Acknowledgments

I wish to thank the organizers of KAON 2016, in particular Cristina Lazzeroni.

6. References

[1] V. Cirigliano, G. Ecker, H. Neufeld, A. Pich and J. Portoles, Rev. Mod. Phys. **84**, 399 (2012) doi:10.1103/RevModPhys.84.399 [arXiv:1107.6001 [hep-ph]].
[2] A. J. Buras, arXiv:1609.05711 [hep-ph].
[3] W. A. Bardeen, A. J. Buras and J. M. Gerard, Nucl. Phys. B **293** (1987) 787. doi:10.1016/0550-3213(87)90091-5; A. J. Buras, J. M. Gérard and W. A. Bardeen, "Large N Approach to Kaon Decays and Mixing 28 Years Later: $\Delta I = 1/2$ Rule, $\hat{B}_K$ and ΔM_K", Eur. Phys. J. C **74**, 2871 (2014)
[4] J. M. Gerard, Acta Phys. Polon. B **21** (1990) 257.
[5] A. J. Buras, M. Gorbahn, S. Jager and M. Jamin, JHEP **1511**, 202 (2015) doi:10.1007/JHEP11(2015)202 [arXiv:1507.06345 [hep-ph]].
[6] E. Coluccio Leskow, G. D'Ambrosio, D. Greynat and A. Nath, Phys. Rev. D **93**, no. 9, 094031 (2016) doi:10.1103/PhysRevD.93.094031 [arXiv:1603.09721 [hep-ph]].
[7] G. D'Ambrosio and D. N. Gao, JHEP **0010**, 043 (2000) doi:10.1088/1126-6708/2000/10/043 [hep-ph/0010122]; L. Cappiello and G. D'Ambrosio, Phys. Rev. D **75**, 094014 (2007) [arXiv:hep-ph/0702292].
[8] J. R. Batley *et al.* [NA48/2 Collaboration], Eur. Phys. J. C **68**, 75 (2010) [arXiv:1004.0494 [hep-ex]].
[9] J. Bijnens, G. Colangelo, G. Ecker and J. Gasser, arXiv:hep-ph/9411311 and references there
[10] L. Cappiello, O. Cata, G. D'Ambrosio and D. -N. Gao, Eur. Phys. J. C **72**, 1872 (2012) [arXiv:1112.5184 [hep-ph]].
[11] L. M. Sehgal and M. Wanninger, Phys. Rev. D **46**, 1035 (1992) [Erratum-ibid. D **46**, 5209 (1992)]; P. Heiliger and L. M. Sehgal, Phys. Rev. D **48**, 4146 (1993) [Erratum-ibid. D **60**, 079902 (1999)]; J. K. Elwood, M. B. Wise, M. J. Savage, Phys. Rev. **D52**, 5095 (1995). [hep-ph/9504288]; J. K. Elwood, M. B. Wise, M. J. Savage, J. W. Walden, Phys. Rev. **D53**, 4078-4081 (1996). [hep-ph/9506287].
[12] E. Abouzaid *et al.* [KTeV Collaboration], Phys. Rev. Lett. **96**, 101801 (2006) [hep-ex/0508010]; A. Lai *et al.* [NA48 Collaboration], Eur. Phys. J. C **30**, 33 (2003).
[13] The Review of Particle Physics (2016) C. Patrignani et al. (Particle Data Group), Chin. Phys. C, 40, 100001 (2016)
[14] H. Pichl, Eur. Phys. J. C **20**, 371 (2001) [hep-ph/0010284].
[15] S. R. Gevorkyan and M. H. Misheva, Eur. Phys. J. C **74**, no. 5, 2860 (2014) [arXiv:1403.1053 [hep-ph]].
[16] Brigitte Bloch-Devaux On behalf of the NA48/2 collaboration "*Study of the K to ? ?0 e+ e- decay with NA48/2 @ CERN*": these proceedings.
[17] G. D'Ambrosio, G. Ecker, G. Isidori and J. Portoles, JHEP **9808**, 004 (1998) doi:10.1088/1126-6708/1998/08/004 [hep-ph/9808289].
[18] G. Isidori and R. Unterdorfer, JHEP **0401**, 009 (2004) [hep-ph/0311084].
[19] G. Ecker and A. Pich, Nucl. Phys. B **366**, 189 (1991);
[20] G. D'Ambrosio, D. Greynat and G. Vulvert, Eur. Phys. J. C **73**, 2678 (2013) [arXiv:1309.5736 [hep-ph]].

Study of the $K^{\pm} \to \pi^{\pm}\pi^{0}e^{+}e^{-}$ decay by the NA48/2 experiment at CERN

B Bloch-Devaux[1,2]

[1] Dipartimento di Fisica Sperimentale del Università di Torino, Torino, Italy

E-mail: brigitte.bloch@cern.ch

Abstract. The first observation of about 5000 candidates, with a 5% background contamination, of the rare decay $K^{\pm} \to \pi^{\pm}\pi^{0}e^{+}e^{-}$is reported by the NA48/2 experiment at CERN. From the analysis of 1.7×10^{11} kaon decays collected in 2003–2004, the preliminary branching ratio in the full kinematic region is measured to be $\mathcal{B}(K^{\pm} \to \pi^{\pm}\pi^{0}e^{+}e^{-}) = (4.22 \pm 0.15) \times 10^{-6}$. The observed value is in perfect agreement with theoretical predictions based on Chiral Perturbation Theory.

1. Introduction

Kaon decays have played a major role in establishing the quark mixing flavour structure of the Standard Model. Radiative kaon decays are dominated by long-distance effects but can still bring information on short-distance physics processes when considered in a detailed Dalitz plot analysis. This is the case of the $K^{\pm} \to \pi^{\pm}\pi^{0}e^{+}e^{-}$decay, never observed so far, which proceeds through virtual photon exchange followed by internal conversion into an electron-positron pair, i.e. $K^{\pm} \to \pi^{\pm}\pi^{0}\gamma^{*} \to \pi^{\pm}\pi^{0}e^{+}e^{-}$.

The virtual γ^{*} can be produced by two different mechanisms: Inner Bremsstrahlung (IB), where the γ^{*} is emitted by one of the charged mesons in the initial or final state and Direct Emission (DE) where the γ^{*} is radiated off at the weak vertex of the intermediate state. Consequently, the differential decay rate consists of three terms: the dominant long-distance IB contribution (pure electric part E), the DE component (electric E and magnetic M parts),

[2] for the NA48/2 Collaboration: G. Anzivino, R. Arcidiacono, W. Baldini, S. Balev, J.R. Batley, M. Behler, S. Bifani, C. Biino, A. Bizzeti, B. Bloch-Devaux, G. Bocquet, N. Cabibbo, M. Calvetti, N. Cartiglia, A. Ceccucci, P. Cenci, C. Cerri, C. Cheshkov, J.B. Chèze, M. Clemencic, G. Collazuol, F. Costantini, A. Cotta Ramusino, D. Coward, D. Cundy, A. Dabrowski, P. Dalpiaz, C. Damiani, M. De Beer, J. Derré, H. Dibon, L. DiLella, N. Doble, K. Eppard, V. Falaleev, R. Fantechi, M. Fidecaro, L. Fiorini, M. Fiorini, T. Fonseca Martin, P.L. Frabetti, L. Gatignon, E. Gersabeck, A. Gianoli, S. Giudici, A. Gonidec, E. Goudzovski, S. Goy Lopez, M. Holder, P. Hristov, E. Iacopini, E. Imbergamo, M. Jeitler, G. Kalmus, V. Kekelidze, K. Kleinknecht, V. Kozhuharov, W. Kubischta, G. Lamanna, C. Lazzeroni, M. Lenti, L. Litov, D. Madigozhin, A. Maier, I. Mannelli, F. Marchetto, G. Marel, M. Markytan, P. Marouelli, M. Martini, L. Masetti, E. Mazzucato, A. Michetti, I. Mikulec, M. Misheva, N. Molokanova, E. Monnier, U. Moosbrugger, C. Morales Morales, D.J. Munday, A. Nappi, G. Neuhofer, A. Norton, M. Patel, M. Pepe, A. Peters, F. Petrucci, M.C. Petrucci, B. Peyaud, M. Piccini, G. Pierazzini, I. Polenkevich, Yu. Potrebenikov, M. Raggi, B. Renk, P. Rubin, G. Ruggiero, M. Savrié, M. Scarpa, M. Shieh, M.W. Slater, M. Sozzi, S. Stoynev, E. Swallow, M. Szleper, M. Valdata-Nappi, B. Vallage, M. Velasco, M. Veltri, S. Venditti, M. Wache, H. Wahl, A. Walker, R. Wanke, L. Widhalm, A. Winhart, R. Winston, M.D. Wood, S.A. Wotton, A. Zinchenko, M. Ziolkowski.

Content from this work may be used under the terms of the Creative Commons Attribution 3.0 licence. Any further distribution of this work must maintain attribution to the author(s) and the title of the work, journal citation and DOI.
Published under licence by IOP Publishing Ltd

and their interference. The interference term collects the different contributions, IBE, IBM and EM. In the related $K^{\pm} \to \pi^{\pm}\pi^{0}\gamma$ mode the interference consists only of the IBE term[1], because the remaining interferences are P-violating, and cancel out upon angular integration. In the $K^{\pm} \to \pi^{\pm}\pi^{0}e^{+}e^{-}$mode, the IBM term is a novel interference contribution. For this reason the $K^{\pm} \to \pi^{\pm}\pi^{0}e^{+}e^{-}$ decay offers interesting short and long-distance parity-violating observables that can be studied in the 3D space that now includes the non-zero $q^2 = M_{ee}^2$ kinematic variable. There are only few theoretical publications related to the $K^{\pm} \to \pi^{\pm}\pi^{0}e^{+}e^{-}$ mode [2][3][4]. The authors of [2] were able to predict, on the basis of the NA48/2 measurement of the magnetic and electric terms in the $K^{\pm} \to \pi^{\pm}\pi^{0}\gamma$ decay mode [5], the branching ratio of the single components. No experimental observation has so far been reported.

2. The NA48/2 experiment

The NA48/2 experiment at the CERN SPS was specifically designed for charge asymmetry measurements in the $K^{\pm} \to 3\pi$ decay modes [6]. Large sample of charged kaon ($K^{\pm}$) decays were collected during the 2003–2004 data taking period. The experiment beam line had been designed to deliver simultaneous narrow momentum band K^{+} and K^{-} beams originating from primary 400 GeV/c protons extracted from the CERN SPS and impinging on a beryllium target. Secondary beams with central momenta of 60 GeV/c and a momentum band of ±3.8% (rms) were selected and brought to a common beam axis. The beam kaons decayed in a fiducial decay volume contained in a 114 m long cylindrical evacuated tank. The momenta of charged decay products were measured in a magnetic spectrometer, housed in a tank filled with helium and placed after the decay volume. The spectrometer was composed of four drift chambers (DCH) and a dipole magnet providing in the horizontal plane a momentum kick $\Delta p = 120$ MeV/c to charged particles. The momentum resolution achieved was $\sigma_p/p = (1.02 \oplus 0.044 \cdot p)\%$ (p in GeV/c). A hodoscope (HOD) consisting of two planes of plastic scintillators segmented into vertical and horizontal strip-shaped counters followed the spectrometer. The HOD surface was logically subdivided into 16 exclusive square regions producing fast signals used to trigger the detector readout on charged track topologies. Further downstream was a liquid krypton electromagnetic calorimeter (LKr), an almost homogeneous ionization chamber with an active volume of 7 m^3 of liquid krypton, $27X_0$ deep, segmented transversally into 13248 projective $\sim 2\times2$ cm^2 cells and with no longitudinal segmentation. The energies of photons and electrons were measured with a resoltion $\sigma_E/E = (3.2/\sqrt{E} \oplus 9.0/E \oplus 0.42)\%$ (E in GeV). An iron/scintillator hadronic calorimeter and muon detectors were located further downstream. A dedicated two-level trigger was used to collect three track decays with a very high efficiency. A detailed description of the detector can be found in [7].

3. First observation of the $K^{\pm} \to \pi^{\pm}\pi^{0}e^{+}e^{-}$ decay

The signal mode is selected simultaneously with the normalization mode ($K^{\pm} \to \pi^{\pm}\pi^{0}_{D}, \pi^{0}_{D} \to \gamma e^{+}e^{-}$) chosen to have a final state topology ($\pi^{\pm}\gamma e^{+}e^{-}$) differing from the signal ($\pi^{\pm}\gamma\gamma e^{+}e^{-}$) by only one photon while satisfying similar kinematic constraints on the reconstructed π^0 and kaon masses.

3.1. Candidate selections and background estimates

Both signal and normalization candidates are reconstructed from three charged tracks (two same-sign tracks and one opposite-charge track) forming a common vertex in the fiducial decay volume (the vertex charge is then $q_{vtx} = \pm1$). The tracks are required to be in time within 5 ns of each other using the HOD time of the hit associated to the track and their impact point should lie within the geometrical acceptance of the drift chambers (radial distance to the beam line axis, monitored by fully reconstructed $K^{\pm} \to \pi^{\pm}\pi^{+}\pi^{-}$ ($K_{3\pi}$) events, larger than 12 cm). The track momenta are required to be in the range (2–60) GeV/c and any track-to-track distance at

DCH1 should be larger than 2 cm to suppress photon conversions to an e^+e^- pair in the upstream material. The first level trigger requires coincidences of hits in the two HOD planes in at least two of the 16 square regions. To avoid inefficiencies of geometrical origin, configurations where the three considered tracks, extrapolated to the HOD front face from positions and directions in DCH4, have their impact points in a single quadrant are rejected. This affects 2.3% of the signal simulated events and is negligible for the normalisation events while it restores high and similar trigger efficiencies for both modes.

Isolated energy clusters without associated track in the LKr, in time within 5 ns with the vertex time (mean of the three track times) are identified as the photon candidates. The minimum photon energy required is 2 GeV. The photon four-momenta are reconstructed assuming they originate from the 3-track vertex.

In the signal selection, the two-photon invariant mass is required to be within ± 15 MeV/c^2 from the nominal PDG[8] π^0 mass. The reconstructed invariant mass of the $\pi^\pm\pi^0e^+e^-$system is required to be within $\pm$ 45 MeV/c^2 from the nominal PDG $K^\pm$ mass. In the normalization selection, the same cuts applies to the invariant mass of the γe^+e^- system (π^0_D) and to the $\pi^\pm\pi^0_D$ system, respectively. The chosen ranges are loose cuts with respect to the achieved resolution for each of the reconstructed masses, minimize the dependence on possible resolution effects and maximize both acceptances.

The total reconstructed momentum is required to be in the range (54–66) GeV/c, compatible with the beam momentum and both transverse positions, of the 3-track vertex and of the reconstructed kaon at the LKr front face (defined as the energy weighted sum of the non-deviated tracks and photon(s) positions), to be consistent with the beam position as monitored by $K_{3\pi}$ events.

Kinematic identification is used instead of standard electron-pion separation based on the E/p ratio of the energy deposited in LKr and momentum measured in the spectrometer. This choice reduces the impact of particle mis-identification and increases both acceptances. The procedure assigns an electron mass to the single track with its charge opposite to q_{vtx}. The remaining electron-pion ambiguity for the two other same-sign tracks is then solved by testing the two mass hypotheses against the full selection.

The last requirement relies on the strong correlation between the reconstructed π^0 and kaon masses. This correlation can be seen in Figure 1a (Figure 2a) for the normalization (signal) mode, respectively. When selecting events within a band defined as $|M_{\pi^0} - 0.42 \cdot M_K + 73.2| < 6$ MeV/c^2 (masses in MeV/c^2), more than 99% of the normalisation simulated events and 96.5% of the signal simulated events are kept.

Two main sources of background are contributing to the signal final state: $K^\pm \to \pi^\pm\pi^0\pi^0_D$ ($K_{3\pi D}$) when one of the photons is lost, and $K^\pm \to \pi^\pm\pi^0_D(\gamma)$ ($K_{2\pi D}$), when an extra photon combines with the Dalitz photon to mimic a $\pi^0 \to \gamma\gamma$ decay. An additional suppression of the $K_{3\pi D}$ background events is obtained by requiring the squared invariant mass of the $\pi^+\pi^0$ system to be greater than 0.12 GeV$^2/c^4$, exploiting the larger phase space available in the signal mode. To reject further $K_{2\pi D}$ background contamination, each of the two possible invariant masses $M_{ee\gamma}$ are required to be more than 7 MeV/c^2 away from the nominal mass of the neutral pion. The background processes contributing to the normalization mode are $K_{\mu 3D}(K^\pm \to \mu^\pm\nu\pi^0_D)$ and K_{e3D} ($K^\pm \to e^\pm\nu\pi^0_D$) where the π^0_D is correctly reconstructed but the muon (electron) is given a pion mass and therefore can sometimes satisfy all other selection criteria.

Samples of $16.8 \times 10^6 K_{2\pi D}$ candidates and 5076 signal candidates have been selected from a subset of a 1.7×10^{11} kaon decay exposure in 2003–2004. The background estimates from simulation amount to (15182 $\pm$ 173) $K_{\mu 3D}$ events and (10334 $\pm$ 140) K_{e3D} events in the normalisation mode, corresponding to a total background contamination of 0.15%. In the signal mode, the background estimates from simulation amount to (159 $\pm$ 8) events from $K_{3\pi D}$ and (130 $\pm$ 24) events from $K_{2\pi D}$, adding up to a background contamination of 5.7%. Reconstructed

π^0_D and $\pi^\pm\pi^0_D$ mass distributions are displayed in Figures 1b,c for the selected normalisation candidates. Reconstructed $\gamma\gamma$ and $\pi^\pm\pi^0e^+e^-$masses are shown in Figures 2b,c for the signal candidates. Expected background and normalization (signal) simulations, normalized to the number of observed candidates, show a good agreement with the data distributions.

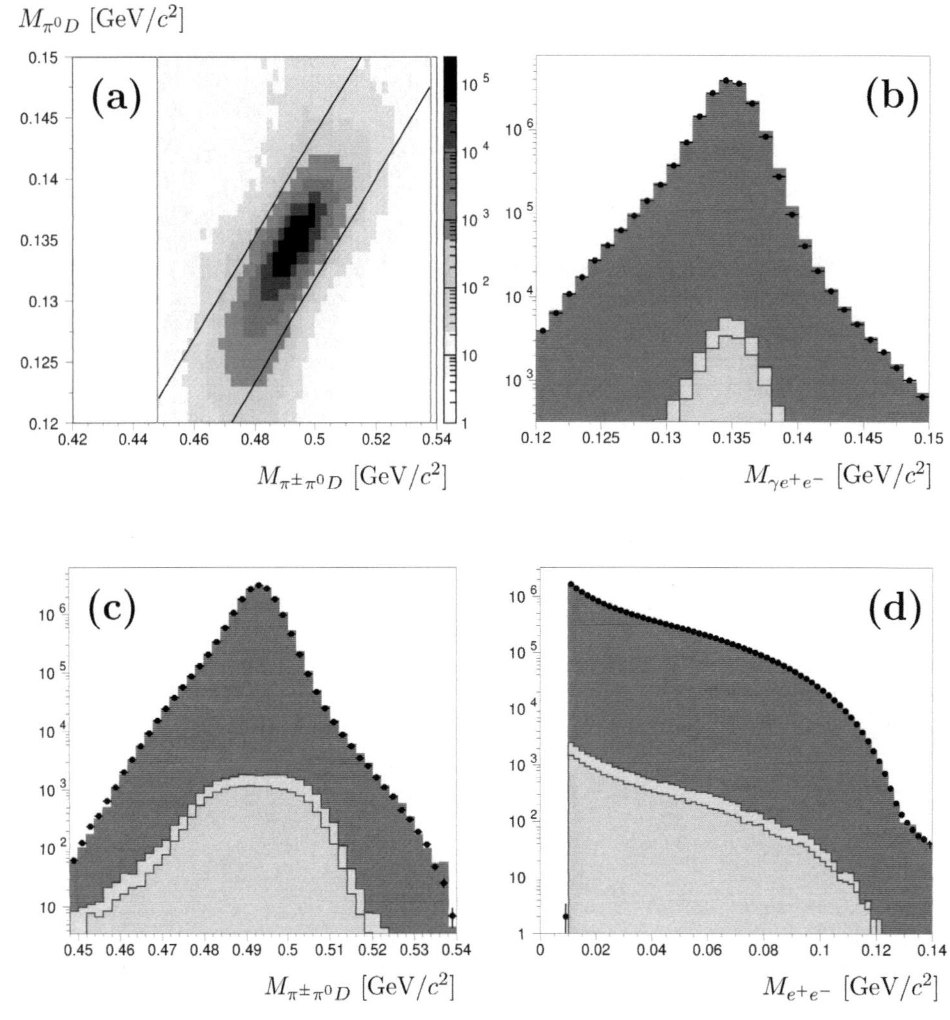

Figure 1. Normalization candidates: **(a)** Reconstructed simulated signal events in the plane $(M_{\pi^\pm\pi^0D}, M_{\pi^0D})$. The vertical lines shows the $M_{\pi^\pm\pi^0D}$ selected range and the slanted band corresponds to the kinematic constraint. **(b)** Reconstructed invariant mass M_{π^0D} distribution. **(c)** Reconstructed invariant mass $M_{\pi^\pm\pi^0D}$ distribution. **(d)** Reconstructed e^+e^- invariant mass distribution. The selection requires $M_{e^+e^-} > 10$ MeV/c^2 well above threshold where the acceptance is steeply falling. In the invariant mass plots, full dots correspond to data candidates, stacked histograms are, from bottom to top, the expected $K_{\mu 3D}$ and K_{e3D} backgrounds estimated from simulation and normalization best simulation including radiative effects.

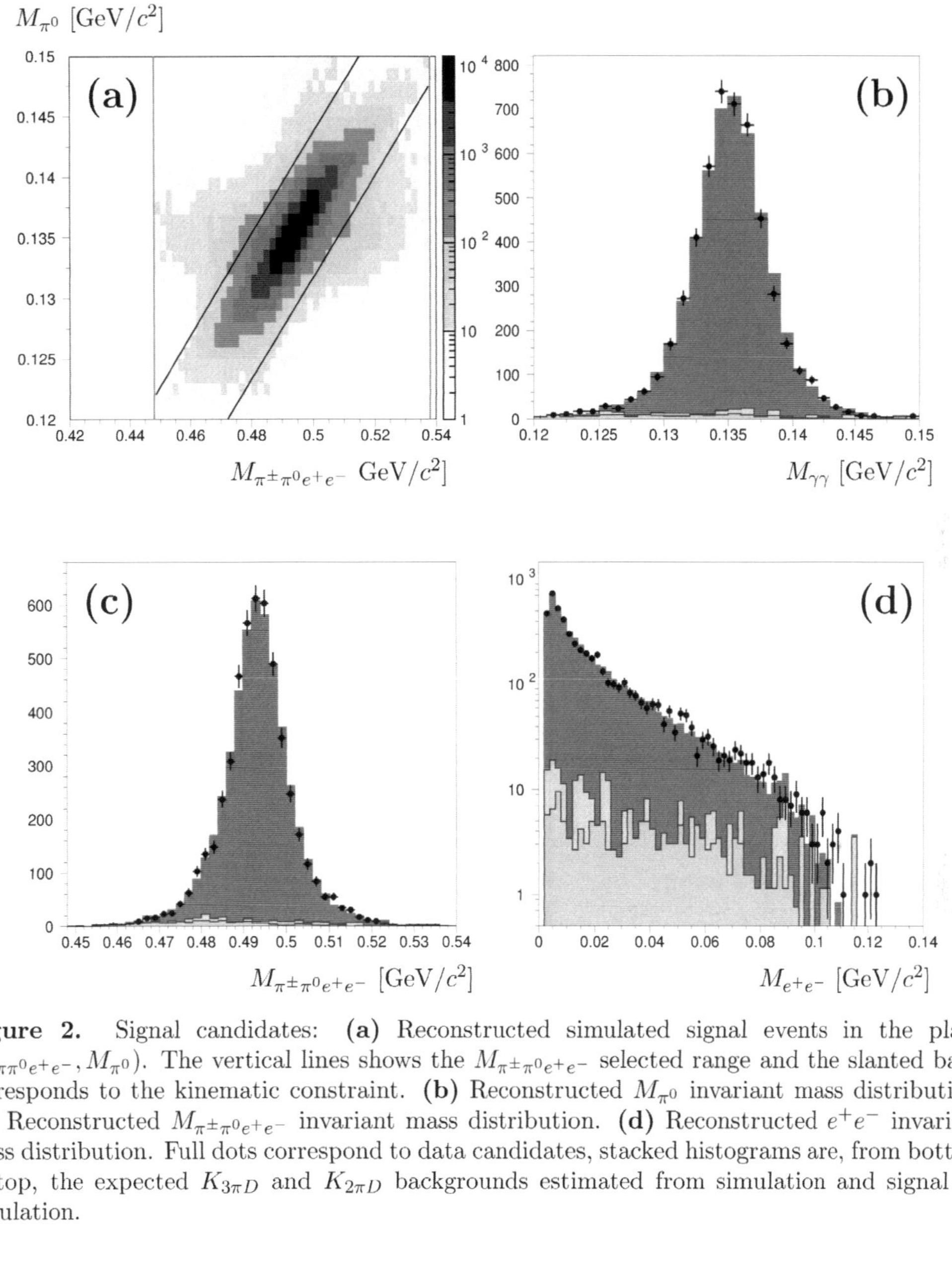

Figure 2. Signal candidates: **(a)** Reconstructed simulated signal events in the plane $(M_{\pi\pi^0e^+e^-}, M_{\pi^0})$. The vertical lines shows the $M_{\pi^\pm\pi^0e^+e^-}$ selected range and the slanted band corresponds to the kinematic constraint. **(b)** Reconstructed M_{π^0} invariant mass distribution. **(c)** Reconstructed $M_{\pi^\pm\pi^0e^+e^-}$ invariant mass distribution. **(d)** Reconstructed e^+e^- invariant mass distribution. Full dots correspond to data candidates, stacked histograms are, from bottom to top, the expected $K_{3\pi D}$ and $K_{2\pi D}$ backgrounds estimated from simulation and signal IB simulation.

3.2. Branching ratio measurement

The Branching Ratio of the $K^\pm \to \pi^\pm\pi^0e^+e^-$decay mode is obtained using the expression:

$$\mathcal{B}(K^\pm \to \pi^\pm\pi^0e^+e^-) = \frac{(N_S - N_{Bs})}{(N_N - N_{Bn})}\frac{A_N\epsilon_N}{A_S\epsilon_S}\mathcal{B}(N) \quad (1)$$

where $N_{S,Bs,N,Bn}$ are the number of signal (5076), background to signal (289 ± 25), $K_{2\pi D}$ (16.8×10^6) and corresponding background (25517 ± 223) events. $A_{S,N}$ and $\epsilon_{S,N}$ are the acceptances and trigger efficiencies of the signal and normalization modes. The trigger efficiency (ϵ_N), is measured using control data samples. It is also evaluated on simulated samples and found to be well reproduced (ϵ_N = 98.4%). Because of the very limited data statistics, simulated signal samples are therefore used to evaluate the signal trigger efficiency (ϵ_S = 99.2%) . The normalization mode branching ratio $\mathcal{B}(N) = (2.427 \pm 0.073) \times 10^{-3}$ is obtained from the PDG[8] world average. The acceptances of the signal, the normalization and the background channels are computed using a GEANT3-based [9] Monte Carlo (MC) simulation which includes full detector geometry and material description, stray magnetic fields, accurate beam line geometry, local detector imperfections and time variations of the above throughout the data taking period. The acceptance A_N (4.08%) is computed using the implementation of $K^{\pm} \rightarrow \pi^{\pm}\pi^0$ according to [10] followed by π^0_D according to the most recent calculations [11], including the best current implementation of radiative effects. The agreement between data and simulation can be seen from the $M_{e^+e^-}$ distribution of Figure 1d. The MC simulations for the different $\pi\pi^0e^+e^-$ contributions IB, DE, and the electric interference IBE, have been generated separately according to the theoretical description given in [2] neglecting the magnetic interference in the present preliminary result. The signal acceptance A_S (0.67%) has been obtained from a weighted average of the single component acceptances, using as weights the relative contributions with respect to IB computed in [2]:

$$A_S = \frac{A_{IB} + A_{DE} \cdot w_{DE} + A_{INT} \cdot w_{INT}}{1 + w_{DE} + w_{INT}} \tag{2}$$

To take into account the E,M measurement uncertainties in [5], the weights entering the total signal acceptance were varied accordingly resulting in a 0.25% relative change quoted as the systematic uncertainty due to the acceptance modeling. As radiative corrections to the $\pi\pi^0e^+e^-$ mode are not considered in [2], the signal MC simulation included both the classical Coulomb attraction/repulsion between charged particles and the real photon(s) emission as implemented in the PHOTOS package. The preliminary result for the total branching ratio is obtained:

$$\mathcal{B}(K^{\pm} \rightarrow \pi^{\pm}\pi^0e^+e^-) = (4.22 \pm 0.06_{stat.} \pm 0.04_{syst.} \pm 0.13_{ext.}) \times 10^{-6} \tag{3}$$

where systematic errors include uncertainties on acceptance, trigger efficiencies and radiative corrections. The external error originating from the normalisation mode branching ratio uncertainty is the dominant error in the present measurement obtained with an overall precision of about 3%.

In summary, the first experimental measurement of the branching ratio $\mathcal{B}(K^{\pm} \rightarrow \pi^{\pm}\pi^0e^+e^-)$ $= (4.22 \pm 0.15) \times 10^{-6}$ is in very good agreement with the theoretical predictions [2] of 4.19×10^{-6} without Isopsin breaking corrections and 4.10×10^{-6} including Isospin breaking corrections. The NA48/2 data sample analyzed has no sensitivity to the DE and INT contributions to the M_{ee} spectrum within the current statistics (Figure 2d). It will be difficult to perform a full Dalitz plot analysis without a proper description of the radiative effects, particularly relevant in a final state with two electron/positron.

References

[1] Christ N 1967 Phys. Rev. **159** 1292
[2] Cappiello L, Cata O, D'Ambrosio G and Gao D N 2012 Eur. Phys. J. C **72** 1872
[3] Pichl H 2001 Eur. Phys. J. C **20** 371
[4] Gevorkyan S R and Misheva M H 2014 Eur. Phys. J. C **74** 2860
[5] Batley J R *et al.* [NA48/2 Collaboration] 2010 Eur. Phys. J. C **68** 75
[6] Batley J R *et al.* [NA48/2 Collaboration] 2007 Eur. Phys. J. C **52** 875
[7] Fanti V *et al.* 2007 Nucl. Instrum. Methods A **574** 443
[8] Patrignani C *et al.* [Particle Data Group] 2016 Chin. Phys. C **40** 100001
[9] GEANT3 Detector Description and Simulation Tool 1994 CERN Program Library W5013
[10] Gatti C 2006 Eur. Phys. J. C **45** 417
[11] Husek T, Kampf K and Novotny J 2015 Phys. Rev. D **92** 054027

IOP Conf. Series: Journal of Physics: Conf. Series **800** (2017) 012030 doi:10.1088/1742-6596/800/1/012030

$K \to \pi \ell^+ \ell^-$: Status and update

Jorge Portolés

IFIC, Universitat de València – CSIC, Apt. Correus 22085, E-46071 València, Spain

E-mail: Jorge.Portoles@ific.uv.es

Abstract. $K \to \pi \ell^+ \ell^-$ decays, $\ell = e, \mu$, will be one of the goals of future kaon facilities, like NA62 (CERN) and K^{0}TO (J-PARC). We review briefly the theoretical status of those processes.

1. Introduction

Suppressed (rare) kaon decays in the Standard Model (SM) constitute an interesting benchmark where to look for New Physics. Common sense and logic tell us that, in any observable, in order to look for physics beyond the SM one needs to have a reliable SM prediction and this is not obvious in many kaon decays. Indeed low-energy strong interacting effects, of difficult implementation, may be involved in those determinations [1].

$K \to \pi \nu \overline{\nu}$ are atypical, poorly measured, decays that are dominated by short-distance interactions [2, 3] and, accordingly, have a very accurate prediction within the SM. This is the reason why they are the main goal of present dedicated facilities: $K^+ \to \pi^+ \nu \overline{\nu}$ at NA62 (CERN) and $K_L \to \pi^0 \nu \overline{\nu}$ at K^{0}TO (J-PARC) [4]. These devices should be able, in addition, to get samples of other related processes like $K^+ \to \pi^+ \ell^+ \ell^-$ ($\ell = e, \mu$) at NA62 or $K_L \to \pi^0 \gamma\gamma$ at K^{0}TO (as we will comment the latter is relevant to determine the CP-conserving contribution to $K_L \to \pi^0 \ell^+ \ell^-$).

In this note we summarize the theoretical status of the $K \to \pi \ell^+ \ell^-$ decays. These are $\Delta S = 1$ weak neutral current transitions in the SM and, hence, very suppressed . Contrarily to the case above, these decays are fairly dominated by strongly interacting low-energy dynamics which is always harder to determine. As in many other processes of analogous features, Chiral Perturbation Theory (ChPT) is the appropriate tool to handle these decays [5, 6]. They were first studied within this framework by Ecker, Pich and De Rafael [7, 8] when they computed the dominant octet ($\Delta I = 1/2$) contributions at the leading $\mathcal{O}(p^4)$ in ChPT. This set up a systematic approach that would pave the way for later improvements, as we will see. The subdominant contributions of $SU(3)$ 27-representation operators, at $\mathcal{O}(p^4)$, have been carried out in Ref. [9].

$K^+, K_S \to \pi \ell^+ \ell^-$ decays are dominated by single virtual-photon exchange, $K \to \pi \gamma^*$. However this contribution is CP violating for $K_L \to \pi^0 \ell^+ \ell^-$ and, consequently, the latter is driven by substantially different dynamics. The theoretical status of the first processes will be set forth in Section 1. Meanwhile Section 2 will be dedicated to the description of the K_L decay.

A more detailed account of $K \to \pi \ell^+ \ell^-$ decays and their corresponding bibliography can be looked up in Ref. [1].

Content from this work may be used under the terms of the Creative Commons Attribution 3.0 licence. Any further distribution of this work must maintain attribution to the author(s) and the title of the work, journal citation and DOI.
Published under licence by IOP Publishing Ltd

2. $K^+, K_S \to \pi \ell^+ \ell^-$

CP allowed $K(k) \to \pi(p)\ell^+\ell^-$, $\ell = e, \mu$, decays are dominated by single virtual photon exchange $K \to \pi\gamma^*$. This is the case for K^+ and K_S decays. Hence the amplitude is determined by one electromagnetic transition form factor in the presence of nonleptonic weak interactions [7, 10]:

$$i \int d^4x \, e^{iqx} \langle \pi(p)| T \left\{ J^{\mu}_{em}(x) \, \mathcal{L}_{\Delta S=1}(0) \right\} |K_j(k)\rangle = \frac{G_F M_K^2}{(4\pi)^2} V_j(z) \left[z(k+p)^{\mu} - (1-r_{\pi}^2) q^{\mu} \right], \quad (1)$$

with $j = +, S$, to distinguish both decays, and $q = k-p$, $z = q^2/M_K^2$, $r_P = M_P/M_K$. Here J^{μ}_{em} is the electromagnetic current and $\mathcal{L}_{\Delta S=1}$ the strangeness changing nonleptonic weak Lagrangian. The spectrum in the dilepton invariant mass is then given by:

$$\frac{d\Gamma_j}{dz} = \frac{G_F^2 \alpha^2 M_K^5}{12\pi(4\pi)^4} \lambda^{3/2}(1, z, r_{\pi}^2) \sqrt{1 - 4\frac{r_{\ell}^2}{z}} \left(1 + 2\frac{r_{\ell}^2}{z}\right) |V_j(z)|^2, \quad (2)$$

with $\lambda(a,b,c) = (a+b-c)^2 - 4ab$ and $4r_{\ell}^2 \leq z \leq (1-r_{\pi})^2$.

The $V_j(z)$ form factor can be determined within the ChPT framework, i.e. in a model-independent way. Because gauge invariance, $V_j(z) = 0$ at $\mathcal{O}(p^2)$ in the chiral expansion. Hence the leading contribution is $\mathcal{O}(p^4)$ that was computed in Ref. [7]. Although a complete study to next-to-leading $\mathcal{O}(p^6)$ has not been performed yet, the dominant unitarity corrections from $K \to \pi\pi\pi$, shown in Fig. 1, were studied in Ref. [10], that constitute the present *state-of-the-art* (there is also a tiny contribution from a kaon running in the loop, that obviously has no absorptive part). The full result can be decomposed in a polynomial contribution (linear in z

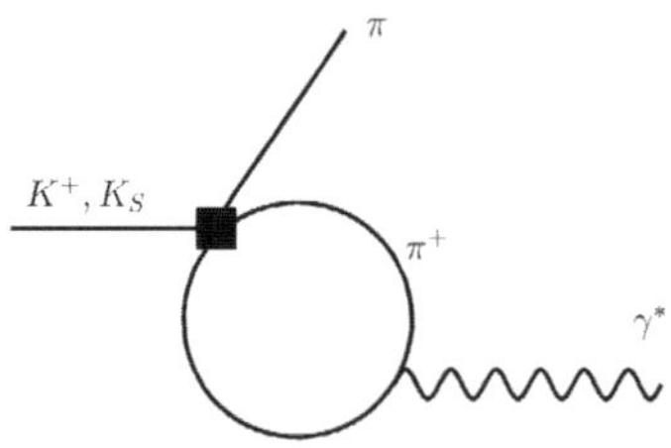

Figure 1. $K \to \pi\pi\pi$ contribution to $K^+, K_S \to \pi\gamma^*$.

up to this order) plus a unitarity loop correction :

$$V_j(z) = a_j + b_j z + V_j^{\pi\pi}(z), \qquad (j = +, S). \quad (3)$$

The polynomial part is, unfortunately, not fully given by chiral symmetry. It translates, mainly, into the Low Energy Couplings (LECs) of the ChPT framework, and it encodes contributions from higher energy spectra that have been integrated out in the process to reach the low-energy effective theory. a_j gets contributions starting at $\mathcal{O}(p^4)$ while b_j, but for a tiny $\mathcal{O}(p^4)$ kaon loop input in Fig. 1, starts at $\mathcal{O}(p^6)$. As we will comment we have, up to now, only model-dependent theoretical predictions for these parameters.

The unitarity corrections in $V_j(z)$ are given by :

$$V_j^{\pi\pi}(z) = \overbrace{\frac{\alpha_j + \beta_j(z-z_0)/r_{\pi}^2}{G_F M_K^2 r_{\pi}^2}}^{K\to\pi\pi\pi} \overbrace{\left[\frac{4}{9} - \frac{4}{3z} + \frac{4}{3z}\left(1 - \frac{z}{4}\right) G(z)\right]}^{\text{1-loop}} \overbrace{\left[1 + \frac{z}{r_{\rho}^2}\right]}^{F_V(z)}, \quad (4)$$

where $z_0 = r_{\pi}^2 + 1/3$. We now explain the structure of the unitarity corrections:

- The interaction of the vertex $K \to \pi\pi\pi$ in Fig. 1 is driven by the α_j and β_j constants that are expressed in terms of the parameterizations of Ref. [11]:

$$\begin{aligned} \alpha_+ &= \beta_1 - \frac{1}{2}\beta_3 + \sqrt{3}\gamma_3\,, \qquad & \beta_+ &= 2(\xi_1 + \xi_3 - \xi_3') \\ \alpha_S &= -\frac{4}{\sqrt{3}}\gamma_3\,, & \beta_S &= \frac{8}{3}\xi_3'. \end{aligned} \tag{5}$$

  It can be noticed that α_+, β_+ are given by both $\Delta I = 1/2, 3/2$ transitions, while α_S, β_S come only from $\Delta I = 3/2$ transitions, because these are the only ones that drive the CP conserving contribution to $K_S \to \pi^0\pi^+\pi^-$. As a consequence $V_S^{\pi\pi}(z)$ is fairly suppressed in comparison with $V_+^{\pi\pi}(z)$.
- The 1-loop function $G(z)$ is given in Ref. [1].
- The $\pi^+\pi^-\gamma^*$ vertex in the one-loop diagram is dressed through the vector form factor in the Resonance Chiral Theory framework [12]:

$$F_V(z) = \frac{M_\rho^2}{M_\rho^2 - M_K^2 z} \simeq 1 + \frac{z}{r_\rho^2}\,, \tag{6}$$

If, as expected, $a_j, b_j \sim \mathcal{O}(1)$, the polynomial contribution fairly dominates over the unitarity-cut loop corrections given by $V_j^{\pi\pi}(z)$. Hence these decays are very sensitive to the chiral LECs. Several models have been considered for their computation. One of them comes to mind immediately from the comparison of the electromagnetic form factor in Eq. (6) and the weak polynomial term in $V_j(z)$ (3). One may consider that this later term could also be dominated by a resonance transition (VMD). This would predict:

$$\left\{ \quad a_j\, b_j > 0, \qquad\qquad \frac{b_j}{a_j} = \frac{M_K^2}{M_\rho^2} \simeq 0.4 \quad \right\}^{\text{VMD model}}. \tag{7}$$

Other models have also been employed: weak deformation model [13], minimal hadronic approximation [14], Padé type [15] or large-N_C considerations [16]. Recently RCB and UKQCD collaborations [17, 18] have studied these decays within lattice QCD and have produced:

$$a_+ = 1.4 \pm 0.7, \qquad\qquad b_+ = 0.7 \pm 0.8\,, \tag{8}$$

that are in good agreement with the VMD model prediction (7). It has to be noted that this is a first try in the lattice. They are working with unphysically heavy pion and kaon masses and, accordingly, an extrapolation to the physical point is needed. We should expect an improvement from the lattice in the near future.

The values of a_j, b_j, can be extracted from the experimental data by using our model-independent description of $V_j(z)$ in Eq. (3) as they are the only unknowns. The present situation, given by NA48/1 and NA48/2 is depicted in Table 1. While in K^+ decay both, branching and spectrum, have been determined, in the K_S decay we only have the branching ratio. As a result only the a_S parameter has been resolved in the neutral decay, while we have results for both a_+ and b_+. By comparing the prediction of the VMD model in Eq. (7) with the phenomenological results for a_+ and b_+ in Table 1 we notice that the only agreement is in the common sign of both parameters but not in their ratio. Hence one could consider either that the model is not correct or that higher order chiral corrections are not negligible. This later hypothesis would be rather unexpected.

The spectra of $K^+ \to \pi^+e^+e^-$ decay is shown in Figure 2 while the one of $K^+ \to \pi^+\mu^+\mu^-$ is shown in Figure 3. In those figures we also show the theoretical prediction as given by the

Process	BR $\times 10^9$	a_j	b_j	b_j/a_j
$K^+ \to \pi^+ e^+ e^-$ [19]	314 ± 10	-0.578 ± 0.016	-0.779 ± 0.066	~ 1.35
$K^+ \to \pi^+ \mu^+ \mu^-$ [20]	96.2 ± 2.5	-0.575 ± 0.039	-0.813 ± 0.145	~ 1.41
$K_S \to \pi^0 e^+ e^-$ [21]	$5.8^{+2.9}_{-2.4}$	$\lvert 1.06 \rvert^{+0.26}_{-0.21}$	-	-
$K_S \to \pi^0 \mu^+ \mu^-$ [22]	$2.0^{+1.5}_{-1.2}$	$\lvert 1.54 \rvert^{+0.40}_{-0.32}$	-	-

Table 1. Experimental determinations of $K \to \pi \ell^+ \ell^-$ by NA48/1 (K_S) and NA48/2 (K^+).

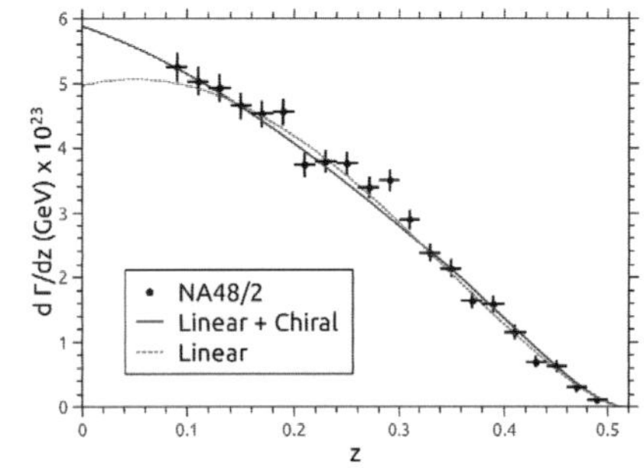

Figure 2. Spectra for $K^+ \to \pi^+ e^+ e^-$ in the dilepton invariant mass. NA48/2 data from [19].

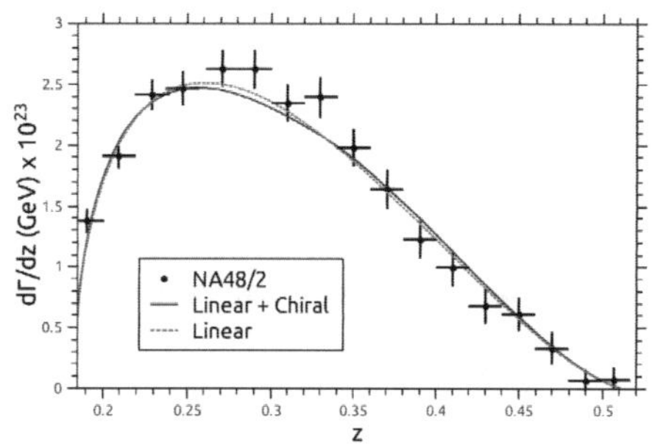

Figure 3. Spectra for $K^+ \to \pi^+ \mu^+ \mu^-$ in the dilepton invariant mass. NA48/2 data from [20].

full $V_+(z)$ in Eq. (3) (Linear + Chiral) and keeping the polynomial part only (Linear) for the a_+ and b_+ parameters in Table 1. As can be noticed this latter contribution dominates and the role of the unitarity loop is rather small.

In addition to the dominant $K^\pm \to \pi^\pm \gamma^*$ amplitude, the form factor $V_+(z)$ receives a short-distance contribution from the operator Q_{7V} from the effective Lagrangian [1]. Although negligible in the branching ratios and spectra, its interference with the long-distance amplitude leads to a CP-violating charge asymmetry in $K^\pm \to \pi^\pm \ell^+ \ell^-$ [8], defined by:

$$\Delta\Gamma \equiv \frac{\Gamma(K^+ \to \pi^+ e^+ e^-) - \Gamma(K^- \to \pi^- e^+ e^-)}{\Gamma(K^+ \to \pi^+ e^+ e^-) + \Gamma(K^- \to \pi^- e^+ e^-)} \sim 0.07 \, \mathrm{Im}\, \lambda_t, \tag{9}$$

where $\lambda_q = V_{qd} V_{qs}^*$. For the numerical determination we have used that $\mathrm{Im}\, b_+/\mathrm{Im}\, a_+ \simeq M_K^2/M_\rho^2$ (VMD model). With $\mathrm{Im}\, \lambda_t \sim \eta \lambda^5 A^2 \sim 10^{-4}$, this asymmetry is tiny within the SM [10, 23]. Experimentally the present bound is given by $\Delta\Gamma < 0.021$ at 90% C.L. [19].

An interesting observable could be the unintegrated charge asymmetry [10] defined by:

$$\delta\Gamma(z) \equiv \frac{\frac{d\Gamma}{dz}(K^+ \to \pi^+ e^+ e^-) - \frac{d\Gamma}{dz}(K^- \to \pi^- e^+ e^-)}{\Gamma(K^+ \to \pi^+ e^+ e^-) + \Gamma(K^- \to \pi^- e^+ e^-)}. \tag{10}$$

In order to skip the background produced by the Dalitz decay of a π^0, namely $K^\pm \to \pi^\pm \pi_D^0$, $\pi_D^0 \to e^+ e^- \gamma$, a cut in the experimental data excludes events with $z < 0.08$. The unintegrated asymmetry, as seen in Fig. 4, stays a higher values.

3. $K_L \to \pi^0 \ell^+ \ell^-$

The decay $K_L \to \pi^0 \gamma^*$ is CP violating. As a consequence the dynamics of the $K_L \to \pi^0 \ell^+ \ell^-$ decay is more involved than the processes above. Indeed both long-distance and short-distance

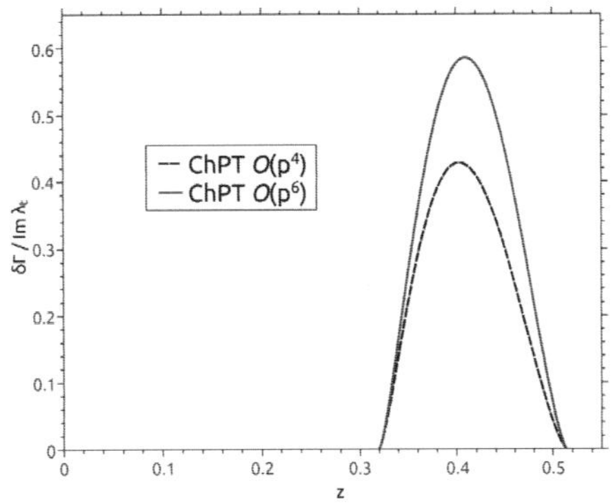

Figure 4. ChPT predictions for the unintegrated charge asymmetry for the process $K^+ \to \pi^+ e^+ e^-$ in the z variable.

contributions drive their observables. The structure of the amplitude includes, in addition to the vector form factor $V_L(z)$, other form factors, namely scalar, pseudoscalar and axial-vector, may contribute to this decay [24].

There are three dominating contributions:

- An indirect CP-violating transition due to the K^0-$\overline{K}^0$ mixing, proportional to $V_S(z)$ in Eq. (3):
$$V_L^{\rm ind}(z) = \pm \varepsilon\, V_S(z)\,. \tag{11}$$
- A direct CP-violating contribution arising from the four-fermion effective operators Q_{7V} and Q_{7A} given in [1]. The contributions to the form factors are proportional to $\mathrm{Im}\,\lambda_t$ and the semileptonic form factors of $\langle \pi(p_\pi)|\overline{s}\gamma_\mu u|K(p_K)\rangle$ [24].
- A CP-conserving contribution from $K_L \to \pi^0\gamma\gamma \to \pi^0\ell^+\ell^-$ [25], as given by the Feynman diagram in Figure 5.

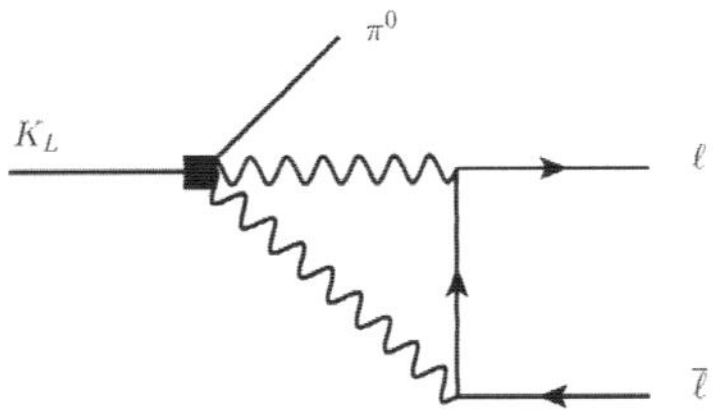

Figure 5. CP-conserving contribution to $K_L \to \pi^0\ell^+\ell^-$.

The addition of both CP-violating contributions gives (using the VMD model to eliminate b_S):

$$\begin{aligned}
10^{12} \times \mathrm{BR}(K_L \to \pi^0 e^+ e^-)\Big|_{\rm CPV} &= 15.7\,|a_S|^2 \pm 6.2\,|a_S| \left(\frac{\mathrm{Im}\,\lambda_t}{10^{-4}}\right) + 2.4 \left(\frac{\mathrm{Im}\,\lambda_t}{10^{-4}}\right)^2 , \\
10^{12} \times \mathrm{BR}(K_L \to \pi^0 \mu^+ \mu^-)\Big|_{\rm CPV} &= 3.7\,|a_S|^2 \pm 1.6\,|a_S| \left(\frac{\mathrm{Im}\,\lambda_t}{10^{-4}}\right) + 1.0 \left(\frac{\mathrm{Im}\,\lambda_t}{10^{-4}}\right)^2 .
\end{aligned} \tag{12}$$

Process	Theory CP-violating	Theory CP-conserving	Experiment (90% C. L.)
$K_L \to \pi^0 e^+ e^-$	3.1 ± 0.9	~ 0	< 28 [26]
$K_L \to \pi^0 \mu^+ \mu^-$	1.4 ± 0.5	0.52 ± 0.16	< 38 [27]

Table 2. Comparison between theoretical and experimental results for BR($K_L \to \pi^0 \ell^+ \ell^-$). The numbers correspond to BR$\times 10^{11}$. We assume positive interference between both CP-violating contributions (theoretically preferred [28]).

It can be seen that for $|a_S| \sim 1$ (as the phenomenological results of Table 1 suggest) the indirect CP-violating contribution should be dominating in the electronic case while for the muon both are alike.

In Table 2 we compare the present experimental bounds with both CP-violating and CP-conserving pieces. The dominating contribution to the latter is helicity suppressed ($\propto M_\ell$) and, therefore, it is negligible in the electron case. However, although smaller than the CP-violating, the CP-conserving contribution is not tiny in $K_L \to \pi^0 \mu^+ \mu^-$. It is interesting to note how close to the present experimental bounds are the SM predictions. We are almost there.

Acknowledgments
This work has been partially supported by MEC (Spain) under grant FPA2014-53631-C2-1-P], by the Spanish Centro de Excelencia Severo Ochoa Programme [SEV-2014-0398] and by the Generalitat Valenciana [PrometeoII/2013/007].

References

[1] Cirigliano V, Ecker G, Neufeld H, Pich A and Portolés J 2012 *Rev. Mod. Phys.* **84** 399 (*Preprint* 1107.6001)
[2] Buchalla G and Buras A J 1994 *Nucl. Phys.* **B412** 106–142 (*Preprint* hep-ph/9308272)
[3] Buras A J, Gorbahn M, Haisch U and Nierste U 2005 *Phys. Rev. Lett.* **95** 261805 (*Preprint* hep-ph/0508165)
[4] Komatsubara T K 2012 *Prog. Part. Nucl. Phys.* **67** 995–1018 (*Preprint* 1203.6437)
[5] Gasser J and Leutwyler H 1984 *Ann. Phys.* **158** 142
[6] Gasser J and Leutwyler H 1985 *Nucl. Phys.* **B250** 465
[7] Ecker G, Pich A and de Rafael E 1987 *Nucl. Phys.* **B291** 692
[8] Ecker G, Pich A and de Rafael E 1988 *Nucl. Phys.* **B303** 665
[9] Ananthanarayan B and Sentitemsu Imsong I 2012 *J. Phys.* **G39** 095002 (*Preprint* 1207.0567)
[10] D'Ambrosio G, Ecker G, Isidori G and Portolés J 1998 *JHEP* **08** 004 (*Preprint* hep-ph/9808289)
[11] Kambor J, Missimer J H and Wyler D 1991 *Phys. Lett.* **B261** 496–503
[12] Ecker G, Gasser J, Leutwyler H, Pich A and de Rafael E 1989 *Phys. Lett.* **B223** 425
[13] Ecker G, Pich A and de Rafael E 1990 *Phys. Lett.* **B237** 481
[14] Friot S, Greynat D and de Rafael E 2004 *Phys. Lett.* **B595** 301–308 (*Preprint* hep-ph/0404136)
[15] Dubničková A Z, Dubnička S, Goudzovski E, Pervushin V and Sečansky M 2008 *Phys. Part. Nucl. Lett.* **5** 76–84 (*Preprint* hep-ph/0611175)
[16] Coluccio Leskow E, D'Ambrosio G, Greynat D and Nath A 2016 *Phys. Rev.* **D93** 094031 (*Preprint* 1603.09721)
[17] Christ N H, Feng X, Portelli A and Sachrajda C T (RBC, UKQCD) 2015 *Phys. Rev.* **D92** 094512 (*Preprint* 1507.03094)
[18] Christ N H, Feng X, Juttner A, Lawson A, Portelli A and Sachrajda C T 2016 (*Preprint* 1608.07585)
[19] Batley J R *et al.* (NA48/2) 2009 *Phys. Lett.* **B677** 246–254 (*Preprint* 0903.3130)
[20] Batley J R *et al.* (NA48/2) 2011 *Phys. Lett.* **B697** 107–115 (*Preprint* 1011.4817)
[21] Batley J R *et al.* (NA48/1) 2003 *Phys. Lett.* **B576** 43–54 (*Preprint* hep-ex/0309075)
[22] Batley J R *et al.* (NA48/1) 2004 *Phys. Lett.* **B599** 197–211 (*Preprint* hep-ex/0409011)
[23] D'Ambrosio G and Gao D N 2002 *JHEP* **0207** 068 (*Preprint* hep-ph/0203087)
[24] Isidori G, Smith C and Unterdorfer R 2004 *Eur. Phys. J.* **C36** 57–66 (*Preprint* hep-ph/0404127)
[25] Heiliger P and Sehgal L M 1993 *Phys. Rev.* **D47** 4920–4938
[26] Alavi-Harati A *et al.* (KTeV) 2004 *Phys. Rev. Lett.* **93** 021805 (*Preprint* hep-ex/0309072)
[27] Alavi-Harati A *et al.* (KTeV) 2000 *Phys. Rev. Lett.* **84** 5279–5282 (*Preprint* hep-ex/0001006)
[28] Buchalla G, D'Ambrosio G and Isidori G 2003 *Nucl. Phys.* **B672** 387–408 (*Preprint* hep-ph/0308008)

Prospects for the search of $K^0_S \to \pi^+\pi^-e^+e^-$ at LHCb

C. Marin Benito on behalf of the LHCb collaboration

Universitat de Barcelona, Physics Faculty, Marti i Franques, 1, Barcelona 08028, Spain

E-mail: carla.marin.benito@cern.ch

Abstract. The feasibility of observing the $K^0_S \to \pi^+\pi^-e^+e^-$ decay at LHCb is studied using simulated and real data. During the Run I of LHC, the yield of events expected per fb^{-1} of pp collisions at $\sqrt{s} = 8$ TeV is found to be $N_{RunI}(K^0_S \to \pi^+\pi^-e^+e^-) = 120^{+280}_{-100}$. A dedicated trigger selection has been developed for the 2016 data-taking. A large signal yield, $N_{Upgrade}(K^0_S \to \pi^+\pi^-e^+e^-) = (5.0 \pm 0.3) \times 10^4$ per fb^{-1}, is expected in the LHC upgrade phase. Pseudoexperiments have been run to asses the feasibility of discovering evidence for the observation of the signal already in the Run I data-set.

1. Introduction

LHCb has proved to be very competitive in the search for rare strange decays, the most recent proof of this being the new results and prospects presented at this conference. Among the decays of interest one can find $K^0_S \to \ell^+\ell^-\ell'^+\ell'^-$ where $\ell^{(\prime)}$ can be either an electron or a muon.

These decays have never been observed and no experimental limits are available in the literature [1]. The predicted branching fractions in the Standard Model (SM) are of the order [2]:

$$\begin{aligned} \mathcal{B}(K^0_S \to e^+e^-e^+e^-) &\sim 10^{-10} \\ \mathcal{B}(K^0_S \to \mu^+\mu^-e^+e^-) &\sim 10^{-11} \\ \mathcal{B}(K^0_S \to \mu^+\mu^-\mu^+\mu^-) &\sim 10^{-14} \end{aligned} \quad (1)$$

The introduction of possible New Physics (NP) terms in the amplitude calculation could lead to dominant contributions. Therefore, any experimental deviation from the SM predicted values would be a hint of contributions from NP. Moreover, the measurement of the time interference of $K^0_L \to \ell^+\ell^-\ell'^+\ell'^-$ with $K^0_S \to \ell^+\ell^-\ell'^+\ell'^-$ would allow stringent CKM tests.

For those decays containing electrons in the final state the reconstruction of the electrons is the most challenging issue at LHCb, due to the low momentum they have in this decay and the large energy loss they suffer by bremsstrahlung.

Preliminary analysis on simulated data allow to study the expected mass resolution and displacement of the mass peak due to the energy loss of the electrons. For $K^0_S \to \mu^+\mu^-e^+e^-$, $K^0_S \to \pi^+\pi^-e^+e^-$ events are used as a proxy given the similar topology. A simple fit with a double sided Crystal-Ball [3] to $K^0_S \to e^+e^-e^+e^-$ and $K^0_S \to \pi^+\pi^-e^+e^-$ simulated events is shown in Figure 1.

In addition to the mass resolution, the expected overlap with the copious background $K^0_S \to \pi^+\pi^-e^+e^-$ with the two pions misidentified as electrons or muons is also checked. For this, $K^0_S \to \pi^+\pi^-e^+e^-$ simulated events are reconstructed changing the mass hypothesis of the two pions to two electrons and two muons, respectively. Figure 2 shows these distributions

Content from this work may be used under the terms of the Creative Commons Attribution 3.0 licence. Any further distribution of this work must maintain attribution to the author(s) and the title of the work, journal citation and DOI.
Published under licence by IOP Publishing Ltd

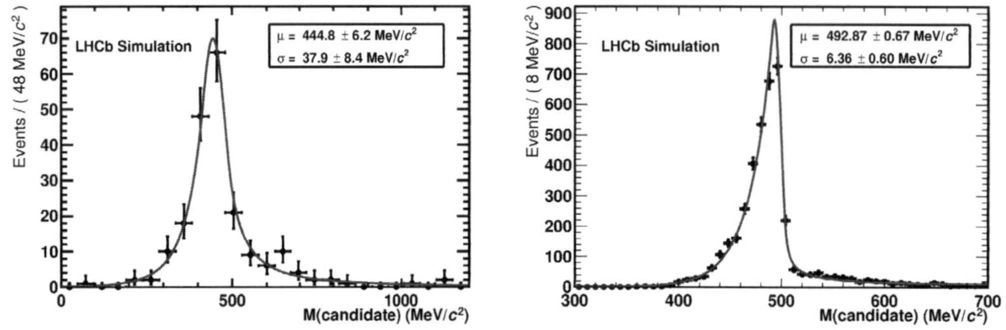

Figure 1. Mass distributions for reconstructed $K_S^0 \to e^+e^-e^+e^-$ (left) and $K_S^0 \to \pi^+\pi^-e^+e^-$ (right) simulated events. The continuos line shows a fit with a double sided Crystal-Ball function.

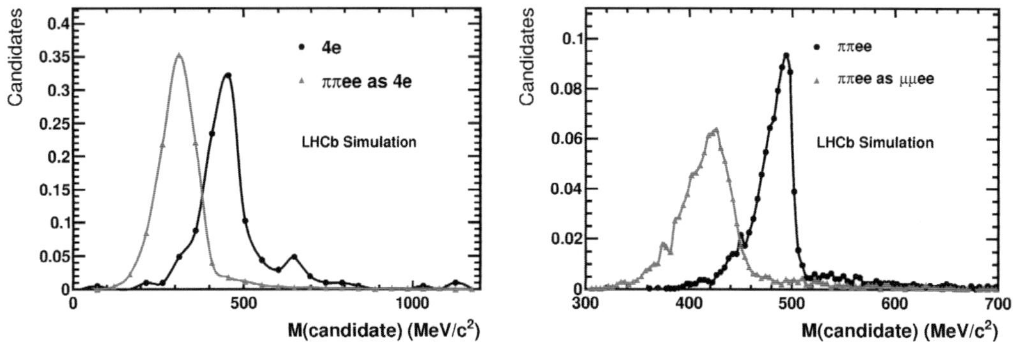

Figure 2. Mass distributions for reconstructed $K_S^0 \to e^+e^-e^+e^-$ and $K_S^0 \to \pi^+\pi^-e^+e^-$ simulated events under the $e^+e^-e^+e^-$ mass hypothesis (left) and reconstructed $K_S^0 \to \pi^+\pi^-e^+e^-$ simulated events under the $\pi^+\pi^-e^+e^-$ and $\mu^+\mu^-e^+e^-$ mass hypothesis (right).

overlaid with the $K_S^0 \to e^+e^-e^+e^-$ and $K_S^0 \to \pi^+\pi^-e^+e^-$ reconstructed mass. It can be seen that the signal peaks can be discriminated from the $K_S^0 \to \pi^+\pi^-e^+e^-$ one thanks to the good mass resolution. Nevertheless, the tails of the distributions overlap and therefore it becomes necessary to have this decay under complete control before being able to set a limit on the signal channels.

In addition, the $K_S^0 \to \pi^+\pi^-e^+e^-$ decay is the most obvious candidate as normalisation channel for the leptonic 4-body final states, given the similar topology to the signal channels with four tracks including a di-electron in the final state. Moreover, this channel is interesting by itself as a place to look for new light dark matter states decaying dominantly to a pair of leptons that would peak in the di-electron invariant mass [4]. An example could be the search for dark photons using a strategy similar to the one proposed in [5], which has become of great interest after recent hints of a light new boson decaying to a e^+e^- pair [6]. The feasibility of observing the $K_S^0 \to \pi^+\pi^-e^+e^-$ decay at LHCb is presented in this study.

The NA48 collaboration has observed and measured the $\mathcal{B}(K_S^0 \to \pi^+\pi^-e^+e^-)$. The current world average for this observable, $\mathcal{B}(K_S^0 \to \pi^+\pi^-e^+e^-) = (4.79 \pm 0.15) \times 10^{-5}$ [1], is used in this study to asses the feasibility of observation at LHCb.

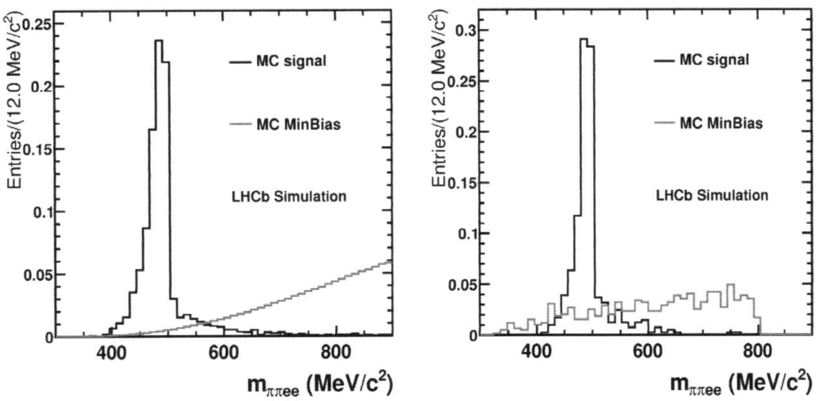

Figure 3. Invariant $\pi^+\pi^-e^+e^-$ mass distribution for $K^0_S \to \pi^+\pi^-e^+e^-$ (grey) and minimum bias (red) reconstructed MC candidates before (left) and after (right) the offline selection.

2. $K^0_S \to \pi^+\pi^-e^+e^-$ at LHCb

This study is based on simulated data reproducing the conditions of the 2012 LHCb data-taking. Moreover, some studies are also performed on a data sample corresponding to 2 fb^{-1} of integrated luminosity at $\sqrt{s} = 8$ TeV collected by LHCb during 2012. The efficiency of the reconstruction, the selection and the trigger is obtained from the MC sample. This allows to extract the expected signal yield while the background level is estimated from data. Possible improvements on the trigger side are studied for Run II of the LHC and beyond and the expected signal yield is recomputed taking them into account. Finally, we asses the feasibility of observing this decay at LHCb already with Run I data.

2.1. Reconstruction, selection and trigger

The reconstruction performance for this decay is evaluated by requiring the reconstructed particles to match the MC generated signal. The efficiency of this matching is found to be $\epsilon^{sig}_{reco} = (0.134 \pm 0.002)$ %. The $\pi\pi ee$ invariant mass distribution for reconstructed signal and background MC is shown in Figure 3 left. An offline selection based on linear cuts is defined exploiting the topology of the decay. Loose momentum and transverse momentum and large displacement from the interaction vertex are combined with track quality and particle identification requirements. The full selection is detailed in [7] and the invariant mass distribution for candidates fulfilling it can be found in the right plot of Figure 3. The selection efficiency for signal and minimum bias MC samples is $\epsilon^{sig}_{sel} = (10.1 \pm 0.5) \times 10^{-2}$ and $\epsilon^{bkg}_{reco,sel} = (2.95 \pm 0.12) \times 10^{-5}$, respectively. In the second case, the reconstruction efficiency is included here.

No specific trigger selection was present during the Run I data-taking to select $K^0_S \to \pi^+\pi^-e^+e^-$ decays. Thus, the LHCb trigger efficiency to select this decay is evaluated on signal and minimum bias simulated events that satisfy the offline selection as an OR of all the physics selections that were present in the trigger in Run I [8]. The total trigger efficiency is found to be $\epsilon^{sig}_{trig} = (0.24^{+0.56}_{-0.20})$ % and $\epsilon^{bkg}_{trig} < 0.51$ % for signal and background, respectively. In the second case, since no events are selected in the MC sample a limit is set on the efficiency assuming the number of selected events is < 3 at 90 % CL.

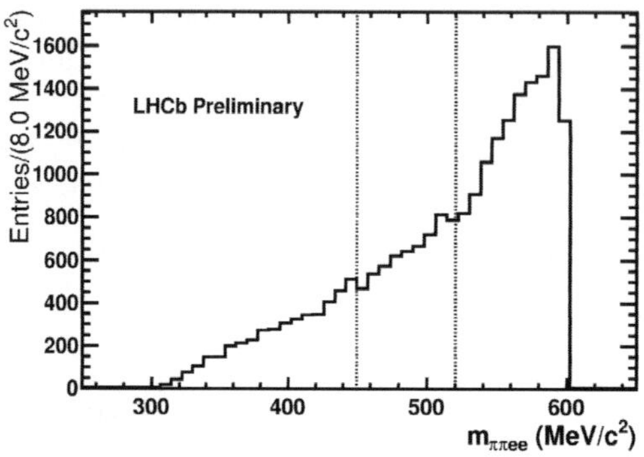

Figure 4. Invariant $\pi^+\pi^-e^+e^-$ mass distribution for candidates satisfying the offline selection and trigger requirements. The signal region, $450 - 520$ MeV/c^2, is delimited by dashed lines.

2.2. Expected yields in Run I

The expected yields are obtained in a tight mass window, $450 - 520$ MeV/c^2, selecting (76.9 ± 1.8) % of the signal and (4.33 ± 0.02) % of the background. The expected signal yield can be computed from the following expression:

$$N^{sig}_{exp} = N(\mathrm{K^0_S/\,fb^{-1}}) \cdot \mathcal{B}(\mathrm{K^0_S} \to \pi^+\pi^-\mathrm{e^+e^-}) \cdot \epsilon^{sig} \quad (2)$$

where $N(\mathrm{K^0_S/\,fb^{-1}}) \sim 10^{13}$ is the number of $\mathrm{K^0_S}$ mesons decaying inside the LHCb acceptance [9], $\mathcal{B}(\mathrm{K^0_S} \to \pi^+\pi^-\mathrm{e^+e^-}) = (4.79 \pm 0.15) \times 10^{-5}$ [1] is the world average branching ratio for this decay and $\epsilon^{sig} = \epsilon^{sig}_{reco} \cdot \epsilon^{sig}_{sel} \cdot \epsilon^{sig}_{trig} \cdot \epsilon^{sig}_{mass}$. The expected signal yield per fb^{-1} of LHCb data at 8 TeV is found to be:

$$N^{sig}_{exp} = 120^{+280}_{-100}$$

where the uncertainty is dominated by the MC statistics.

A similar technique can be used to estimate the expected background yield from MC:

$$N^{bkg}_{exp} = \sigma_{tot} \cdot \epsilon^{bkg} \quad (3)$$

where $\sigma_{tot} = (94.6 \pm 0.3)\ mb$ is the total cross-section simulated in the MC minimum bias sample and $\epsilon^{bkg} = \epsilon^{bkg}_{reco,sel} \cdot \epsilon^{bkg}_{trig} \cdot \epsilon^{bkg}_{mass}$. The expected background yield after the full selection is:

$$N^{bkg}_{exp} < 6.1 \times 10^5$$

per fb^{-1} at 90 % CL.

2.3. Studies on data

The full selection defined in Section 2.1 is applied to the 2012 LHCb dataset corresponding to an integrated luminosity of 2 fb^{-1}. The resulting $\pi^+\pi^-e^+e^-$ invariant mass distribution is shown in Figure 4. Since no evidence for a signal peak is observed, this contribution is assumed to be negligible. With this assumption the background level in the signal region amounts to:

$$N^{bkg}_{obs} \sim 6 \times 10^3$$

which is compatible with the expectations from the MC study.

2.4. Expected yields in Run II and beyond

A new trigger selection has been developed in order to inclusively select K^0_S decays into a di-electron pair. The requirements follow the offline selection presented in Section 2.1. This configuration started running in the LHCb trigger at the beginning of the 2016 data-taking.

The efficiency of the new configuration is estimated on the MC sample in 2012 conditions since no sample in Run II conditions is available. The total trigger efficiency with the new selection is $\epsilon^{sig}_{trig} = (0.24^{+0.56}_{-0.20})$ %.

Assuming no other improvements have been achieved in Run II and also neglecting the increase in the K^0_S production cross section and flight distance from 8 to 13 TeV, the expected yield per fb^{-1} in Run II is:

$$N^{sig}_{RunII} = 120^{+280}_{-100}$$

where the uncertainty is dominated by the MC statistics.

Although the Run I and Run II trigger selections have the same efficiency selecting the signal channel, they use complementary approaches. Thus combining both would lead to an increase in the trigger efficiency for this decay.

Beyond the LHC Run II, during its upgrade phase, the LHCb trigger will be fully based on software [10]. This will allow to have many exclusive selections exploiting the full event reconstruction at trigger level. In the ideal scenario a signal efficiency of $\sim$ 100% can be achieved. With this assumption, the yield of $K^0_S \to \pi^+\pi^-e^+e^-$ per fb^{-1} is expected to be:

$$N^{sig}_{upgrade} = (5.0 \pm 0.3) \times 10^4$$

2.5. Observation feasibility with Run I data

Taking into account the expected signal, $N^{sig}_{exp} \sim 240$, and observed background, $N^{bkg}_{obs} \sim 6 \times 10^3$, in the 2012 LHCb dataset it might be possible to observe this decay with the current data sample by means of a high-efficiency large-rejection further selection. A multivariate (MVA) selection would be appropriate in this case. Other dedicated tools for the selection of low momentum electrons could help to further discriminate signal from background.

In order to study the possibility of an observation or evidence in a more quantitative way, a study based on pseudoexperiments has been performed. It is based on the expected signal and observed background yields in the signal region, together with the corresponding mass Probability Density Functions (PDFs). A detailed description of the study is presented in [7]. The main result can be found in Figure 5, where the signal efficiency vs. background rejection curves needed to obtain a 3σ evidence or 5σ observation of the decay are represented. Notice that the working point for the selection presented in Section 2.1, corresponding to 100% signal efficiency and 0% background rejection, is found way below the curves, confirming the selected data sample is not statistically significant. However, both curves are well within the usual discrimination achieved by standard MVA selections, which reinforces the conclusion that an evidence or observation could be possible using only Run I data.

3. Conclusions

The feasibility of observing the $K^0_S \to \pi^+\pi^-e^+e^-$ decay at LHCb has been studied using MC and data samples in Run I conditions. The expected signal yield per fb^{-1} of data at 8 TeV is found to be:

$$N^{exp}_{RunI}(K^0_S \to \pi^+\pi^-e^+e^-) = 120^{+280}_{-100} \quad (4)$$

where the uncertainty is dominated by the MC statistics.

No hint of signal is observed in the 2012 LHCb dataset corresponding to an integrated luminosity of 2 fb^{-1} while the total background amounts to:

$$N^{obs}_{RunI}(bkg) \sim 6 \times 10^3 \quad (5)$$

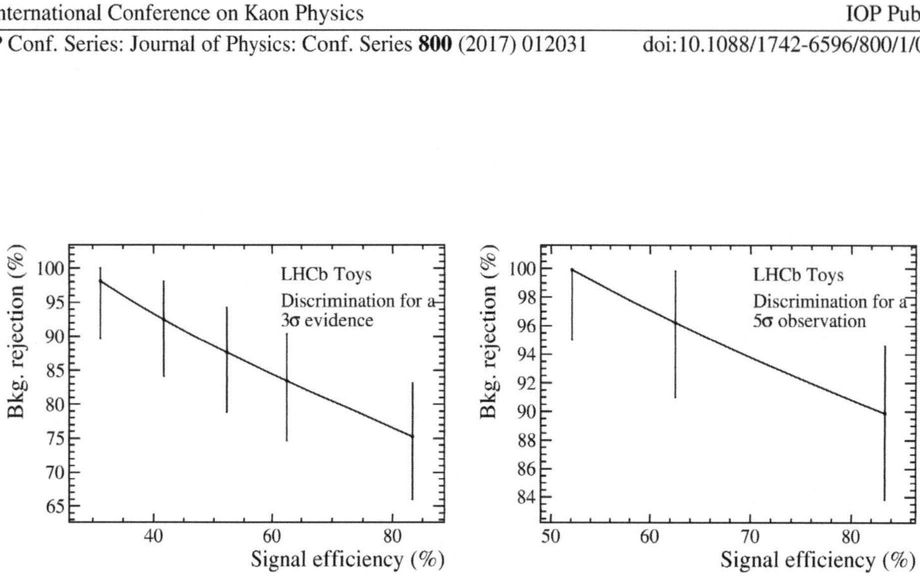

Figure 5. Signal efficiency vs. background rejection curves needed to obtain a 3σ evidence (left) or 5σ observation (right) of $K^0_S \to \pi^+\pi^-e^+e^-$ using the 2012 LHCb dataset. The error bars correspond to 68% C.L. variation in the toys for each expected signal yield.

A dedicated trigger selection has been developed for the 2016 data-taking. Its efficiency is expected to be the same as in Run I. Since a complementary approach is used, the combination of both could improve the efficiency for Run II.

In the LHC upgrade phase with a full software trigger at LHCb, a signal efficiency of $\sim 100\%$ can be achieved. Under this assumption the signal yield per fb^{-1} is expected to be:

$$N_{Upgrade}(K^0_S \to \pi^+\pi^-e^+e^-) = (5.0 \pm 0.3) \times 10^4$$

Pseudoexperiments have been run with the expected signal and observed background yields in the 2012 data-set to obtain the signal efficiency vs background rejection curves needed in order to asses an evidence or observation for this decay. Both are well within the usual discrimination achieved by standard MVA selections, which reinforces the conclusion that an evidence or observation could be possible using only Run I data. During the upgrade phase the large expected signal yield would allow to set stringent constrains in the $K^0_S \to \ell^+\ell^-\ell'^+\ell'^-$ decays and search for peaks in the di-electron invariant mass distribution of the signal.

References

[1] Particle Data Group, K. A. Olive *et al.*, *Review of particle physics*, Chin. Phys. **C38** (2014) 090001.
[2] G. D'Ambrosio, D. Greynat, and G. Vulvert, *Standard Model and New Physics contributions to K_L and K_S into four leptons*, Eur. Phys. J. **C73** (2013), no. 12 2678, arXiv:1309.5736.
[3] J. E. Gaiser, *Charmonium Spectroscopy From Radiative Decays of the J/ψ and ψ'*, PhD thesis, SLAC, 1982.
[4] G. Isidori, *Future prospects in kaon physics: some remarks*, Rare'n'Strange Workshop, CERN.
[5] P. Ilten, J. Thaler, M. Williams, and W. Xue, *Dark photons from charm mesons at LHCb*, Phys. Rev. **D92** (2015), no. 11 115017, arXiv:1509.06765.
[6] A. J. Krasznahorkay *et al.*, *Observation of Anomalous Internal Pair Creation in Be8: A Possible Indication of a Light, Neutral Boson*, Phys. Rev. Lett. **116** (2016), no. 4 042501, arXiv:1504.01527.
[7] C. Marin Benito, L. Garrido Beltran, and X. Cid Vidal, *Feasibility study of $K^0_S \to \pi^+\pi^-e^+e^-$ at LHCb* , Tech. Rep. LHCb-PUB-2016-016. CERN-LHCb-PUB-2016-016, CERN, Geneva, Oct, 2016.
[8] A. Puig, *The LHCb trigger in 2011 and 2012*, LHCb-PUB-2014-046.
[9] LHCb collaboration, R. Aaij *et al.*, *Search for the rare decay $K^0_S \to \mu^+\mu^-$*, JHEP **01** (2013) 090, arXiv:1209.4029.
[10] LHCb collaboration, *LHCb Trigger and Online Technical Design Report*, CERN-LHCC-2014-016. LHCb-TDR-016.

IOP Conf. Series: Journal of Physics: Conf. Series **800** (2017) 012032 doi:10.1088/1742-6596/800/1/012032

Search for Dark Photon at NA48/2, and measurement of π^0 form factor

Mauro Raggi[1]

Sapienza Università di Roma and INFN Roma, Piazzale Aldo Moro 5 , Roma Italy

E-mail: mauro.raggi@uniroma1.infn.it

Abstract. The NA48/2 experiment at CERN performed a search for the dark photon (A') via the decay chain $\pi^0 \to \gamma A'$, $A' \to e^+e^-$. Using a sample of $\sim 1.7 \cdot 10^7$ neutral pions tagged from $K^\pm \to \pi^\pm \pi^0_D$ and $K^\pm \to \mu^\pm \pi^0_D \nu$ decays collected in 2003-04 no signal is observed. The limits in the plane dark photon mixing parameter ε^2 versus its mass $m_{A'}$ are reported. The NA62-RK experiment collected, during the 2007 data taking, a large sample of neutral pion Dalitz decays with a trigger optimised for electrons final states with respect to NA48/2 one. We report the measurement of the π^0 electromagnetic transition form factor (TFF) slope parameter (a) based on a sample of $\sim 1 \cdot 10^6$ π^0_D decays. The preliminary measured value $a = (3.70 \pm 0.53_{stat} \pm 0.36_{syst}) \cdot 10^{-2}$ is compatible with theoretical expectations and previous measurements.

1. Introduction

The NA48/2 beam line was designed to deliver simultaneous narrow momentum band K^+ and K^- beams with central momenta of 60 GeV/c derived from the 400 GeV/c protons extracted from the CERN SPS. The beam kaons decayed in an 114 m long fiducial volume contained in a cylindrical vacuum tank. The momenta of charged decay products were measured by a magnetic

[1] for the NA48/2 and NA62 Collaborations: F. Ambrosino, A. Antonelli, G. Anzivino, R. Arcidiacono, W. Baldini, S. Balev, J.R. Batley, M. Behler, S. Bifani, C. Biino, A. Bizzeti, B. Bloch-Devaux, G. Bocquet, V. Bolotov, F. Bucci, N. Cabibbo, M. Calvetti, N. Cartiglia, A. Ceccucci, P. Cenci, C. Cerri, C. Cheshkov, J.B. Chèze, M. Clemencic, G. Collazuol, F. Costantini, A. Cotta Ramusino, D. Coward, D. Cundy, A. Dabrowski, G. D'Agostini, P. Dalpiaz, C. Damiani, H. Danielsson, M. De Beer, G. Dellacasa, J. Derré, H. Dibon, D. Di Filippo, L. DiLella, N. Doble, V. Duk, J. Engelfried, K. Eppard, V. Falaleev, R. Fantechi, M. Fidecaro, L. Fiorini, M. Fiorini, T. Fonseca Martin, P.L. Frabetti, A. Fucci, S. Gallorini, L. Gatignon, E. Gersabeck, A. Gianoli, S. Giudici, A. Gonidec, E. Goudzovski, S. Goy Lopez, E. Gushchin, B. Hallgren, M. Hita-Hochgesand, M. Holder, P. Hristov, E. Iacopini, E. Imbergamo, M. Jeitler, G. Kalmus, V. Kekelidze, K. Kleinknecht, M. Koval, V. Kozhuharov, W. Kubischta, V. Kurshetsov, G. Lamanna, C. Lazzeroni, M. Lenti, E. Leonardi, L. Litov, N. Lurkin, D. Madigozhin, A. Maier, I. Mannelli, F. Marchetto, G. Marel, M. Markytan, P. Marouelli, M. Martini, L. Masetti, P. Massarotti, E. Mazzucato, A. Michetti, I. Mikulec, M. Misheva, N. Molokanova, E. Monnier, U. Moosbrugger, C. Morales Morales, M. Moulson, S. Movchan, D.J. Munday, M. Napolitano, A. Nappi, G. Neuhofer, A. Norton, T. Numao, V. Obraztsov, V. Palladino, M. Patel, M. Pepe, A. Peters, F. Petrucci, M.C. Petrucci, B. Peyaud, R. Piandani, M. Piccini, G. Pierazzini, I. Polenkevich, I. Popov, Yu. Potrebenikov, M. Raggi, B. Renk, F. Retière, P. Riedler, A. Romano, P. Rubin, G. Ruggiero, A. Salamon, G. Saracino, M. Savrié, M. Scarpa, V. Semenov, A. Sergi, M. Serra, M. Shieh, S. Shkarovskiy, M.W. Slater, M. Sozzi, T. Spadaro, S. Stoynev, E. Swallow, M. Szleper, M. Valdata-Nappi, P. Valente, B. Vallage, M. Velasco, M. Veltri, S. Venditti, M. Wache, H. Wahl, A. Walker, R. Wanke, L. Widhalm, A. Winhart, R. Winston, M.D. Wood, S.A. Wotton, O. Yushchenko, A. Zinchenko, M. Ziolkowski.

Content from this work may be used under the terms of the Creative Commons Attribution 3.0 licence. Any further distribution of this work must maintain attribution to the author(s) and the title of the work, journal citation and DOI.
Published under licence by IOP Publishing Ltd

spectrometer consisting of four drift chambers (DCHs), two upstream and two downstream of a dipole magnet. The magnet provided a horizontal transverse momentum kick to charged particles of 120 MeV/c. Each DCH was made of eight planes of sense wires. Downstream of the spectrometer a plastic scintillator hodoscope producing fast trigger signals and providing precise time measurements of charged particles was placed. Further downstream was a liquid krypton electromagnetic calorimeter (LKr), an almost homogeneous ionization chamber with an active volume of 7 m^3, $27X_0$ deep, segmented transversally into 13248 projective 2x2 cm^2 cells and with no segmentation along the beam axis. The calorimeter information is used for photon energy measurements and charged particle identification. An iron/scintillator hadronic calorimeter and muon detectors were located further downstream. A description of the detector can be found in [1]. The NA62-RK experiment collected data during 2007, using the NA48/2 detector, aiming at measuring the ratio R_K of the rates of the leptonic kaon decays. The experiment used modified beam central momentum of 74 GeV/c and different trigger conditions optimized to collect electrons.

2. Search for the Dark Photon in π^0 decays

The hypothesis that the dark matter can be hiding in the so-called "dark sectors" has become more and more popular in recent years, attracting an increasing interest both in experimental searches and theoretical studies[3]. In this set of models, the Standard Model (SM) and the dark sector are connected by "mediators", particles with quantum numbers of both SM and dark sector forces. In the most general $U(1)$ extension of the SM the mediator can kinetically mix with the photon [2] and is commonly called "dark photon" (DP). Due to the mixing with the standard model photon, the DP can be directly produced in collisions of an electron or proton with a target, and in the decays of mesons (π^0, η, ϕ) containing a γ in the final state[4]. The DP decay modes depend on its mass and on the hypothesis on the dark sector particle mass spectrum. The most commonly used are to consider the DP the lightest state in the dark sector and that it decays to SM particles, "visible decays". NA48/2 performed a search for the dark photon in π^0 decays exploiting the process $\pi^0 \to \gamma A'$, $A' \to e^+e^-$. During the 2003 and 2004 data taking periods NA48/2 accumulated the world largest sample of tagged π^0 decays, $\sim 5 \cdot 10^{10}$, originating from $K^\pm \to \pi^\pm\pi^0$ and $K^\pm \to \mu^\pm\pi^0\nu_\mu$ kaon decays.

The predicted branching fraction of the π^0 decay to $\gamma A'$ is expressed in term of the dark sector coupling strength ε and the DP mass $m_{A'}$ as follows [6]:

$$\mathcal{B}(\pi^0 \to \gamma A') = 2\varepsilon^2 \left(1 - \frac{m_{A'}^2}{m_{\pi^0}^2}\right)^3 \mathcal{B}(\pi^0 \to \gamma\gamma), \tag{1}$$

with a strong kinematic suppression of the decay rate for DP masses approaching m_{π^0}. In the mass range of interest for the present analysis, $2m_e < m_{A'} < m_{\pi^0}$, under the hypothesis of visible decays, the $\mathcal{B}(A' \to e^+e^-)$ is ≈ 1. The sensitivity of the NA48/2 data set is unfortunately limited by the $\pi_D^0 \to e^+e^-\gamma$ decay, which represents an irreducible background producing a signature identical to that of the DP. In fact, in the accessible parameter range ($m_{A'} > 10$ MeV/c^2 and $\varepsilon^2 > 5 \times 10^{-7}$), the maximum DP mean path does not exceed 10 cm and the DP can not be distinguished from genuine π_D^0 decays which decay promptly. On the other hand, the NA48/2 3-track vertex reconstruction does not introduce significant acceptance losses for DP events as the typical resolution on the vertex longitudinal coordinate is ≈ 1 m.

The full NA48/2 data sample is used in the present analysis. The event selection criteria developed are very similar for both $K^\pm \to \pi^\pm\pi_D^0$ ($K_{2\pi D}$) and $K^\pm \to \mu^\pm\pi^0\nu_\mu$ ($K_{\mu 3D}$) tagged π^0 decays. Both require a three-track vertex reconstructed in the fiducial decay region and two opposite-sign electrons ($e^\pm$) candidate tracks. Charged particle identification is based on the ratio of energy deposition in the LKr calorimeter to the momentum measured by the spectrometer. In addition, a single isolated LKr energy deposition cluster is required as the photon candidate. The number of π_D^0 decay candidates reconstructed with the joint π^0 decay selection is $1.69 \cdot 10^7$. The reconstructed invariant mass of the $\pi^\pm\pi_D^0$ system and the squared missing mass of the $\mu^\pm\pi_D^0$ (the neutrino invariant mass squared) are presented for the corresponding decay modes in the left column of (Fig. 1). The width of the DP is expected to be very small, due to the ϵ^2 suppression of the coupling to SM particles. For this reason the characteristic signature of its decay into lepton pair is a very narrow peak in the di-electron invariant mass spectra, that are shown in the right column of (Fig. 1) for both the selected data samples.

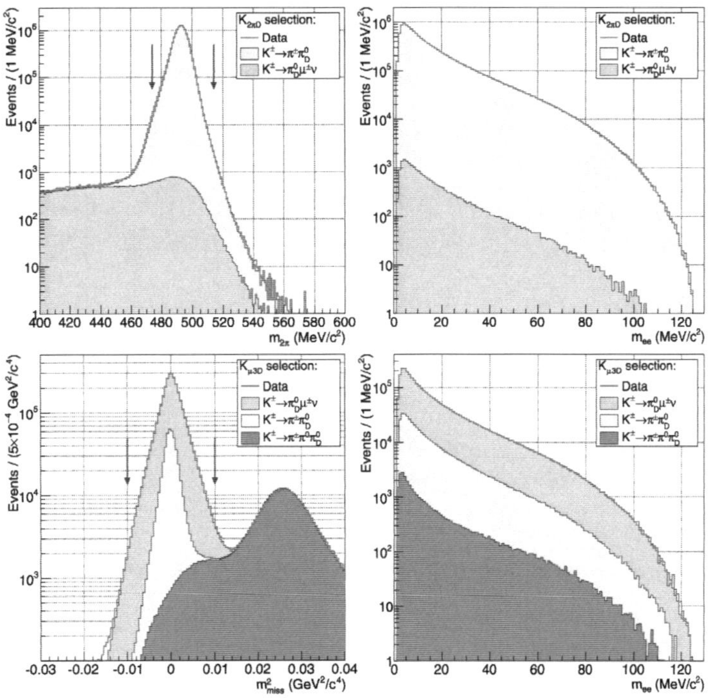

Figure 1. Left) Reconstructed $\pi^\pm\pi_D^0$ invariant mass (top) and $\mu^\pm\pi_D^0\nu$ missing mass (bottom) distributions of the data and simulated samples. The selection condition is illustrated with blue arrows.
Right) Invariant di-lepton mass distributions for the selected data samples. A dark photon signal would manifest itself as a spike in the m_{ee} distributions.

A scan for a DP signal is performed with 404 tentative DP masses in the range 9 MeV/c^2 < $m_{A'}$ <120 MeV/c^2, limited at the lower boundary by the reduced accuracy of the π_D^0 background simulation at low e-e mass, and on the higher boundary by the kinematic suppression of the $\pi^0 \to \gamma A'$ decay in Eq. 1. For each considered DP mass value, the number of data candidates N_{obs} passing the joint DP selection, is compared to the estimated number of background events N_{exp} evaluated with MC simulations. The local statistical significance of the DP signal for each mass hypothesis is presented in Fig. 2. Its value never exceeds 3σ, therefore no DP signal is observed. For each DP mass value the upper limits at 90% CL on the number of A' candidates

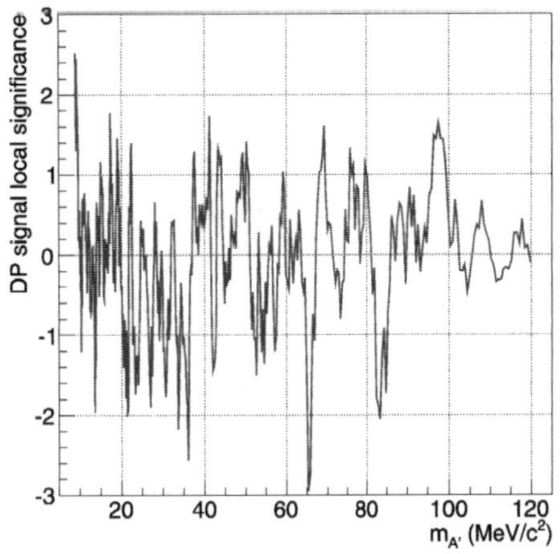

Figure 2. Local significance of the DP signal for each A' mass hypothesis. Neighbouring values are strongly correlated since the mass step of the scan is much smaller than the width of the signal window[5].

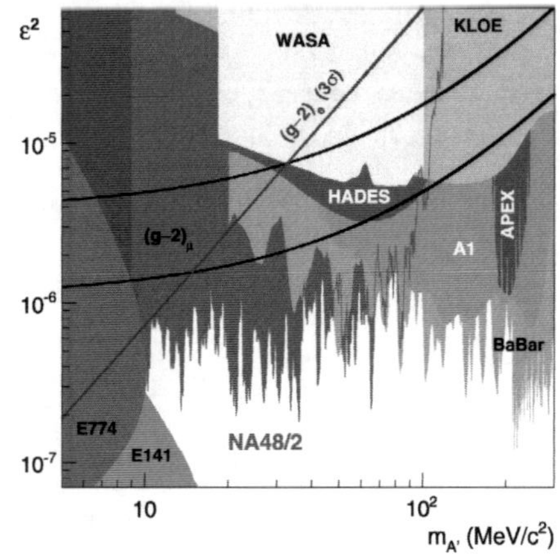

Figure 3. The NA48/2 upper limits at 90% CL on the mixing parameter ε^2 versus the DP mass $m_{A'}$, compared to the other exclusion limits[5].

(N_{DP}), computed from N_{obs}, N_{exp} and δN_{exp} using the frequentist Rolke-Lpez method[7], is translated into the corresponding branching fraction limit $\mathcal{B}(\pi^0 \to \gamma A')$ using the relation:

$$\mathcal{B}(\pi^0 \to \gamma A') = \frac{1}{N_K \epsilon} \frac{N_{\rm DP}}{\left[\mathcal{B}(K_{2\pi})A(K_{2\pi}) + \mathcal{B}(K_{\mu 3})A(K_{\mu 3}) + 2\mathcal{B}(K_{3\pi})A(K_{3\pi})\right]}. \quad (2)$$

where A($K_{2\pi}$,$K_{\mu 3}$,$K_{3\pi}$) are the acceptances of the joint DP selection for the corresponding K decay followed by the prompt $A' \to ee$ decay, ϵ is the trigger efficiency, and $N_K = (1.57 \pm 0.05) \times 10^{11}$[5] is the total number on kaon decays collected by NA48/2. Upper limits at 90% CL on the mixing parameter ε^2, calculated from Eq. 1 for each DP mass value $m_{A'}$, are shown in Fig. 3 together with a collection of present constraints[5]. The band in between the two black lines, where the inconsistency of theoretical and experimental values of the muon (g-2) reduces to less than 2 standard deviations, is also shown as well as the region above the blue line excluded by the electron (g-2) measurement.

The search presented represents the most stringent upper limit on the mixing parameter ε^2 in the mass range 9-70 MeV/c^2. In combination with other experimental limits, under the assumption that DP couples to quarks and decays to SM particles only, this result completes the exclusion of the region of parameter space which could solve the g-2 anomaly, ruling out the DP as a possible explanation.

3. Measurement of the π^0 Form Factor slope

The neutral pion, composed of only u and d quarks, is the lightest of all the mesons and is, therefore, a particularly interesting field to study the low energy behaviour of the SM strong interaction. The π^0 decays electromagnetically to a pair of photons with a Branching Fraction of $(98.823 \pm 0.034)\%$ while in almost all the remaining cases decays into an electron-positron pair and a photon. In this decay mode $\pi^0 \to e^+e^-\gamma$, one of the two photons acquires an off-shell mass greater than ~1 MeV and then decays into an electron-positron pair. In the theoretical treatment of this decay the most common approach is to factorize the differential decay width into a point-like part and a Form Factor (FF), encoding the information about the particle structure without an explicit description of the underlying physics.

3.1. The π^0 Transition Form Factor slope

The π^0_D decay width is parameterised in term of two independent kinematic variables $x = (M_{e^+e^-}/m_{\pi^0})^2$, $y = 2p_{\pi^0}(p_{e+} - p_{e-})/(m^2_{\pi^0}(1-x))$ as follows:

$$\frac{1}{\Gamma(\pi^0_{\gamma\gamma})}\frac{d^2\Gamma(\pi^0_D)}{dxdy} = \frac{\alpha}{4\pi}\frac{(1-x)^3}{x}(1+y^2+\frac{r^2}{x})|F(x)|^2(1+\delta(x,y)) \quad (3)$$

where $r^2 = (2m_e/m_{\pi^0})^2$. The first term represents the point-like approximation, F(x), the Transition From Factor (TFF), while the last term $(1+\delta(x,y))$ the radiative corrections due to the interaction of the two electrons in the final state.

In the allowed kinematic region for the π^0_D decay, F(x) is expected to vary slowly and can be described with a linear approximation:

$$F(x) = 1 + ax \quad (4)$$

where a is the so-called TFF slope parameter. In the vector meson dominance (VMD) approach, first proposed by Gell-Mann and Zachariasen [8], $F(x)$ is dominated by ρ and ω mesons contributions, resulting in a predicted value of $a \approx m^2_{\pi^0}(m_\rho^{-2} + m_\omega^{-2})/2 \approx 0.03$.

3.2. NA62-RK Transition Form Factor slope measurement

NA62-RK experiment has performed the most precise measurement to date of the form factor slope a based on a sample 1.05×10^6 π^0_D decays collected during the 2007 data taking period. The effect of changing the value of the parameter a is very small in the distribution of the di-electron invariant mass, therefore a proper treatment of the radiative corrections and a large unbiased data sample are necessary to perform a precise measurement. The radiative corrections to the π^0_D decay were originally computed in NA62 by interpolating the table of points in the (x,y) plane provided by the Mikaelian and Smith paper[9]. Recent revision of the original calculation by Husek et al. [10] provided the NA62 experiment with a code for the exact $\delta(x,y)$ computation in any phase space point. The new calculation also adds to the original one the one-loop one-photon irreducible contribution, and is able to produce extra radiated photons, which can play a significant role in the determination of experimental acceptance.

Since the π^0 electromagnetic TFF does not depend on the y Dalitz variable, we can integrate the differential decay width in Eq.3 over y and obtain the slope parameter only from the reconstructed distribution of the x Dalitz variable:

$$\frac{1}{\Gamma(\pi^0_{\gamma\gamma})}\frac{d^2\Gamma(\pi^0_D)}{dx} = \frac{2\alpha}{3\pi}\frac{(1-x)^3}{2x}\left(1+\frac{r^2}{2x}\right)\sqrt{\left(1-\frac{r^2}{x}\right)}(1+\delta(x,y))(1+ax)^2 \quad (5)$$

The extraction of the TFF slope a is performed by comparing the x data distribution with different MC spectra obtained varying the value of the TTF slope parameter. The best value for a is chosen by selecting the MC sample showing the best agreement with the x data distribution.

A tagging procedure for π^0_D decays similar to the one already described for the DP searches has been used to select the NA62-RK data sample. Dalits decays of the π^0 are tagged using $K_{2\pi D}$ and $K_{\mu 3D}$ kaon decays profiting of the trigger conditions optimised for electrons in the NA62-RK experiment. In addition, the total energy deposit in the calorimeter (of e^+, e^- and

γ) has to exceed 14 GeV and at least one of the leptons tracks is required to have momentum above 5.5 GeV/c. These cuts are intended in order to just select the region of high trigger efficiency where the agreement with the MC is satisfactory. Since the acceptance of events in the region $x \leq 0.01$ is not well reproduced in the simulation, the reconstructed x variable has to be greater than 0.01. After all cuts, a data sample of 1.05×10^6 candidates was selected. Candidate events are divided into 20 equipopulous bins and a χ^2 minimisation is used to select

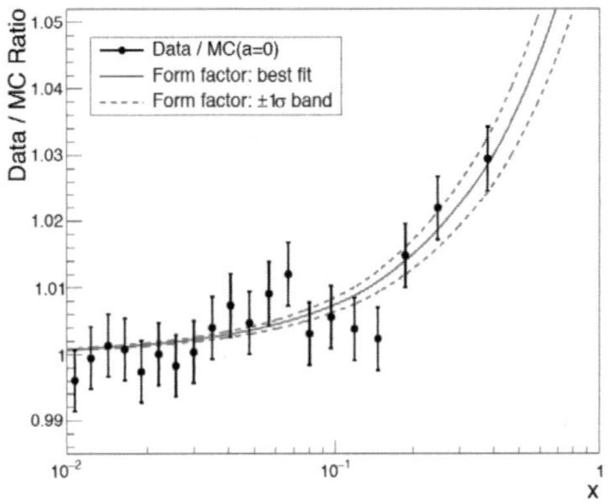

Figure 4. Result of the transition from factor slope fit. Black dots represent data/MC ratio with no form factor slope a=0.

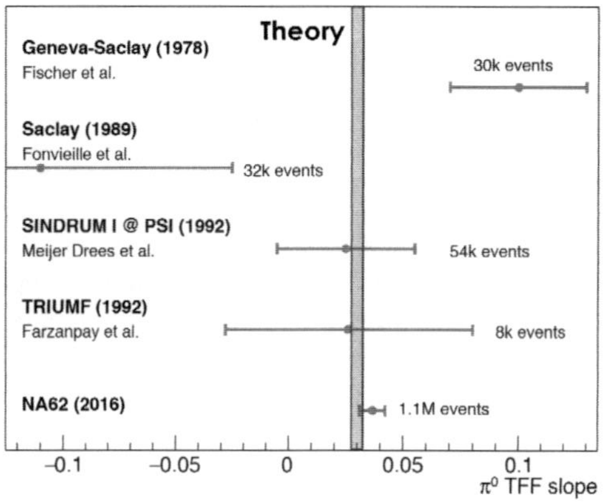

Figure 5. Comparison of the NA62-RK preliminary measurement of the TFF slope with other experiments and theoretical expectation band.

the best the data/MC histogram comparison. The preliminary value of a obtained by the fit is:

$$a = (3.70 \pm 0.53_{stat} \pm 0.36_{syst}) \times 10^{-2} = (3.70 \pm 0.64) \times 10^{-2} \quad (6)$$

with a χ^2/d.o.f = 52.5/49. An illustration of the fit to the data Motecarlo ratio is shown Fig. 4. The solid red line represents the form factor best fit while the dashed red lines shows the $\pm 1\sigma$ uncertainty band. The NA62-RK preliminary result improves the precision on the π^0 TFF in the time-like momentum region compared with previous measurements [11, 12, 14, 13] as shown in Fig. 5 and is in good agreement with theoretical expectations. The analysis is at an advanced state and the final result is expected to be submitted for publication in the near future.

References

[1] V. Fanti *et al.* [NA48 Collaboration], *Nucl. Instrum. Meth. A* **574**, 433 (2007).
[2] B. Holdom, Phys. Lett. **B166**, 196 (1986).
[3] J. Alexander *et al.*, arXiv:1608.08632 [hep-ph].
[4] M. Raggi and V. Kozhuharov, Riv. Nuovo Cim. **38**, no. 10, 449 (2015).
[5] J. R. Batley *et al.* [NA48/2 Collaboration], Phys. Lett. B **746**, 178 (2015)
[6] B. Batell, M. Pospelov and A. Ritz, Phys. Rev. **D80**, 095024 (2009).
[7] W. A. Rolke and A. M. Lopez, Nucl. Instrum. Meth. A **458**, 745 (2001)
[8] Gell-Mann M. and Zachariasen F. Phys. Rev.bf 124 1961 953.
[9] K.O. Mikaelian and J. Smith, Phys. Rev. **D5**, 1763 (1972).
[10] T. Husek, K. Kampf and J. Novotny, Phys. Rev. D **92**, no. 5, 054027 (2015)
[11] J. Fischer *et al.*, Phys. Lett. **73B**, 359 (1978).
[12] H. Fonvieille *et al.*, Phys. Lett. B **233**, 65 (1989).
[13] R. Meijer Drees *et al.* [SINDRUM-I Collaboration], Phys. Rev. D **45**, 1439 (1992).
[14] F. Farzanpay *et al.*, Phys. Lett. B **278**, 413 (1992).

Prospects for the search of $K^0_S \to \pi^0\mu^+\mu^-$ at LHCb

Miriam Lucio Martinez

Universidade de Santiago de Compostela, Physics Faculty, 15782 Santiago de Compostela, Spain

E-mail: miriam.lucio.martinez@cern.ch

Abstract. The sensitivity of the LHCb experiment to $\mathcal{B}(K^0_S \to \pi^0\mu^+\mu^-)$ is analyzed in light of the 2011, 2012 and 2016 data and the oportunity of having a full software trigger with the LHCb upgrade. Two strategies are considered: the full reconstruction of the decay products and the partial reconstruction using only the dilepton pair and kinematic constraints. In both cases, the sensitivity achieved can surpass the world's current best measurement. Both approaches could be statistically combined to further improve the result.

1. Physics Motivation

Kaons play a major role in particle physics, both for Standard Model (SM) and for New Physics searches. If this New Physics stands at energies higher than the TeV its effect in quark flavour physics would be visible only with new sources of Flavour Violation not originating from the Yukawa couplings (non Minimal Flavour Violating, non-MFV). In this scenario, the $s \to d$ decay processes (see Fig. 1) play a central role. This is because they have the strongest CKM suppression factor of all quark transitions, which makes them particularly sensitive to non-MFV sources. Among these transitions, the decay $(K^0_L \to \pi^0\mu^+\mu^-)$ is shown to be sensitive to, for example, models with extra dimensions [1]. However, the potential for this decay to constrain scenarios beyond the Standard Model is limited by the large SM uncertainty on its branching fraction prediction [1]:

$$\mathcal{B}(K^0_L \to \pi^0\mu^+\mu^-)_{\rm SM} = 1.4 \pm 0.3 \times 10^{-11} \quad (1)$$

$$\mathcal{B}(K^0_L \to \pi^0\mu^+\mu^-)_{\rm SM} = 0.9 \pm 0.2 \times 10^{-11} \quad (2)$$

The two values correspond to two theoretical solutions, depending on whether constructive or destructive interference between the contributing waves is present. The large theoretical uncertainty on $\mathcal{B}(K^0_L \to \pi^0\mu^+\mu^-)_{\rm SM}$ arises from the large uncertainty on one of the parameters involved in the calculation. An improved measurement of $\mathcal{B}(K^0_S \to \pi^0\mu^+\mu^-)$ will reduce this uncertainty. The only measurement of $\mathcal{B}(K^0_S \to \pi^0\mu^+\mu^-)$ was performed by the NA48 experiment at CERN [2], which obtained:

$$\mathcal{B}(K^0_S \to \pi^0\mu^+\mu^-) = (2.9^{+1.5}_{-1.2}({\rm stat}) \pm 0.2({\rm syst})) \times 10^{-9}. \quad (3)$$

Current and planned kaon experiments are not expected to improve $\mathcal{B}(K^0_L \to \pi^0\mu^+\mu^-)$. Thus, a sensitivity study is performed to assess if it could be done at LHCb [4].

Content from this work may be used under the terms of the Creative Commons Attribution 3.0 licence. Any further distribution of this work must maintain attribution to the author(s) and the title of the work, journal citation and DOI.
Published under licence by IOP Publishing Ltd

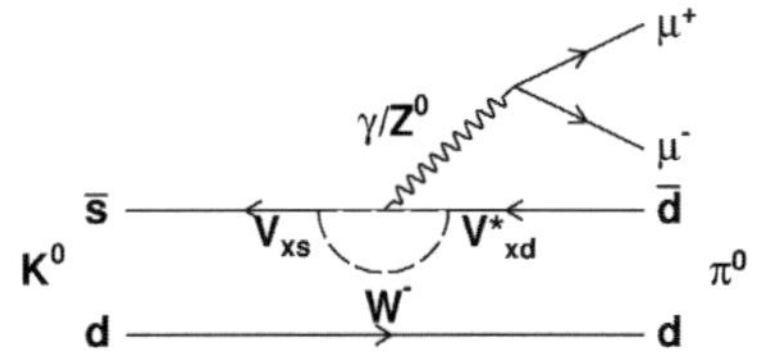

Figure 1. Feynman diagram of the process $K^0_S \to \pi^0\mu^+\mu^-$.

2. Analysis Strategy

Decays of the K^0_S in LHCb are characterized by decay vertices separated from the interaction point[1], and with tracks having an average transverse momentum significantly lower than those from the decay of b- or c-hadrons. Muon candidates are combined into $\mu^+\mu^-$ pairs. Then a π^0 can be added to the dimuon pair to make a fully reconstructed K^0_S decay. However, due to the kinematic constraints the K^0_S mass resolution does not depend much on the information of the reconstructed π^0. Indeed, a peaking distribution is found using only the constraints on the π^0 mass and the K^0_S momentum, with an estimate of the typical value of the π^0 momentum ($\sim$ 10GeV/c) (Fig. 2). Therefore, since the reconstruction efficiency of the π^0 is limited, events without reconstructed π^0's are also considered. This leads to two independent analyses: one for the events in which all decay products are considered (hereafter FULL) and one in which only the dimuon pair is used (hereafter PARTIAL). More details on the procedure can be found in [4]. For the FULL analysis, data from Run I is used, corresponding to a luminosity of 3 fb^{-1}.

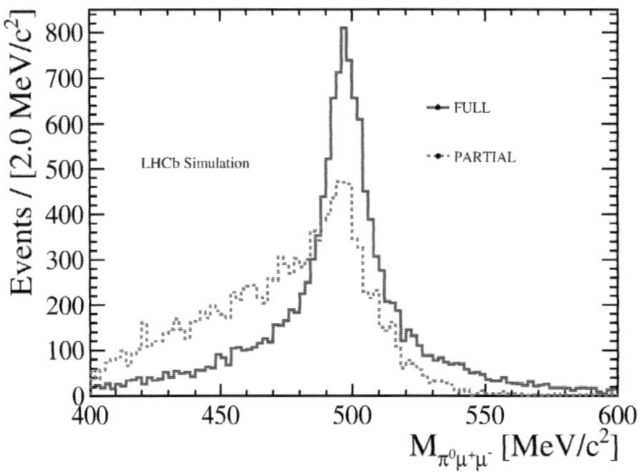

Figure 2. Comparison between the FULL (solid red) and PARTIAL (dashed green) kaon candidate mass distributions.

In the PARTIAL case, data from 2016 is analyzed, with an integrated luminosity of 0.3 fb^{-1}.

[1] The K_S at LHC typically decays after traversing tens of centimeters to several meters.

3. Reconstruction and selection

Pairs of muon candidates are reconstructed combining opposite-charged tracks with hits in the vertex locator, tracker stations, and muon chambers. In addition, the tracks are required to be separated by at least 6σ (impact parameter significance) from any $p-p$ collision point in the event. Tracks with transverse momentum lower than 80 MeV/c are ignored. Neutral pion candidates are reconstructed from γ candidate pairs that correspond to two independent clusters in the calorimeter. Each photon candidate is required to have a transverse momentum of at least 200 MeV/c. The pion candidate is required to have a mass within 30 MeV/c of the world average π^0 mass. The mass resolution is then improved by constraining the π^0 candidate mass to the world average π^0 mass, and by constraining the three-momentum vector of the K^0_S to point back to the production vertex. The PARTIAL selection does not require any information about a reconstructed π^0.

A multivariate classifier (MVA) is used to separate signal from combinatorial background. It is trained with simulated events and a fraction of the data not used afterwards in the analysis. The MVA input variables contain information about the geometrical properties of the events, kinematics, track quality, and muon identification quality. First, they are divided into continuous variables and discrete variables. The continuous variables are gaussianized, decorrelated, and gaussianized again. Then the gaussianized and the discrete variables are inputs for a boosted decision tree (BDT) training. The BDT response for signal and background for both FULL and PARTIAL are shown in Fig. 3. The analysis is then performed in the BDT region [0.6,1.0], with

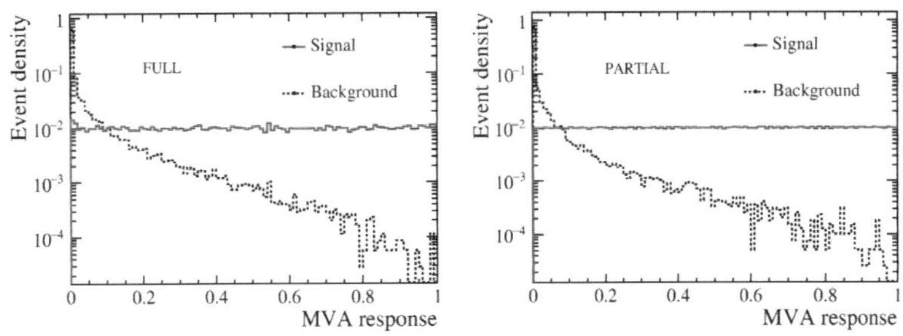

Figure 3. BDT response both for signal (solid red) and background (dashed black). Right: FULL channel. Left: PARTIAL channel. Signal and background are normalized to the same area.

a signal efficiency of 40% for both FULL and PARTIAL and a background rejection of 99.3% for FULL and 99.8% for PARTIAL.

4. Background sources

Several sources of background are investigated to assess their relevance:

- $K^0_S \to \pi^+\pi^-$ decays, where both pions are misidentified as muons, and in the case of the FULL category, combined with a random π^0 from the underlying event. These decays have a mass larger than that of the K^0_S and do not enter the fit region, except for potential residual tails that effectively add up to the combinatorial background. No evidence for $K^0_S \to \pi^+\pi^-$ background is seen for the BDT region fitted.
- $K^0 \to \mu^+\mu^-\gamma\gamma$ decays. This background was considered in the NA48 analysis [2]. However, its contribution at LHCb is found to be negligible. Further details on this study are given in [4].

- $K^0_L \to \pi^+\pi^-\pi^0$ decays. Since there is no evidence of this background in the data, it is neglected. Including a $K^0_L \to \pi^+\pi^-\pi^0$ component to the observed background does not change significantly the sensitivity estimates. The K^0_S counterpart has a branching fraction of 3.5×10^{-7} and thus is about four orders of magnitude smaller than $K^0_L \to \pi^+\pi^-\pi^0$. In general, no sign of a resonant structure in the $\pi^+\pi^-\pi^0$ is seen on data.
- Combinatorial background. It is considered to be composed by random combination of tracks, including those generated by pseudo-random combinations of hits during the pattern recognition. It has a monotonic shape across the studied invariant mass range.

5. Fit model

Only events in the BDT range [0.6,1] are considered in the fit to the data. A simultaneous fit to the mass distribution across four equally-sized independent bins of the BDT response is performed. The combinatorial background is described with an exponential PDF for both FULL and PARTIAL analysis, with independent floating yields and decay constants in each BDT bin. The signal model is an Hypathia distribution [3] with different configurations for FULL and PARTIAL (see Fig. 4). The signal model parameters are independent in each BDT bin and are obtained from simulation. The fractions of signal events allotted to each BDT bin are also fixed from values obtained from simulation. The total signal yield is left floating free in the simultaneous fit. The signal yield is floated in the fit to the data. It is measured to be compatible with zero within one sigma for the FULL case and two sigma for the PARTIAL case. The fit projections to the FULL and PARTIAL data are shown in Fig. 5.

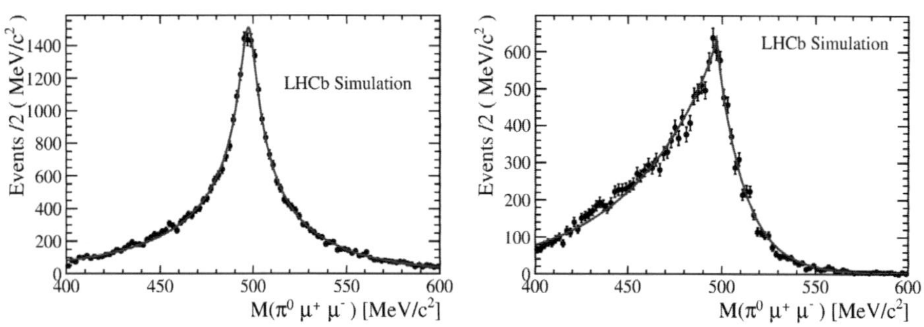

Figure 4. Invariant mass distribution fot $\pi^0\mu^+\mu^-$ events (black dots). The fit result is shown in blue for FULL (left) and PARTIAL (right) categories.

6. Expected sensitivity

The expected statistical precision on $\mathcal{B}(K^0_S \to \pi^0\mu^+\mu^-)$ for multiple values of the integrated luminosity up to 100 fb^{-1} is estimated. The expected background yield is extrapolated from the current data fit result, where the signal yield is consistent with zero. The background yield is scaled linearly for larger integrated luminosities. For each integrated luminosity in the studied range, sets of pseudo-experiments are generated with the above background expectations, and with a signal yield expectation of

$$N_{sig} = \frac{\mathcal{B}(K^0_S \to \pi^0\mu^+\mu^-)\epsilon(K^0_S \to \pi^0\mu^+\mu^-)}{\mathcal{B}(K^0_S \to \pi^+\pi^-)\epsilon(K^0_S \to \pi^+\pi^-)} N(K^0_S \to \pi^+\pi^-)\frac{\mathcal{L}_{fut}}{\mathcal{L}_{curr}}, \quad (4)$$

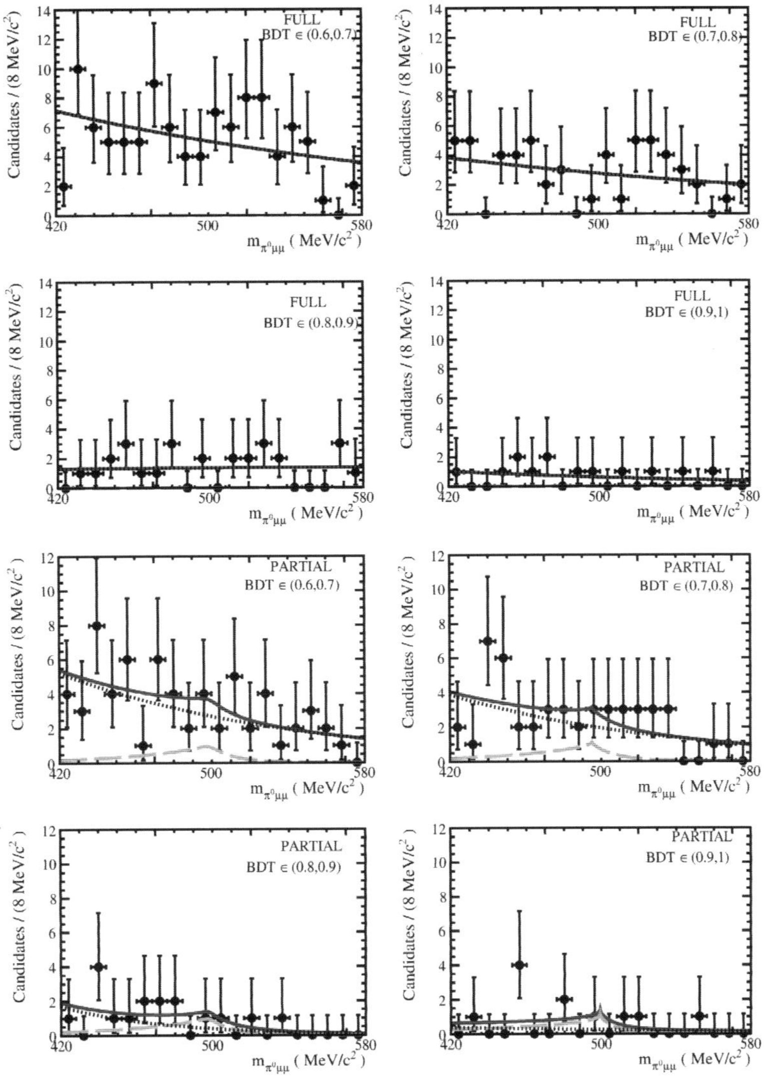

Figure 5. Fit to data for FULL (top) and PARTIAL (bottom) categories. The magenta dashed line shows the signal contribution, the dotted black line the background, and the solid blue line the prediction from the total fit model.

where L_{fut} and L_{curr} are the future and current luminosities, respectively, and $K_S^0 \to \pi^+\pi^-$ is used as normalisation channel. The models described in the previous section are fitted to each pseudo-experiment with a floating $\mathcal{B}(K_S^0 \to \pi^0\mu^+\mu^-)$. The background model parameters used are the ones obtained from the fit to the data. The statistical uncertainties are obtained as the variations of $\mathcal{B}(K_S^0 \to \pi^0\mu^+\mu^-)$ that deviate from the minimum of the log-likelihood profile by half a unit. Finally, the uncertainties are averaged across the set of pseudo-experiments for a given integrated luminosity. The resulting sensitivity curves are shown in Fig. 6. Since both of them enter values for σ_{STAT} lower than those measured by NA48 (horizontal orange lines) [2], it can be seen that the analyses of both PARTIAL and FULL categories can lead to a precision better than NA48 for the LHCb upgrade if a trigger efficiency above $\approx 50\%$ (corresponding to $\varepsilon^{TRIG}\mathcal{L} \approx 50\text{fb}^{-1}$) can be maintained.

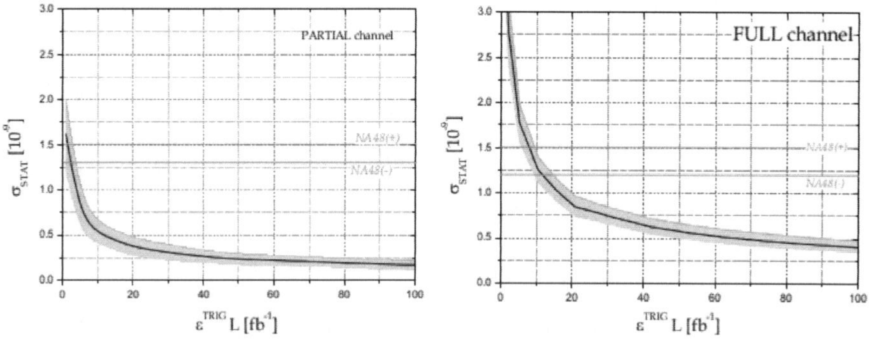

Figure 6. Expected precision on $\mathcal{B}(K^0_S \to \pi^0\mu^+\mu^-)$ for the FULL (top) and PARTIAL (bottom) channels, as a function of the integrated luminosity times trigger efficiency, $L \times \varepsilon^{TRIG/SEL}$. The horizontal orange lines represent the statistical errors measured by NA48 [2], both positive (+) and negative (-)

7. Summary and Outlook

A precise measurement of the $K^0_S \to \pi^0\mu^+\mu^-$ branching fraction is crucial for a precise $\mathcal{B}(K^0_L \to \pi^0\mu^+\mu^-)$ SM theoretical prediction and the search for physics beyond the SM. The sensitivity of the LHCb experiment to $\mathcal{B}(K^0_S \to \pi^0\mu^+\mu^-)$ was studied using the 3 fb^{-1} of data recorded at 7 and 8 TeV center-of-mass energy during 2011 and 2012, and on 0.3 fb^{-1} of data recorded at 13 TeV center-of-mass energy during 2016. Full and partial decay reconstruction algorithms were considered, provided that the efficiency for reconstructing kaons is higher for the LHCb upgrade. The sensitivity study was performed using pseudo-experiments by extrapolating signal yield results based on the currently available data to expected future integrated luminosities. If a trigger efficiency of at least 50% can be assured in the future, LHCb can determine $\mathcal{B}(K^0_S \to \pi^0\mu^+\mu^-)$ with a precision significantly better than that of NA48.

References

[1] M. Bauer, S. Casagrande, U. Haisch and M. Neubert, JHEP **1009** (2010) 017 doi:10.1007/JHEP09(2010)017 [arXiv:0912.1625 [hep-ph]].

[2] J. R. Batley *et al.* [NA48/1 Collaboration], Phys. Lett. B **599** (2004) 197 doi:10.1016/j.physletb.2004.08.058 [hep-ex/0409011].

[3] D. Martnez Santos and F. Dupertuis, Nucl. Instrum. Meth. A **764** (2014) 150 doi:10.1016/j.nima.2014.06.081 [arXiv:1312.5000 [hep-ex]].

[4] V. G. Chobanova,X. Cid Vidal,J. P. Dalseno,M. Lucio Martinez,D. Martinez Santos and V. Renaudin, *Sensitivity of LHCb and its upgrade in the measurement of* $\mathcal{B}(K^0_S \to \pi^0\mu^+\mu^-)$ Tech. Rep. LHCb-PUB-2016-017. CERN-LHCb-PUB-2016-017, CERN, Geneva, Oct. 2016

Dispersive analysis of $K_S \to \gamma\gamma$ and $K_S \to \gamma\ell^+\ell^-$

Lewis C. Tunstall,[0,1] **Gilberto Colangelo**[1] **and Ramon Stucki**[1]
[1]Albert Einstein Centre for Fundamental Physics, Institute for Theoretical Physics, University of Bern, Sidlerstrasse 5, CH–3012 Bern, Switzerland

E-mail: tunstall@itp.unibe.ch

Abstract. We calculate the decay rates for $K_S \to \gamma\gamma$ and $K_S \to \gamma\ell^+\ell^-$ ($\ell = e$ or μ) within a dispersive framework in which the weak Hamiltonian carries momentum. Given input from $K_S \to \pi\pi$ and $\gamma\gamma^{(*)} \to \pi\pi$, we solve the once-subtracted dispersion relations numerically and find that final-state $\pi\pi$ interactions generate sizeable corrections to the predictions from 3-flavour chiral perturbation theory. Our analysis predicts $\text{BR}(K_S \to \gamma\gamma) = (2.34 \pm 0.31) \times 10^{-6}$, $\text{BR}(K_S \to \gamma e^+e^-) = (4.38 \pm 0.57) \times 10^{-8}$, and $\text{BR}(K_S \to \gamma\mu^+\mu^-) = (1.45 \pm 0.27) \times 10^{-9}$.

1. Motivation

The advent of high-statistics kaon experiments such as NA62 [1], KLOE-2 [2] and (somewhat surprisingly) LHCb [3], presents an opportunity for theorists to re-examine the Standard Model (SM) background for various rare decays. In order to identify areas where progress may be made, it is useful to adopt a classification system of the possible decay modes:

- ☺ **The good:** so-called golden modes like $K \to \pi\nu\bar{\nu}$, where the dominant effect is of short-distance nature and the corresponding hadronic matrix element is largely determined by a single form factor;[1]
- 😐 **The bad:** CP-violating decays like $K_L \to \pi^0\ell^+\ell^-$ ($\ell = e$ or μ), where short- and long-distance effects come in roughly equal measure;
- ☹ **The ugly:** non-leptonic decays like $K_S \to \pi\pi$ and $K_L \to 3\pi$ which are dominated by long-distance contributions. In such cases, non-perturbative methods are essential; especially to make sense of long-standing puzzles like the $\Delta I = 1/2$ rule (which for the purposes of this talk is assumed to be exact).

In between the above trio are decays where a separation of the short- and long-distance effects can be achieved with varying degree of success. Since kaon decays occur at low energies, a systematic analysis can be undertaken within χPT_3, where amplitudes are expanded as an asymptotic series in powers of $O(m_K)$ momentum and light quark masses $m_{u,d,s} = O(m_K^2)$. In general, our ability to obtain precise predictions from the effective field theory is influenced by two main factors:

(i) hadronic uncertainties are parametrised in terms of low-energy constants (LECs) like the pion decay constant F_π, whose values are not fixed by chiral symmetry alone. In an ideal

[0] Speaker
[1] However, as remarked at this conference [4], "good" does not necessarily imply "trivial": e.g. for $K \to \pi\nu\bar{\nu}$ there are also bi-local, long-distance contributions, whose evaluation requires input from lattice QCD.

Content from this work may be used under the terms of the Creative Commons Attribution 3.0 licence. Any further distribution of this work must maintain attribution to the author(s) and the title of the work, journal citation and DOI.
Published under licence by IOP Publishing Ltd

world, these LECs would be extracted from a combination of input from experiment and lattice QCD. However, for the weak radiative decays which interest us, this reality appears to be far into the future;

(ii) at energies above the $\pi\pi$ threshold, final-state interactions (FSI), especially in the 0^{++} channel [5–8], can spoil the convergence of the χPT$_3$ expansion. In these cases, chiral-perturbative methods must be abandoned in favour of non-perturbative methods based on unitarity, analyticity, and crossing symmetry.

Dispersion relations offer a means to address items (i) and (ii) within a model-independent framework. In this talk we consider the application of these methods to $K_S \to \gamma\gamma^*$ transitions [11], with the two-fold aim of (a) extending the predictions from LO χPT$_3$ to fully account for effects due to FSI, and (b) providing a reliable estimate for the theory uncertainty. For the pure radiative decay, the chiral prediction [12–14] for the rate

$$\mathrm{BR}(K_S \to \gamma\gamma)_{\chi\mathrm{PT}_3} = 2.0 \times 10^{-6} \tag{1}$$

is in reasonable agreement with the experimental average

$$\mathrm{BR}(K_S \to \gamma\gamma) = (2.63 \pm 0.17) \times 10^{-6}\,, \tag{2}$$

while the predictions [15] for the leptonic modes are typically expressed in terms of the ratios

$$\left.\frac{\Gamma(K_S \to \gamma\ell^+\ell^-)}{\Gamma(K_S \to \gamma\gamma)}\right|_{\chi\mathrm{PT}_3} = \begin{cases} 1.6 \times 10^{-2} & (\ell = e) \\ 3.8 \times 10^{-4} & (\ell = \mu) \end{cases}\,. \tag{3}$$

Although these decays have not yet been measured, they may lie within reach of LHCb's kaon programme or the KLOE-2 experiment, which is projected to be sensitive down to K_S branching ratios of $O(10^{-9})$. As shown below, the dispersive analysis yields sizeable corrections to the predictions from LO χPT$_3$. Our results support the generally accepted view that getting FSI under good control is essential if long-distance contributions to 0^{++} channels are to be reliably determined.[2]

2. Dispersive framework for $K_S \to \gamma\gamma^*$

In two-body decays, the kinematics is completely fixed and thus a representation of the amplitude in terms of Mandelstam variables s, t, u is not immediately forthcoming. One popular approach is to promote the mass of the decaying particle to a dynamical variable "s" and solve the corresponding dispersion integral. Although seductively simple, this prescription is subject to an inherent ambiguity associated with the fact that there are an *infinite number of ways* in which to extrapolate the mass off-shell [19, 20].

In order to avoid the arbitrariness induced by off-shell extrapolations, an alternative approach was proposed by Büchler *et al.* [21] in the study of $K_S \to \pi\pi$. The key idea in that work was to allow the $\Delta S = 1$ weak Hamiltonian $\mathcal{H}_w$ to inject momentum in the amplitude $\langle\pi\pi|\mathcal{H}_w|K_S\rangle$, and solve the resulting set of coupled dispersion integrals in the crossed s, t, u channels. Adapted to $K_S \to \gamma\gamma^*$ transitions (Fig. 1), this prescription yields a tensor decomposition of the amplitude[3]

$$A_{\mu\nu}(k, q_1, q_2) \supset \big(q_1 \cdot q_2 g_{\mu\nu} - q_{2\mu}q_{1\nu}\big)A_{\gamma\gamma^*}(s, t, u, q_2^2) \tag{4}$$

[2] A topical example concerns the SM prediction for ϵ'/ϵ, where the recent lattice result from RBC/UKQCD [16] is in tension with experiment at approximately the 3σ confidence level. However, that same calculation produces a $\pi\pi$ scattering phase-shift $\delta_0^0(m_K^2)$ that falls $\approx 3\sigma$ short of model-independent analyses based on the Roy equations. This suggests that the effects from FSI are not (yet) fully accounted for in the lattice result; see e.g. [17, 18] for further details.

[3] The most general tensor decomposition actually involves three scalar functions, but two of them vanish if contributions from D waves and higher are neglected; see main text.

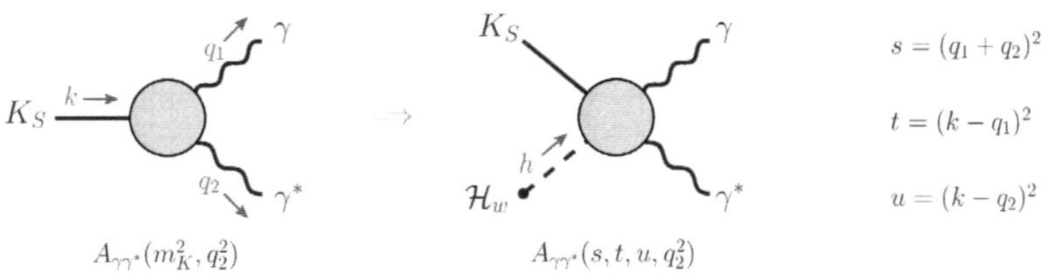

Figure 1. For the physical decay amplitude $A_{\gamma\gamma^*}$ (left), the kinematics are completely fixed by the external momenta. Allowing the weak Hamiltonian $\mathcal{H}_w$ to carry momentum (right) promotes $A_{\gamma\gamma^*}$ to a function of s, t, u, with the physical amplitude recovered by taking the limit $h \to 0$.

that involves a scalar function $A_{\gamma\gamma^*}$ which depends on s, t, u and is free from kinematic zeros and singularities. A straightforward one-loop calculation determines the LO χPT$_3$ result for $A_{\gamma\gamma^*}$ in this framework, and reproduces the predictions listed in eqs. (1) and (3) once the physical limit $h \to 0$ is taken.

Our goal, however, is to construct a dispersive representation for $A_{\gamma\gamma^*}$, which in general requires an analysis of all possible intermediate states $\pi\pi$, 4π, KK, ... in all three channels. This is rather complicated, so instead we make use of a simplification which has proven to be particularly effective for other scattering processes at low energies, namely to neglect the contributions to discontinuities coming from D waves and higher. As in those cases, we expect that at the physical point $h \to 0$, the contributions to the S wave coming from discontinuities in the t and u channels are negligible.

In the s-channel, the leading contribution to the discontinuity of $A_{\gamma\gamma^*}$ is due to the $\pi\pi$ intermediate state. The unitarity relation for this contribution reads

$$\text{disc}_s\, A_{\gamma\gamma^*}(s, q_2^2) = -\alpha\sqrt{1 - 4m_\pi^2/s}\left\{\frac{A_{\pi\pi}(s)[h^0_{++}(s, q_2^2)]^*}{2(s - q_2^2)}\right\}, \tag{5}$$

so we write down a once-subtracted dispersion relation

$$A_{\gamma\gamma^*}(s, s_0, q_2^2) = a_{\gamma\gamma^*}(s_0, q_2^2) + \frac{s - s_0}{\pi}\int_{4m_\pi^2}^{\infty} dz\, \frac{\text{disc}_s\, A_{\gamma\gamma^*}(z, q_2^2)}{(z - s_0)(z - s - i\epsilon)}, \tag{6}$$

where the subtraction constant $a_{\gamma\gamma^*}$ is needed to ensure convergence of the integral. Evidently, knowledge of $a_{\gamma\gamma^*}$ and the discontinuity (5) completely fixes the prediction for $A_{\gamma\gamma^*}$. Concerning quantities which enter in the latter, h^0_{++} is a helicity partial wave corresponding to the $\gamma\gamma^* \to \pi\pi$ subprocess, while $A_{\pi\pi}$ is the $K_S \to \pi\pi$ amplitude, whose dispersive representation [21] given to good approximation by

$$A_{\pi\pi}(s) \simeq \sqrt{2}a_{\pi\pi}[1 + E(X)s/m_K^2]\Omega_0^0(s)\,. \tag{7}$$

Here $a_{\pi\pi}$ is a subtraction constant whose value is fixed from the physical $K_S \to \pi\pi$ decay amplitude at $s = m_K^2$, while the quantity

$$E(X) = \frac{3m_K^2(1 + X)}{m_K^2 - m_\pi^2(4 + 3X)} \tag{8}$$

parametrises the NLO corrections in terms of the parameter X. An appealing feature of this representation is that the effects from FSI are fully accounted for in terms of the calculable Omnès factor

$$\Omega_0^0(s) = \exp\left(\frac{s}{\pi}\int_{4m_\pi^2}^{\infty} dz\, \frac{\delta_0^0(z)}{z(z - s - i\epsilon)}\right). \tag{9}$$

3. Dispersive treatment of $K_S \to \gamma\gamma$

With a dispersive representation of the $K_S \to \gamma\gamma^*$ amplitude at hand (6), our first application concerns the case where both photons are on-shell. Here, the subtraction constant $a_{\gamma\gamma}$ is fixed by matching to χPT$_3$ well below the $\pi\pi$ threshold, where the typically large corrections due to FSI are entirely absent. In particular, we match at the kinematic point $s_0 = -0.098$ GeV^2, where the χPT$_3$ amplitude vanishes and thus the dependence on $a_{\gamma\gamma}$ is eliminated — up to $O(p^6)$ corrections which we fold into our final estimate of the systematic uncertainty.

Regarding the input for $h^0_{0,++}$, we use data from the coupled-channel analysis of $\gamma\gamma \to \pi\pi$ performed by García-Martín and Moussallam [22]. Since the determination of $h^0_{0,++}$ in this analysis is expected to be reliable up to $s \lesssim 2$ GeV^2, it is necessary to impose a cutoff Λ in our dispersion integral. We have checked that at the physical point $s = m_K^2$, the cutoff dependence is very mild, and we take $\Lambda = 1.2$ GeV as a benchmark value.

The resulting energy dependence of the real and imaginary parts of $A_{\gamma\gamma}$ is shown in Fig. 2. As expected, the dispersive representation agrees with LO χPT$_3$ below the $\pi\pi$ threshold, but for $s > 4m_\pi^2$, the effects from FSI distort the amplitude, producing a significant enhancement (suppression) of the real (imaginary) part. These effects lead to an enhanced prediction for the branching ratio

$$\text{BR}(K_S \to \gamma\gamma) = \frac{m_K^3}{64\pi} \frac{|A_{\gamma\gamma}(m_K^2)|^2}{\Gamma(K_S)_{\text{tot}}} = (2.34 \pm 0.31) \times 10^{-6} \tag{10}$$

which brings the SM and experiment (2) into much better agreement. The quoted uncertainty is due to four main sources of systematic error which enter in our analysis:

- the variation $X = \pm 0.3$ based on the expected convergence pattern of χPT$_3$ below the $\pi\pi$ threshold;
- $O(p^6)$ corrections to the subtraction constant $a_{\gamma\gamma}$, which we estimate by shifting the value of the subtraction point s_0 by 30%;
- lack of knowledge about the phase of the $K_S \to \pi\pi$ sub-amplitude in the inelastic region. We account for this by comparing two Omnès factors which exhibit "dip" and "non-dip" behaviour in this kinematic region;
- contributions from the high-energy region $\Lambda > 1.2$ GeV. These we estimate by guiding the phase of Ω^0_0 to π and fixing the helicity partial wave to a constant value $|h_{0,++}| \approx 4$.

Combined in quadrature, the final uncertainty has turned out to be remarkably modest.

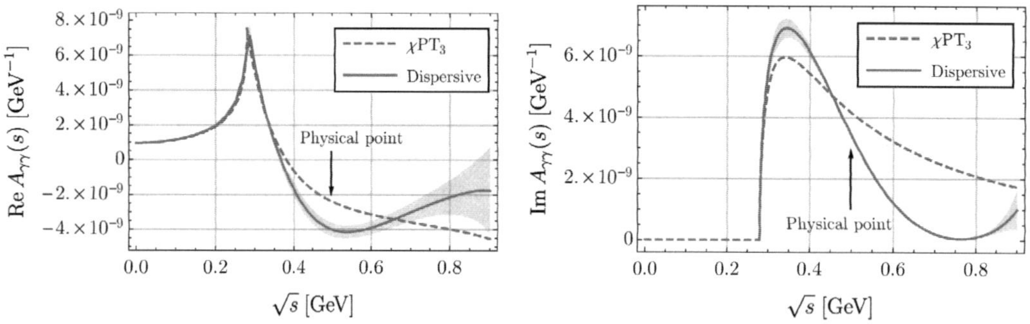

Figure 2. Energy dependence of the real (left) and imaginary (right) parts of the $K_S \to \gamma\gamma$ amplitude. The blue band in the dispersive result corresponds to the systematic uncertainty.

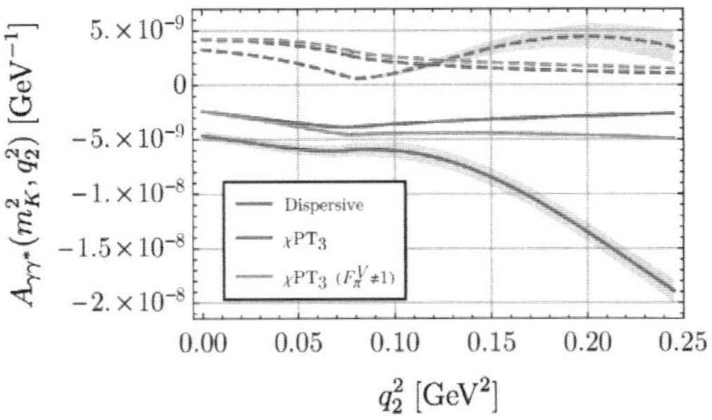

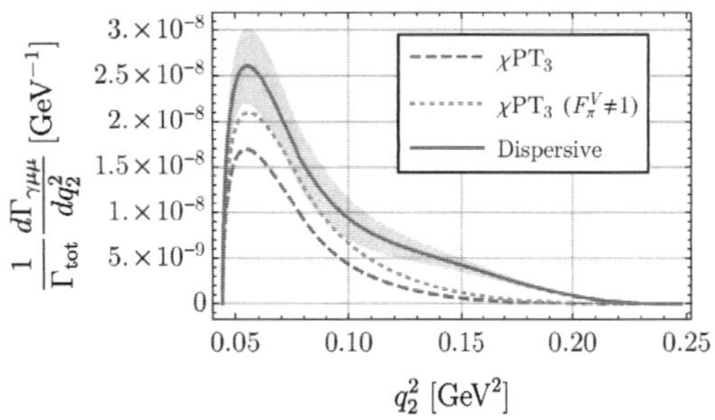

Figure 3. Dependence of the $K_S \to \gamma\gamma^*$ amplitude (left) and $K_S \to \gamma\mu^+\mu^-$ differential decay rate (right) on the photon momentum q_2^2. In the left figure, the real parts are denoted by the solid curves, while the imaginary parts are dashed. The bands on the dispersive results correspond to the systematic uncertainty.

4. Dispersive treatment of $K_S \to \gamma\ell^+\ell^-$

We now consider the $K_S \to \gamma\gamma^*$ transition, where the photon momentum lies in the physical region $4m_\ell^2 \leq q_2^2 \leq m_K^2$ relevant for $K_S \to \gamma\ell^+\ell^-$ decays. In this case, we subtract the dispersion integral (6) at $s_0 = 0$, and fix the subtraction constant $a_{\gamma\gamma^*}$ by matching to the χPT$_3$ amplitude. Regarding input for the discontinuity, we use the pion form factor and helicity partial wave $h_{0,++}$ obtained from Moussallam's single-channel analysis of $\gamma\gamma^* \to \pi\pi$ [23]. Strictly speaking, the input for $h_{0,++}$ is only valid for energies $\sqrt{s} \lesssim 0.8$ GeV, however, we have checked that increasing the cutoff to $\Lambda = 1.2$ GeV does not lead to a difference of more than $\approx 7\%$ in the resulting prediction for $A_{\gamma\gamma^*}$. Since this small change is covered by our estimate of the systematic uncertainty, we take the larger cutoff as a benchmark value in our numerics

For fixed $s = m_K^2$, we evaluate the dispersion integral numerically and show the q_2^2 dependence of the amplitude in the left plot of Fig. 3. As shown in the figure, when $q_2^2 < 4m_\pi^2$, the effect of FSI resembles that seen in $K_S \to \gamma\gamma$ (Fig. 2), with the real (imaginary) parts enhanced (suppressed) relative to χPT$_3$. However, as q_2^2 increases above the $\pi\pi$ threshold, the pion vector form factor F_π^V becomes progressively more important, and both real and imaginary parts in the dispersive amplitude are enhanced relative to LO χPT$_3$.

Turning to the $K_S \to \gamma\ell^+\ell^-$ decay rates, in Fig. 3 we compare the χPT$_3$ prediction for the differential decay rate involving muons against our dispersive result. Evidently, the corrections are large for $q_2^2 \gtrsim 0.05$: this can be inferred from the q_2^2 behaviour shown in the left plot in Fig. 3. We also see that, for this mode, the dominant source of the enhancement is due to the pion form factor.

The resulting branching fractions are given in Table 1, where the uncertainties are determined as in Sec. 2, except for the subtraction constant: here we keep s_0 fixed and vary the χPT$_3$ amplitude by 30%. In both cases, the corrections are sizeable: for the electron mode we see a shift of $O(50\%)$, while in the muon mode we have a shift of $O(100\%)$. The origin of these shifts are different in each case. For the electron mode, the phase space is peaked near the origin $q_2^2 = 0$, so the role of F_π^V is suppressed and the dominant effect is due to FSI. On the other hand, as shown in the right plot of Fig. 3, the enhancement in the muon mode is due to both FSI and the form factor.

Input	BR($K_S \to \gamma e^+e^-$)	BR($K_S \to \gamma\mu^+\mu^-$)
χPT$_3$	3.09×10^{-8}	7.25×10^{-10}
χPT$_3$ ($F_\pi^V \neq 1$)	3.17×10^{-8}	9.97×10^{-10}
This work	$(4.38 \pm 0.57) \times 10^{-8}$	$(1.45 \pm 0.27) \times 10^{-9}$

Table 1. Predictions for the branching ratio of $K_S \to \gamma\ell^+\ell^-$. The second row indicates the effect of including the pion vector form factor F_π^V in the χPT$_3$ amplitude.

5. Remarks

Chiral perturbation theory and lattice QCD are two of the main tools which allow a systematic calculation of long-distance contributions to kaon decays, but getting FSI under good control in either of these approaches is challenging. Dispersion relations offer a different, complementary methodology to the previous two, which addresses specifically the treatment of FSI. If one can match the dispersive and the chiral representation, and solve the dispersion relation, one can usually obtain much better control over FSI effects. In this talk, we have presented a first step in this direction by introducing a dispersive framework for $K_S \to \gamma\gamma$ and $K_S \to \gamma\ell^+\ell^-$. In both cases we find the corrections due to FSI are large with respect to LO χPT$_3$. The extension of this framework to the more challenging $K_S \to \ell^+\ell^-$ decays is currently in progress.

Acknowledgements

LCT thanks the conference organisers for the opportunity to present the work discussed in this article. We also thank Marc Knecht and Peter Stoffer for helpful discussions regarding the systematic uncertainties in our analysis. Financial support from the Swiss National Science Foundation is gratefully acknowledged.

References

[1] G. Collazuol [NA62 Collaboration], PoS EPS **-HEP2009** (2009) 260.
[2] G. Amelino-Camelia *et al.*, Eur. Phys. J. C **68** (2010) 619 [arXiv:1003.3868].
[3] R. Aaij *et al.* [LHCb Collaboration], JHEP **1301**, 090 (2013) [arXiv:1209.4029].
[4] G. Isidori, these proceedings [URL: http://tinyurl.com/kaonsummary].
[5] A. Neveu and J. Scherk, Annals Phys. **57** (1970) 39.
[6] T. N. Truong, Acta Phys. Polon. B **15** (1984) 633.
[7] T. N. Truong, Phys. Lett. B **207** (1988) 495.
[8] A. Dobado, M. J. Herrero and T. N. Truong, Phys. Lett. B **235** (1990) 134.
[9] J. R. Pelaez, Phys. Rept. **658**, 1 (2016) [arXiv:1510.00653].
[10] I. Caprini, G. Colangelo and H. Leutwyler, Phys. Rev. Lett. **96** (2006) 132001 [arXiv:hep-ph/0512364].
[11] G. Colangelo, R. Stucki and L. C. Tunstall, Eur. Phys. J. C (to be published) [arXiv:1609.03574].
[12] G. D'Ambrosio and D. Espriu, Phys. Lett. B **175** (1986) 237.
[13] J. L. Goity, Z. Phys. C **34** (1987) 341.
[14] V. Cirigliano, G. Ecker, H. Neufeld, A. Pich and J. Portoles, Rev. Mod. Phys. **84** (2012) 399 [arXiv:1107.6001].
[15] G. Ecker, A. Pich and E. de Rafael, Nucl. Phys. B **303** (1988) 665.
[16] Z. Bai *et al.* [RBC and UKQCD Collaborations], Phys. Rev. Lett. **115**, 212001 (2015) [arXiv:1505.07863].
[17] G. Colangelo, talk at NA62 Kaon Physics Handbook MITP Workshop, 11-22 January, 2016 [URL: http://tinyurl.com/NA62colangelo].
[18] A. Pich, talk at NA62 Kaon Physics Handbook MITP Workshop 11-22 January, 2016 [URL: http://tinyurl.com/NA62pich].
[19] G. Colangelo, Nucl. Phys. Proc. Suppl. **106** (2002) 53 [arXiv:hep-lat/0111003].
[20] M. Büchler, G. Colangelo, J. Kambor and F. Orellana, Phys. Lett. B **521** (2001) 29 [arXiv:hep-ph/0102289].
[21] M. Büchler, G. Colangelo, J. Kambor and F. Orellana, Phys. Lett. B **521** (2001) 22 [arXiv:hep-ph/0102287].
[22] R. García-Martín and B. Moussallam, Eur. Phys. J. C **70** (2010) 155 [arXiv:1006.5373].
[23] B. Moussallam, Eur. Phys. J. C **73** (2013) 2539 [arXiv:1305.3143].

Measurement of $BR(\mathrm{K}^{+} \to \mu^{\pm}\nu_{\mu}\mathrm{e}^{+}\mathrm{e}^{-})$ with NA48/2 at CERN

Gia Khoriauli*

Talk given at the KAON2016 conference (14-17 September 2016, Birmingham, UK) on behalf of the NA48/2 Collaboration: G. Anzivino, R. Arcidiacono, W. Baldini, S. Balev, J.R. Batley, M. Behler, S. Bifani, C. Biino, A. Bizzeti, B. Bloch-Devaux, G. Bocquet, N. Cabibbo, M. Calvetti, N. Cartiglia, A. Ceccucci, P. Cenci, C. Cerri, C. Cheshkov, J.B. Chèze, M. Clemencic, G. Collazuol, F. Costantini, A. Cotta Ramusino, D. Coward, D. Cundy, A. Dabrowski, P. Dalpiaz, C. Damiani, M. De Beer, J. Derré, H. Dibon, L. DiLella, N. Doble, K. Eppard, V. Falaleev, R. Fantechi, M. Fidecaro, L. Fiorini, M. Fiorini, T. Fonseca Martin, P.L. Frabetti, L. Gatignon, E. Gersabeck, A. Gianoli, S. Giudici, A. Gonidec, E. Goudzovski, S. Goy Lopez, M. Holder, P. Hristov, E. Iacopini, E. Imbergamo, M. Jeitler, G. Kalmus, V. Kekelidze, K. Kleinknecht, V. Kozhuharov, W. Kubischta, G. Lamanna, C. Lazzeroni, M. Lenti, L. Litov, D. Madigozhin, A. Maier, I. Mannelli, F. Marchetto, R. Marchevski, G. Marel, M. Markytan, P. Marouelli, M. Martini, L. Masetti, E. Mazzucato, A. Michetti, I. Mikulec, N. Molokanova, E. Monnier, U. Moosbrugger, C. Morales Morales, D.J. Munday, A. Nappi, G. Neuhofer, A. Norton, M. Patel, M. Pepe, A. Peters, F. Petrucci, M.C. Petrucci, B. Peyaud, M. Piccini, G. Pierazzini, I. Polenkevich, Yu. Potrebenikov, M. Raggi, B. Renk, P. Rubin, G. Ruggiero, M. Savrié, M. Scarpa, M. Shieh, M.W. Slater, M. Sozzi, S. Stoynev, E. Swallow, M. Szleper, M. Valdata-Nappi, B. Vallage, M. Velasco, M. Veltri, S. Venditti, M. Wache, H. Wahl, A. Walker, R. Wanke, L. Widhalm, A. Winhart, R. Winston, M.D. Wood, S.A. Wotton, A. Zinchenko, M. Ziolkowski

*Johannes Gutenberg University of Mainz, Staudingerweg 7, 55128 Mainz, Germany

E-mail: khoriaul@uni-mainz.de

Abstract. The exclusive branching ratio of the radiative decay $\mathrm{K}^{\pm} \to \mu^{\pm}\nu_{\mu}\mathrm{e}^{+}\mathrm{e}^{-}$ is measured using data collected by the NA48/2 experiment at CERN in 2003-2004. The measurement is performed in the phase-space region of the mass of the electron-positron pair, $m_{\mathrm{e}^{+}\mathrm{e}^{-}} \geq 140\,\mathrm{MeV}/c^{2}$. This is the most accurate measurement of the exclusive $BR(\mathrm{K}^{\pm} \to \mu^{\pm}\nu_{\mu}\mathrm{e}^{+}\mathrm{e}^{-})$. The result, $BR(\mathrm{K}^{\pm} \to \mu^{\pm}\nu_{\mu}\mathrm{e}^{+}\mathrm{e}^{-} \mid m_{\mathrm{e}^{+}\mathrm{e}^{-}} \geq 140\,\mathrm{MeV}/c^{2}) = (7.81 \pm 0.23) \times 10^{-8}$ is compatible with the theoretical value, 8.51×10^{-8}, calculated using the Chiral Perturbation Theory framework at next-to-leading order.

1. Introduction

The branching ratios of the radiative four-lepton Kaon decays, $\mathrm{K}^{\pm} \to l^{\pm}\nu_{l}l'^{+}l'^{-}$, where, l, $l' = \mathrm{e}$, μ and $l = l'$ or $l \neq l'$, are precisely known in the Standard Model (SM) already at the tree level. However, their differential branching ratios in the high kinematic tail of the invariant mass of the $l'^{+}l'^{-}$ pair, $m_{l'^{+}l'^{-}}$, are particularly interesting for two reasons: a) it is possible to have a precise and unambiguous calculation of the low-energy QCD effects within the framework of the Chiral Perturbation Theory (ChPT) at next-to-leading order; b) the suppression of the SM branching ratio in the high $m_{l'^{+}l'^{-}}$ region makes the decays sensitive

Content from this work may be used under the terms of the Creative Commons Attribution 3.0 licence. Any further distribution of this work must maintain attribution to the author(s) and the title of the work, journal citation and DOI.
Published under licence by IOP Publishing Ltd

to a new physics phenomenon, which can contribute in the production of the $l'^{+}l'^{-}$ pair via some new particle with a non-zero mass [1, 2, 3].

In the following work the exclusive branching ratio, $BR(\mathrm{K}^{\pm} \to \mu^{\pm}\nu_{\mu}\mathrm{e}^{+}\mathrm{e}^{-})$ is measured in the kinematic region of $m_{\mathrm{e}^{+}\mathrm{e}^{-}} \geq 140$ MeV/c^2. The corresponding theoretical value, $BR(\mathrm{K}^{\pm} \to \mu^{\pm}\nu_{\mu}\mathrm{e}^{+}\mathrm{e}^{-} \mid m_{\mathrm{e}^{+}\mathrm{e}^{-}} \geq 140\ \mathrm{MeV}/c^2) = 8.51 \times 10^{-8}$, is calculated [4] using the ChPT framework. The calculation includes next-to-leading order loop corrections, which lead to the Structure Dependent (SD) contribution (see Figure 1c) in the branching ratio. The Inner Bremsstrahlung (IB) contribution is mainly determined by the leading order amplitudes of the initial- and final-state radiation (see Figures 1a and 1b) and dominates if a calculation is done in the full phase space. Table 1 shows results of theoretical calculation [4] taking only IB or IB, SD

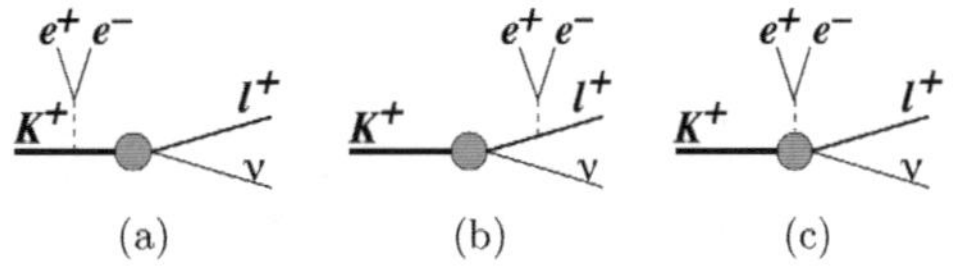

Figure 1: Tree level Inner Bremsstrahlung diagrams (a, b) and Structure Dependent diagram (c), which arises only at the next-to-leading order.

and their interference (INT) contributions into account. Comparison is made for both the full and the interesting $m_{\mathrm{e}^{+}\mathrm{e}^{-}} \geq 140$ MeV/c^2 phase space regions. The SD and INT contributions have no significant impact on the inclusive branching ratio. In contrast, their contribution is about 40% in the exclusive branching ratio in the $m_{\mathrm{e}^{+}\mathrm{e}^{-}} \geq 140$ MeV/c^2 region.

Table 1: Comparison of branching ratios obtained with only IB and with all IB+SD+INT contributions included in the calculation of the ChPT form factors. Results for both the full phase space region and the high mass region of the $\mathrm{e}^{+}\mathrm{e}^{-}$ pair are presented.

	IB	IB+SD+INT
Full phase space	2.49×10^{-5}	2.49×10^{-5}
$m_{\mathrm{e}^{+}\mathrm{e}^{-}} \geq 140$ MeV/c^2	4.98×10^{-8}	8.51×10^{-8}

The first measurement of $BR(\mathrm{K}^{\pm} \to \mu^{\pm}\nu_{\mu}\mathrm{e}^{+}\mathrm{e}^{-} \mid m_{\mathrm{e}^{+}\mathrm{e}^{-}} \geq 140\,\mathrm{MeV}/c^2) = (12.3 \pm 3.2) \times 10^{-8}$ was carried out at CERN in 1976 [5]. The mean value is higher than the theoretical one but the difference is not significant due to the large uncertainty of the measurement. A more precise measurement of the exclusive branching ratio but with a more severe cut on the phase space was performed later by the E865 experiment at BNL in 2002 [6]. The result, $BR(\mathrm{K}^{\pm} \to \mu^{\pm}\nu_{\mu}\mathrm{e}^{+}\mathrm{e}^{-} \mid m_{\mathrm{e}^{+}\mathrm{e}^{-}} \geq 145\,\mathrm{MeV}/c^2) = (7.06 \pm 0.16\,(stat.) \pm 0.26\,(syst.)) \times 10^{-8}$ is in agreement with the ChPT expectation.

2. The NA48/2 Experiment

The main goal of the fixed-target experiment NA48/2 was to measure the direct CP violation in the $\mathrm{K}^{\pm} \to \pi^{\pm}\pi^{+}\pi^{-}$ and $\mathrm{K}^{\pm} \to \pi^{\pm}\pi^{0}\pi^{0}$ decays [7]. The experiment was operating on the primary beam of 400 GeV/c momentum protons supplied by the SPS accelerator facility at CERN. A secondary beam of simultaneous and collinear K^{+} and K^{-} with the momentum (60 ± 3.7) GeV/c was extracted after the Beryllium target. Kaon decays in vacuum inside the 114 m long fiducial volume were used for measurements. The magnetic spectrometer was placed downstream after the decay volume to measure charged tracks and was operating in Helium at the atmospheric pressure. It consisted of four drift chambers (DCH) and a dipole magnet installed in between the 2nd and 3rd DCHs. The dipole magnet was providing 120 MeV/c horizontal momentum kick. The momentum resolution of the spectrometer was $\sigma_p/p = (1.02 \oplus 0.044 \cdot p)\%$, where, p is given in GeV/c. The magnetic spectrometer was followed downstream by the charged track hodoscope (HOD) to trigger events with charged tracks and measure their time. Its resolution was about 200 ps. The HOD was followed by the $27X_0$ deep, almost homogeneous electromagnetic calorimeter (LKr) filled with liquid Krypton. The LKr had excellent energy and spatial resolution. As an example, $\sigma_{\rm E}/{\rm E} = 1.4\%$ and $\sigma_x = \sigma_y = 1.5$ mm at E = 10 GeV. The LKr was followed by the hadronic calorimeter and the muon veto system (MUV). The MUV consisted of three orthogonal planes of scintillator strips to measure time and coordinates of tracks. Each MUV plane was preceded with a 80 cm thick iron wall for absorbing hadrons. An additional 40 cm thick iron wall was placed after the last MUV plane to stop hadrons scattered backward from the material at the rear of the beamline. A complete description of the NA48/2 beamline and the detector can be found in [8].

3. Measurement of $BR(\mathrm{K}^{\pm} \to \mu^{\pm}\nu_{\mu}\mathrm{e}^{+}\mathrm{e}^{-} \mid m_{\mathrm{e}^{+}\mathrm{e}^{-}} \geq 140\ \mathrm{MeV}/c^2)$

The signal $\mathrm{K}^{\pm} \to \mu^{\pm}\nu_{\mu}\mathrm{e}^{+}\mathrm{e}^{-}$ decay (Figure 2a) has a similar three-track event topology as the $\mathrm{K}^{\pm} \to \pi^{\pm}\pi^{+}\pi^{-}$ decay. Therefore, its measurement is based on the NA48/2 main physics trigger for selection of the three-track events. In total, order of 10^{11} $\mathrm{K}^{\pm}$ decays were recorded during the whole operation of the experiment using the main trigger.

3.1. Signal event selection

Events with the reconstructed good-quality vertices of three tracks within the 98 m long fiducial volume are selected. The total charge of the three tracks is required to be ±1 in order to avoid events with the fake three-track vertices. All tracks are required to be within the geometrical acceptance of the DHCs, HOD, LKr and MUV detectors. Two out of the three tracks have to be identified as electrons using the LKr calorimeter. The third track is identified as muon if it behaves as a minimum ionising particle (MIP) in the LKr and is detected by the MUV. Further selection criteria are applied to reject background events. The requirement $m_{\mathrm{e}^{+}\mathrm{e}^{-}} \geq 140\ \mathrm{MeV}/c^2$ helps to completely get rid of the background from the $\mathrm{K}^{\pm} \to \pi^{\pm}\pi_0(\pi_0 \to \mathrm{e}^{+}\mathrm{e}^{-}\gamma)$ and $\mathrm{K}^{\pm} \to \mu^{\pm}\nu_{\mu}\pi_0(\pi_0 \to \mathrm{e}^{+}\mathrm{e}^{-}\gamma)$ decays. A cut on the invariant mass of the muon-neutrino pair, $\mathrm{M}_{\mu\nu_{\mu}} > m_{\pi^{\pm}} + 3 \times \sigma(\mathrm{M}_{\mu\nu_{\mu}}) = 170\ \mathrm{MeV}/c^2$, is applied to suppress the $\mathrm{K}^{\pm} \to \mathrm{e}^{+}\mathrm{e}^{-}\pi^{\pm}(\pi^{\pm} \to \mu^{\pm}\nu_{\mu})$ decay background from 10% down to a negligible level. Here, $\mathrm{M}_{\mu\nu_{\mu}} = \sqrt{(\mathrm{P}_{\mathrm{K}^{\pm}} - \mathrm{P}_{\mathrm{e}^{+}\mathrm{e}^{-}})^2}$ and the resolution $\sigma(\mathrm{M}_{\mu\nu_{\mu}}) = 10\ \mathrm{MeV}/c^2$ in the peak region of $m_{\pi^{\pm}}$ is found from Monte-Carlo simulation. The requirement $\mathrm{M}_{\mu\nu_{\mu}} > 170\ \mathrm{MeV}/c^2$ decreases the signal acceptance by 11%.

3.2. Data driven estimation of background

Diagrams of the background decays that are not completely suppressed with the event selection cuts are shown in Figures 2b - 2e. They contribute in the signal selection due to particle misidentification or/and detection inefficiency. The $\mathrm{K}^{\pm} \to \pi^{\pm}\pi^{0}\pi^{0}$ decay ($\mathrm{K}_{3\pi^0}$), where $\pi^{\pm}$ decays further into the $\mu^{\pm}\nu_{\mu}$ pair (Figure 2b) and only two electrons out of four electrons and two photons are detected, mimics a signal event. The $\mathrm{K}^{\pm} \to \pi^{\pm}\pi^{+}\pi^{-}$ decay ($\mathrm{K}_{3\pi}$) can mimic the signal if one of the pions decays into $\mu\nu_{\mu}$ and the remaining pions are both misidentified as electrons (Figure 2c). The misidentification probability is small but given the large branching ratio of the $\mathrm{K}_{3\pi}$ decay the double-misidentified pions as electrons are still non-negligible background. The $\mathrm{K}^{\pm} \to \mathrm{e}^{\pm}\nu_{\mathrm{e}}\pi^{+}\pi^{-}$ decay ($\mathrm{K}_{\mathrm{e}4}$) can contribute in the signal selection in two different ways. In one scenario one of the pions is misidentified as an electron and another one as a muon (Figure 2d). In the second scenario one of the pions is misidentified as an electron while the second pion decays into a $\mu^{\pm}\nu_{\mu}$ pair (Figure 2e). Data driven approach is used to

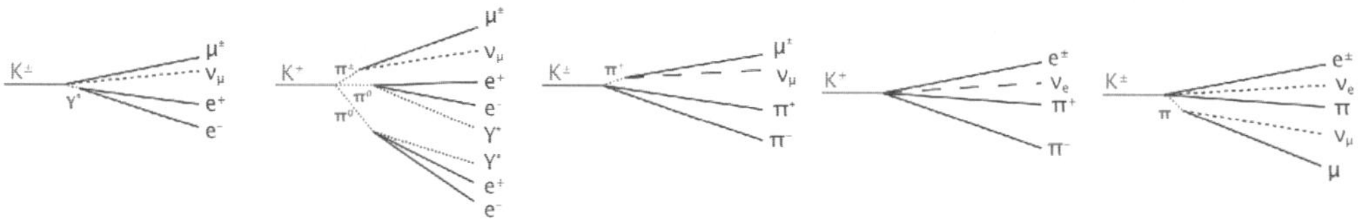

(a) (b) (c) (d) (e)

Figure 2: Diagrams of signal (a) and $K_{3\pi^0}$ (b), $K_{3\pi}$ (c) and K_{e4} (d, e) background decays.

estimate contribution of the irreducible background decays, $K_{3\pi^0}$, $K_{3\pi}$ and K_{e4} into the final selection of events. The idea of the data-driven method lies in selecting events with same-sign pair of electrons while other selection cuts are the same as for the signal event selection. The muon charge has to have the opposite sign to the same-sign electron pair. The misidentification probability of pions as electrons and the detector and trigger inefficiencies and acceptance do not depend on the electric charge. Therefore, the expected number of background events of a given type in the signal region (opposite-sign ee-pair) is equal to the observed number of background events of the given type in the control region (same-sign ee-pair) multiplied by a corresponding combinatorial factor. The combinatorial factor for the $K_{3\pi^0}$ background (Figure 2b) is 2, since there are twice more possibilities of having e^+e^- pairs selected than the same sign pairs, e^+e^+ or e^-e^-. For other irreducible background decays (Figures 2c-2e), the corresponding combinatorial factors are equal to 1. Figure 3 shows distributions of squared missing mass, defined as $M^2_{miss} = (P_{K^\pm} - P_{\mu^\pm e^+ e^-})^2$, before the last selection cut of events is applied. The dashed vertical lines in red represent the final selection cut. This cut helps to remove a large fraction of the estimated background (solid blue histogram), which tends to contribute in the higher M^2_{miss} region due to misidentified pions as electrons. Figure 4 shows a distribution of a dimensionless observable, $z = (m_{e+e-}/m_{K^\pm})^2$, in data (black histogram) after the final selection. In total, 1663 events are selected while the total background contribution is estimated to be about 3%. The z-distribution is compared with two different Monte-Carlo simulated samples of the signal. The shape of the z-distribution in the signal sample simulated with only IB amplitudes (red histogram) does not match the distribution in data. In contrast, the distribution in the signal sample produced with all IB+SD+INT contributions included in the Monte-Carlo simulation (solid cyan histogram) correctly describes the measured spectrum. In both Figure 3 and Figure 4 the simulated signal samples are normalised to the number of selected events in data minus the estimated number of events of the total background.

3.3. Signal normalisation

The total number of Kaon decays in the 98 m long decay volume is measured by using the $K^\pm \to \pi^\pm\pi^+\pi^-$ decay. Background-free selection of this decay events can be performed by selecting the three-track events (same as for the signal selection) and applying two-sided cut (as shown in Figure 5 with dashed vertical lines in red) on a distribution of the mass of the three tracks around the mass peak of K^+. Order of 10^9 $K^\pm \to \pi^\pm\pi^+\pi^-$ decay events are selected. The acceptance of the $K^\pm \to \pi^\pm\pi^+\pi^-$ decay channel, $(24.04 \pm 0.01)\%$, is found from Monte-Carlo simulation. Using the number of selected $K^\pm \to \pi^\pm\pi^+\pi^-$ events and the acceptance and the known $BR(K^\pm \to \pi^\pm\pi^+\pi^-)$, the total number of the $K^\pm$ decays is calculated to be $(1.56 \pm 0.01) \times 10^{11}$. Similarity of the event topologies of the normalisation $K^\pm \to \pi^\pm\pi^+\pi^-$ and the signal decays leads to the first order cancelation of some common systematic effects such as detector and trigger inefficiencies, etc.

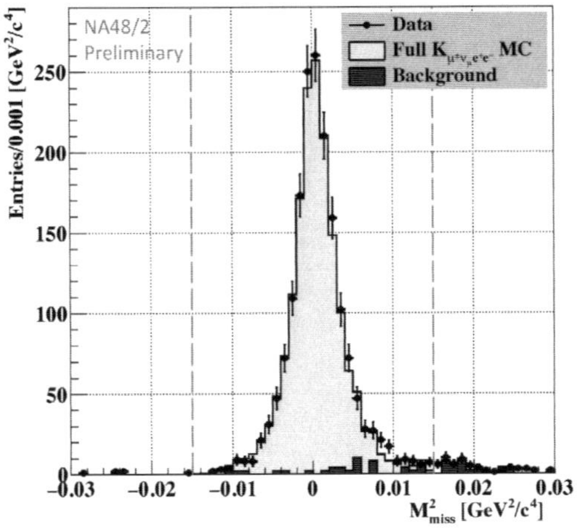

Figure 3: M^2_{miss} - distribution in selected data (black), in signal sample from full Monte-Carlo simulation (solid cyan) and in total background estimate (solid blue).

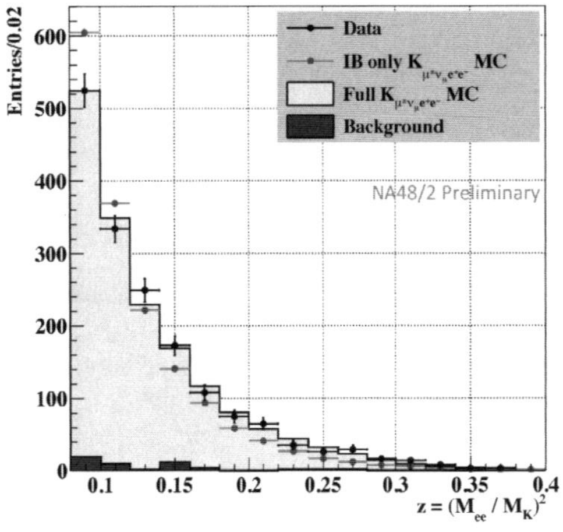

Figure 4: z-distribution after final selection of events in data (black) and in signal samples from three level (red) and full (solid cyan) Monte-Carlo simulation.

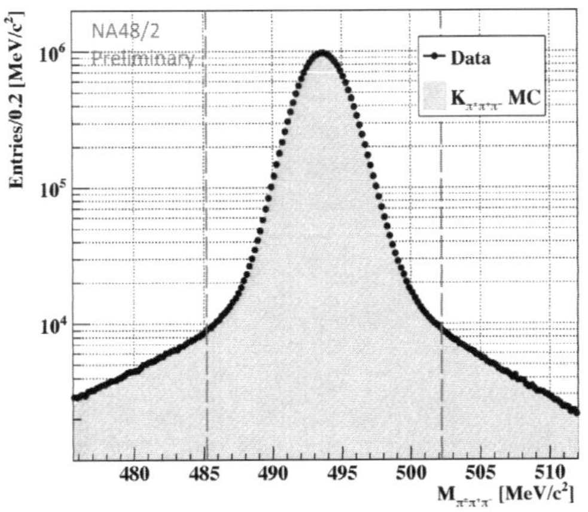

Figure 5: Distribution of mass of three tracks after selection of $K^\pm \to \pi^\pm\pi^+\pi^-$ decay candidates in data (black) and in simulated sample (solid cyan).

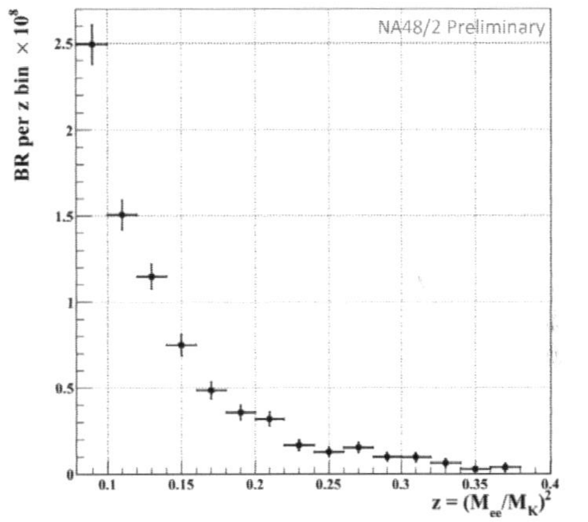

Figure 6: Branching ratio ($\times 10^8$) of signal decay measured separately in each z-bin of interesting phase space region $m_{e^+e^-} \geq 140\ \mathrm{MeV}/c^2$.

3.4. Results

The signal acceptance is found to depend on the z-region according to Monte-Carlo simulation and to vary from 12% to 15%. In order to properly account for the variation of the acceptance the signal decay branching ratio is measured in 15 bins of z. The result of the measurement is presented in Figure 6. The total exclusive branching ratio is the sum of the measured branching ratios over all bins of z and is equal to $BR(K^\pm \to \mu^\pm\nu_\mu e^+e^- \mid m_{e^+e^-} \geq 140\ \mathrm{MeV}/c^2) = (7.81 \pm 0.21\,(stat.) \pm 0.08\,(syst.) \pm 0.06\,(ext.)) \times 10^{-8}$. The uncertainty of the measurement due to the uncertainty of $BR(K^\pm \to \pi^\pm\pi^+\pi^-)$ is presented separately. It is referred as an external

Table 2: List of the sources of the uncertainty of the measurement. The list is divided into three groups of the statistical, systematical and external uncertainty sources.

Uncertainty source	$\Delta BR/BR$ [$\times 10^2$]
Data statistics	2.54
Normalisation channel statistics	0.02
Radiation corrections	0.70
Background statistics	0.62
Background systematics	0.30
Trigger efficiency	0.54
Muon identification efficiency	0.13
Electron identification efficiency	0.04
Signal channel acceptance	0.12
Normalisation channel acceptance	0.03
$BR(\mathrm{K}^\pm \to \pi^\pm\pi^+\pi^-)$	0.72
Total	2.88

uncertainty. Table 2 presents a detailed list of all uncertainties of the measurement. The data statistical uncertainty is the main source of the total uncertainty. The main contributors in the systematical uncertainties are the precision of the knowledge of $BR(\mathrm{K}^\pm \to \pi^\pm\pi^+\pi^-)$ and modelling of the radiative corrections. The uncertainty due to the latter source is estimated in a conservative way by switching on/off the radiative corrections in Monte-Carlo simulation of the signal. The uncertainties of the background estimation and the trigger efficiency also have significant impact on the precision. The other sources of uncertainty are negligible.

4. Conclusions

Model independent measurement of the exclusive branching ratio of the radiative decay $\mathrm{K}^\pm \to \mu^\pm\nu_\mu\mathrm{e}^+\mathrm{e}^-$ has been performed in the region $m_{\mathrm{e}^+\mathrm{e}^-} \geq 140$ MeV/c^2. In total, $(1.56\pm0.01)\times10^{11}$ events of $\mathrm{K}^\pm$ decays collected by the NA48/2 experiment at CERN were used for the measurement. The result is in agreement with the ChPT prediction. The uncertainty of the measurement is dominated by the statistical component.

References

[1] Carlson C E and Rislow B C 2014 Constraints to new physics models for the proton charge radius puzzle from the decay $\mathrm{K}^+ \to \mu^+\nu\mathrm{e}^+\mathrm{e}^-$ *Phys. Rev.* D **89** 035003
[2] Barger V *et al* 2012 Constraint on parity-violating muonic forces *Phys. Rev. Lett.* **108** 081802
[3] Batell B *et al* 2016 The Leptonic Higgs Portal *Prerpint* arXiv:1606.04943 [hep-ph]
[4] Bijnens J, Ecker G and Gasser J 1993 Radiative semileptonic Kaon decays *Nucl. Phys.* B **396** 81-118
[5] Diamant-Berger A M *et al* 1976 Study of some rare decays of K^+ meson *Physics Letters* B **62** Issue **4** 485-490
[6] Poblaguev A A *et al* 2002 Experimental study of the radiative decays $\mathrm{K}^+ \to \mu^+\nu\mathrm{e}^+\mathrm{e}^-$ and $\mathrm{K}^+ \to \mathrm{e}^+\nu\mathrm{e}^+\mathrm{e}^-$ *Phys. Rev. Lett.* **89** 061803
[7] The NA48/2 Collaboration 2007 Search for direct CP violating charge asymmetries in $\mathrm{K}^\pm \to \pi^\pm\pi^+\pi^-$ and $\mathrm{K}^\pm \to \pi^\pm\pi^0\pi^0$ decays *Eur. Phys. J.* C **52** 875
[8] The NA48/2 Collaboration 2007 The beam and detector for the NA48 neutral kaon CP violation experiment at CERN *Nucl. Instrum. Meth.* A **574** 433

Probing CPT symmetry in transitions with entangled neutral kaons

A Di Domenico

Dipartimento di Fisica, Sapienza Università di Roma,
and I.N.F.N. Sezione di Roma, P.le A. Moro, 2, I-00185 Rome, Italy

E-mail: antonio.didomenico@roma1.infn.it

Abstract. A new kind of *CPT* test for transitions in the neutral kaon system is presented, where the exchange of *in* and *out* states (and *CP* conjugation), required for a direct and genuine *CPT* test, is performed exploiting the entanglement of the kaon pair produced at a ϕ-factory. Using this method it would be possible for the first time to directly test the *CPT* symmetry in transition processes between meson states, rather than comparing masses, lifetimes, or other intrinsic properties of particle and anti-particle states. The proposed test in the neutral kaon system is very clean and fully robust, and might shed light on possible new *CPT* violating mechanisms, or further improve the precision of the present experimental limits. It could be implemented at the DAΦNE facility in Frascati, where the KLOE-2 experiment can reach a statistical sensitivity of $\mathcal{O}(10^{-3})$ on the newly proposed observable quantities.

1. Introduction

CPT symmetry, i.e. the symmetry under the combination of charge conjugation (C), parity (P), and time reversal (T) transformations, at present appears to be the only discrete symmetry of Quantum Mechanics respected in Nature. This result has a very solid theoretical foundation in the well known *CPT* theorem [1, 2, 3, 4], ensuring exact *CPT* invariance for any quantum field theory formulated on flat space-time assuming (1) Lorentz invariance, (2) Locality, and (3) Unitarity (i.e. conservation of probability).

A violation of the *CPT* symmetry would have a dramatic impact on our present theoretical picture and would definitely constitute an unambiguous signal of a New Physics framework, thus strongly motivating both experimental searches and theoretical studies on this subject.

CPT invariance has been confirmed by all present experimental tests, particularly in the neutral kaon system where strong limits have been set to a variety of possible *CPT* violation effects which might arose in a quantum gravity scenario [5, 6, 7, 8, 9, 10]. The best limits on the δ parameter expressing *CPT* violation in the K^0-$\bar{\mathrm{K}}^0$ mixing matrix, i.e. in the *standard* Weisskopf-Wigner approach [11] are obtained in the CPLEAR experiment for $\Re\delta$ [12], and using the Bell-Steinberger relation for $\Im\delta$ [13, 14], yielding a stringent limit on the difference of mass terms for K^0 and $\bar{\mathrm{K}}^0$: $|m_{\mathrm{K}^0} - m_{\bar{\mathrm{K}}^0}| < 4 \times 10^{-19}$ GeV at 95% c.l. [5].

The *CPT* violating probe has been, however, often limited to a difference of masses (and other intrinsic properties) for a particle and its anti-particle, i.e. to diagonal mass terms. In many physical phenomena the perturbing effect does not appear at first order in perturbation theory: it would be sufficient that the perturbation breaks a symmetry of the non-perturbed states. This vanishing effect at first order for the diagonal elements, like e.g. the case of the

Content from this work may be used under the terms of the Creative Commons Attribution 3.0 licence. Any further distribution of this work must maintain attribution to the author(s) and the title of the work, journal citation and DOI.
Published under licence by IOP Publishing Ltd

electric dipole moment for T violation, is not present for transitions (non-diagonal elements) [15].

Here a new kind of *CPT* test for transitions in the neutral kaon system is discussed where the exchange of *in* and *out* states (and *CP* conjugation), required for a direct and genuine *CPT* test, is performed exploiting the entanglement of the kaon pair produced at a ϕ-factory (see Ref. [16] for a more detailed discussion). This methodology has been proposed for a direct test of the T symmetry in the same context [17], similarly to the one adopted for the performed test in the B meson system at B-factories [18, 19, 20, 21, 15]. The decay is not an essential ingredient for a non-vanishing effect and it is only used for filtering the appropriate initial and final states of the neutral kaon transition [22]. Explicitly, in the standard Weisskopf-Wigner approach to this system, our *CPT*-violating effects can be connected to the $\Re\delta$ parameter, a genuine *CPT*-violating effect independent of $\Delta\Gamma = \Gamma_S - \Gamma_L$, with Γ_S and Γ_L the widths of the physical states.

2. Direct test of *CPT* symmetry with neutral kaons

In order to implement a direct test of the *CPT* symmetry, the Einstein-Podolsky-Rosen (EPR) entanglement of neutral mesons produced at a ϕ-factory is exploited. In fact in this case the initial state of the neutral kaon pair produced in $\phi \to \mathrm{K}^0\bar{\mathrm{K}}^0$ decay can be rewritten in terms of any pair of orthogonal states $|\mathrm{K}_+\rangle$ and $|\mathrm{K}_-\rangle$:

$$|i\rangle = \frac{1}{\sqrt{2}}\{|\mathrm{K}^0\rangle|\bar{\mathrm{K}}^0\rangle - |\bar{\mathrm{K}}^0\rangle|\mathrm{K}^0\rangle\} = \frac{1}{\sqrt{2}}\{|\mathrm{K}_+\rangle|\mathrm{K}_-\rangle - |\mathrm{K}_-\rangle|\mathrm{K}_+\rangle\} . \tag{1}$$

Here one can consider the states $|\mathrm{K}_+\rangle$, $|\mathrm{K}_-\rangle$ defined as follows: $|\mathrm{K}_+\rangle$ is the state filtered by the decay into $\pi\pi$ ($\pi^+\pi^-$ or $\pi^0\pi^0$), a pure $CP = +1$ state; analogously $|\mathrm{K}_-\rangle$ is the state filtered by the decay into $3\pi^0$, a pure $CP = -1$ state. Their orthogonal states correspond to the states which cannot decay into $\pi\pi$ or $3\pi^0$, defined, respectively, as

$$\begin{aligned} |\widetilde{\mathrm{K}}_-\rangle &\propto [|\mathrm{K_L}\rangle - \eta_{\pi\pi}|\mathrm{K_S}\rangle] \\ |\widetilde{\mathrm{K}}_+\rangle &\propto [|\mathrm{K_S}\rangle - \eta_{3\pi^0}|\mathrm{K_L}\rangle] , \end{aligned} \tag{2}$$

with $\eta_{\pi\pi} = \langle\pi\pi|T|\mathrm{K_L}\rangle/\langle\pi\pi|T|\mathrm{K_S}\rangle$ and $\eta_{3\pi^0} = \langle 3\pi^0|T|\mathrm{K_S}\rangle/\langle 3\pi^0|T|\mathrm{K_L}\rangle$. With these definitions of states, it can be shown that the condition of orthogonality $\langle\mathrm{K}_-|\mathrm{K}_+\rangle = 0$, (i.e. $|\mathrm{K}_+\rangle \equiv |\widetilde{\mathrm{K}}_+\rangle$ and $|\mathrm{K}_-\rangle \equiv |\widetilde{\mathrm{K}}_-\rangle$) corresponds to assume negligible direct *CP* (or *CPT*) violation contributions, assumption quite well satisfied for neutral kaons [16]. The validity of the $\Delta S = \Delta Q$ rule is also assumed, so that the two flavor orthogonal eigenstates $|\mathrm{K}^0\rangle$ and $|\bar{\mathrm{K}}^0\rangle$ are identified by the charge of the lepton in semileptonic decays, i.e. a $|\mathrm{K}^0\rangle$ can decay into $\pi^-\ell^+\nu$ and not into $\pi^+\ell^-\bar{\nu}$, and vice-versa for a $|\bar{\mathrm{K}}^0\rangle$[1].

Thus, exploiting the perfect anticorrelation of the states implied by Eq. (1), it is possible to have a "flavor-tag"or a "CP-tag", i.e. to infer the flavor (K^0 or $\bar{\mathrm{K}}^0$) or the *CP* (K_+ or K_-) state of the still alive kaon by observing a specific flavor decay ($\pi^+\ell^-\nu$ or $\pi^-\ell^+\bar{\nu}$) or CP decay ($\pi\pi$ or $\pi^0\pi^0\pi^0$) of the other (and first decaying) kaon in the pair.

In this way one can experimentally access – for instance – the transition $\mathrm{K}^0 \to \mathrm{K}_+$, taken as reference, and the $\mathrm{K}_+ \to \mathrm{K}^0$, $\bar{\mathrm{K}}^0 \to \mathrm{K}_+$ and $\mathrm{K}_+ \to \bar{\mathrm{K}}^0$ transitions, i.e. the T, *CP* and *CPT* conjugated transitions, respectively. Specifically for the *CPT* symmetry test, one can *directly* compare the probabilities for the reference transition and the *CPT* conjugated one through the

[1] It is important to underline that both these assumptions can be relaxed for some specific observables, as discussed below

following ratios of probabilities:

$$\begin{aligned} R_{1,CPT}(\Delta t) &= P\left[\mathrm{K}_+(0) \to \bar{\mathrm{K}}^0(\Delta t)\right]/P\left[\mathrm{K}^0(0) \to \mathrm{K}_+(\Delta t)\right] \\ R_{2,CPT}(\Delta t) &= P\left[\mathrm{K}^0(0) \to \mathrm{K}_-(\Delta t)\right]/P\left[\mathrm{K}_-(0) \to \bar{\mathrm{K}}^0(\Delta t)\right] \\ R_{3,CPT}(\Delta t) &= P\left[\mathrm{K}_+(0) \to \mathrm{K}^0(\Delta t)\right]/P\left[\bar{\mathrm{K}}^0(0) \to \mathrm{K}_+(\Delta t)\right] \\ R_{4,CPT}(\Delta t) &= P\left[\bar{\mathrm{K}}^0(0) \to \mathrm{K}_-(\Delta t)\right]/P\left[\mathrm{K}_-(0) \to \mathrm{K}^0(\Delta t)\right] \ . \end{aligned} \tag{3}$$

The measurement of any deviation from the prediction $R_{i,CPT}(\Delta t) = 1$ imposed by *CPT* invariance is a signal of *CPT* violation.
It is worth noting that for $\Delta t = 0$:

$$R_{1,CPT}(0) = R_{2,CPT}(0) = R_{3,CPT}(0) = R_{4,CPT}(0) = 1 \ , \tag{4}$$

i.e. the *CPT*-violating effect is built in the time evolution of the system, and it is absent at $\Delta t = 0$, within our approximations.
For $\Delta t \gg \tau_S$, assuming the presence of *CPT* violation only in the mass matrix ($\delta \neq 0$) and nothing else, one gets:

$$R_{2,CPT}(\Delta t \gg \tau_S) \simeq 1 - 4\Re\delta \tag{5}$$

$$R_{4,CPT}(\Delta t \gg \tau_S) \simeq 1 + 4\Re\delta \ , \tag{6}$$

i.e. the *CPT*-violating effect built in the time evolution reaches a "plateau" regime and dominates in this limit. It is a *genuine* effect because $\Re\delta$ does not depend on $\Delta\Gamma$ as an essential ingredient [15, 22].

At a ϕ-factory one can define two observable ratios:

$$R^{\mathrm{exp}}_{2,CPT}(\Delta t) \equiv \frac{I(\ell^-, 3\pi^0; \Delta t)}{I(\pi\pi, \ell^-; \Delta t)} \tag{7}$$

$$R^{\mathrm{exp}}_{4,CPT}(\Delta t) \equiv \frac{I(\ell^+, 3\pi^0; \Delta t)}{I(\pi\pi, \ell^+; \Delta t)} \ , \tag{8}$$

where $I(f_1, f_2; \Delta t)$ are the double decay rates into decay products f_1 and f_2 as a function of the difference of kaon decay times Δt [17, 23], with f_1 occurring before f_2 decay for $\Delta t > 0$, and viceversa for $\Delta t < 0$.
They are related to the $R_{i,CPT}(\Delta t)$ ratios defined in eqs (3) as follows, for $\Delta t \geq 0$:

$$\begin{aligned} R^{\mathrm{exp}}_{2,CPT}(\Delta t) &= R_{2,CPT}(\Delta t) \times D_{CPT} \\ R^{\mathrm{exp}}_{4,CPT}(\Delta t) &= R_{4,CPT}(\Delta t) \times D_{CPT} \end{aligned} \tag{9}$$

whereas for $\Delta t < 0$ one has:

$$\begin{aligned} R^{\mathrm{exp}}_{2,CPT}(\Delta t) &= R_{1,CPT}(|\Delta t|) \times D_{CPT} \\ R^{\mathrm{exp}}_{4,CPT}(\Delta t) &= R_{3,CPT}(|\Delta t|) \times D_{CPT} \ , \end{aligned} \tag{10}$$

with

$$D_{CPT} = \frac{|\langle 3\pi^0|T|\mathrm{K}_-\rangle|^2}{|\langle \pi\pi|T|\mathrm{K}_+\rangle|^2} = \frac{\mathrm{BR}\left(\mathrm{K_L} \to 3\pi^0\right)\Gamma_L}{\mathrm{BR}\left(\mathrm{K_S} \to \pi\pi\right)\Gamma_S} \ , \tag{11}$$

where the last r.h.s. equality holds with a high degree of accuracy, at least $\mathcal{O}(10^{-7})$. The value of D_{CPT} can be therefore directly evaluated from branching ratios and lifetimes.

The explicit expressions of ratios (7) and (8) (neglecting higher order terms in small parameters and for not too large negative Δt) are:

$$\begin{aligned} R_{2,CPT}^{\text{exp}}(\Delta t) &= \frac{P[\text{K}^0(0) \to \text{K}_-(\Delta t)]}{P[\text{K}_-(0) \to \bar{\text{K}}^0(\Delta t)]} \times D_{CPT} \\ &\simeq |1 - 2\delta|^2 \left|1 + 2\delta e^{-i(\lambda_S - \lambda_L)\Delta t}\right|^2 \times D_{CPT} \ , \end{aligned} \tag{12}$$

$$\begin{aligned} R_{4,CPT}^{\text{exp}}(\Delta t) &= \frac{P[\bar{\text{K}}^0(0) \to \text{K}_-(\Delta t)]}{P[\text{K}_-(0) \to \text{K}^0(\Delta t)]} \times D_{CPT} \\ &\simeq |1 + 2\delta|^2 \left|1 - 2\delta e^{-i(\lambda_S - \lambda_L)\Delta t}\right|^2 \times D_{CPT} \ . \end{aligned} \tag{13}$$

The expected behavior of the observables $R_{2,CPT}^{\text{exp}}(\Delta t)$ and $R_{4,CPT}^{\text{exp}}(\Delta t)$ as a function of Δt, and without the approximations of eqs.(12) and (13), is shown in figure 1, where – for visualization purposes – the probabilities involved have been evaluated fixing the *CPT* violating parameters $\Re\delta$ and $\Im\delta$ to a value different from zero, and equal to their present uncertainties [5], i.e. $\Re\delta = 3.3 \times 10^{-4}$ and $\Im\delta = 1.6 \times 10^{-5}$. In figure 2 a zoom of the $\Delta t > 0$ region, where the "plateau" regimes (5) and (6) dominate, is shown. Experimentally, this is the most interesting and statistically most populated region, where the best sensitivity to *CPT* violation effects can be reached by the KLOE-2 experiment [16, 17, 24].

The presence of direct *CP* violation contributions in the decay amplitudes, even though in principle could mimic *CPT* violation effects, turns out to be totally irrelevant for the plateau region $\Delta t \gg \tau_S$ (see detailed description in Ref. [16]). The effect of a possible violation of the $\Delta S = \Delta Q$ rule is also not affecting the *CPT* test in the same region with the double ratio defined as:

$$\frac{R_{2,CPT}^{\text{exp}}(\Delta t \gg \tau_S)}{R_{4,CPT}^{\text{exp}}(\Delta t \gg \tau_S)} = 1 - 8\Re\delta - 8\Re x_- \ . \tag{14}$$

with x_- describing *CPT* violation in the $\Delta S \neq \Delta Q$ semileptonic decay amplitudes. Therefore the double ratio (14) constitutes one of the most robust observables for our proposed *CPT* test. It is independent of D_{CPT}, and in the limit $\Delta t \gg \tau_S$ it exhibits a pure and genuine *CPT* violating effect, even without the assumptions of the validity of the $\Delta S = \Delta Q$ rule and of negligible contaminations from direct *CP* violation.

The KLOE-2 experiment at the DAΦNE facility aims at reaching a total integrated luminosity $L \geq 5$ fb^{-1} in the next couple of years, and could make a precise measurement of the two observable ratios $R_{2,CPT}^{\text{exp}}(\Delta t)$ and $R_{4,CPT}^{\text{exp}}(\Delta t)$. By considering a large Δt interval in the statistically most populated region, e.g. $0 \leq \Delta t \leq 300\ \tau_S$, a statistical sensitivity on the double ratio (14) of (3.0×10^{-3}) and (2.1×10^{-3}) is obtained for $L = 5$ and 10 fb^{-1}, respectively [16]. Once translated into an uncertainty on $\Re\delta$, these results will improve its present limit.

3. Conclusions

A novel *CPT* test has been studied in the neutral kaon system based on the direct comparison of a transition probability with its *CPT* reverse transition. The appropriate preparation and detection of *in* and *out* states in both the reference and the reverse processes is made by exploiting

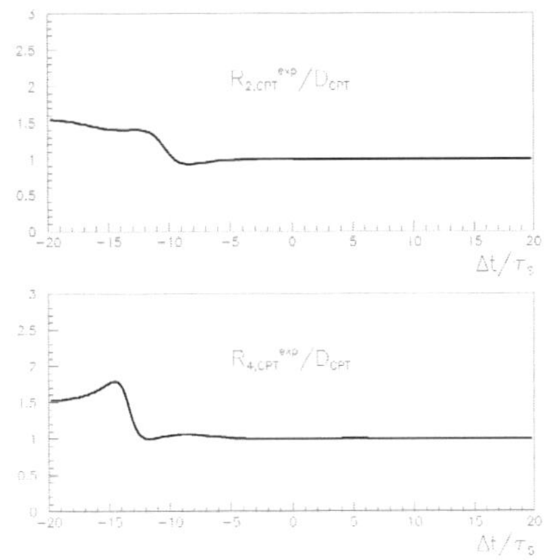

Figure 1. The ratios $R^{\rm exp}_{2,CPT}(\Delta t)$ and $R^{\rm exp}_{4,CPT}(\Delta t)$ as a function of Δt. For visualization purposes the *CPT* violating parameters have been fixed to the values $\Re\delta = 3.3 \times 10^{-4}$ and $\Im\delta = 1.6 \times 10^{-5}$.

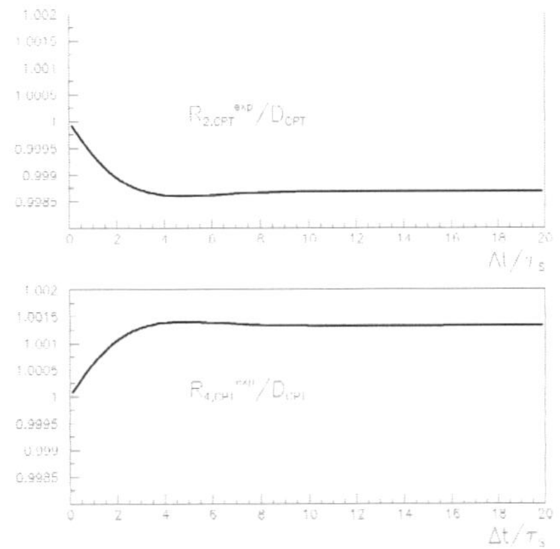

Figure 2. A zoom of the plots shown in Fig.1 in the region $0 \leq \Delta t \leq 20\tau_S$.

the EPR entanglement of neutral kaons produced in a ϕ-factory and using their decays as filtering measurements of the kaon states only.

Possible spurious effects induced by CP violation in the decay and/or a violation of the $\Delta S = \Delta Q$ rule have been shown to be either negligible or disentangled by the dependence with the time evolution. The proposed measurement is thus fully robust leading to definite conclusions on CPT Violation.

Acknowledgements
The author would like to warmly thank C. Lazzeroni, and all the organizing committee for the invitation to the KAON 2016 conference, and the very pleasant stay in Birmingham.

References
[1] G. Lüeders, *Proof of the TCP Theorem*, Ann. Phys. (NY) **2**, 1 (1957), reprinted in Ann. Phys. (NY) **281**, 1004 (2000)
[2] W. Pauli, *Exclusion principle, Lorentz group and reflexion of space-time and charge* in *Niels Bohr and the development of physics*, edited by W. Pauli, Pergamon, London, 1955, p.30
[3] J. S. Bell, *Time reversal in field theory*, Proc. R. Soc. London A **231**, 479 (1955)
[4] R. Jost, *Eine Bemerkung zum CPT-theorem*, Helv. Phys. Acta **30**, 409 (1957)
[5] K. A. Olive et al. (Particle Data Group), *Review of particle physics*, Chin. Phys. **C 38**, 090001 (2014)
[6] F. Ambrosino et al., KLOE collaboration, *First observation of quantum interference in the process* $\phi \to K_S K_L \to \pi^+\pi^-\pi^+\pi^-$*: A test of quantum mechanics and CPT symmetry*, Phys. Lett. B **642**, 315 (2006)
[7] A. Di Domenico and the KLOE collaboration, *Search for CPT violation and decoherence effects in the neutral kaon system*, J. Phys. Conf. Ser. **171**, 012008 (2009)
[8] A. Di Domenico and the KLOE collaboration, *CPT Symmetry and Quantum Mechanics Tests in the Neutral Kaon System at KLOE*, Found. Phys. **40** , 852 (2010)
[9] D. Babusci et al., KLOE-2 Collaboration, *Test of CPT and Lorentz symmetry in entangled neutral kaons with the KLOE experiment*, Phys. Lett. B **730**, 89 (2014)
[10] R. Adler et al., CPLEAR collaboration, *Tests of CPT symmetry and quantum mechanics with experimental data from CPLEAR* Phys. Lett. B **364**, 239 (1995)
[11] V. Weisskopf and E. P. Wigner, *Über die natürliche linienbreite in der strahlung des harmonischen oszillators*, Z. Phys. **63**, 54 (1930)
[12] A. Angelopoulos et al., CPLEAR collaboration, *A determination of the CPT violation parameter Re(δ) from the semileptonic decay of strangeness-tagged neutral kaons*, Phys. Lett. B **444**, 52 (1998)
[13] J. S. Bell and J. Steinberger, *Weak interactions of kaons*, Proc. of the Oxford Int. Conf. on Elementary Particles (1965)
[14] F. Ambrosino et al., KLOE collaboration, *Determination of CP and CPT violation parameters in the neutral kaon system using the Bell-Steinberger relation and data from the KLOE experiment*, JHEP **12**, 011 (2006)
[15] J. Bernabeu and F. Martinez-Vidal, *Colloquium: Time-reversal violation with quantum-entangled B mesons*, Rev. Mod. Phys. **87**, 165 (2015)
[16] J. Bernabeu, A. Di Domenico, P. Villanueva-Perez, *Probing CPT in transitions with entangled neutral kaons*, JHEP **10**, 139 (2015)
[17] J. Bernabeu, A. Di Domenico, P. Villanueva-Perez, *Direct test of time reversal symmetry in the entangled neutral kaon system at a* ϕ*-factory*, Nucl. Phys. B. **868**, 102 (2013)
[18] M. C. Banuls and J. Bernabeu, *CP, T and CPT versus temporal asymmetries for entangled states of the* B_d*-system*, Phys. Lett. B **464**, 117 (1999)
[19] M. C. Banuls and J. Bernabeu, *Studying indirect violation of CP, T and CPT in a B-factory*, Nucl. Phys. B **590**, 19 (2000)
[20] J. Bernabeu, F. Martínez-Vidal, P. Villanueva-Pérez, *Time reversal violation from the entangled* $B^0\bar{B}^0$ *system*, JHEP **08**, 064 (2012)
[21] J. P. Lees *et al.*, BABAR Collaboration, *Observation of Time-Reversal Violation in the* B^0 *Meson System*, Phys. Rev. Lett. **109**, 211801 (2012)
[22] L. Wolfenstein, *The search for direct evidence for time reversal violation*, Int. J. Mod. Phys. E **8**, 501 (1999)
[23] A. Di Domenico (Editor) , *Handbook on Neutral Kaon Interferometry at a* ϕ*-factory*, Frascati Phys. Ser. **43** (2007)
[24] G. Amelino-Camelia et al., *Physics with the KLOE-2 experiment at the upgraded DAΦNE*, Eur. Phys. J. C **68**, 619 (2010)

Prospects for an experiment to measure BR($K_L^0 \to \pi^0\nu\bar{\nu}$) at the CERN SPS

M Moulson, for the NA62-KLEVER Project
INFN Laboratori Nazionali di Frascati, 00044 Frascati RM, Italy
E-mail: moulson@lnf.infn.it

Abstract. Precise measurements of the branching ratios for the $K \to \pi\nu\bar{\nu}$ decays can provide unique constraints on CKM unitarity and, potentially, evidence for new physics. It is important to measure both decay modes, $K^+ \to \pi^+\nu\bar{\nu}$ and $K_L^0 \to \pi^0\nu\bar{\nu}$, since different new physics models affect the rates for each channel differently. We are investigating the feasibility of performing a measurement of BR($K_L^0 \to \pi^0\nu\bar{\nu}$) using a high-energy secondary neutral beam at the CERN SPS in a successor experiment to NA62. The planned experiment would reuse some of the NA62 infrastructure, including possibly the NA48 liquid-krypton calorimeter. The mean momentum of K_L^0 mesons decaying in the fiducial volume is 70 GeV; the decay products are boosted forward, so that less demanding performance is required from the large-angle photon veto detectors. On the other hand, the layout poses particular challenges for the design of the small-angle vetoes, which must reject photons from K_L^0 decays escaping through the beam pipe amidst an intense background from soft photons and neutrons in the beam. We present some preliminary conclusions from our feasibility studies, summarizing the design challenges faced and the sensitivity obtainable for the measurement of BR($K_L^0 \to \pi^0\nu\bar{\nu}$).

1. Introduction

The $K \to \pi\nu\bar{\nu}$ decays are flavor-changing neutral current (FCNC) processes that probe the $s \to d\nu\bar{\nu}$ transition via the Z-penguin and box diagrams shown in figure 1. They are highly GIM suppressed and their Standard Model (SM) rates are very small. For several reasons, the

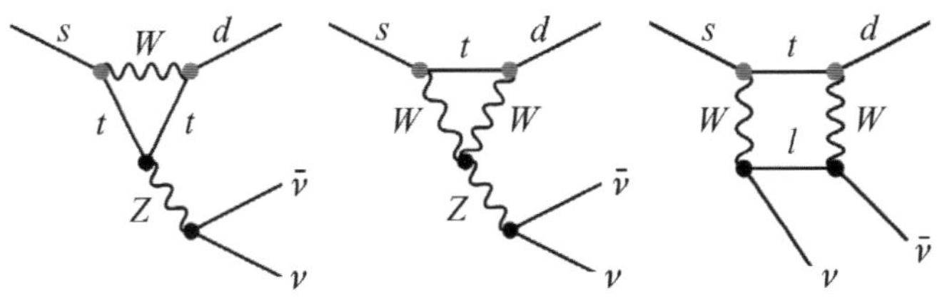

Figure 1. Diagrams contributing to the process $K \to \pi\nu\bar{\nu}$.

SM calculation for their branching ratios (BRs) is particularly clean (see [1] for a review):

- The loop amplitudes are dominated by the top-quark contributions. The neutral decay violates CP; its amplitude involves the top-quark contribution only.

Content from this work may be used under the terms of the Creative Commons Attribution 3.0 licence. Any further distribution of this work must maintain attribution to the author(s) and the title of the work, journal citation and DOI.
Published under licence by IOP Publishing Ltd

- The hadronic matrix element for these decays can be obtained from the precise experimental measurement of the K_{e3} rate.
- There are no long-distance contributions from processes with intermediate photons.

In the SM, $\mathrm{BR}(K^+ \to \pi^+ \nu\bar{\nu}) = (8.4 \pm 1.0) \times 10^{-11}$ and $\mathrm{BR}(K_L^0 \to \pi^0 \nu\bar{\nu}) = (3.4 \pm 0.6) \times 10^{-11}$ [2]. The uncertainties are entirely dominated by the CKM inputs, which in this case are from tree-level observables. Without the parametric errors from this source, the uncertainties would be just 0.30×10^{-11} (3.5%) and 0.05×10^{-11} (1.5%), respectively. Because of corrections for lighter-quark contributions to the amplitudes, the intrinsic uncertainty is slightly larger for the charged channel.

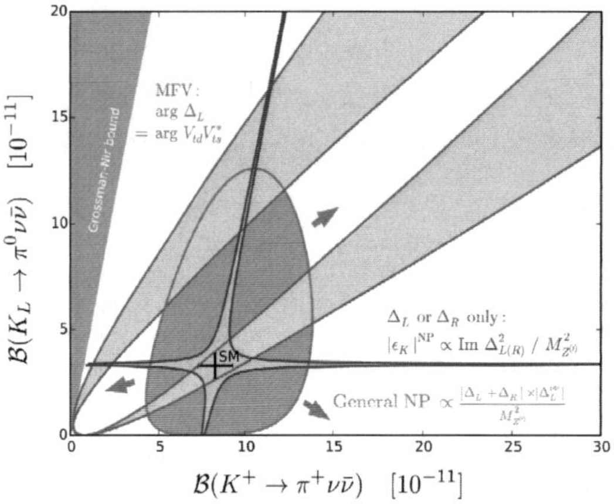

Figure 2. Schematic illustration of correlations between BRs for $K^+ \to \pi^+ \nu\bar{\nu}$ and $K_L^0 \to \pi^0 \nu\bar{\nu}$ expected in different new-physics scenarios, from [3].

Because the SM rates are small and predicted very precisely, the BRs for these decays are sensitive probes for new physics. In general, $\mathrm{BR}(K_L^0 \to \pi^0 \nu\bar{\nu})$ and $\mathrm{BR}(K^+ \to \pi^+ \nu\bar{\nu})$ are differently sensitive to modifications from a given new-physics scenario. If one or both BRs is found to differ from its SM value, it may be possible to characterize the physical mechanism responsible, as schematized in Figure 2, from [3]. For example, if the pattern of flavor-symmetry breaking from new physics were the same as in the SM (minimal flavor violation), the K_L^0 and K^+ BRs would lie along the band of correlation shown in green. If the new interaction were to couple to only left-handed or only right-handed quark currents, as expected for example in models with modified Z couplings or littlest Higgs models with T parity, the BRs would lie along one of the branches shown in blue. New physics without these constraints, as expected for example in models with large extra dimensions, could modify the K^+ and K_L^0 BRs in an arbitrary way, as illustrated in red.

The BR for the decay $K^+ \to \pi^+ \nu\bar{\nu}$ has been measured by Brookhaven experiment E787 and its successor, E949. The combined result from the two generations of the experiment, obtained with seven candidate events, is $\mathrm{BR}(K^+ \to \pi^+ \nu\bar{\nu}) = 1.73^{+1.15}_{-1.05} \times 10^{-10}$ [4]. The goal of the NA62 experiment at the CERN SPS [5] is to measure $\mathrm{BR}(K^+ \to \pi^+ \nu\bar{\nu})$ with a precision of about 10%. NA62 is currently running and is expected to collect ~ 100 signal events by the end 2018.

The decay $\mathrm{BR}(K_L^0 \to \pi^0 \nu\bar{\nu})$ has never been measured. The KOTO experiment at J-PARC has a good chance of observing it. The experiment makes use of a tightly collimated, low-energy neutral beam (peak momentum 1.4 GeV) and compact, hermetic detector. From a brief pilot run in 2013, KOTO obtained the limit $\mathrm{BR}(K_L^0 \to \pi^0 \nu\bar{\nu}) < 5.1 \times 10^{-8}$ (90% CL) [6]. The experiment has been running since 2015. From the preliminary analysis of 5% of the data collected so far,

KOTO has reached a single-event BR sensitivity of 5.9×10^{-9}; background levels are still under evaluation [7]. By the end of 2015, the beam power reached 42 kW; it is expected to gradually increase to 100 kW. If this can be done, the experiment should reach single-event sensitivity for the SM BR by about 2019. In the longer term, KOTO strongly intends to upgrade the beam and experiment to perform a measurement with ~100 event sensitivity. A plan to do so was outlined in the original 2006 KOTO proposal, but there is no official Step 2 proposal yet. The Step 2 measurement would require construction of a new neutral beamline, a complete rebuild of the detector, and extension of the experimental hall; data taking would not be expected to begin before 2025.

2. KLEVER

Given the both the importance and the difficulty of the measurement of BR($K_L^0 \to \pi^0\nu\bar{\nu}$), an experiment making use of a technique complementary to that used by KOTO is well motivated. We are evaluating the feasibility of an experiment at the CERN SPS to measure BR($K_L^0 \to \pi^0\nu\bar{\nu}$) using a high-energy beam. This makes photon vetoing significantly easier, but increases considerably the size of the detector, and in particular, the volume to be covered with photon vetoes. On the other hand, since the photons from background K_L^0 are boosted forwards, the coverage of the large-angle photon vetoes may not need to extend beyond 100 mrad in the polar angle. The experiment could reuse some of the NA62 experimental infrastructure, and possibly the NA48 liquid-krypton calorimeter [8]. Given the time needed for R&D and construction, in addition to the constraints from the LHC running schedule, the natural target date for the experiment to turn on would be at the start of LHC Run 4, in early 2026.

2.1. Beam

Assuming the SM BR for $K_L^0 \to \pi^0\nu\bar{\nu}$ and an acceptance of 10% for decays within the experiment's fiducial volume (FV), 3×10^{13} K_L^0 decays would be required for the observation of 100 signal events. Extraction of the 400 GeV proton beam onto a $1\lambda_{\text{int}}$ beryllium target to produce a tightly collimated (0.28 μsr) neutral beam at an angle of 2.4 mrad (chosen to optimize the ratio of K_L^0 decays in the FV to neutrons in the beam) would result in a K_L^0 flux of 2.8×10^{-5} per proton on target (pot). About 2.2% of the K_L^0 mesons in the beam would decay in a 50-m FV with an average momentum of 70 GeV. The required integrated proton flux is then 5×10^{19} pot. Assuming that this is delivered in 5 years[1] the required primary beam intensity would be 2×10^{13} protons per pulse (ppp) with one pulse every 16.8 s. The sixfold increase in intensity relative to NA62 is made necessary by the tight beam collimation and long K_L^0 lifetime.

The maximum intensity that the SPS can deliver to the North Area is about 4×10^{13} ppp, but apart from competition for protons from the SPS, a primary beam intensity of 2×10^{13} ppp is not currently available on any of the North Area targets. Running with an intensity this high would require comprehensive upgrades to the current beamline cavern and experimental area, as well as the nearly 2 km long beam transport from the extraction point in the SPS to the production target of the experiment. An alternative could be to site the experiment at the North Area Beam Dump Facility under discussion in the context of the SHiP experiment. Either way, detailed solutions and meaningful cost estimates will require studies in collaboration with the CERN Accelerator and Technology Sector.

The layout of the secondary beamline is illustrated in figure 3. The 400-GeV beam is incident on the target at a downward angle of 2.4 mrad. Three collimators are used to obtain a neutral beam with an opening angle of 0.3 mrad: a moveable dump collimator at $z = 15$ m from the target, the defining collimator at $z = 60$ m, and a final collimator at $z = 105$ m with an aperture just larger than the angular limit of the background from the dump collimator, to

[1] 100 effective days of running per year, similar to a Snowmass year.

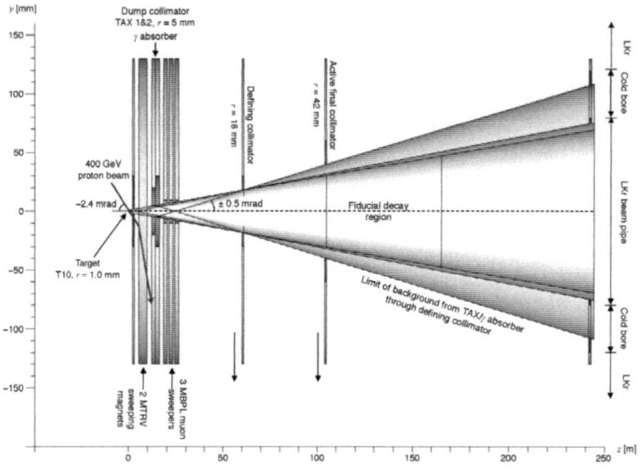

Figure 3. Layout of the neutral beamline.

remove particles showering on the edges of the defining collimator. The 45 m length between the defining and final collimators is equivalent to nearly 9 K_S^0 lifetimes at the mean beam momentum. Vertical sweeping magnets are located just upstream of the dump collimator, and horizontal muon-sweeping magnets are just downstream. A photon converter of high-Z material is placed in the center of the dump collimator, to reduce the flux of photons in the neutral beam.

The K_L^0, photon, and neutron fluxes in this beamline have been simulated with Fluka and Geant4, with the former used for the beam-target interaction and the latter used for propagation of secondaries through the beamline. The simulation predicts 290 MHz of K_L^0 in the beam, about 50% more than the parameterizations used to estimate the intensity requirement would suggest. The simulation also predicts 230 MHz of photons with $E > 5$ GeV and 20 MHz of photons with $E > 30$ GeV, assuming a 30-mm iridium absorber in the dump collimator. More problematic is the fact that the simulation predicts 3 GHz of neutrons in the beam, which places stringent requirements on the insensitivity of the small-angle calorimeter to neutrons.

2.2. Detector

The experimental signature of the $K_L^0 \to \pi^0 \nu \bar{\nu}$ decay consists of two photons with unbalanced transverse momentum in an event in which there are no additional particles. The only available kinematic constraint is on the two-photon invariant mass: $M(\gamma\gamma) = m_{\pi}^0$. The most dangerous background is from $K_L^0 \to \pi^0 \pi^0$ decays (which are about 3×10^7 times more abundant than the signal) with two lost photons. The centerpiece of the experiment is the NA48 liquid-krypton calorimeter, which is used both to reconstruct the π^0 for the signal decay and to veto any other photons present. The invariant mass constraint is imposed to obtain the z position of the $\gamma\gamma$ vertex. The layout of the detector elements is schematically illustrated in figure 4. The largest elements are about 3 m in diameter.

The vacuum volume begins 25 m upstream of the final collimator. Decays in this region for which two photons pass through the final collimator can give rise to background. The final collimator itself is thus an active detector, with the collimating surfaces made of LYSO surrounded by an electromagnetic calorimeter extending out to a radius of 1 m to provide upstream veto coverage.

The FV extends from $z = 105$ m (the position of the final collimator) to $z = 155$ m. The front face of the sensitive part of the LKr calorimeter is at z = 241.5 m. The 90-m distance from the downstream end of the FV to the LKr significantly aids with background rejection, since most

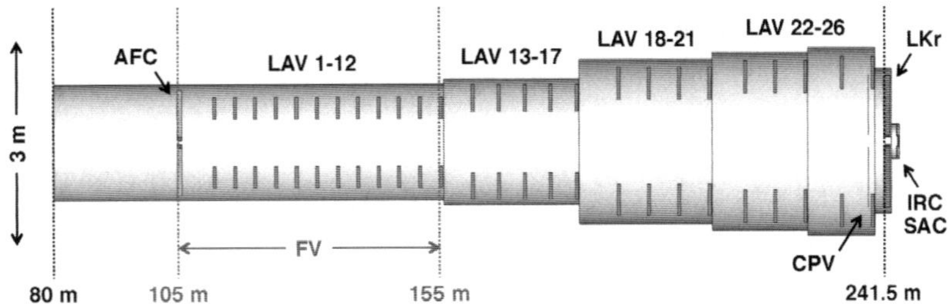

Figure 4. Schematic illustration of the experimental layout. The scales for transverse and longitudinal dimensions are in the ratio 10:1.

$K_L^0 \to \pi^0\pi^0$ decays with lost photons occur just upstream of the LKr. Even for $K_L^0 \to \pi^0\pi^0$ events in which the two detected photons are from different π^0s, the vertex is almost always reconstructed downstream of the FV.

Besides the LKr, there are a total of 26 ring-shaped large-angle photon veto (LAV) stations in five different sizes, placed at intervals of 4 to 6 m to guarantee coverage out to $\theta = 100$ mrad. The LAV detectors are based on the lead/scintillating-tile design of the large-angle photon vetoes for the (canceled) CKM experiment at Fermilab [9].

The small-angle vetoes (IRC and SAC) on the downstream side of the LKr intercept photons from K_L^0 decays that pass through the beam pipe. The small-angle calorimeter (SAC) squarely intercepts the neutral beam. The intermediate-ring calorimeter (IRC) is a ring-shaped detector between the SAC and LKr that intercepts photons from downstream decays that make it through the calorimeter bore at slightly larger angles. Because of the high rates of neutrons and photons in the beam, the design of these detectors, in particular the SAC, is one of the most challenging aspects of the experiment. The task is made a little easier by the fact that photons from K_L^0 decays in the SAC acceptance have very high energy. The SAC can be blind to photons with $E < 5$ GeV, and need only have very high efficiency (99.99%) for photons with $E > 30$ GeV. The current design calls for a compact tungsten/silicon-pad sampling calorimeter with a crystal metal absorber. The coherent interaction of photons with the crystal lattice reduces the radiation length of the absorber, and hence the ratio X_0/λ_{int} (see, e.g., [10]), resulting in a calorimeter with better transparency to neutrons. The effect is greatest for high-energy photons; for 30 GeV photons incident at an angle of 2 mrad, the pair-production cross section is enhanced by a factor of about three. Preliminary Geant4 simulations indicate that perhaps 20% of the neutrons in the beam will interact in the SAC; information on the transverse and longitudinal shower spread and pulse shape information will be used to obtain additional neutron discrimination power.

In addition to the photon vetoes, the experiment makes use of charged-particle veto (CPV) detectors and a hadronic calorimeter downstream of the LKr (not shown in figure 4), to reject background from the copious K_L^0 decays into charged particles.

2.3. Expected performance

The rejection power for background from $K_L^0 \to \pi^0\pi^0$ events has been studied using a fast Monte Carlo, with statistics equivalent to those from five years of data taking. Signal candidates are required to have exactly two clusters on the LKr and no hits on any other detector. The decay vertex position in z is obtained from the separation between the clusters on the calorimeter and the cluster energies, and is required to be inside the FV. There is a substantial amount of background from upstream decays, but this can be eliminated by requiring that the clusters on the LKr lie outside of a radius of 35 cm from the beam axis. Figure 5 shows the distribution of the

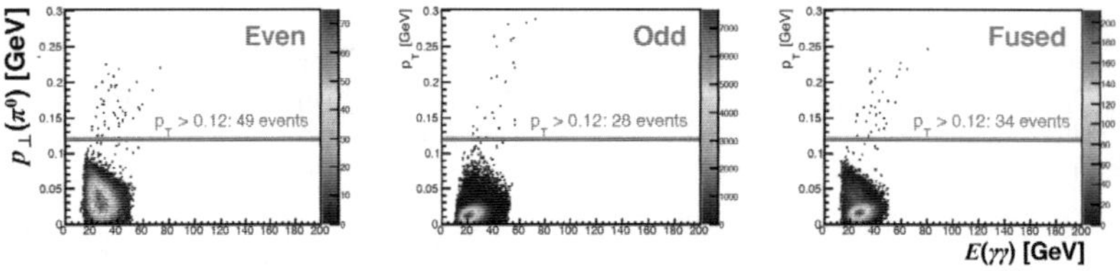

Figure 5. Distribution in $(p_{\perp\gamma\gamma}, E_{\gamma\gamma})$ plane of $K_L^0 \to \pi^0\pi^0$ events passing all selection cuts, from fast simulation: even events (left), odd events (center), and events with fused clusters (right).

remaining events in the plane of pair transverse momentum $p_{\perp\gamma\gamma}$ vs. pair energy $E_{\gamma\gamma}$ for events in which both photons come from the same π^0 (even pairing), different π^0s (odd pairing), or in which one or both clusters contain overlapping LKr hits (fused). If $p_{\perp\gamma\gamma}$ for signal candidates is required to be larger than 0.12 GeV, 111 background events remain. Similar high-statistics studies of $K_L^0 \to \pi^0\pi^0\pi^0$ and $K_L^0 \to \gamma\gamma$ backgrounds give 15 and 0 counts, respectively ($K_L^0 \to \gamma\gamma$ is effectively eliminated by the $p_\perp$ cut). The acceptance for $K_L^0 \to \pi^0\nu\bar{\nu}$ decays in the fiducial volume is about 9% (the $p_{\perp\gamma\gamma}$ cut retains 78% of signal events). In five years, 97 signal events are expected to be collected. The photon converter in the dump collimator may absorb or scatter as much as 35% of the beam. This effect is not simulated; taking it into account, at the SM BR, about 60 signal events per year are expected, together with a similar number of background events ($S/B \sim 1$).

3. Conclusions

The present results are highly preliminary. In particular, high-statistics background studies have been carried out only for a few channels. Some potential backgrounds, such $nn \to nn\pi^0$ from interactions of beam neutrons on residual gas, need to be investigated. The detector concepts require validation, and the suitability of the LKr performance, in particular concerning photon detection efficiency and time resolution, needs to be verified. The possibility of adding charged-particle tracking to the experiment, which would provide more complete final-state reconstruction for efficiency estimation and systematic control, and which would allow significant expansion of the physics program, is under investigation. Finally, the available intensity and siting options at CERN need to be worked out. With these caveats, our preliminary design studies indicate that an experiment to measure BR($K_L^0 \to \pi^0\nu\bar{\nu}$) can be performed at the SPS during LHC Run 4 (2026-2029). These ideas are under development in the context of the CERN Physics Beyond Colliders initiative [11], with the intention to move towards an official proposal within the next two years.

References

[1] Cirigliano V *et al.* 2012 *Rev. Mod. Phys.* **84** 399
[2] Buras A *et al.* 2015 *JHEP* **1511** 033
[3] Buras A, Buttazzo D and Knegjens R 2015 *JHEP* **1511** 166
[4] E949 Collaboration, Artamonov A *et al.* 2009 *Phys. Rev. D* **79** 092004
[5] Hahn F *et al.* 2010 NA62 technical design document `http://cds.cern.ch/record/1404985`
[6] KOTO Collaboration, Ahn J *et al.* 2016 arXiv:1609.03637
[7] KOTO Collaboration, Shiomi K *et al.* these proceedings
[8] NA48 Collaboration, Fanti V *et al.* 2007 *Nucl. Instrum. Meth. A* **574** 433
[9] Ramberg E, Cooper P and Tschirhart R 2004 *IEEE Trans. Nucl. Sci.* **51** 2201
[10] Bak J *et al.* 1988 *Phys. Lett. B* **202** 615
[11] `https://indico.cern.ch/event/523655/`

The KLOE-2 experiment at DAΦNE

A. Passeri on behalf of the KLOE-2 Collaboration

INFN sezione Roma Tre, via della Vasca Navale 84, 00146 Roma, Italy

antonio.passeri@roma3.infn.it

Abstract. The KLOE-2 experiment is steadily taking data since November 2014 at the upgraded DAΦNE ϕ-factory, in Frascati Laboratories, collecting up to now about 2.4 fb^{-1} at a rate around 10 pb^{-1}/day. Performances of the upgraded KLOE-2 detector, data taking conditions, data quality monitoring and physics perspectives are here discussed.

1. Introduction

KLOE-2 is the continuation of the KLOE experiment [1], upgraded with state-of-the-art detectors to improve its performances and extend its physics reach [2]. KLOE-2 has already collected an integrated luminosity of 2.4 fb^{-1} at the upgraded DAΦNE e^+e^- collider and aims to collect more than 5 fb^{-1}.

DAΦNE implements the crab-waist collision scheme [3] which allows to increase the collider's specific luminosity while keeping under control the beam-beam effects. During the KLOE-2 run the best achieved peak luminosity up to now is $L_{peak} = 2.2\times10^{32}$ $cm^{-2}s^{-1}$ and the maximum daily integrated luminosity is $L_{max} = 13$ pb^{-1}/day, while the average up-time of the machine is 80%.

2. The KLOE-2 detector

The KLOE detector consists in a large Drift Chamber [4] and a Calorimeter [5], both immersed in a 0.5 T axial magnetic field. The cylindrical Drift Chamber, with stereo wire geometry, provides ~ 150 μm spatial resolution in the bending plane and ~2 mm along the beam line, and allows to reconstruct charged particle tracks with high momentum resolution ($\sigma_p/p = 0.4\%$) while the Pb-scintillating fibers calorimeter, which ensures a 98% solid angle coverage, allows to reconstruct clusters with excellent time ($\sigma_t = 54$ ps/$\sqrt{E(GeV)} \oplus$ 100 ps) and good energy ($\sigma_E/E = 5.7\%/\sqrt{E(GeV)}$) resolutions. DC and EMC performances turn out to be very stable in time, despite the very different operational conditions of the present KLOE-2 data taking, in terms of machine background, with respect to the KLOE run.

Along with the pre-existing KLOE detector, new detectors have been installed inside the interaction region and along the beam lattice in order to enhance track and vertex reconstruction, photon and electron/positron detection needed to accomplish the KLOE-2 physics program: a novel cylindrical GEM Inner Tracker (IT), new calorimeters (QCALTs and CCLATs) to enlarge the angular acceptance of the apparatus and four tagging stations (LETs and HETs) for $\gamma\gamma$-physics studies.

2.1. The Inner Tracker

To improve the resolution on decay vertices close to the IP, reconstructed from low-momentum charged secondaries, the Inner Tracker (IT) has been inserted in the free space between the beam pipe and the DC inner wall, at 25 cm from IP. The IT is the first ever built and operated cylindrical GEM detector, with a total material budget below 2% of radiation length and < 0.5% dead area, allowing to

Content from this work may be used under the terms of the Creative Commons Attribution 3.0 licence. Any further distribution of this work must maintain attribution to the author(s) and the title of the work, journal citation and DOI.
Published under licence by IOP Publishing Ltd

minimize the multiple scattering of low- momentum tracks and the probability of photon conversions. The resolution on vertices close to the IP is expected to improve by about a factor 3 [6].

The IT is composed by four concentric Cylindrical GEM (CGEM) [7] detectors, with radii from 13 cm, to preserve the K_SK_L quantum interference, up to 20.5 cm, due to the constraint from DC inner wall. Each layer, with a total active length of 70 cm, is a triple-GEM detector with 5 concentric cylindrical electrodes: a cathode, to set the drift field, 3 GEM foils acting as multiplication stages, and an anode/readout plane. The anode plane is a multi-layer circuit which implements X-V stereo readout with a 650 μm pitch for a total amount of 30000 front-end channels.

The IT is operated with a Ar:*i*C4H10 90:10 gas mixture at a nominal gain of 12000 to limit the discharge probability, measured with α-particles. The measured efficiency [8] with cosmic-ray muon tracks reconstructed by DC is 98% for the single-view and 95% for the two-views.

The IT operational parameters with colliding beams have been optimized as a function of the beam currents and background conditions, which are carefully monitored through the current measured on the IT inner layer, the total current measured in the DC and the background level measured on both calorimeter's end-caps. Online monitoring of IT temperature is also available to keep operation safe, as well as offline software tools for checking detector status, occupancy and performances [8].

The DC excellent track reconstruction has been exploited for IT alignment and calibration. Two effects determine a focusing or a defocusing of the electron cloud, depending on the impact parameter of the track: a non-radial track effect and the presence of the KLOE-2 magnetic field. The first, due to the non-zero angle between the impinging track and the radial direction of ionization electrons motion, induces a shift and a spread of the reconstructed hit position on the readout plane. The second, due to the Lorentz force acting on the signal electron cloud, further shifts and spreads the reconstructed position. These effects must be studied and measured independently: cosmic-ray muon runs acquired without magnetic field have been used to evaluate the non-radial correction, while the magnetic field influence has been investigated using cosmic-ray muon data and Bhabha scattering events. After applying alignment and calibration corrections the preliminary IT-DC Bhabha tracking residuals distributions show a resolution of ~400 μm, which is well within expectations, as displayed in figure 1 for IT layer 2. Further alignment and calibration refinements will definitely improve this first result.

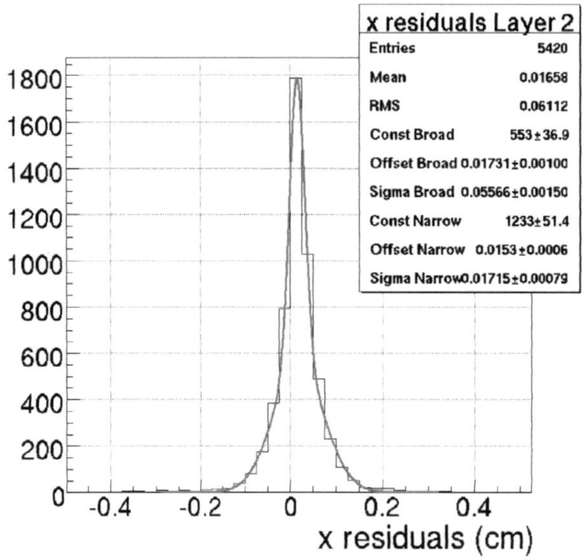

Figure 1. Residuals between the DC extrapolated tracks and the IT reconstructed ones on the IT layer 4 surface.

2.2. *The low angle calorimeters*

Two new Quadrupole CALorimeters with Tiles, QCALTs [9], have been installed around the DAΦNE low-β quadrupoles, at both sides of the IP, with the goal of improving the acceptance for rejecting KL → $3\pi^0$ background events in CP-violating KL→$2\pi^0$ decays. Each calorimeter, 1 m long, consists in a dodecagonal structure, arranged as a sampling of 5 layers of 5 mm thick scintillator plates, alternated with 3.5 mm thick tungsten plates, for a total of ~ $5X_0$ thickness. The active part of each plane is divided into 20 tiles of 5 X 5 cm^2 area with 1 mm diameter wavelength shifter fibers embedded in circular grooves. Each fiber is then optically connected to a SiPM, for a total of 2400 channels. QCALT calorimeters have a ~2 mm resolution along the beam axis and a time resoultion of ~1ns. Time calibration and channel-by-channel equalization is performed using cosmic-ray muon runs acquired with and without magnetic field. The KLOE calorimeter is used as timing reference.
In order to enlarge the angular acceptance for particles coming from the IP from 20° to 10°, with the aim of improving multi-photon detection in rare decays such as $K_S \rightarrow \gamma\gamma$, $\eta \rightarrow \pi^0\gamma\gamma$ and $K_S \rightarrow 3\pi^0$, two identical Crystal CALorimeters with Timing (CCALTs) [9] have been intalled very close to the IP, near the first focusing quadrupoles of DAΦNE. Each CCALT module is made of 4 aluminum shells, with projective geometry, containing 4 LYSO crystals readout by SiPM. Tests performed at the Frascati Beam Test Facility allowed to measure an energy resolution better than 5% and a time resolution of about 49 ns (120 ps) at 100 MeV (500 MeV) energy. Similar performances are measured with cosmic-ray muons, used also for calibration and equalization of CCALT channels.
Such detector provides also fast luminosity monitor to the DAΦNE operations.

2.3. *The γ-γ taggers*

Low Energy Taggers (LETs) and High Energy Taggers (HETs) have been installed with the aim of detecting electrons and positrons scattered in $e^+e^- \rightarrow e^+e^-\gamma^*\gamma^* \rightarrow e^+e^-X$ reactions, which deviate from the equilibrium orbit during the propagation along the accelerator lattice with energy below 510 MeV. Two identical LET stations [10] have been placed symmetrically at 1 m on both sides of the IP, in order to tag electrons and positrons with energy 160 < E < 400 MeV. Each LET station consists in an array of 5 x 4 LYSO crystals, each read out by a Silicon photomultiplier (SiPM) with 1.5 x 1.5 cm^2 section and 12 cm length, pointing to the average direction of the arriving particles (~11° with respect to the beam line). The two stations are rotated by an angle of 17° with respect to the horizontal plane in order to maximize the number of collected positrons and electrons. The energy resolution of the LET calorimeters, less than 10% in the energy range 150 < E < 400 MeV, has been measured with electrons of energy between 50 and 500 MeV at the Frascati Beam Test Facility and it well matches the requirements. Equalization of the LET crystals response and time calibration have been performed selecting minimum ionizing particles (MIPs) as high-momentum tracks from cosmic-ray muons. The absolute energy scale calibration is performed with radiative Bhabha scattering events ($e^+e^- \rightarrow e^+e^-\gamma$), with the photon and one lepton reconstructed in the KLOE main detector and the other one detected in the LET.

The HET stations [11] are position detectors for measuring the deviation of $e^\pm$ from the main beam orbit. Together with time information, this measurement allows to tag scattered electrons and positrons with energy greater than 420 MeV in $\gamma\gamma$ processes. The two HET detectors are placed 11 m away from the IP, in symmetrical positions. The sensitive area of each HET station is a set of twenty-eight 3x5x6 mm^3 plastic scintillators, with an additional larger one used for coincidence purposes, which are placed at different distances from the beam-line. The fired scintillator provides a measurement of the distance between the impinging particle and the beam.

The KLOE-2 apparatus is synchronized with the DAΦNE bunch crossing, which occurs every 2.7 ns. Since HET stations are far from the main detector installation, its time synchronization with KLOE-2 detectors is obtained using dedicated runs of radiative Bhabha scattering events. After calibration the HET taggers nicely reproduce the DAΦNE bunch time structure thanks to the good time resolution of the HET detectors.

3. Data taking and data quality

KLOE-2 data taking started in November 2014 and is going on steadily up to now with just a few months summer shutdown stops which were exploited to perform programmed maintenance and upgrade operations to detector and accelerator components. DAFNE performances in terms of delivered luminosity and machine background levels have constantly improved during this period, allowing the KLOE-2 detector to operate with increasingly higher beam currents and peak luminosity.

By june 2016 DAΦNE delivered an integrated luminosity of 3.0 fb^{-1}, out of which 2.4 fb^{-1} were acquired by KLOE-2. The maximum peak luminosity reached 2.2×10^{32} cm^{-2} s^{-1}, while the best daily integrated luminosity reached 13 pb^{-1} (11 pb^{-1} acquired). DAFNE implements a topping-up injection scheme, which allows to keep the average beam currents between 1.2÷1.6 A (electrons) and 0.7÷1.0 A (positrons). KLOE-2 detector provides instantaneous luminosity measurement, by counting the large angle Bhabha events, and machine backgrounds monitoring via 3 different observables: the current drawn by the IT inner layer, the current drawn by the DC and the counting rate of the inner regions of the calorimeter end-caps. All of these 3 background probes are continuously monitored and threshold values have been defined such to allow the detector to operate safely and to collect good quality data.

Data quality is monitored both online and offline by checking beam parameters ($\sqrt{s}$, IP position and ϕ boost) and event classification counters stability. Benchmark physics channels have also been selected such as $K_{S,L} \to \pi^+\pi^-$, $\phi \to \eta\gamma$ ($\eta \to \gamma\gamma$ or $\eta \to 3\pi^0$): thanks to a careful analysis a general good agreement with previous KLOE data distributions is observed despite the increased levels of the machine background.

To assess the actual beam energy scale an energy scan around the ϕ peak has been performed by shifting the central value of the DAΦNE radiofrequency. The event rates in different ϕ decay channels have been measured and normalized to the large angle Bhabha event rate. As shown in figure 2, event rates are nicely fitted by the ϕ lineshape, including radiative corrections and beam energy spread effect.

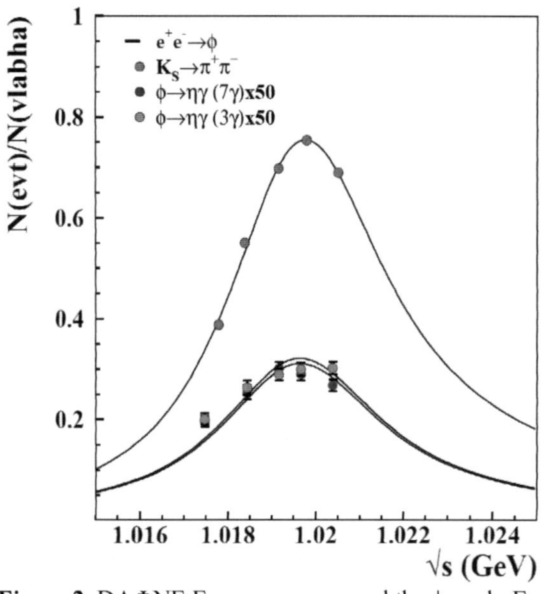

Figure 2. DAΦNE Energy scan around the ϕ peak. Event counting is normalized to the number of "very large angle Bhabhas" ("vlabhas"), i.e. with $\theta > 55^{\circ}$

4. Perspectives and conclusions

While KLOE-2 data taking campaign is continuing, physics analysis has started according to the guidelines described in [2]. Rare neutral kaon decays, in particular K_S ones, are being studied: with 5 fb^{-1} total integrated luminosity a pure data sample of about $5x10^9$ tagged K_S decays will be available. Neutral kaon interferometry is also a key analysis subject, as KLOE-2 offers a unique environment to test fundamental symmetries such as CPT and Lorentz invariance with entangled particles' pairs. $\gamma\gamma$ taggers will also provide clean data to study low energy QCD effects in light mesons decay widths and form factors. The study of η and η' decays and the search for dark photon signatures complete the physics goals of the present run.

Acknowledgements

We thank the DAΦNE team for their efforts in maintaining low background running conditions and their collaboration during all data taking. We want to thank our technical staff: G.F. Fortugno and F. Sborzacchi for their dedication in ensuring efficient operation of the KLOE computing facilities; M. Anelli for his continuous attention to the gas system and detector safety; A. Balla, M. Gatta, G. Corradi, G. Papalino and F.Budano for electronics maintenance; M. Santoni, G. Paoluzzi and R. Rosellini for general detector support; C. Piscitelli for his help during major maintenance periods. This work was supported in part by the EU Integrated Infrastructure Initiative Hadron Physics Project under contract number RII3-CT- 2004-506078; by the European Commission under the 7th Framework Programme through the 'Research Infrastructures' action of the 'Capacities' Programme, Call: FP7-INFRASTRUCTURES-2008-1, Grant Agreement No. 227431; by the Polish National Science Centre through the Grants No. 2011/03/N/ST2/02652, 2013/08/M/ST2/00323, 2013/11/B/ST2/04245, 2014/14/E/ST2/00262, 2014/12/S/ST2/00459.

References

[1] Bossi F., De Lucia E., Franzini J. L., Miscetti S., Palutan M. and KLOE Collaboration, Nuovo Cimento, 30 (2008) 10;
[2] Amelino-Camelia G. et al., Eur. Phys. J. C, 68 (2010) 619;
[3] M.Zobov et al., Phys. Rev. Lett. 104, 174801 (2010) 1;
[4] Adinolfi M. et al., Nucl. Instrum. Meth. A, 488 (2002) 51;
[5] Adinolfi M. et al., Nucl. Instrum. Meth. A, 482 (2002) 364;
[6] Archilli F. et al., arXiv:1002.2572v1 and LNF-10/3(P), INFN-LNF, Frascati, 2010;
[7] Bencivenni G., Domenici D., Nucl. Instrum. Meth. A, 581 (2007) 581;
[8] Di Cicco A. and Morello G., Acta Phys. Pol. B, 46 (2015) 73;
[9] Happacher F. and Martini M., Acta Phys. Pol. B, 46 (2015) 87;
[10] Babusci D. et al., Acta Phys. Pol. B, 46 (2015) 87;
[11] Babusci D. et al., Acta Phys. Pol. B, 46 (2015) 81;

Prospects for exotics and LFV at NA62

P Petrov [1]

Centre for Cosmology, Particle Physics and Phenomenology (CP3)
Université catholique de Louvain, Louvain-la-Neuve, Belgium

E-mail: plamen.petrov@uclouvain.be

Abstract. The NA62 experiment at the CERN SPS is designed to measure the branching ratio of the ultra-rare decay $K^+ \to \pi^+\nu\bar{\nu}$ with 10% precision. For this purpose the experiment aims to collect of the order of 100 $K^+ \to \pi^+\nu\bar{\nu}$ events which would require at least $10^{13}K^+$ decays in the fiducial volume of the detector. The large sample of K^+ decays that will be available at NA62 allows the experiment to carry out a rich program for rare and forbidden K^+ and π^0 decays, including lepton flavour and/or number violating modes, sterile neutrinos, exotic particles (e.g. Dark Photons). NA62's potential for such searches in K^+ and π^0 decays is discussed, with initial sensitivity estimates. In addition, plans for continuing the experiment after 2021 (Long Shutdown 2) are presented, including in particular the possibility to operate in a beam-dump mode for exploring various Dark Matter scenarios.

[1] for the NA62 Collaboration:
G. Aglieri Rinella, R. Aliberti, F. Ambrosino, R. Ammendola, B. Angelucci, A. Antonelli, G. Anzivino, R. Arcidiacono, I. Azhinenko, S. Balev, M. Barbanera, J. Bendotti, A. Biagioni, L. Bician, C. Biino, A. Bizzeti, T. Blazek, A. Blik, B. Bloch-Devaux, V. Bolotov, V. Bonaiuto, M. Boretto, M. Bragadireanu, D. Britton, G. Britvich, M.B. Brunetti, D. Bryman, F. Bucci, F. Butin, J. Calvo, E. Capitolo, C. Capoccia, T. Capussela, A. Cassese, A. Catinaccio, A. Cecchetti, A. Ceccucci, P. Cenci, V. Cerny, C. Cerri, B. Checcucci, O. Chikilev, S. Chiozzi, R. Ciaranfi, G. Collazuol, A. Conovaloff, P. Cooke, P. Cooper, G. Corradi, E. Cortina Gil, F. Costantini, F. Cotorobai, A. Cotta Ramusino, D. Coward, G. D'Agostini, J. Dainton, P. Dalpiaz, H. Danielsson, J. Degrange, N. De Simone, D. Di Filippo, L. Di Lella, S. Di Lorenzo, N. Dixon, N. Doble, B. Dobrich, V. Duk, V. Elsha, J. Engelfried, T. Enik, N. Estrada, V. Falaleev, R. Fantechi, V. Fascianelli, L. Federici, S. Fedotov, M. Fiorini, J. Fry, J. Fu, A. Fucci, L. Fulton, S. Gallorini, S. Galeotti, E. Gamberini, L. Gatignon, G. Georgiev, A. Gianoli, M. Giorgi, S. Giudici, L. Glonti, A. Goncalves Martins, F. Gonnella, E. Goudzovski, R. Guida, E. Gushchin, F. Hahn, B. Hallgren, H. Heath, F. Herman, T. Husek, O. Hutanu, D. Hutchcroft, L. Iacobuzio, E. Iacopini, E. Imbergamo, O. Jamet, P. Jarron, E. Jones, T. Jones K. Kampf, J. Kaplon, V. Kekelidze, S. Kholodenko, G. Khoriauli, A. Khotyantsev, A. Khudyakov, Yu. Kiryushin, A. Kleimenova, K. Kleinknecht, A. Kluge, M. Koval, V. Kozhuharov, M. Krivda, Z. Kucerova, Yu. Kudenko, J. Kunze, G. Lamanna, G. Latino, C. Lazzeroni, G. Lehmann-Miotto, R. Lenci, M. Lenti, E. Leonardi, P. Lichard, R. Lietava, V. Likhacheva, L. Litov, R. Lollini, D. Lomidze, A. Lonardo, M. Lupi, N. Lurkin, K. McCormick, D. Madigozhin, G. Maire, C. Mandeiro, I. Mannelli, G. Mannocchi, A. Mapelli, F. Marchetto, R. Marchevski, S. Martellotti, P. Massarotti, K. Massri, P. Matak, E. Maurice, M. Medvedeva, A. Mefodev, E. Menichetti, E. Minucci, M. Mirra, M. Misheva, N. Molokanova, J. Morant, M. Morel, M. Moulson, S. Movchan, D. Munday, M. Napolitano, I. Neri, F. Newson, A. Norton, M. Noy, G. Nuessle, T. Numao, V. Obraztsov, A. Ostankov, S. Padolski, R. Page, V. Palladino, G. Paoluzzi, C. Parkinson, E. Pedreschi, M. Pepe, F. Perez Gomez, M. Perrin-Terrin, L. Peruzzo, P. Petrov, F. Petrucci, R. Piandani, M. Piccini, D. Pietreanu, J. Pinzino, I. Polenkevich, L. Pontisso, Yu. Potrebenikov, D. Protopopescu, F. Raffaelli, M. Raggi, P. Riedler, A. Romano, P. Rubin, G. Ruggiero, V. Russo, V. Ryjov, A. Salamon, G. Salina, V. Samsonov, C. Santoni, G. Saracino, F. Sargeni, V. Semenov, A. Sergi, M. Serra, A. Shaikhiev, S. Shkarovskiy, I. Skillicorn, D. Soldi, A. Sotnikov, V. Sugonyaev, M. Sozzi, T. Spadaro, F. Spinella, R. Staley, A. Sturgess, P. Sutcliffe, N. Szilasi, D. Tagnani, S. Trilov, M. Valdata-Nappi, P. Valente, M. Vasile, T. Vassilieva, B. Velghe, M. Veltri, S. Venditti, P. Vicini, R. Volpe, M. Vormstein, H. Wahl, R. Wanke, P. Wertelaers, A. Winhart, R. Winston, B. Wrona, O. Yushchenko, M. Zamkovsky, A. Zinchenko.

Content from this work may be used under the terms of the Creative Commons Attribution 3.0 licence. Any further distribution of this work must maintain attribution to the author(s) and the title of the work, journal citation and DOI.
Published under licence by IOP Publishing Ltd

1. Introduction

NA62 [1] is a fixed target experiment located in the North Area of the CERN SPS which makes use of an unseparated 75 GeV/c positive secondary beam produced by 400 GeV/c protons from the SPS impinging on a beryllium target. The secondary beam has a 6% component of K^+ which are allowed to decay in-flight inside a 65 m long fiducial decay volume (see Figure 1). The goal of the NA62 experiment is to measure the branching ratio of the $K^+ \to \pi^+\nu\bar{\nu}$ decay, a theoretically very clean process that is considered to be a sensitive probe for physics beyond the Standard Model (SM). While the SM prediction $BR(K^+ \to \pi^+\nu\bar{\nu}) = (9.11 \pm 0.72) \times 10^{-11}$ [2] is quite precise it is not matched by the current experimental state-of-art $BR(K^+ \to \pi^+\nu\bar{\nu}) = (17.3^{+11.5}_{-10.5}) \times 10^{-11}$ [3], based on only 7 candidate events. The high uncertainty of the experimental result motivates the new measurement by the NA62 which plans to collect of the order of 100 $K^+ \to \pi^+\nu\bar{\nu}$ events. Considering the SM decay rate and assuming a 10% signal acceptance, the kaon flux should correspond to at least 10^{13} K^+ decays in the fiducial volume. Furthermore, since $BR(K^+ \to \pi^+\pi^0) = 20.69\%$, NA62 will be exposed to $\sim 10^{12}$ tagged π^0 decays.

These large samples of kaon and pion decays provide an opportunity to perform searches for a range of novel phenomena and forbidden decay modes with an unprecedented precision. In particular, NA62 is well positioned to improve the current limits in the searches for K^+ decays with explicit violation of the conservation of lepton flavour (LFV) or number (LNV). These include the LNV modes: $K^+ \to \pi^-\mu^+\mu^+$, $K^+ \to \pi^-e^+e^+$ and $K^+ \to \pi^-\mu^+e^+$; and the LFV: $K^+ \to \pi^+\mu^\pm e^\mp$. NA62 can also search for heavy neutrinos produced in $K^+ \to l^+\nu_N (l = e, \mu)$, where ν_N does not decay inside the fiducial volume of the detector.

The exploitation of rare and forbidden decay physics beyond the flagship mode will rely critically on the implementation of dedicated trigger chains. The NA62 trigger is divided into three logical levels: hardware L0 trigger that reduces the $\sim$ 10 MHz input rate at nominal beam to $\sim$ 1 MHz using trigger algorithms based on multiplicity signals from the ring-imaging Cerenkov (RICH), the charged-particle hodoscope (CHOD) and the muon-veto (MUV3) detectors, and the energy deposited in the LKr calorimeter; software L1 and L2 triggers that reduce the input rate further by using the K^+ beam tagging detector (CEDAR), the large-angle photon vetoes (LAV), reconstructed tracks and decay topologies using the STRAW spectrometer, as well as refined information from all other detectors, including those used at L0. Depending on the beam intensity, the trigger bandwidth at L0 for final states other than "single positive track + missing energy" ($\pi\nu\bar{\nu}$) is limited to a few 100 KHz.

The NA62 detector and the adjoining beamline have been commissioned up to the designed beam intensity in pilot runs in 2014 and 2015, and at present the experiment is taking data for physics. The physics program of NA62 is approved until 2018 when the Long Shutdown 2 (LS2) period starts and the SPS will not provide beam to the North Area. The SPS will resume in 2021 and the NA62 collaboration is considering various proposals for rare and exotic searches that could be done with minimal modifications of the existing apparatus. A particularly interesting possibility, also discussed here, is to operate the detector and the beam-line in a "beam-dump" mode, similar to the SHiP proposal [4], to search for MeV-GeV mass hidden-sector candidates, such as Dark Photons (A'), heavy neutral leptons (HNL), axion-like particles (ALP), etc.

2. LNV and LFV in $K \to \pi ll$ decays

The apparent lepton flavour and/or number conservation observed so far in particle physics is conveniently accommodated in the SM if the neutrino masses are zero. The discovery of neutrino oscillations has led to a conclusion that the lepton flavour symmetry is approximate rather than exact. It also implies that the neutrinos have non-zero masses which, in addition to constituting a non-SM phenomenon by itself, opens the questions about the possible Majorana nature of the neutrino as well as the origin of the small observed neutrino mass scale. Simply

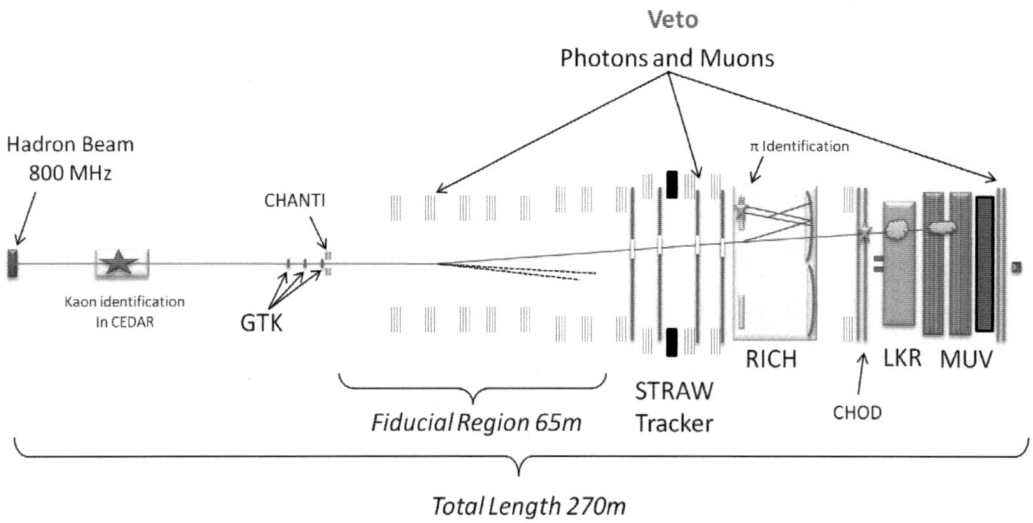

Figure 1. Schematic view of the NA62 experiment, excluding the beam line.

Table 1. Status of searches for LFV and LNV K^+ decays for which limits can potentially be improved by NA62.

Decay mode	SM violation	Limits on BR	Experiment
$K^+ \to \pi^-\mu^+\mu^+$	LNV	8.6×10^{-11}	CERN NA48/2 [5]
$K^+ \to \pi^-e^+e^+$	LNV	6.4×10^{-10}	BNL E865/CERN NA48/2 [6]
$K^+ \to \pi^-\mu^+e^+$	LNV	5.0×10^{-10}	BNL E865/CERN NA48/2 [6]
$K^+ \to \pi^+\mu^-e^+$	LFV	5.2×10^{-10}	BNL E865/CERN NA48/2 [6]
$K^+ \to \pi^+\mu^+e^-$	LFV	1.3×10^{-11}	BNL E777/E865 [7]

incorporating massive neutrinos into the SM results in a prediction of LFV and LNV decays with an unobservably low branching ratios. Therefore, an observation of any of the $K^+ \to \pi^{\pm}ll$ processes in Table 1 would serve as a clear indication of physics beyond the SM, such as supersymmetry, Little Higgs models, extra dimensions, Z' vector bosons, etc.

The LNV decays $K^+ \to \pi^-l^+l'^+$ could be mediated by heavy Majorana neutrino exchange in the same way that could lead to the neutrinoless nuclear double β decay. One extension of the SM that could provide simultaneously explanation for neutrino oscillations, Dark Matter and the baryon asymmetry of the Universe, the so called νMSM [8], introduces three heavy right-handed sterile neutrinos with Majorana mass terms. The first, with low mass, could be a dark matter candidate. The others, with masses ranging from 100 MeV/c^2 to few GeV/c^2, could be responsible for the standard neutrino masses through the Seesaw mechanism and introduce extra CP violating phases, which can account for the observed baryon asymmetry. The heavy neutrinos can be produced in charged kaon decays as long as their masses are within the kinematically accessible range of a few ~ 100 MeV. Provided that the heavy neutrinos are short lived in order to decay within the detector, NA62 is in excellent position to look for $K^+ \to \pi^-l^+l'^+$ decays mediated by them.

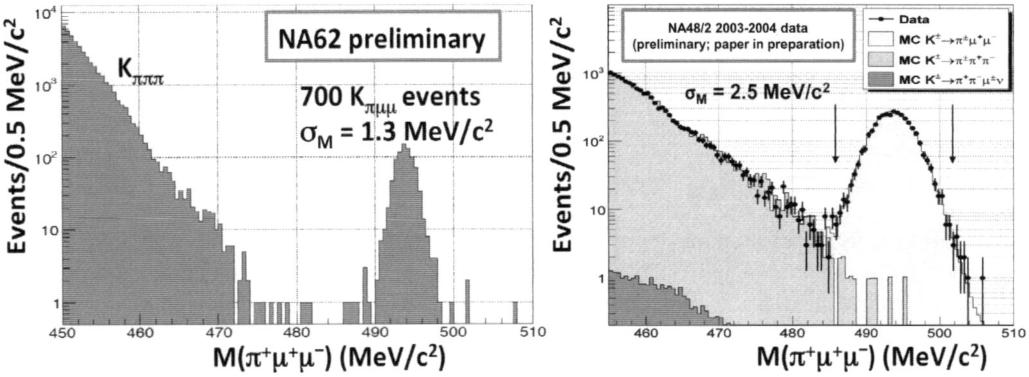

Figure 2. $K^+ \to \pi^+\mu^+\mu^-$ invariant mass distribution as measured by the NA62 (left), using 2016 data collected with a dedicated $\pi\mu\mu$ trigger chain, and by NA48/2 (right) [5].

Recent anomalies in the semileptonic decays $B \to K^\star\mu^+\mu^-$ (P_5' observable) and $B_s^0 \to \phi\mu^+\mu^-$ (LHCb [9, 10]), in $H \to \mu\tau$ (ATLAS/CMS [11]), and the tensions in the ratio ε'/ε [12] have hinted at New Physics, consistent with MFV [13], that could produce LFV in $K^+ \to \pi^+\mu^\pm e^\mp$ decays. Furthermore, the parameter space relevant for the explanation of the B-meson anomalies lies within the sensitivities of the NA62 experiment. Of course pushing down the limits on the $K^+ \to \pi^+\mu^\pm e^\mp$ modes depends critically on the availability of trigger bandwidth so that the corresponding triggers are used without any downscaling.

The trigger chains that are used at the moment are based on multiple tracks final states with additional conditions to identify the lepton pairs (multiplicity in MUV3 and energy deposits in the LKr) in order to reject the high-rate $K_{3\pi}$ contribution. Figure 2 shows a comparison between the invariant mass distribution of the $K^+ \to \pi^+\mu^+\mu^-$ candidates (a lepton number conserving decay with $BR = (9.4 \pm 0.6) \times 10^{-8}$) from a recent NA48/2 measurement (right) and using NA62 data collected during the 2016 run with a dedicated trigger for $\pi\mu\mu$ final states (left). The mass resolution at NA62 is better by a factor ~ 2 as compared to NA48/2 thanks to the new spectrometer and the higher B-field of the dipole magnet (0.36T). The tail towards higher values of invariant mass from $K_{3\pi}$ seen in the NA48/2 measurement is eliminated. Thus in the period until the LS2 it may be possible for NA62 to push the limits on the branching ratios of the LNV and LFV $K^+ \to \pi ll$ decays down to $\sim 10^{-12}$. Furthermore, it is possible that the searches for LNV and LFV would be continued in the period after the LS2, when the $K \to \pi\nu\bar{\nu}$ measurement will be finished and more trigger and acquisition bandwidth will be available.

3. Heavy neutrinos in K_{l2} decays

In 2015 NA62 collected a large sample of ~700M minimum-bias triggered events which allows to carry out a search for HNL production in $K^+ \to l^+\nu$ decays. The search is model independent and it is based on scanning for a peak in the missing mass distributions of $K_{\mu 2}$ and K_{e2} type decays. The preliminary analysis shows that the background can be kept low and with the current data sample NA62 can improve the upper limits on the two mixing matrix elements $|U_{\mu N}|^2$ and $|U_{eN}|^2$ in the mass ranges $250 < m_N < 375$ GeV$/c^2$ and $180 < m_N < 450$ GeV$/c^2$.

4. Hidden sector candidates in "beam dump" mode

Despite the discovery of the Higgs boson, it is clear that the SM is not a complete theory as it fails to explain a number of observed phenomena, including neutrino masses and oscillations, the baryon asymmetry of the Universe, Dark Matter, etc. Various models of Hidden sectors that attempt to solve these puzzles predict some yet unobserved particles which mediate interactions with the SM sector, i.e HNL [8, 14], Dark Photon A' (vector) [15]), ALPs (pseudoscalar coupled to two photons) [16], etc. Searches for such heavy exotic particles can be carried out at the LHC. However, it is possible that we have not observed any of these potential DM candidates due to their extremely feeble interactions with the SM sector rather than due to their heavy masses. Fixed target (or "beam-dump") experiments are particularly useful to search for new weakly coupled, long-lived particles in the MeV to GeV range. The high intensity at the SPS and the large production of charm mesons with the 400 GeV proton beam allow accessing a wide variety of light long-lived exotic particles. While an experiment dedicated to such exotic searches is proposed at the SPS, the SHIP experiment [4], in the shorter term, after the LS2 in 2021, there is a possibility to operate NA62 in "beam dump" mode [17, 18]. The Beryllium target used by NA62 is followed by two 1.6 m long, water-cooled, beam-defining copper collimators (TAX) which can act as a dump ($\sim 10.7\lambda_I$).

ALPs can be produced directly in the beam dump in the fusion of two photons coming from the proton-nucleus interaction (Primakoff production [19]). Although the transverse momenta of the produced ALPs are small the detector is placed far away from the target (> 100 m) and therefore covers only a small angle from the production point. Taking the detector acceptance into account and assuming that all background can be suppressed, even with a rather modest beam-time of a few weeks, NA62 would have a sizeable discovery potential for ALPs in the mass range of $\sim 30 - 200$ MeV (Figure3, left).

Dark Photons can be produced in decays of mesons created in the beam dump, assuming A' is coupled to quarks, or in hard Bremsstrahlung from the beam protons ($\gamma^* N \to N'A'$). NA62 can look for visible decays of A' to e^+e^- pairs, dominant decay channel for $2m_e < M_{A'} < 210$ MeV, and $\mu^+\mu^-$, which is as frequent as e^+e^- for heavier A'. Figure 3 (middle) shows the expected sensitivities assuming 2×10^{18} protons on target (equivalent of 2 years of running) and zero background , and taking into account the trigger efficiency and detector acceptance. NA62 could be sensitive to even larger phase space as only production in the Beryllium target is considered in this estimate, and not in the TAX.

Concerning the HNL, they can be produced in decays of mesons created in the dump and NA62 can search for them in the 60 m long decay volume of the detector. Unlike the heavy neutrino search done in $K^+ \to l^+\nu$ decays the searches for visible decays of the HNL are model dependent. Possible final states to look for are $e^\pm\pi^\mp$ (for the mass range of $140 < M_{A'} < 250$ MeV/c^2), $\mu^\pm\pi^\mp$ ($M_{A'} > 250$ MeV/c^2) and $e^+e^-\nu$ ($2m_e < M_{A'} < 140$ MeV/c^2). The expected sensitivity is shown in Figure 3 (right).

During the 2015 and 2016 runs NA62 collected several hours of data at different intensities with a closed TAX. In addition, during the 2016 run, in normal data-taking conditions, the trigger configuration included chains for exotic searches which were ran in parallel with the main trigger: RICH & 2×CHOD quad. & 2×MUV3 hits ($\mu\mu$ trigger) and RICH & 2×CHOD quad. & MUV3 hit & $E_{LKr} > 10$ GeV ($\pi\mu$ trigger). The collected data is used for feasibility studies and tests of the zero background hypothesis. The preliminary analysis indicates that zero background is achievable [18].

5. Summary

NA62 is designed to make a precise measurement of the $BR(K^+ \to \pi^+\nu\bar{\nu})$. The high-intensity kaon beam provides an opportunity to extend the physics program beyond the flagship mode

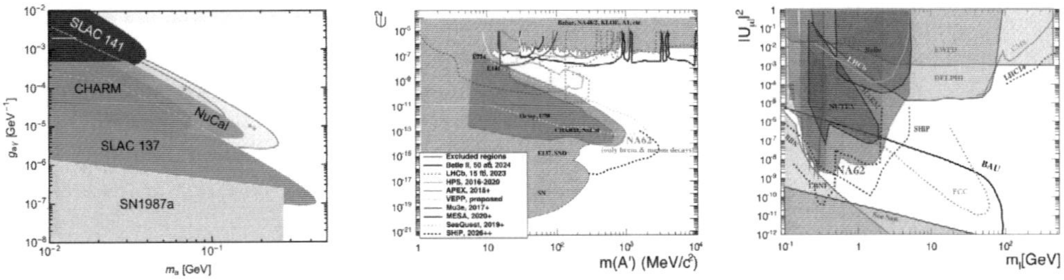

Figure 3. (left) Projected sensitivity of NA62 in the (m_{ALP}, ALP-γ coupling) parameter space for a number of protons on target equivalent to one day, or 1.3×10^{16} protons on target (marked with *), and equivalent to one month (marked with **) [17]; (middle) NA62 sensitivity (90% CL) to A' in the ($m_{A'}$, A' coupling to γ) parameter space considering 2 years of data-taking (2×10^{18} protons on the Beryllium target); (right) NA62 sensitivity (90% CL) to HNL in the (m_l, HNL coupling to μ) parameter space considering 2 years of data-taking [18]; Regions covered by previously performed experiments and astrophysical constraints are overlayed in the plots.

and search for a number of rare and forbidden decays, including the LNV and LFV $K^+ \to \pi^\pm ll$ decays and heavy neutrinos in $K^+ \to l^+\nu$. The detector and the beamline were commissioned during runs in 2014 and 2015, and in 2016 the experiment is taking data for physics with a number of dedicated triggers for exotic decays which are ran in parallel with the main trigger. NA62 is currently approved to take data until the end of 2018 when the LS2 of the SPS/LHC starts. Without major modifications of the existing apparatus NA62 can operate in beam-dump mode using one of the beam-defining collimators as a dump. The experiment has a potential to search for weakly-interacting sub-GeV Dark Matter candidates. Such a run is currently being considered for the period after the LS2 in 2021.

References

[1] Hahn F *et al.* 2010 NA62 Document 10-07 `http://cds.cern.ch/record/1404985`
[2] Buras A *et al.* 2015 *JHEP* **11** 033
[3] Artamonov A *et al.* 2008 *Phys. Rev. Lett.* **101** 191802
[4] Alekhin S *et al.* 2016 *Rep. Prog. Phys.* **79** 124201
[5] Massri K 2016 *Proc. Rencontres de Moriond* arXiv:1607.04216 [hep-ex]
[6] Appel R*et al.* 2000 *Phys. Rev. Lett.* **85**
[7] Sher A*et al.* 2005 *Phys. Rev.* D **72**
[8] M Shaposhnikov *et al.* 2005 *Phys. Lett.* B **631**
[9] Aaij R *et al.* 2013 *Phys. Rev. Lett.* **111** 191801
[10] Aaij R *et al.* 2015 *JHEP* **09** 179
[11] Khachatryan V *et al.* 2015 *Phys. Lett.* B **749** 337
[12] Buras A 2016 *JHEP* **1604** 071 arXiv:1601.00005 [hep-ph]
[13] Crivellin A *et al.* 2016 *Phys. Rev.* D **93** 074038
[14] Atre A *et al.* 2009 *JHEP* **0905** 030 arXiv:0901.3589 [hep-ph]
[15] Holdom B 1986 *Phys.Lett.* B **166** 196
[16] Rouven E *et al.* arXiv:1311.0029 [hep-ph] YITP-SB-36
[17] Döbrich B *et al.* 2016 *JHEP* **1602** 018
[18] Spadaro T 2016 *Physics beyond colliders workshop @ CERN (7 Sep 2016)*
[19] Primakoff H *et al.* 1966 *Phys. Rev.* **152**

IOP Conf. Series: Journal of Physics: Conf. Series **800** (2017) 012040 doi:10.1088/1742-6596/800/1/012040

Recent results on exclusive hadronic cross sections measurements at *BaBar*

Roger Barlow

The University of Huddersfield, Queensgate, Huddersfield HD1 3DH, UK

E-mail: R.Barlow@hud.ac.uk

Abstract. The *BaBar* Collaboration has an intensive program studying hadronic cross sections in low-energy e^+e^- annihilations, accessible via initial-state radiation. Our measurements allow significant improvements in the precision of the predicted value of the muon anomalous magnetic moment. These improvements are necessary for shedding light on the current 3 sigma difference between the predicted and the experimental values. We have published results on a number of processes with two to six hadrons in the final state, and other final state are currently under investigation. We report here on the most recent results obtained by analysing the entire *BaBar* dataset, including the $\pi^+\pi^-\pi^0\pi^0$, $\pi^+\pi^-\eta$, $K^0_S K^0_L \pi^0$, $K^0_S K^0_L \pi^0\pi^0$, $K^0_S K^0_L \eta$, and other final states.

1. PEP-II and *BaBar*: more than just a B factory

The *BaBar* experiment collected 0.5 ab^{-1} of data over $\sim$ 10 years, at centre of mass energies on or just below the mass of the $\Upsilon(4S)$ meson. Although the primary objective was the study of B meson properties – which it accomplished very successfully – the accelerator also produced significant numbers of charm quarks and τ leptons, resulting in many *BaBar* τ and charm publications.

However the detector can also be used for the study of light quark pair production at low energies (from threshold up to $\sqrt{s} \sim$ 3GeV) through the process of Initial State Radiation (ISR), in which a Bremsstrahlung photon is emitted from the electron or positron, leading to an e^+e^- collision at reduced energy This is shown in Figure 1, as a Feynman diagram and with a typical event. The clearly-separated photon balances the hadronic system.

The effective centre of mass energy is given by

$$s' = s(1 - 2E_\gamma/\sqrt{s})$$

and is measured on an event by event basis. It is not as precisely known as the full CMS energy s, so some unfolding is needed, but this effect is not large. Event rates are low, but the enormous size of the *BaBar* data sample means that numbers of events are large enough.

Photons can also be produced by electromagnetic radiation from final state particles (FSR), and the two amplitudes can in principle interfere. However the amplitude falls with energy so much that this effect is very small.

The apparatus and its performance have been described previously [1]. In this analysis the Electromagnetic calorimeter, with 6,580 CsI(Tl) crystals, is of paramount importance as it enables the ISR photon to be clearly distinguished and its energy accurately measured. It

Content from this work may be used under the terms of the Creative Commons Attribution 3.0 licence. Any further distribution of this work must maintain attribution to the author(s) and the title of the work, journal citation and DOI.
Published under licence by IOP Publishing Ltd

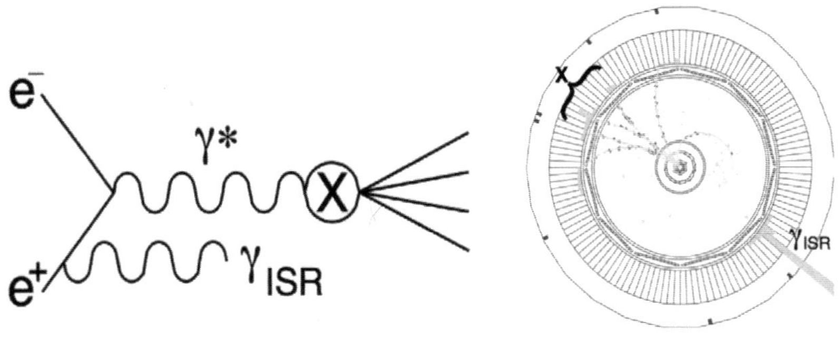

Figure 1. The ISR process: Feynman diagram and an observed event

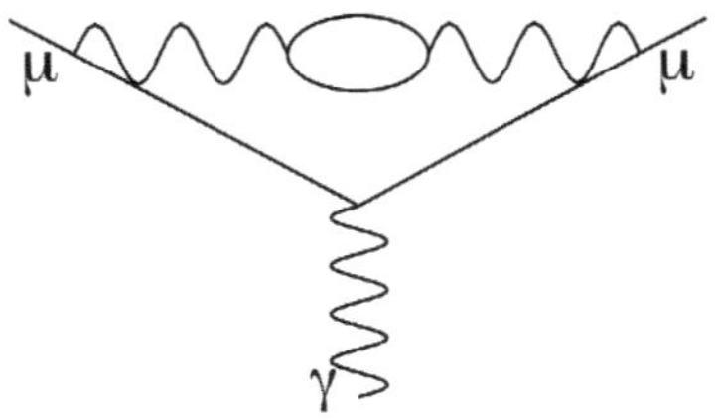

Figure 2. Radiative corrections to $g-2$

also provides excellent reconstruction for π^0 and η mesons. The DIRC ring imaging Cherenkov detector is also significant as it provides superb K/π separation.

2. The $g-2$ puzzle

One motivation for exploring this region of low energy e^+e^- scattering is the reported anomaly in the measurement of the magnetic moment of the muon. The value $a_\mu = \frac{g-2}{2}$ is measured[2] as $(11\ 659\ 209.1 \pm 5.4 \pm 3.3) \times 10^{-10}$ but the Standard Model calculation[3] gives $(11\ 659\ 180.4 \pm 5.1) \times 10^{-10}$. This is a 3.4 standard deviation 'tension' between experiment and the Standard Model. Although such effects have a history of disappearance, it deserves serious investigation and the BNL muon storage ring has been moved to FNAL where measurements will reduce the experimental statistical and systematic errors.

This must be matched by improvement in the theoretical errors. The difference from the Dirac value of 2 is due to higher order corrections to the $\mu - \mu - \gamma$ vertex, including the loop correction shown in Figure 2.

The corrections from QED and weak interactions are calculable, but if the particle in the loop is a quark then the amplitude cannot be calculated perturbatively as the strong coupling constant is large at this low energy. While lattice calculations offer interesting possibilities, in practice this correction needs to be obtained from experimental measurements. This can be done using the optical theorem which links the $\gamma - X - \gamma$ loop correction to the cross section for the reaction $e^+e^- \to \gamma^* \to X$.

One thus needs to measure the value of the ratio $R = \sigma(hadrons)/\sigma(\mu^+\mu^-)$ as a function of

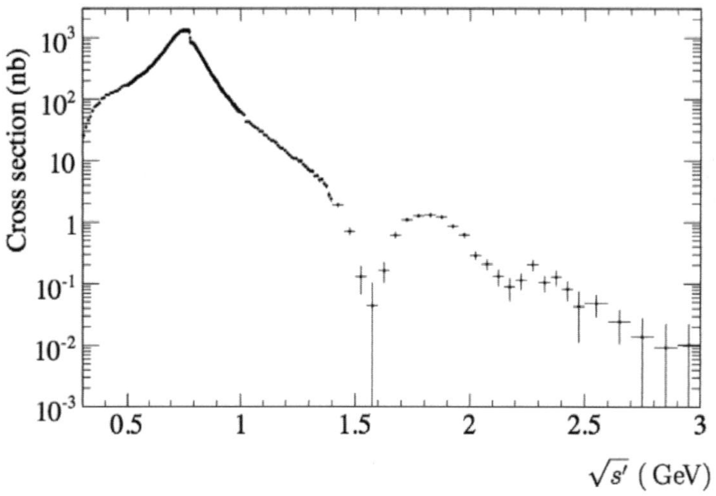

Figure 3. Cross section for the reaction $e^+e^- \to \pi^+\pi^-$ as a function of energy. Taken from [4]

energy, with the range below 2 GeV being of greatest importance. Although this is an inclusive quantity, it has to be measured through the different exclusive channels, which are then summed, as these have very different experimental efficiencies.

3. The reaction $e^+e^- \to \pi^+\pi^-$

The most important contribution to a_μ is the $\pi^+\pi^-$ channel, and results are shown in Figure 3. The spectrum is dominated by the broad $\rho(770)$ resonance, and other resonances are needed to fit the data at high energies.

Use of data driven techniques drives the systemaic uncertainties below the 1% level. For example, the effective luminosity is obtained from measuring $\mu\mu\gamma$ events rather than relying on Monte Carlo simulations (though these have been performed, and are consistent). Efficiencies for particle identification - which uses information from the DIRC and other detectors, and also the results of a kinematic fit - are determined by Monte Carlo data corrected by using clearly identifiable control channels.

This analysis also includes the next order correction $\pi^+\pi^-\gamma\gamma$ events, where the second γ may be visible or it may have been emitted along the beam pipe.

From this we obtain the pion contribution to vacuum polarisation (from threshold up to 1.8 GeV) of $(514.1 \pm 2.2 \pm 3.1) \times 10^{-10}$. The relative statistical error is large above the 1.8 GeV limit, but the absolute error is small, and for these high energies a QCD calculation can be used.

4. The reaction $e^+e^- \to K^+K^-$

A similar analysis has been performed where the two tracks are identified as kaons [5]. The cross section is shown in Figure 4. It is dominated by the narrow ϕ resonance. We have fitted the mass and the width of the peak, and the results agree with standard values [6]. Results are in excellent agreement with previous low energy experiments at CMD-2 [7] and SND [8], but our statistics are significantly higher than theirs. This analysis has also been used to extract the kaon form factor.

Integration gives the K^+K^- contribution to a_μ as $(22.93 \pm 0.18 \pm 0.22 \pm 0.03) \times 10^{-10}$ where the 3rd error is a systematic arising from the parametrisation of the ϕ meson.

IOP Conf. Series: Journal of Physics: Conf. Series **800** (2017) 012040 doi:10.1088/1742-6596/800/1/012040

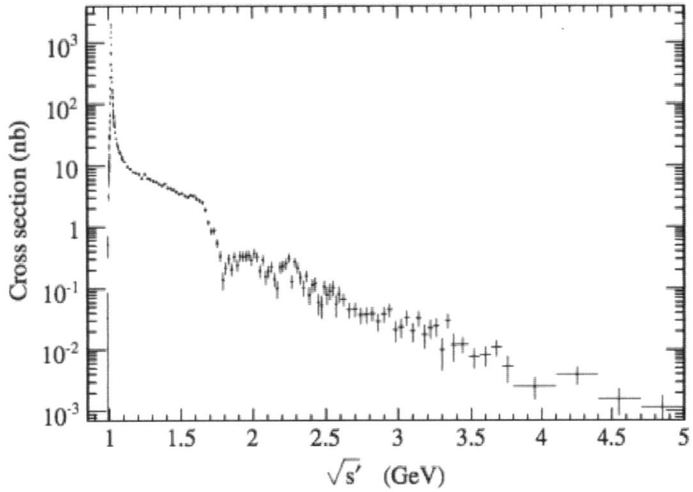

Figure 4. Cross section for the reaction $e^+e^- \to K^+K^-$ as a function of energy. Taken from [5]

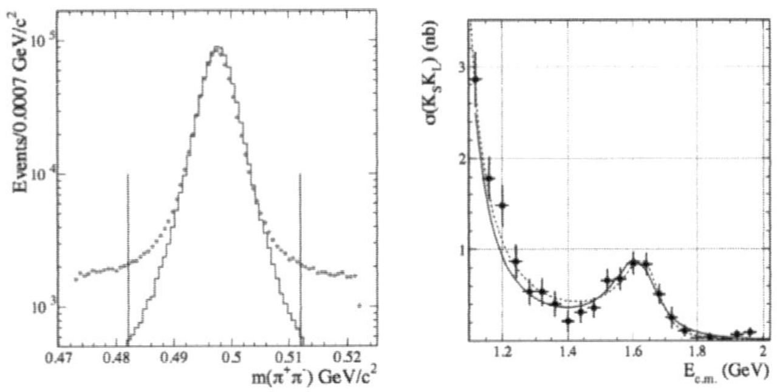

Figure 5. (a) Left: the invariant mass distribution of K_S^0 candidates. (b) Right: The cross section for the reaction $e^+e^- \to K_S^0K_L^0$ as a function of energy. Taken from [9]

5. The reactions $e^+e^- \to K_S^0K_L^0, K_S^0K_L^0\pi^+\pi^-, K_S^0K_S^0\pi^+\pi^-$ and $K_S^0K_S^0K^+K^-$

The analysis has been extended to neutral kaons [9]. K_S^0 particles are easy to identify, from pairs of $\pi^+\pi^-$ tracks forming a distinct vertex. The mass distribution is shown in Figure 5a. Note the scale of the x axis and the logarithmic y axis. K_L^0 particles are harder. They do not decay in the detector but interact to give a cluster in the EM calorimeter, which appears similar to a photon cluster. However the efficiency can be comprehensively studied from $\phi \to K_S^0K_L^0$ decays.

The $K_S^0K_L^0$ reaction is dominated by ϕ, like K^+K^-, and again fitted results agree with standard values. The $\phi'(1680)$ appears as a positive peak, shown in close-up in Figure 5b, whereas for K^+K^- it produces an edge, pointing to differences in the phases of the ϕ and ϕ' in the two channels.

The analysis also gives first measurements of the channels $e^+e^- \to K_S^0K_L^0\pi^+\pi^-$, which is dominated by K^* production with some $\phi\pi\pi$, and of $K_S^0K_S^0\pi^+\pi^-$ and $K_S^0K_S^0K^+K^-$. Full details are in the reference [9].

IOP Conf. Series: Journal of Physics: Conf. Series **800** (2017) 012040 doi:10.1088/1742-6596/800/1/012040

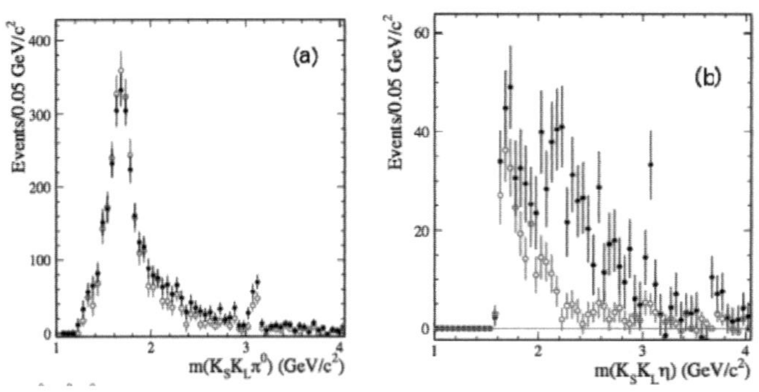

Figure 6. : The mass spectra for (a) Left: $e^+e^- \to K^0_S K^0_L \pi^0$, where the red circles show the $KK^*(892)$ contribution and (b) Right $e^+e^- \to K^0_S K^0_L \eta$, where the red circles show the $\phi\eta$ contribution.. Taken from [10]

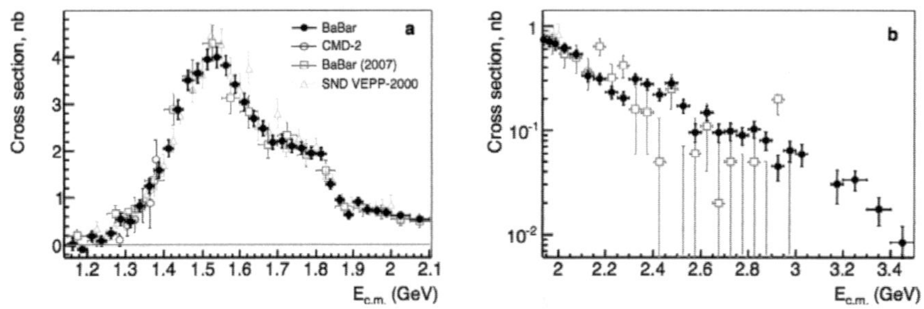

Figure 7. : The cross section for $e^+e^- \to \pi^+\pi^-\eta$ as a function of energy, with a close-up of the high mass region. Taken from [11]

6. The reactions $e^+e^- \to K^0_S K^0_L \pi^0, K^0_S K^0_L \eta$ and $K^0_S K^0_L \pi^0\pi^0$

Recently the analysis has been extended to states with neutral pions and kaons[10]. Some result are shown in Figure 6.

The $K^0_S K^0_L \pi^0$ system is dominated by $KK^*(892)$ (red circles) with a small $\phi\pi^0$ contribution, and a J/ψ signal. The $K^0_S K^0_L \eta$ system is dominated by ϕ (red circles). Again, a small J/ψ peak appears

The measurement of the $K^0_S K^0_L \pi^0\pi^0$ state gives the total contribution of $KK\pi\pi$ to a_μ as $(0.85 \pm 0.05) \times 10^{-10}$ - previously it was $(1.35 \pm 0.39) \times 10^{-10}$. This is an order of magnitude improvement in the error, as the channels are now all measured with no reliance on isospin relations.

7. The reaction $e^+e^- \to \pi^+\pi^-\eta$

The $\pi^+\pi^-\eta$ channel has also been studied [11]. The η is detected through the decay mode $\eta \to \gamma\gamma$ so the event comprises 3 photons and two tracks, and the η peak is very clean, with low backgrounds. The regions $E_{\gamma\gamma}$ above and below are treated separately.

The cross section is shown as a function of energy in Figure 7. Data from previous experiments, and an earlier BABAR result, are shown for comparison. The reults agree, but the precision of the new data is clearly superior.

IOP Conf. Series: Journal of Physics: Conf. Series **800** (2017) 012040 doi:10.1088/1742-6596/800/1/012040

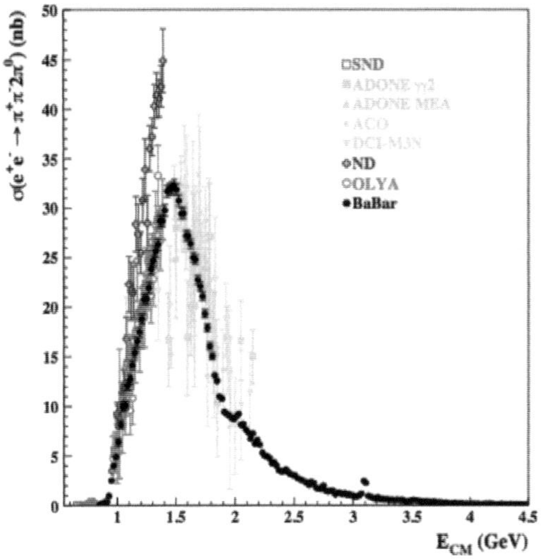

Figure 8. : The cross section for $e^+e^- \to \pi^+\pi^-\pi^0\pi^0$ as a function of energy, together with results from previous experiments. Taken from [12]

The cross section is well described by models including contributions from the $\rho(770)$, $\rho(1450)$ and $\rho(1700)$. There is a peak at 3.1 GeV (removed from the plot) which can be used to measure the branching ration $BR(J/\psi \to \pi^+\pi^-\eta) = (0.042 \pm 0.008)\%$.

8. The reaction $e^+e^- \to \pi^+\pi^-\pi^0\pi^0$

Finally, we have performed an analysis of the four pion $\pi^+\pi^-\pi^0\pi^0$ state[12]. This used the full 431 fb^{-1} sample of e^+e^- collisions, and 150,000 signal events were selected. Results are shown in Figure 8. The superiority of the BaBar data (black) over previous experiments is clear.

Our precision gives a significant improvement to the evaluation of the a_μ theory uncertainty for this channel, which was one of the larger contributors to the the previous total theory error. We now have a contribution to a_μ $(18.1 \pm 0.1 \pm 0.6) \times 10^{-10}$ in the range 0.85 to 1.8 GeV. The uncertainties correspond to a total relative precision of 3.2%. Hence, the relative precision of the final BaBar measurement alone is a factor 2.5 higher than the precision of the world data set without BaBar.

9. Conclusions

Use of the ISR technique with an identified photon has enabled BaBar to do precision studies in low energy e^+e^- annihilation, over a large number of channels.The high luminosity and the particle identification and energy/momentum measurement more than compensate for any drawbacks, and the data quality and quantity surpasses the results of the low energy storage rings.

These measurements have been used to improve the uncertainty on the theoretical calculation of the muon $g-2$ value. This shows that, if the currently observed anomaly persists when new results are obtained from the forthcoming run at FNAL, this will not be explainable by uncertainties in the hadronic vacuum polarisation.

References

[1] B. Aubert et al. (BABAR Collaboration), Nucl. Instrum. Methods Phys. Res., Sect. A **479**, 1 (2002).
[2] G.W. Bennett et al. (Muon g-2 Collaboration), Phys. Rev. D**73**, 072003 (2006)
[3] M. Davier, Nucl. Phys. B Proc. Suppl. **169**, 288 (2007)
[4] J. P. Lees *et al* (BABAR Collaboration), *Precise measurement of the* $e^+e^- \to \pi^+\pi^-\gamma$ *cross section with the initial-state radiation method at BaBar*, Phys. Rev. D **86**, 032013 (2012).
[5] J. P. Lees *et al* (BABAR Collaboration), *Precision measurement of the* $e^+e^- \to K^+K^-\gamma$ *cross section with the initial-state radiation method at BaBar*, Phys. Rev. D **88** 032013 (2013)
[6] C. Patrignani et al. (Particle Data Group), *The Review of Particle Physics*, Chin. Phys. C, **40**, 100001 (2016)
[7] R.R. Akhmetshin et al. (CMD-2 Collaboration), Phys. Lett. B **669**, 217 (2008)
[8] M. N. Achasov et al. (SND Collaboration), Phys. Rev. D **63**, 072002 (2001)
[9] J. P. Lees *et al* , (*BaBar* Collaboration), *Cross sections for the reactions* $e^+e^- \to K^0_S K^0_L$, $K^0_S K^0_L \pi^+\pi^-$, $K^0_S K^0_S \pi^+\pi^0$, *and* $K^0_S K^0_S K^+K^-$ *from events with initial-state radiation*, Phys. Rev. D **89**, 092002 (2014)
[10] J. P. Lees *et al* , (*BaBar* Collaboration), *Cross sections for the reactions* $e^+e^- \to K^0_S K^0_L \pi^0, K^0_S K^0_L \eta$ *and* $K^0_S K^0_L \pi^- 0\pi^0$ *from events with initial-state radiation*, Proc 38th Int. Conf. on High Energy Physics (ICHEP2016), Chicago (2016)
[11] J. P. Lees *et al* , (*BaBar* Collaboration), *Study of the process* $e^+e^- \to \pi^+\pi^-\eta$ *using initial state radiation*, Proc 38th Int. Conf. on High Energy Physics (ICHEP2016), Chicago (2016)
[12] J. P. Lees *et al* , (*BaBar* Collaboration), *Measurement of the energy dependent* $e^+e^- \to \pi^+\pi^-\pi^0\pi^0$ *cross section using initial state radiation*, Proc 38th Int. Conf. on High Energy Physics (ICHEP2016), Chicago (2016)

Upgrade of In-Beam Charged Particle Detector for the KOTO Experiment

I Kamiji[1] **and K Nakagiri**[1]

[1]Department of Physics, Kyoto University, Kitashirakawa Oiwake-cho, Sakyo-ku, Kyoto-shi, Kyoto, Japan

E-mail: `i.kamiji@scphys.kyoto-u.ac.jp`

Abstract. The KOTO experiment aims at measuring the branching ratio of the $K_L \to \pi^0 \nu\bar{\nu}$ decay. Since the decay is highly suppressed in the Standard Model, it is sensitive to new physics beyond the standard model. In the KOTO experiment, two photons from the π^0 in the final state are detected by an electromagnetic calorimeter, and the absence of extra particles except for the escaping neutrinos is ensured by the surrounding veto counters. As one of the veto counters, an in-beam charged particle detector, composed of 3-mm thick plastic scintillators, covered the beam hole of the calorimeter. It was operating in a high-rate environment with a single counting rate of 10 MHz, which caused 10% loss to the $K_L \to \pi^0 \nu\bar{\nu}$ signal. To reduce the hits by neutral particles, the detector has been upgraded to a thin-gap wire chamber. Keeping a high detection efficiency over 99.6% for penetrating charged particles, a 65% reduction of the single counting rate was achieved, corresponding to 40% recovery of acceptance for $K_L \to \pi^0 \nu\bar{\nu}$.

1. Introduction

1.1. KOTO experiment

The KOTO experiment aims at discovering a CP-violating rare decay of the long-lived neutral K meson, $K_L \to \pi^0 \nu\bar{\nu}$. The branching ratio has been calculated as $(3.00 \pm 0.30) \times 10^{-11}$ with a small theoretical uncertainty[1] in the Standard Model. This smallness of the branching ratio makes the decay mode sensitive to the contribution of new physics beyond the Standard Model.

Figure 1 shows a cross-sectional view of the detector system used in the KOTO experiment. In this figure, a K_L comes into the detector system and decays to $\pi^0 \nu\bar{\nu}$. The two photons from π^0 are measured with an electromagnetic calorimeter made of undoped cesium iodide crystals, named CsI, and the other detectors ensure no existence of extra particles other than neutrinos.

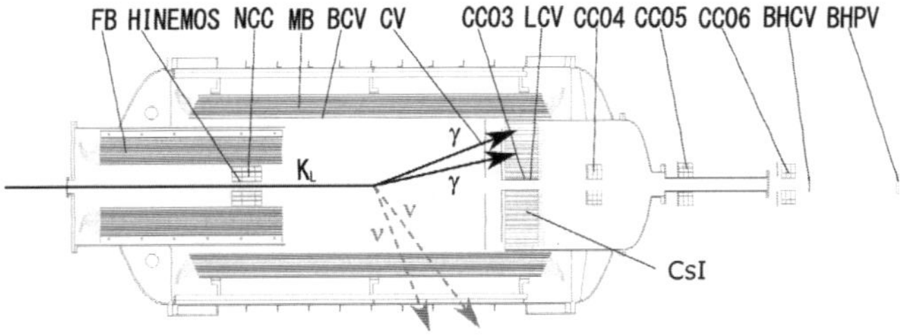

Figure 1. Cross-sectional view of the KOTO detectors.

Content from this work may be used under the terms of the Creative Commons Attribution 3.0 licence. Any further distribution of this work must maintain attribution to the author(s) and the title of the work, journal citation and DOI.
Published under licence by IOP Publishing Ltd

1.2. Upgrade of the in-beam charged particle detector

A charged particle detector, called Beam Hole Charged Veto (BHCV), is located at the downstream of the electromagnetic calorimeter. The role of BHCV is to suppress the $K_L \to \pi^+\pi^-\pi^0$ background where charged pions, π^+ and π^-, escape through the beam hole of the calorimeter as shown in figure 2. Due to the high flux of neutrons and gammas in the K_L beam, its single counting rate reached around 10 MHz when 3-mm-thick plastic scintillators were used as BHCV. Since the measurement of $K_L \to \pi^0\nu\bar{\nu}$ requires no in-time hit in any veto counter, accidental hits in BHCV caused a 10% loss of signal acceptance. To reduce the acceptance loss, BHCV was upgraded to three layers of thin-gap wire chambers, as shown in figure 3, which is more insensitive to neutral particles than the plastic scintillators. Its specification is summarized in table 1.

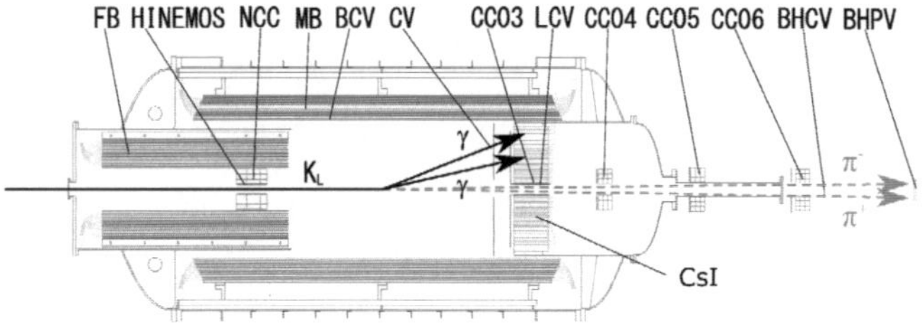

Figure 2. Example of the $K_L \to \pi^+\pi^-\pi^0$ background related to BHCV.

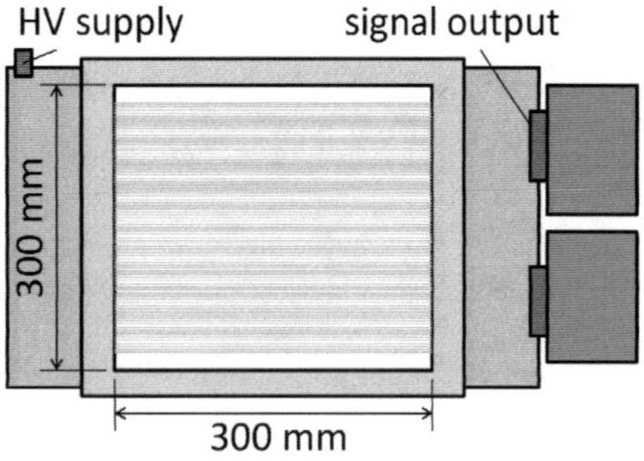

Figure 3. Outline of the wire chamber for BHCV.

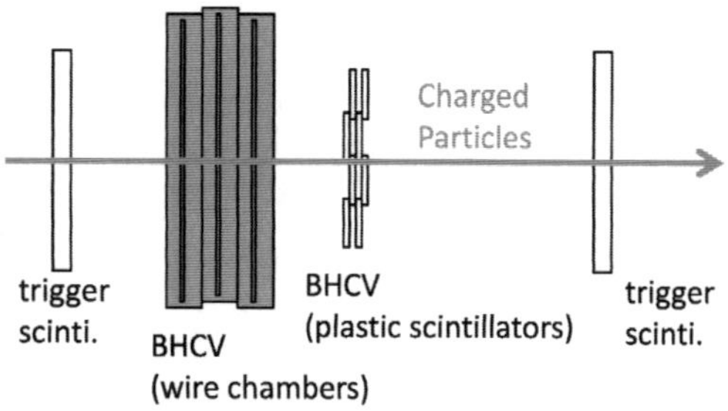

Figure 4. Set up for efficiency measurement. In addition to the plastic sintillators of previous BHCV, two trigger scintillators were used to make coincident trigger.

Table 1. Specification of the wire chamber.

wire diameter/length	50 μm / 30 cm
number of wires	160 (10 wires/channel)
wire-wire spacing	1.8 mm
wire-cathode gap	1.4 mm
cathode plane	graphite-coated kapton film 50-μm thick
gas	CF_4:n-pentane = 55:45
operation voltage	2.7 kV
capacitance per channel	50 pF

2. Study of Detector performance

2.1. Detection efficiency and background suppression

Figure 4 shows a set up of efficiency measurement of the wire chamber for BHCV. To select penetrating charged particles, two additional plastic scintillators were located at the upstream and downstream of BHCV. A distribution of energy deposition in one of the wire chambers is shown in figure 5. The resultant efficiency was over 99.6% in all the three wire chambers. Considering the detector performance of the wire chambers, the background suppression power for $K_L \to \pi^+\pi^-\pi^0$ was estimated by using Geant4 [2] based Monte Carlo simulation. As a result, the background reduction of the wire chambers was the same as the plastic scintillators.

2.2. Single counting rate

The single counting rate was measured in both the plastic scintillators and the wire chambers during physics data taking in 2015. Figure 6 shows the single counting rate with respect to the beam power. The wire chambers had a 65% lower counting rate than that of the plastic scintillators. Although the wire chambers need to use a longer veto timing window of 25 ns to confirm extra particle existence, due to large timing fluctuation coming from drift time of discharged electrons, compared to the 15-ns of the plastic scintillators, the resultant acceptance loss for $K_L \to \pi^0\nu\bar{\nu}$ was reduced by 40%[1].

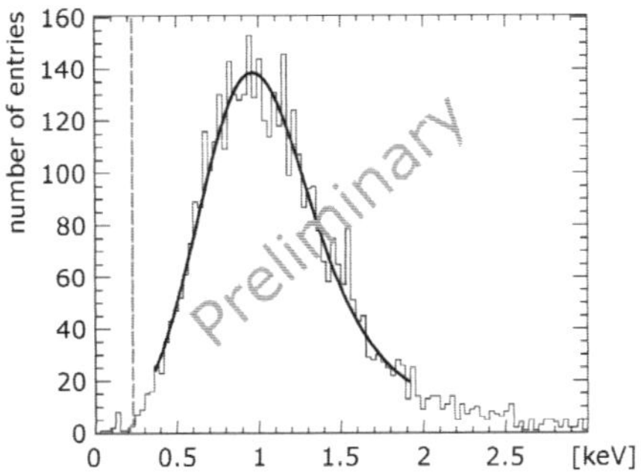

Figure 5. Energy deposit in a single wire chamber. The red dashed line shows a quarter of the peak used as threshold for efficiency calculation.

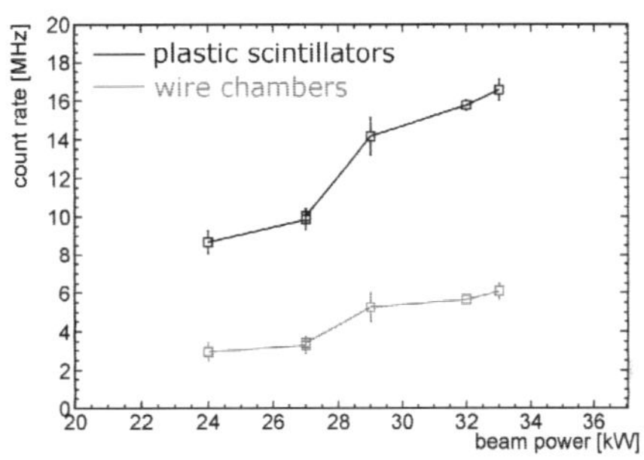

Figure 6. Single counting rate of BHCV with respect to the beam power.

3. Conclusion

In the KOTO experiment, an in-beam charged particle veto, named as BHCV, was upgraded from the plastic scintillators to the three layers of thin-gap wire chambers to suppress the acceptance loss for $K_L \to \pi^0\nu\bar{\nu}$ due to high rate accidental hits by neutral particles. Its single counting rate was reduced by 65% and the resultant acceptance loss of $K_L \to \pi^0\nu\bar{\nu}$ became 40% lower without changing its background reduction ability owing to the high detection efficiency over 99.6% in each single wire chamber.

References

[1] A J Buras, D Buttazzo, J Girrbach-Noe and R Knegjens 2015 *J. High Energy Phys.* **033** 1511.
[2] S AgostinelliA *et al* 2003 *Nucl. Instr. Meth. Phys. Res. A* **506** 250-303.

[1] Acceptance loss due to accidental hits in a veto counter can be calculated by multiplying its counting rate and timing window.

Upgrades of the Data Acquisition System for the KOTO Experiment

Stephanie Su, Jon Ameel, Lucas Beaufore, Brian Beckford, Jacqueline Beechert, Mircea Bogdan, Khalil Bryant, Myron Campbell, Melissa Hutcheson, Roman Lee, Noah McNeal, Jessica Micallef, Joshua Robinson, Christopher Rymph, Hanna Schamis, Yasuyuki Sugiyama, Yasuhisa Tajima, Molly Taylor, Monica Tecchio and Yau Wah

Department of Physics, University of Michigan, Ann Arbor, MI 48109

E-mail: stephsu@umich.edu

Abstract. The KOTO data acquisition system (DAQ) collects detector PMT waveform signals and saves digitized events to permanent storage using frontend ADC modules, two levels of hardware triggers, and a computing farm. The KOTO DAQ system ran stably in 2013 with 24 kW beam power. To maintain high DAQ livetime with increasing beam power, we implemented lossless data compression inside the ADC modules and developed a new L3 computing farm. The upgraded KOTO DAQ system was able to maintain livetime above 80% with 42 kW beam power during the 2015 and 2016 runs. To sustain high DAQ livetime for data taking with beam power of 50 kW and above, an upgrade of our hardware trigger is proposed.

1. Introduction to the KOTO Experiment

The KOTO experiment is located at the J-PARC facility in Tōkai-mura, Ibaraki, Japan. The goal is to measure the branching ratio of the direct CP-violating decay $K_L \to \pi^0 \nu\bar{\nu}$, whose branching ratio is calculated to be $(3.00 \pm 0.30) \times 10^{-11}$ [1] in the Standard Model (SM). The high precision theoretical calculation for the branching ratio enhances the sensitivity towards probing for new physics beyond the SM. The KOTO experiment receives a secondary $\mathrm{K_L}$ neutral beam generated by a 30 GeV primary proton beam hitting on the gold target. The proton beam is slowly extracted in 2 seconds, with 6 seconds repetition. The decay of $K_L \to \pi^0 \nu\bar{\nu}$ is measured by observing the two photons generated by the single π^0 in the calorimeter and nothing else in the other detectors. The KOTO detectors consist of a Cesium Iodide (CsI) calorimeter surrounded by veto detectors sensitive to photons, neutrons, and charged particles.

2. The DAQ System

2.1. The Current System

We built a DAQ system optimized to retain $K_L \to \pi^0 \nu\bar{\nu}$ decays and filter out other $\mathrm{K_L}$ decays which are many orders of magnitude more frequent. The DAQ consists of the ADC modules, two levels of hardware triggers (L1 trigger and L2 trigger), as well as a computing farm (L3 computing farm). The L1 trigger accepts events by requiring a minimum energy deposit in the CsI calorimeter and no energy deposit in the veto detectors. The L2 trigger filters the signal

Content from this work may be used under the terms of the Creative Commons Attribution 3.0 licence. Any further distribution of this work must maintain attribution to the author(s) and the title of the work, journal citation and DOI.
Published under licence by IOP Publishing Ltd

from other neutral decay events by calculating the event Center-of-Energy (CoE), which is a position-weighted energy deposition in the calorimeter[2]. The L3 computing farm manages the event building. An online clustering algorithm is implemented inside the L3 software for potential future online filter usage.

2.2. The ADC Lossless Data Compression

Two types of ADC modules with 125 MHz and 500 MHz sampling rate are used in the DAQ system to accommodate different detector resolutions. The ADC modules shape the analog PMT waveforms into Gaussian waveforms using a 10-pole Bessel function filter, and digitize the waveforms with 8 ns sampling rate. Each waveform has a timing window of 512 ns. A 125 MHz ADC module digitized a detector waveform into 64 samplings and a 500 MHz ADC module digitized that into 256 samplings. A lossless data compression algorithm is used in the ADC firmware to reduce data size and increase the amount of data we can collect with our trigger system. The performance of the lossless data compression effectively increased the DAQ livetime as the beam power and the trigger rate increase, as shown in Fig. 1 below.

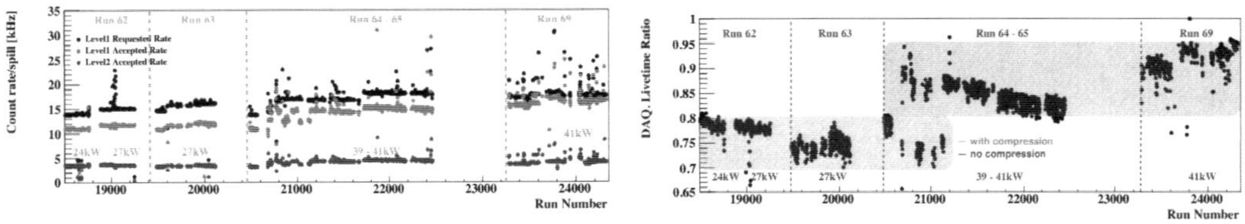

Figure 1. DAQ trigger rate and perfromance in 2015-16 experiment runs.
Left plot: DAQ trigger rates. The trigger rates increase as the beam power increases. The Level 1 (L1) request rate (black points) is the number of triggers generated by the L1 trigger. The L1 accept rate (red points) is the number of triggers being considered for the process in the L2 trigger. L2 accept rate (blue points) is the number of triggers passed the L2 trigger filter.
Right plot: DAQ livetime. The implementation of lossless data compression effectively increased the DAQ livetime, which is defined to be the ratio of L1 accept over L1 request.

2.3. The L3 Computing Farm

We upgraded the L3 computing farm in 2015. The new L3 computing farm consists of 2 head nodes and 47 computer nodes. A full-meshed computer network is implemented using Infiniband hardware with the Message Passing Interface protocol. We divided the computer nodes for event building into two groups - Type 1 and Type 2. Type 1 computer nodes receive the data from the L2 trigger through the Ethernet connection. Based on the spill information and event ID, Type 1 nodes send the data packet to Type 2 nodes via the Infiniband switch. Events in the Type 2 nodes are decompressed and compressed again after rearranging the data packet structure in a fashion more convenient for online event selection and subsequent offline data analysis.

2.4. L2 RPT hardware upgrade

Further increase in the beam power will lower the DAQ livetime ratio due to insufficient hardware resources. We decided to replace the L2 trigger hardware with the Reconfigurable Clustering Element (RCE) Platform Technology (RPT) developed by SLAC. This upgrade is expected to have the capability to process data up to minimum of 100 kW without the lossless compression inside the ADC modules.

The RPT hardware is shown in Fig. 2. It consists of an ATCA blade, called Cluster-On-Board (COB), and Rear Transmission Module (RTM) used to receive data from the detector. Each COB includes two flavors of mezzanine boards: four Data Processing Modules (DPM), each composed of two RCEs with Zynq-7045 FPGA, and one Data Transmission Module (DTM), containing one Zynq-7030 RCE. The RCE comes with both a Real-Time Executive for Multiprocessor Systems (RTEMS) and an ARM Linux operating system loaded on a micro SD card. The communication between each RCE and between each COB has to go through the Clustering Interconnect (CI) Ethernet switch on the COB. A replicated-mesh backplane will be used to assist the full communication between every RCE from different COBs.

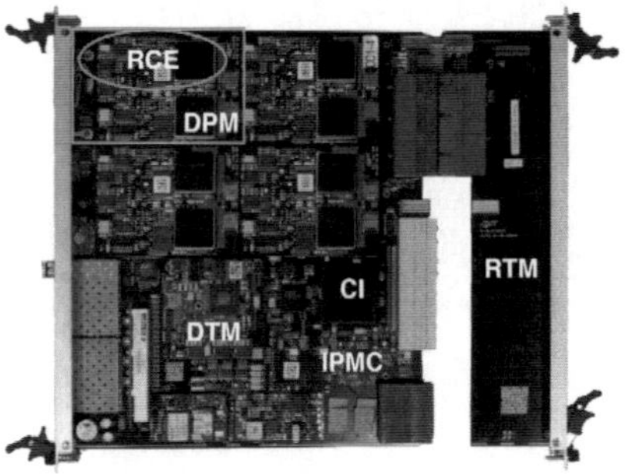

Figure 2. A Cluster-On-Board (COB) and a Rear Transmission Module (RTM). The Clustering Interconnect (CI) Ethernet switch handles all communication traffic between each RCE and is connected to replicated-mesh ATCA backplane. Multiple 10 Gbps links are available on the COB for these communications.

We plan to implement event building inside the new L2 trigger. Having a full event in the trigger system, we will be able to enhance the quality and the flexibility of the L2 trigger rejection. We will use a custom designed RTM to receive data from the ADC modules, as well as to transmit the processed data to the L3 computing farm for further online analysis, filtering, and storage. Each RCE on the DPM will receive event fragments from the ADC modules via the RTM. Each RCE will read the event ID information and via the CI redirect the event fragment to a single RCE where event building will take place. A flexible set L2 trigger filters, such as CoE decision or calorimetric cluster counting, will be implemented to the reject unwanted events. A 1 GB DDR3 memory is available on each RCE for temporary storage during event building and L2 filtering.

3. Summary

The lossless data compression inside the ADC modules increased the DAQ livetime as the beam power increased to 42 kW. To maintain the current DAQ system performance for beam power greater than 50 kW, we are developing an upgrade for the L2 trigger system.

4. Acknowledgments

This material is based upon work supported by the U.S. Department of Energy, Office of Science, Office of High Energy Physics, under Award Numbers DE-SC0007859 and No. DE-SC0009798, and the Science Graduate Student Research Fellowship (SCGSR 2015). The research is supported by High Energy Accelerator Research Organization (KEK), the Ministry of Education, Culture, Sports, Science, and Technology (MEXT) KAKENHI No. 18071006, and the Japan Society for the Promotion of Science (JSPS) KAKENHI No. 23224007.

References

[1] Buras A J, Buttazzo D and Knegjens R 2015 *J. High Energ. Phys.* **2015** 166
[2] Su S, *et al.* 2014 PoS(TIPP2014)203

IOP Conf. Series: Journal of Physics: Conf. Series **800** (2017) 012043 doi:10.1088/1742-6596/800/1/012043

A new cylindrical photon-veto detector for the KOTO experiment

Manabu Togawa for the KOTO collaboration
Department of Physics, Osaka University, Toyonaka 560-0043, Osaka, Japan
E-mail: togawa@champ.hep.sci.osaka-u.ac.jp

Abstract. The KOTO experiment at the J-PARC laboratory seeks to obtain the first observation of the $K_L \to \pi^0\nu\bar{\nu}$ decay. The branching ratio is calculated in the Standard Model (SM) to be 3×10^{-11}. This is a good probe to explore new physics beyond the SM with small theoretical uncertainty of ~2 %. We installed a new barrel-shaped photon veto detector, named the Inner Barrel, inside the KOTO detector to improve detection efficiency of photons from the major background of $K_L \to 2\pi^0$ decay. The Inner Barrel detector is a sampling calorimeter, consisting of 25 layers of 5-mm-thick scintillators and 24 layers of 1-mm-thick lead plates, corresponding to 5 radiation lengths. The volume is 3 m long along the beam direction, and inner and outer diameters are 1.5 m and 1.9 m, respectively. Scintillation light is read out by a photomultiplier at both ends via wavelength shifting fibers. The Inner Barrel was installed in April, 2016 and the performance was demonstrated at the beam time in June. In this paper, the detector design, construction and performance evaluated with the neutral beam are presented.

1. Introduction

The very rare kaon decay $K_L^0 \to \pi^0\nu\bar{\nu}$ is a sensitive probe for direct CP violation in the quark sector. The decay is a Flavor Changing Neutral Current process and is induced through electroweak loop diagrams. The branching ratio is predicted to be 3×10^{-11} in the Standard Model [1], and the theoretical uncertainty is estimated to be only a few percent. The decay is also sensitive to new physics scenarios beyond the Standard Model such as Supersymmetric theories.

The latest experimental result was obtained from the E391a experiment at KEK 12 GeV PS, and the upper limit on the branching ratio was 2.6×10^{-8} (90% *C.L.*)[2]. The goal of the KOTO (K0 at TOkai) experiment is to study $K_L^0 \to \pi^0\nu\bar{\nu}$ at the Standard Model sensitivity [3].

2. An Inner Barrel detector

A schematic cross-sectional view of the KOTO detector is shown in Fig. 1. The KOTO detector subsystems are categorized into two parts. One is the CsI calorimeter, which detects two photons from π^0 decay, and the other is a set of hermetic veto counters, which demand no energy deposit in the $K_L^0 \to \pi^0\nu\bar{\nu}$ events.

One of the main backgrounds is the $K_L^0 \to 2\pi^0$ decay, also shown in Fig. 1; two photons are detected in the CsI calorimeter while other two photons are undetected due to the inefficiency of the veto detectors. The Main Barrel (MB) detector which has large coverage in the transverse region, has only 13.5 X_0. A new detector, the Inner Barrel (IB), was motivated to decrease

Content from this work may be used under the terms of the Creative Commons Attribution 3.0 licence. Any further distribution of this work must maintain attribution to the author(s) and the title of the work, journal citation and DOI.
Published under licence by IOP Publishing Ltd

the inefficiency by adding 5 X_0 another inside the MB. According to the Monte-Carlo (MC) estimation, the amount of $K_L^0 \to 2\pi^0$ will be suppressed by a factor of three

The IB detector is a sampling calorimeter as shown in Fig. 2. It consists of 25 layers of 5-mm-thick scintillators and 24 layers of 1-mm-thick lead plates, corresponding to 5 X_0. The 32 modules were made in a trapezoidal shape and formed as a cylindrical detector. The volume is 3 m long along the beam direction, and inner and outer diameters are 1.5 m and 1.9 m, respectively. Scintillation light is read out by a photomultiplier (Hamamatsu R329-EGP or R7724-100) at both ends via Wave Length Shifting (WLS) Fibers (BCF92).

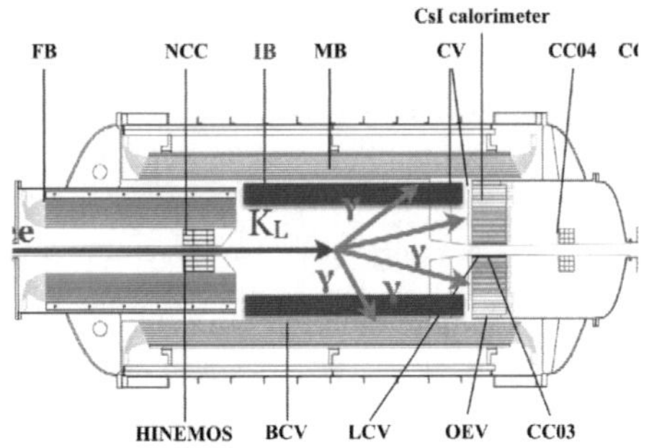

Figure 1. Schematic cross-sectional view of the KOTO detector. The main background event, $K_L \to 2\pi^0$, is also displayed. The new detector is shown as blue color.

Figure 2. Top Left) The WLS fibers are attached in the scintillators. Bottom Left) One module consists of 25 layers of 5 mm thick scintillators and 24 layers of 1 mm lead plates. Right) Formation as cylindrical shape.

3. Module production, construction and insertion to the existing KOTO detector

First, we attached WLS fibers to all 800 scintillators with UV adhesive. After the fibers were glued, we found some cracks in the scintillators caused by uncured adhesive behind fibers. We reproduced new scintillators with fibers for those who have large cracks and also annealed to other scintillators at 80°C for 3 hours to increase chemical resistance based on the result of damage test.

In 2015, we started to make modules as shown in Fig. 3. To bundle the module, we used 0.75 mm-thick stainless band in 9 points. The accuracy of the module production was determined to less than 1 mm. The modules were supported by 8 rings as shown in Fig. 4. All the production and construction processes were made in KEK. The detector was delivered to J-PARC, and then installed in April 2016. To insert the IB in the MB, the IB detector was pulled on the teflon plates attached to the MB and the support rings.

4. Performance check

After installation, the performance of the IB detector was evaluated with the data. Figure 5 shows the timing resolution evaluated with cosmic-rays passing through the MB and IB detectors. We obtained the timing resolutions by comparing relative hit timings between the MB and the IB. The results were almost consistent with the expected values considering the light yield, the decay time of WLS fibers and readout modules.

In May-June 2016, the first physics run with the IB detector was performed. To check the veto response of the IB, we studied events which had four photons in the CsI calorimeter requiring no

IOP Conf. Series: Journal of Physics: Conf. Series **800** (2017) 012043 doi:10.1088/1742-6596/800/1/012043

energy deposit in the MB and the IB. As shown in Fig. 6, we can observe the 498 MeV/c^2 peak of $K_L \to 2\pi^0$ and background contribution mainly from $K_L \to 3\pi^0$. Background contribution of $K_L \to 3\pi^0$ estimated by MC is also plotted and it well reproduces the data.

Figure 3. Module construction.

Figure 4. IB support rings.

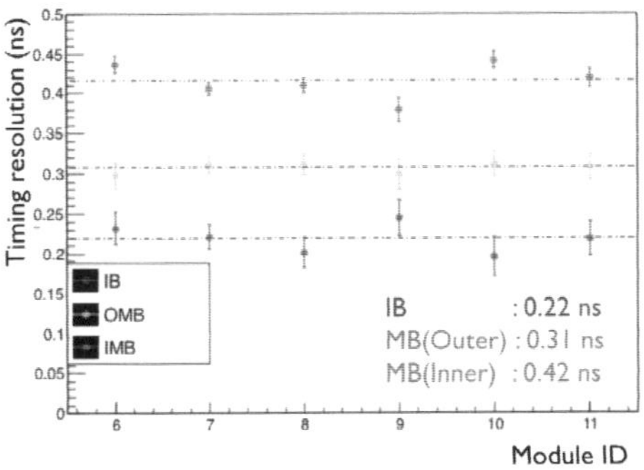

Figure 5. Timing resolution as function of module IDs evaluated with cosmic-rays for several IB (blue) modules. The timing resolutions of the MB inner (red) and outer (green) parts are also plotted.

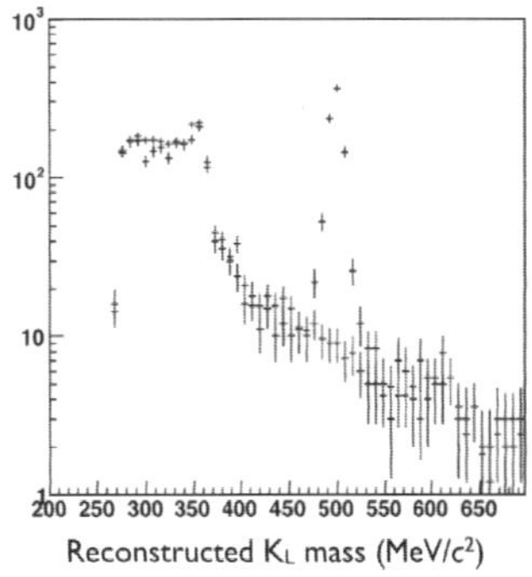

Figure 6. Reconstructed K_L mass using 4 photons in the CsI detector. Selection of no hits in the IB and MB detectors were applied. Black points show 2016 run data, and blue points show $K_L \to 3\pi^0$ MC.

5. Summary

We installed a new photon-veto detector to the KOTO experiment in April 2016. The IB detector is expected to suppress the main background from $K_L^0 \to 2\pi^0$ to be one third. The performances evaluated with the cosmic-ray and the neutral beam data are consistent with expectations and further study is on going.

References

[1] A. J. Buras, D. Buttazzo, J. Girrbach-Noe, R. Knegjens *J. High Energy Phys.* **11**, 033 (2015).
[2] J. K. Ahn *et al.*, *Phys. Rev. D* **81**, 072004 (2010).
[3] J. Comfort *et al. J-PARC E14 Collaboration, Proposal for* $K_L^0 \to \pi^0\nu\bar{\nu}$ *Experiment at J-PARC.* (Available at: http://j-parc.jp/researcher/Hadron/en/pac_0606/pdf/p14-Yamanaka.pdf) (2006).

Beam-edge Photon Detector with Low Sensitivity to Neutrons for the KOTO Experiment

Satoshi Shinohara

Department of Physics, Faculty of Science, Kyoto University Kitashirakawa Oiwake-cho, Sakyo-ku, Kyoto, Japan

E-mail: `shinohara@scphys.kyoto-u.ac.jp`

Abstract. An acrylic Cherenkov counter named Beam Hole Guard Counter (BHGC) was developed as a photon-veto detector to cover the detection gap in the edge of the in-beam photon counter just outside of the beam core in the KOTO experiment. The KOTO experiment took physics data with the BHGC in 2015. We confirmed the stable operation and the expected performance of the BHGC. Therefore, the gap induced background can be rejected with the BHGC.

1. Introduction

The KOTO experiment at the J-PARC aims to observe the CP-violating rare decay $K_L \to \pi^0 \nu\bar{\nu}$. The branching ratio of this decay mode is suppressed in the Standard Model (SM) (3.0×10^{-11}) and its theoretical uncertainty is very small [1]. This decay mode is thus sensitive to new physics beyond the SM.

We identify the signal $K_L \to \pi^0 \nu\bar{\nu}$ decay by detecting a single π^0 reconstructed from two photons, and requiring that there are no observable particles other than the two photons. In order to ensure it, we need a hermetic veto system with high detection efficiency, as shown in Figure 1.

For the detection of particles escaping along the beam, an in-beam photon counter was installed in the downstream of the KOTO detectors, as shown Figure 2.

We found, however, that there is a detection gap in the edge of the in-beam photon counter due to shorter path length. This gap makes 1.9 background events from $K_L \to 2\pi^0$ decays at the SM level sensitivity. For reducing them we designed and made a new photon-veto detector named Beam Hole Guard Counter (BHGC) to cover the gap. Since the neutron rate is high around the beam core, the BHGC needs a high detection efficiency for photons and a low detection efficiency for neutrons.

2. Beam Hole Guard Counter (BHGC)

The BHGC consists of a lead plate to convert photons, and an acrylic plate which acts as Cherenkov radiator. The BHGC has two steps of thresholds to be insensitive to neutrons. The first step is Cherenkov threshold. Charged particles generated from neutrons tend to be slow and cannot make Cherenkov radiation when the velocity is below the Cherenkov threshold. The second step is total reflection condition threshold. The acrylic plate acts as a light guide to photomultiplier tubes (PMTs) attached at the both ends of the acrylic plate. If the speed of

Content from this work may be used under the terms of the Creative Commons Attribution 3.0 licence. Any further distribution of this work must maintain attribution to the author(s) and the title of the work, journal citation and DOI.
Published under licence by IOP Publishing Ltd

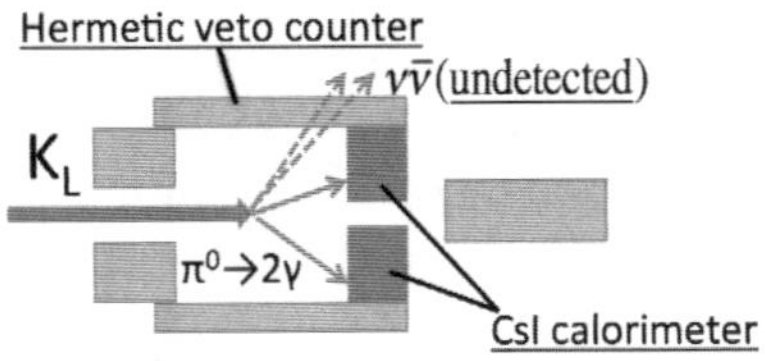

Figure 1. A schematic view of the KOTO detector system and a typical $K_L \to \pi^0 \nu \bar{\nu}$.

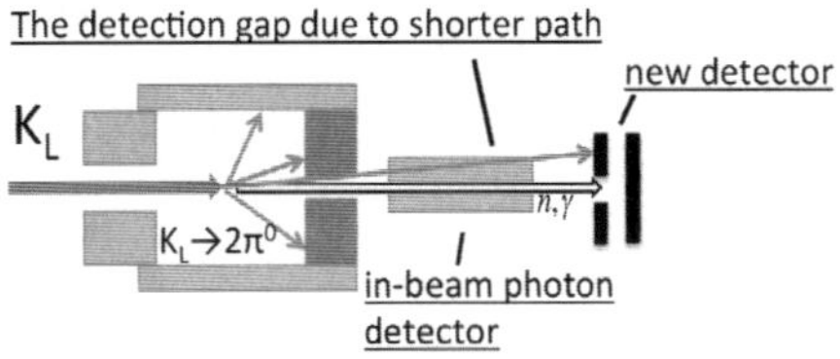

Figure 2. A typical case of $K_L \to 2\pi^0$ background.

the charged particles generated from neutrons exceed the Cherenkov threshold, the Cherenkov radiation is emitted, but the Cherenkov angle tends to be smaller. If the angle is smaller than the total reflection threshold in the acrylic plate, such photons cannot be transported to the edge of the acrylic plate (Figure 3).

3. Physics run with BHGC in 2015

3.1. *Design of BHGC*

The thickness of the lead plate and the position of the BHGC have been optimized. Based on studies with MC simulations, the lead plate thickness was decided to be 9.6 mm, and the module position was decided to be 130 mm away from the beam axis. With this design, 90% of the background events caused by 1 GeV photons going through the detection gap will be rejected.

After the design and construction, the BHGC has been installed in the KOTO detector system in March 2015 (Figure 4).

3.2. *The performance of the BHGC*

We checked and monitored the light yield in each BHGC module with high momentum charged particles. The charged particles were selected by closing beam plug located upstream of the KOTO detectors to stop neutrons and kaons, and by requiring a coincidence with an in-beam photon detector which detects only fast charged particles (β >0.98). As shown in Figure 5, the BHGC light yield was stable within a few percent. This means that there was no radiation damage.

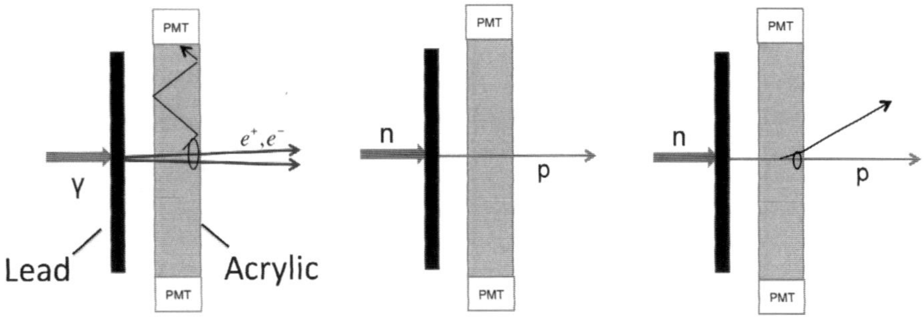

Figure 3. Principle of the BHGC. Left: γ hits BHGC. Middle and right: neutron hits on BHGC. A proton is generated and goes to an acrylic plate normally. Cherenkov threshold for $\beta > 0.67$ (middle), total reflection threshold for $\beta > 0.89$ (right).

The single counting rate agreed with the rate expected from beam photons or K_L decays or neutrons. This agreement shows that the BHGC has low sensitivity to neutrons. We also confirmed that the BHGC response was well reproduced with MC simulations by selecting photon from K_L decays. We finally compared reconstructed π^0 transverse momentum P_t and longitudinal position Z with a MC simulation. The signal (blind) region has been set in terms of those variables, as shown in Figure 6. A MC simulation had one $K_L \to 2\pi^0$ decay event inside the blind region without BHGC veto for the Single Event Sensitivity (S.E.S) of 1.8×10^{-10}. We confirmed that the BHGC rejected this one event inside the blind region.

Figure 4. BHGC viewed from downstream.

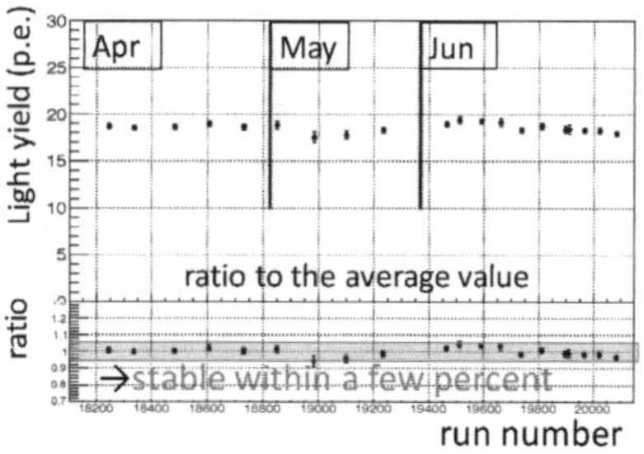

Figure 5. BHGC light yield stability as a function of the run. Top: number pf photoelectrons (p.e.). Bottom: ratio to the average value.

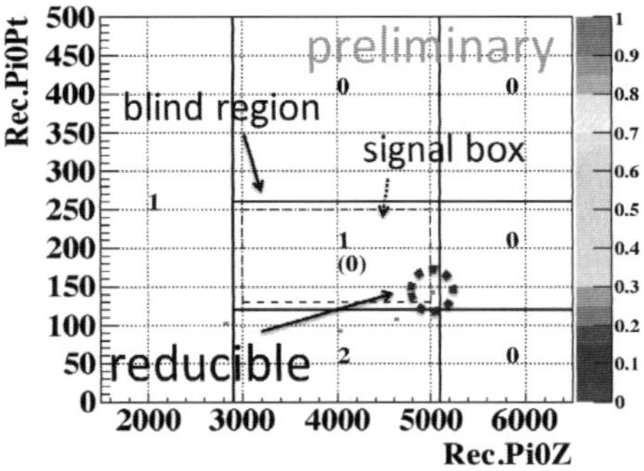

Figure 6. $K_L \to 2\pi^0$ backgrounds in a plane of the reconstructed π^0 P_t and Z position, estimated by a MC simulation.

4. Summary

The BHGC is an acrylic Cherenkov counter developed to cover the detection gap near the beam. Based on the stable operation and the expected performance, 90% of the background events caused by 1 GeV photons passing through the detection gap can be rejected with the BHGC.

5. References

[1] Buras A J, Buttazzo D, Girrbach-Noe J and Knegjens R 2015 *J. High Energy Phys.* **1511** 033.

The Run Control system of the NA62 experiment

Nicolas Lurkin[1,2]

[1]School of Physics and Astronomy, University of Birmingham, Birmingham B15 2TT, UK

E-mail: ncl@hep.ph.bham.ac.uk

Abstract. The NA62 experiment at the CERN SPS aims at measuring the ultra-rare decay $K^+ \to \pi^+\nu\bar{\nu}$ with 10 % accuracy. This can be achieved by detecting about 100 Standard Model events with 10% background in 2–3 years of data taking. The experiment consists of a large number of subsystems dedicated to the detection of the incoming kaon and outgoing pion and also focusing on particle identification and vetoing capabilities. Run Control has been designed to link their trigger and data acquisition systems in a single central application easily controllable by educated but non-expert operators. The application has been continuously evolving over time, integrating new equipments and taking into account requirements and feedback from operation. Future development includes a more automatized system integrating the knowledge acquired during the operation of the experiment.

1. The NA62 experiment

The NA62 experiment at the CERN SPS aims to collect about 100 standard model (SM) events of the ultra-rare kaon decay $K^+ \to \pi^+\nu\bar{\nu}$ with less than 10% background in 2–3 years of data taking. This process is forbidden at tree level and can occur only through a Flavour Changing Neutral Current loop. It is theoretically very clean and the prediction of the branching ratio has been computed within the SM [1] to an exceptionally high degree of precision: $\mathcal{B}(K^+ \to \pi^+\nu\bar{\nu})_{th} = (8.4 \pm 1.0)10^{-11}$. This channel is therefore an excellent probe of the Standard Model and any deviation from this value would indicate a new physics process. A total of 7 events were observed by the BNL E787/E949 experiments [2] including an estimated background of (2.6 ± 0.4) events. The measured branching ratio $\mathcal{B}(K^+ \to \pi^+\nu\bar{\nu})_{exp} = (1.73^{+1.15}_{-1.05}) \times 10^{-10}$ has an uncertainty by far too large to conclude on a possible deviation from the SM predictions known at $\sim$ 10% precision.

To achieve its ambitious goal, NA62 needs to accumulate $\mathcal{O}(10^{13})$ K^+ decays using a secondary 75 GeV/c momentum unseparated hadron beam containing only ~6% K^+. Many different sub-detectors have been developed to identify the signal process and fight the background from other kaon decays and beam accidental coincidences. The current status and achieved sensitivity of the experiment is reported in [3].

2. The NA62 Run Control

Run Control is an application that centrally controls and monitors all equipment involved in the trigger and data acquisition (TDAQ) system. Its purpose is to allow a non-expert operator

[2] On behalf of the NA62 collaboration: Birmingham, Bratislava, Bristol, Bucharest, CERN, Dubna, Fairfax, Ferrara, Florence, Frascati, Glasgow, Liverpool, Louvain, Mainz, Merced, Moscow, Naples, Perugia, Pisa, Prague, Protvino, Rome I, Rome II, San Luis Potos, Sofia, TRIUMF, Turin, Vancouver.

Content from this work may be used under the terms of the Creative Commons Attribution 3.0 licence. Any further distribution of this work must maintain attribution to the author(s) and the title of the work, journal citation and DOI.
Published under licence by IOP Publishing Ltd

to supervise the data taking easily while still giving specialists the possibility to achieve a high level of control of their own equipment.

Technologies and architecture: Operation of the experiment involves other control systems dedicated to hardware equipment (Dectector Control System DCS, Gas System, Cryogenics). A coherent and natural choice was the industrial software "WinCC Open Architecture". This software is the central part of two frameworks - JCOP [4] and UNICOS [5]- already widely used at CERN in the LHC experiments. Some of the components provided by these developer toolkits are of special relevance for NA62:

- DIM (Distributed Interface Management) is the communication layer between Run Control and the numerous equipment distributed across the experiment [6];
- The FSM toolkit manages the finite state machines with SMI++ processes [7] according to their definition (states, transition rules, actions);
- The configuration database tool defines and/or applies sets of parameters from a database;
- The Farm Monitoring and Control manages the PC farm which receives information from individual subdetectors and builds the complete events.

The NA62 TDAQ system is composed of many devices, most of them operating in a specific way. This complication is overcome by internally using a Finite State Machine (FSM) model of each device, therefore presenting a simple and common control to the operator. Little equipment-specific knowledge is integrated in Run Control, while a common generic-device interface is created to transmit commands and information. Simple basic instructions are sent through this channel, and are received by the device control software that is responsible for executing the proper sequence of actions on the hardware. More complex configurations rely on a scheme of flexible XML files where the details of the configuration are again handled by the receiver. Multiple sets of configurations are stored in an Oracle database and can be quickly loaded and distributed to the relevant equipment when needed. This architecture allows Run Control and the controlled systems to evolve independently from each other while keeping a permanent compatibility.

The software is implemented as a hierarchical tree of FSM where the state of each node is either defined by a set of rules summarizing the state of the child-nodes or by the evaluation of logical expressions involving parameters transmitted by the devices.

The external nodes of the tree implement the FSM models representing the specific hardware or software devices (boards, computer, crates and control softwares). Their states are evaluated according to the value of parameters received from the corresponding piece of equipment. The internal nodes represent logical subsystems (subdetectors, PC farm), summarizing the states of their own child-nodes following a set of rules. Finally the top node further aggregates the state of the subsystems to represent the global state of the experiment.

A change of state is always propagated upwards from the device where this change originates to the top node. Conversely, a command can be issued at any node level and is always propagated downwards to all the child-nodes until it reaches a device node (Figure 1). At this point, the command is generated and transmitted to the standardized interface through the network using the DIM protocol.

Infrastructure Run Control is a distributed system spread across different machines in the experiment. The core of the system is located on a dedicated WinCC OA data server on the technical network. A machine bridging the networks hosts the DIM managers which are responsible for loosely binding all devices in the experiment and Run Control. The experiment DAQ continues to run correctly standalone in case of a connectivity problem or if this node is

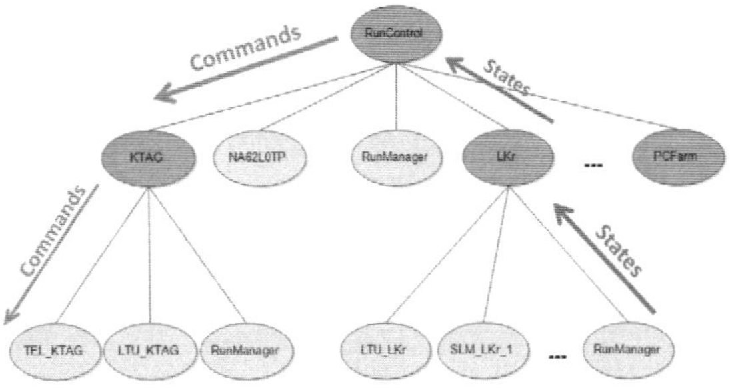

Figure 1. Tree-like hierarchy of units. Green units are logical units and yellow are device units. Commands are propagated downwards and states are propagated upwards and summarized in the logical units.

cut from one of the networks. As soon as communication is re-established, all clients reconnect to the DIM servers and Run Control resumes control over the experiment. The bridge also receives external information from the beam line complex. The user interface itself runs on a different computer in the control room and is remotely connected to the main system.

Future developments The first version of Run Control was successfully deployed for a dry run in 2012 and has continuously evolved since then. New equipment and subdetectors have been delivered and subsequently integrated. The experience acquired during the following years and the feedback from the operators brought Run Control to a good level of reliability and usability.

The next steps in the development are the automatization of known procedures and automatic error detection and recovery. The ELectronic Eye of NA62 (ELENA62) is the core system being developed and already partially deployed. It provides a framework for monitoring functions, interactions with the operator through visual and audio notifications, and confirmation windows. It is highly customizable and configurable as the monitoring and control of specific components of the experiment are provided through plug-in modules.

Among those already in use to increase the efficiency and quality of the data taking are: the PC farm module that handles the PC nodes by restarting the acquisition software after a crash is detected, or that reboots the nodes when a hardware error occurs; two monitoring modules of the beam-line magnets and vacuum in the decay region that provide a fast feedback to the operator; another module that automatizes the start and end of run procedure, additionally guiding the operator when a manual action is necessary.

In conclusion, Run Control has already proved its usability and good reliability, confirming the technological choices made for its design. It has now reached a matured level of development with minimal daily maintenance, allowing the integration of the experience acquired during its operation into a more autonomous system.

References

[1] Buras AJ, Buttazzo D, Girrbach-Noe J and Knegjens R, 2015 JHEP **1511** 033
[2] Artamonov A V *et al.* 2009 Phys. Rev. D **79** 092004
[3] Ruggiero G, in KAON16 proceedings, this conference
[4] Holme O *et al* 2005 Conf. Proc. C **051010** WE2.1
[5] Gayet P *et al.* 2005 ICALEPCS proceedings WE2.2-6I
[6] Gaspar C *et al.* 2001 Comput. Phys. Commun. **140** 102
[7] Franek B and Gaspar C 2004 IEEE Nuclear Science Symposium Conference Record vol **3** 1831

GPU real-time processing in NA62 trigger system

R. Ammendola[1], A. Biagioni[2], S. Chiozzi[3], P. Cretaro[2], S. Di Lorenzo[4,5], R. Fantechi[5], M. Fiorini[6,3], O. Frezza[2], G. Lamanna[7], F. Lo Cicero[2], A. Lonardo[2], M. Martinelli[2], I. Neri[3], P. S. Paolucci[2], E. Pastorelli[2], R. Piandani[5], M. Piccini[8], L. Pontisso[5], D. Rossetti[9], F. Simula[2], M. Sozzi[4,5] and P. Vicini[2]

1 INFN Sezione di Roma Tor Vergata, Roma, Italy
2 INFN Sezione di Roma, Roma, Italy
3 INFN Sezione di Ferrara, Ferrara, Italy
4 Università degli Studi di Pisa, Pisa, Italy
5 INFN Sezione di Pisa, Pisa, Italy
6 Università degli Studi di Ferrara, Ferrara, Italy
7 INFN LNF, Frascati, Italy
8 INFN Sezione di Perugia, Perugia, Italy
9 NVIDIA Corp., Santa Clara, CA, USA

E-mail: gianluca.lamanna@cern.ch

Abstract.
A commercial Graphics Processing Unit (GPU) is used to build a fast Level 0 (L0) trigger system tested parasitically with the TDAQ (Trigger and Data Acquisition systems) of the NA62 experiment at CERN. In particular, the parallel computing power of the GPU is exploited to perform real-time fitting in the Ring Imaging CHerenkov (RICH) detector. Direct GPU communication using a FPGA-based board has been used to reduce the data transmission latency. The performance of the system for multi-ring reconstrunction obtained during the NA62 physics run will be presented.

1. Introduction

General Purpose computing on GPU (GPGPU) is nowadays widespread in scientific areas requiring large processing power such as computational astrophysics, lattice QCD calculations, image reconstruction for medical diagnostics. In a High Energy Physics experiments are using (i.e. ALICE [1]) or are considering to use (i.e. ATLAS [2]) GPUs for high level triggers.

Low level triggers could also benefit from GPUs computing power to build more refined physics-related trigger primitives and to design more selective condition in order to reduce the load to the higher level without sacrificing interesting physics events; Three main aspects must be carefully evaluated in order to establish the real-time capability of this kind of processors: computing power, total processing latency and its stability in time. In the present paper a description is provided of the GPU-based L0 trigger integrating in the experimental setup of the RICH detector of the NA62 experiment in order to reconstruct the ring-shaped hit patterns; we report and discuss results obtained with this system along with the algorithms that will be implemented.

Content from this work may be used under the terms of the Creative Commons Attribution 3.0 licence. Any further distribution of this work must maintain attribution to the author(s) and the title of the work, journal citation and DOI.
Published under licence by IOP Publishing Ltd

2. The NA62 experiment, trigger and RICH detector

The NA62 experiment at CERN [3] has the goal of measuring the branching ratio of the ultra-rare decay of the charged kaon into a pion and a neutrino anti-neutrino pair.

The trigger is a key system to obtain such a result. In its standard implementation, the FPGAs on the readout boards of each sub-detector participating to the L0 trigger, compute simple trigger primitives. The maximum latency allowed for the synchronous L0 trigger was chosen to be rather large in NA62 (up to 1 ms). Such a time budget allows in principle the use of more complex but slower trigger. This would have the benefit of increasing the trigger selectivity for the $K^+ \to \pi^+ \nu\bar{\nu}$ process, and would allow the design of new trigger condition to collect additional physics processes.

The Ring Imaging Cherenkov (RICH) detector is a key element for particle identification in the NA62 experiment and one of the subdetectors responsible for the L0 trigger [4]. Its main purpose is to separate pions from muons in the momentum range from 15 GeV/c to 35 GeV/c and to measure the particle arrival time with better than 100 ps resolution [5]. Thanks to its fast response, it is used in the L0 trigger by providing a multiplicity count. However, much more information would be available from the RICH. Due to the loose relation between hits multiplicity and physics related quantities, evaluating this information at the online trigger stage would allow to build most effective trigger conditions.

3. GPU computing at L0

In a standard approach for GPU computing, data from a detector reach a Network Interface Card (NIC) which copies them periodically on a dedicated area in the PC RAM. Here, a sufficient data load of buffered events is usually prepared and then copied to GPU memory through the PCI express bus. The PC on which the GPU card is plugged, i.e. the host, has the role of starting the GPU "kernel".

In contrast with respect to the applications for which GPUs have been originally designed, in triggering the computing latency is a very important issue. The total latency is essentially composed by two parts: the time to bring the data on the GPU processor and the effective time of processing. With the aim of decreasing the transport latency (and its fluctuations) for L0 applications a direct data transfer protocol from a custom FPGA-based NIC to the GPU has been used. Dedicated pattern recognition algorithms have been developed to exploit the GPU parallel structure. Both these aspects will be discussed below.

NaNet is a low-latency NIC with GPUdirect capability developed at INFN Rome division. Its design comes from the APEnet+ card logic [6] and, in the current implementation, it provides 10 Gb/s communication (NaNet-10) [7]. NaNet board is able to exploit the GPUDirect peer-to-peer (P2P) and RDMA capabilities of NVIDIA Tesla GPUs injecting into their memory an UDP input data stream from the detector front-end. With respect to the previous version (NaNet-1 with 1 Gb/s links) NaNet-10 board is able to cope with higher data rate and allow to receive data from 4 TEL62 RICH readout board [4], providing on-the-fly data processing to perform tasks as data de-compression and event fragments coalescing.

A fast ring finding algorithm is a crucial point to allow the use of RICH in on-line trigger selection. The processing time must be small enough to cope with high input data rate. The parallel structure of the GPUs helps in increasing the computing throughput both allowing the parallel processing of several event concurrently and exploiting pattern recognition parallel algorithms. We focused in particular on two multi-rings standalone pattern recognition algorithms based only on geometrical considerations (as no information from other detectors is available at this level). In case of the first algorithm, named Histogram, the XY plane is divided into a grid and a histogram is created with distances from the grid points and the PMT hits. Rings are identified looking at distance bins whose contents exceed a predefined threshold value. The second algorithm, i.e. Almagest [8], is based on Ptolemy Theorem. Both these algorithms

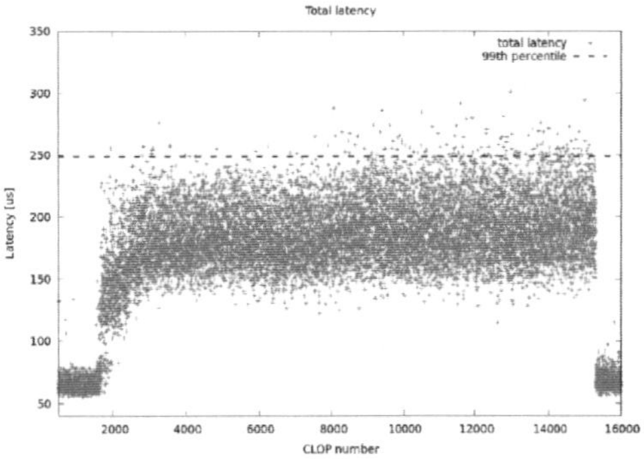

Figure 1. Total latency during an entire burst.

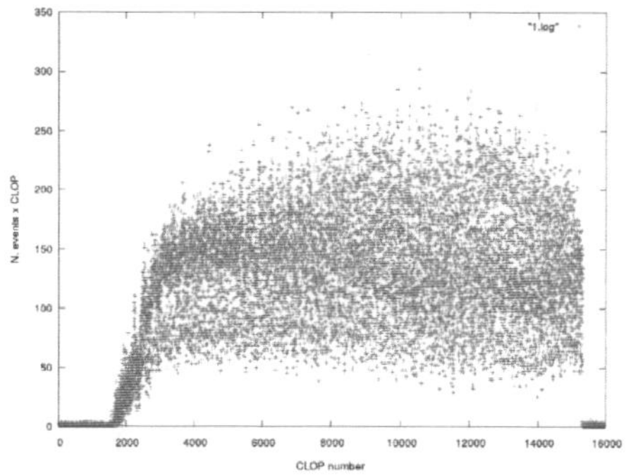

Figure 2. Number of events per CLOP during the burst.

are used for pattern matching, once the number of rings and points belonging to them have been found, it is possible to apply Crawford‘s method [9] to obtain centre coordinates and radii with good spatial resolution.

4. Results

The whole system, equipped with the NaNet-10 board, has been tested during the normal NA62 data taking, with data coming from the four RICH TEL62 readout boards.

The latency measurement is done by exploiting an hardware clock on the FPGA synchronized with the NA62 general clock through a TTC [10] interface. The resolution of this clock is 25 ns. The clock counter is reset at each start of burst (the burst is a bunch of particles produced by the protons extracted from the SPS. The burst duration is about 6 s each 14 s, the duty cycle depends on the SPS extraction scheme).

In fig.1 the total latency including all the components is shown for an entire burst. In fig. the total latency including all the components is shown for an entire burst. The main source of the width of the latency distribution is due to fluctuation in the number of events buffered before the processing: in order to better exploit the parallel structure of the GPU the data are processed in buffers collected in 350 μs, called CLOP (Circular List of Persistent buffers). The latency for processing of a single datagram in the NaNet-10 board, since its receiving in the 10GbE MAC to the completion of the DMA operation towards the GPU memory, accounts for less than 2 μs In 2016, at 1/3 of the nominal intensity, each CLOP contains in average about 150 events (fig.2). Nevertheless, the processing time per event, in the present version of the computing kernel, is quite stable below $1\mu s$ (using a single GPU). The interaction with the present central trigger processor (L0TP) needs a further stage of synchronization. The send of the GPU results, codified in a format compatible with the present NA62 L0 trigger, is done directly from the GPU memory with a low latency RDMA, still through the NaNet-10 board.

5. Conclusions

The GPU system performed as expected in the NA62 2016 run. In the next physics run, the GPU system can be used to provide high quality primitives to increase the purity of the main triggers and to give handle to build selective triggers to enlarge the NA62 trigger menu. Current results for the processing time per event hints to the further development of GPU algorithms and/or to the use of multiple GPU devices to reach an order of magnitude speed-up to cope with the experiment nominal intensity. Additional details on the results shown here can be found in [11].

Acknowledgments
This work was partially funded by the ExaNeSt project (H2020 grant no. 671553). S. Di Lorenzo, M. Fiorini, G. Lamanna, I. Neri, R. Piandani, L. Pontisso,M. Sozzi thank the GAP project, partially supported by MIUR under grant RBFR12JF2Z Futuro in ricerca 2012.

References
[1] D. Rohr, S. Gorbunov, A. Szostak, M. Kretz, T. Kollegger, T. Breitner, T. Alt, Journal of Physics: Conference Series 396 (1) (2012) 012044.
[2] A. T. Delgado, P. C. Muíño, J. A. Soares, R. Gonçalo, J. Baines, T. Bold, D. Emeliyanov, S. Kama, M. Bauce, A. Messina, M. Negrini, A. Sidoti, L. Rinaldi, S. Tupputi, Z. D. Greenwood, A. Elliott, S. Laosooksathit, An evaluation of gpus for use in an upgraded atlas high level trigger, ATL-DAQ-PROC-2015-061, (November 2015).
[3] G. Ruggero on behalf of NA62 Collaboration, in KAON2016 Proceedings
[4] B. Angelucci *et al.*, JINST **9** (2014) no.01, C01055. doi:10.1088/1748-0221/9/01/C01055
[5] D. Aisa *et al.*, Nucl. Instrum. Meth. A **766** (2014) 1. doi:10.1016/j.nima.2014.06.089
[6] R. Ammendola, A. Biagioni, O. Frezza, F. Lo Cicero, A. Lonardo, P. S. Paolucci, D. Rossetti and F. Simula *et al.*, J. Phys. Conf. Ser. **396** (2012) 042059.
[7] R. Ammendola *et al.*, JINST **11** (2016) no.03, C03030. doi:10.1088/1748-0221/11/03/C03030
[8] G. Lamanna, Nucl. Instrum. Meth. A **766** (2014) 241. doi:10.1016/j.nima.2014.05.073
[9] J. F. Crawford, Nucl. Instrum. Meth. **211** (1983) 223. doi:10.1016/0167-5087(83)90575-6
[10] S. Baron Timing, trigger and control (TTC) systems for the LHC, http://ttc.web.cern.ch/ttc/intro.html
[11] A. Lonardo, Proceedings of TWEPP2016 conference (to be published)

A new system of STRAW chambers operating in vacuum for the NA62 experiment

Yu Potrebenikov[1,2]

[1]Joint Institute for Nuclear Research, Dubna, Russia

E-mail: Iouri.Potrebenikov@cern.ch

Abstract. Large 2.15 × 2.15 m^2 area straw tube chambers have been developed to meet the specific requirements of the NA62 experiment. The main goal of the NA62 experiment at the CERN SPS is to measure the branching ratio of the ultra-rare $K^+ \to \pi^+\nu\overline{\nu}$ decay with 10 % accuracy. This can be achieved by detecting about 100 Standard Model events with 10% background in 2 - 3 years of data taking. A low mass tracking system is necessary to achieve a high resolution on kinematic quantities while timing resolution is mandatory to match the outgoing pion with the incoming kaon. Design, construction and first performance are reported.

1. Introduction

High energy particle physics experiments often require large dimension tracking chambers achieving good spatial resolution and built of minimal amount of material to minimize multiple coulomb scattering effects when used as elements of a magnetic spectrometer. Among these detectors are straw tube chambers based on an already well known technique. However, more than twenty years after the first proposal to use such kind of detector, there was no chamber reliably operating in vacuum, except for the chamber of the COSY-TOF collaboration [1], which was placed in vacuum together with its outer frame. The NA62 experiment is located in the CERN North Area and makes use of the primary 400 GeV proton beam from the SPS to produce a secondary unseparated charged hadron beam. The beam line layout and the detector setup have been optimized to study $K^+ \to \pi^+\nu\overline{\nu}$ decays [2] based on decay-in-flight technique of high momentum (75 GeV/c) kaons. Details about the already achieved performance and sensitivity can be found in [3]. A pilot run took place in autumn 2014 to commission the sub-detectors and readout system. The STRAW chamber system was fully commissioned and its performance found to meet the design requirements. In June 2015, NA62 began to take physics data, and will continue throughout 2018.

2. A new type of STRAW tubes

Overpressure inside a straw tube leads to radial and longitudinal deformations. The effect of radial deformation can be minimized for example, by adopting a layout where the straws

[2] for the NA62 CERN-DUBNA STRAW group: N. Azorskiy, L. Glonti, Yu. Gusakov, V. Elsha, T. Enik, S. Kakurin, V. Kekelidze, E. Kislov, A. Kolesnikov, D. Madigozhin, S. Movchan, I. Polenkevich, Yu. Potrebenikov, V. Samsonov, S. Shkarovskiy, S. Sotnikov, H. Danielsson, J. Bendotti, J. Degrange, N. Dixon, P. Lichard, J. Morant, V. Palladino, F. Perez Gomez, G. Ruggiero, M. Vergain.

Content from this work may be used under the terms of the Creative Commons Attribution 3.0 licence. Any further distribution of this work must maintain attribution to the author(s) and the title of the work, journal citation and DOI.
Published under licence by IOP Publishing Ltd

are sufficiently separated from each other. Longitudinal deformation may become critical as it spontaneously destroys the straightness of the straw. An axial pre-tension guarantees straight straws during the operation of the detector and minimizes the need for supporting material. However, classical fabrication of straws by film winding and subsequent gluing of winded layers with elastic glue substantially elongates [4] with time and can completely loosen the initial pretension that again will lead to deformation of straws. As a consequence, ultrasonic welding along the straw axis has been used. Thin-wall tubes (straws) are the elementary components used in the chambers. The 9.82 mm diameter drift tubes are made of 36 μm Mylar film, coated with 50 nm of copper and 20 nm of gold on the inside.They are welded along the straw axis using an ultrasonic welding machine developed at JINR (Figure 1). A straw seam visual quality control and a set of tests allow to produce straws within the strict tolerances defined in [5]. Straw tubes are filled with a mixture of 70% Argon and 30% CO_2 at atmospheric pressure and are themselves operated in vacuum. Gas leakage per meter of straw does not exceed 0.5×10^{-6} litre/min at 1 bar overpressure. A straw tube withstands a pressure of up to 9 bar.

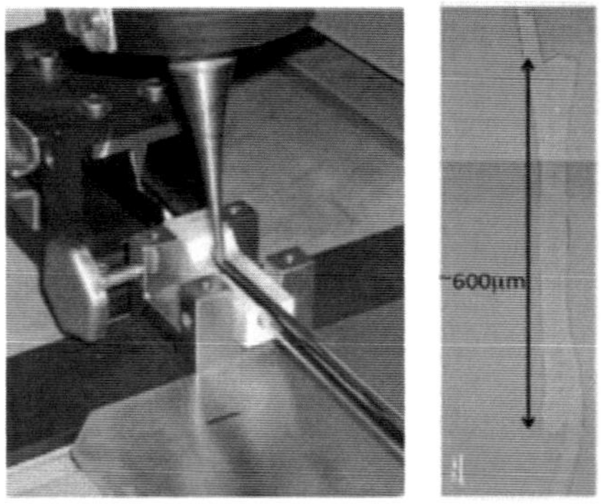

Figure 1. Left: Detail of the straw welding machine. Right: Microscope image showing the cross-section of the weld.

3. Straw chamber assembly and performance

Each straw chamber is composed of two square-shape modules with reinforced sides to withstand the strength of tube pretension (more then 700 kg on each module side). One module contains two views measuring X (0°), Y (90°) coordinates and the second module contains the U (−45°) and V (+45°) views (Figure 2-left). A straw view consists of four straw layers, 112 straws each. Each view has a gap of about 12 cm without straws near the centre, such that, after overlaying the four views, an octagon shaped hole is created for the beam passage. As the beam has an angle of +1.2 mrad and −3.6 mrad in the horizontal plane, upstream and downstream of the spectrometer magnet, respectively, this hole is not centred on the Z axis, but has custom offsets (along the X direction) as a function of the chamber Z-position. Each chamber has a circular active area with an outer diameter of 2.1 m. High detection redundancy is provided through a straw arrangement with four layers per view, which guarantees at least two hits per view (Figure 2-right). Up to 99.98% detection efficiency is obtained within a straw tube diameter of 9.4 mm.

After completion of the straw chamber installation, their actual position relative to the beam axis has been measured by the CERN survey group with an accuracy of 0.3 mm.

The straw readout electronics is based on the CARIOCA chip [6], an 8-channels charge-sensitive preamplifier-shaper-discriminator with base-line restoration and tail cancellation. The gain is 15 mV/fC and the peaking time $\sim 8-10$ ns. The leading time of the straw signal is used

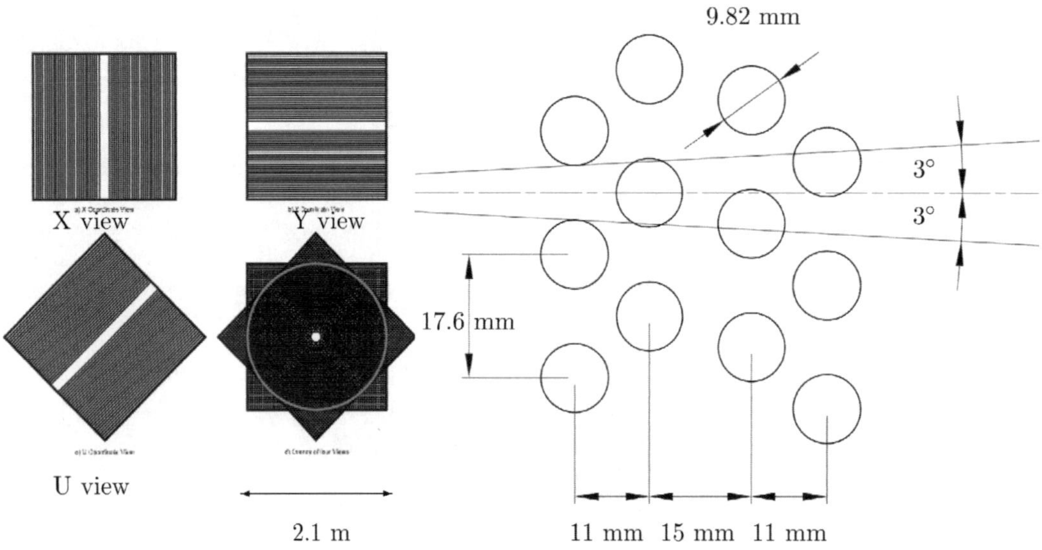

Figure 2. Left: One straw chamber is composed of four views (X, Y, U, V) and each view measures one coordinate. Right: Each view includes four layers of straws providing at least two straws hit per track.

for the track position measurement using radial distance-time correlation. The trailing time is the same for each particle (maximum drift time at the straw wall), and can be used for trigger purpose. Each cell (16 straws) has an individual HV power cable and readout board. The read out board is based on two CARIOCA chips and an ALTERA FPGA (software 16-ch TDC). Data are transmitted to the straw readout board (SRB). Each group of 8 SRBs is installed into a single 9U VME rack (32 SRBs in total).

Four MPOD crates for high and low voltage power supply have been installed and integrated in the Detector Control system of the experiment through a user-friendly interface. On-line monitoring has been developed for quality control of the recorded data.

The experimental drift time distribution was compared with the results of a `GARFIELD` simulation [7] with good agreement. Reconstruction of $K^+ \to \pi^+\pi^+\pi^-$ decays using the straw spectrometer leads to a reconstructed kaon mass resolution of 0.9 MeV/c^2 in line with the design expectations.

In conclusion, a new type a Straw chambers has been developed and built successfully. The performance of the STRAW chambers operating in vacuum has been estimated and meets the requirements of the NA62 experiment in terms of material budget (1.8% of a radiation length in total), space resolution (close to 130 μm) and excellent efficiency.

References

[1] Wintz P [COSY-TOF Collaboration] 2004 AIP Conf. Proc. **698** 789
[2] Anelli G *et al.* 2005 CERN-SPSC-2005-013
[3] Ruggiero G, in KAON16 proceedings, this conference
[4] Kendziora C *et al* FERMILAB- preprint 2002 Pub-02/241-E
[5] The NA62 Collaboration 2010 *NA62: Technical Design Document, NA62-10-07*, available `[online]` `https://cds.cern.ch/record/1404985`
[6] Moraes D. *et al.* IEEE International Symposium on Circuits and Systems 2001 Volume **1** 360
[7] Veenhof R `GARFIELD`, available `[online]` `http://garfield.web.cern.ch/garfield/`

IOP Conf. Series: Journal of Physics: Conf. Series **800** (2017) 012048 doi:10.1088/1742-6596/800/1/012048

The Kaon identification system of the NA62 experiment

E Maurice[1,2]

[1]University of Liverpool

E-mail: emaurice@hep.ph.liv.ac.uk

Abstract. The main goal of the NA62 experiment at the CERN SPS is to measure the branching ratio of the ultra-rare $K^+ \to \pi^+ \nu\overline{\nu}$ decay with 10 % accuracy. This can be achieved by detecting about 100 Standard Model events with 10% background in 2 - 3 years of data taking. NA62 is exposed to a 750 MHz high-energy unseparated charged hadron beam, with a 6 % kaons component, and uses kaon decay-in-flight technique. Precise timing matching of the incident kaon and of the downstream charged track is essential to reject accidental coincidences when working in such a high rate environment. This is achieved by the kaon tagging system KTAG, which identifies kaons with an efficiency higher than 95 % and provides precise time information with a resolution better than 100 ps. KTAG re-uses the Cherenkov radiator and optics of a CEDAR, a ring-focusing Cherenkov detector designed for MHz beam intensity in the 1970s. To reach the required performance, KTAG is equipped with new photon detectors, electronics readout, mechanics, cooling and safety systems.

1. The NA62 CERN Experiment

The NA62 experiment is located at CERN in the North Area and uses the primary 400 GeV proton beam from the SPS to produce a secondary unseparated charged hadron beam. The beam line layout and the detector setup have been optimized to study $K^+ \to \pi^+ \nu\overline{\nu}$ decays [1] based on decay-in-flight technique of high momentum (75 GeV/c) kaons. To reach a sensitivity to branching ratio of $\mathcal{O}(10^{-10})$, the rate of kaons decaying must be $\sim$5 MHz. As only 10 % of the kaons decay in the 65 m fiducial region between the upstream and downstream detectors, a 50 MHz beam of K^+ is required. It is obtained with the secondary beam from the SPS which has a rate of 750 MHz, and consists of 6 % K^+, 24 % p, 70 % π^+. It is therefore essential to provide an accurate timing ($\sim$ 100 ps) of the incoming kaon to precisely match the particles from the decay, and to avoid wrong association below the 1% level. To identify $K^+ \to \pi^+ \nu\overline{\nu}$ signal events and suppress background, NA62 uses techniques based on kinematic properties and particle identification. To reduce the background due to neutral pions (π^0 decaying to photons) and muons in the final state, hermetic coverage enables vetoing of photons and muons. Details about the already achieved performance and sensitivity can be found in [2]. In 2012, a technical run was carried out to study the performances of NA62 sub-detectors and finalize their design, in particular these of KTAG [3]. A pilot run took place in autumn 2014 to commission the sub-detectors and readout system. KTAG was fully commissioned and its performance found to

[2] for the NA62 KTAG group: Birmingham, Bristol and Liverpool Universities, UK.

Content from this work may be used under the terms of the Creative Commons Attribution 3.0 licence. Any further distribution of this work must maintain attribution to the author(s) and the title of the work, journal citation and DOI.
Published under licence by IOP Publishing Ltd

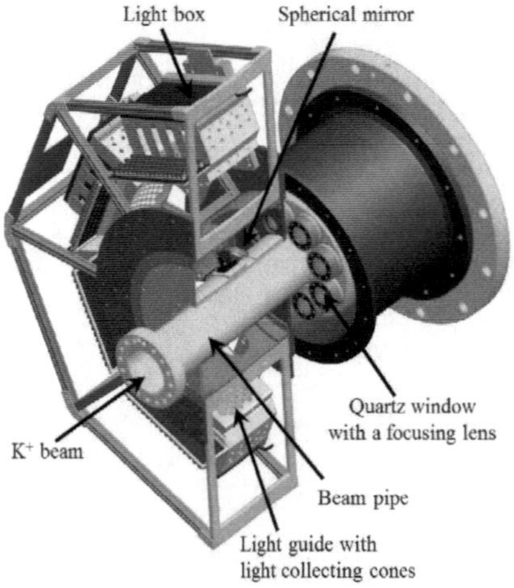

Figure 1. Left: Drawing showing the upstream part of the CEDAR and the KTAG photon detection system. Right: KTAG and CEDAR installed in the NA62 beam line during the 2012 test run.

meet the design requirements. In June 2015, NA62 began to take physics data, and will continue throughout 2018.

2. The Kaon identification system (KTAG)

The Kaon identification system (Figure 1) is instrumental in suppressing the accidental non-kaon background and defines a precise event timestamp with an accuracy better than 100 ps. KTAG identifies the 50 MHz of K^+ within the 750 MHz unseparated beam.

KTAG is based on a ring-focusing Cherenkov counter (CEDAR for Cherenkov Detector with Achromatic Ring focus [4]) equipped with new photon detectors and read-out system. This counter is optimized to detect Cherenkov light emitted by charged particles traveling together in a parallel beam, with an internal optical system focusing the light in concentric rings on the photon detector plane (Figure 2-left). Proper operation requires the intrinsic beam divergence not to exceed 70 μrad in each transverse direction to the beam axis. The pressure of the radiator gas is then used to distinguish the particles of different velocities and so discriminate between kaons, pions and protons, as shown in Figure 2-right.

The Cherenkov light produced in the gas is reflected by a mirror, passes through a corrector lens and finally through a diaphragm. The lens corrects for the optical dispersion in the N_2 gas and ensures that the Cherenkov light, emitted at all wavelengths, arrives at the same radius on the diaphragm plane. The aperture of the diaphragm is adjusted to detect light emitted by particles, as selected by the chosen gas pressure. As a consequence, photons emitted by other particles hit the diaphragm at a different radius and do not illuminate the PMTs. The pressure scan highlighting the kaon operation pressure is shown in Figure 2-right: the optimal N_2 gas pressure is 1.74 bar to select 75 GeV/c kaons. After the diaphragm, the light passes through additional lenses and light guides to finally reach eight light-boxes where the photon detectors (PMT) are located. These PMTs are hosted in an insulated, cooled Faraday enclosure flushed with N_2. Each box contains 48 Hamamatsu PMTS (32 of R9880 type and 16 of R7400 type). The rate of photons on each PMT is reduced to 5 MHz while PMT signals are read out and digitized in dedicated front end electronic boards, enabling KTAG to handle the high beam flux.

Data recorded in 2014–2015 have been analyzed to assess performance of KTAG. This

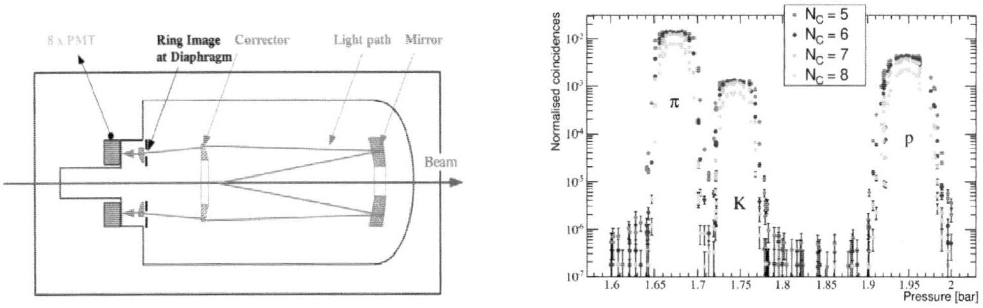

Figure 2. Left: Schematic view of the light path in the CEDAR counter. Right: Result of a pressure scan of the N_2 gas radiator showing the separation between K^+, π^+, p.

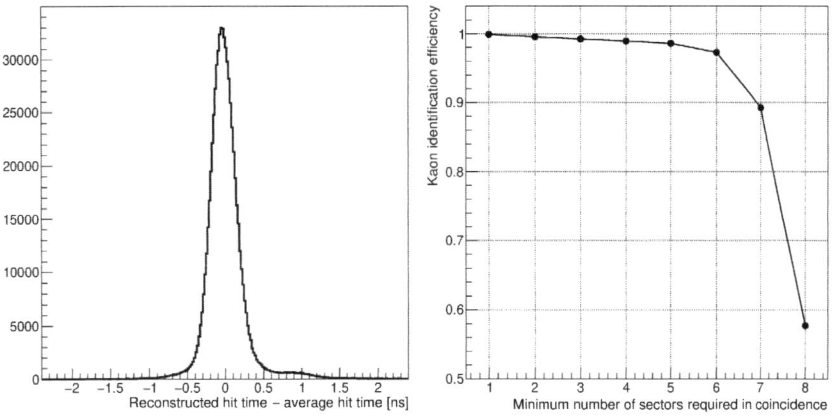

Figure 3. KTAG performance: (Left) Time resolution of a single PMT. (Right) Kaon tagging efficiency as a function of the number of sectors in coincidence.

evaluation uses $K^+ \to \pi^+\pi^0$ decays reconstructed from kinematic properties only. The time resolution of a single PMT signal is 280 ps(Figure 3-left), achieving a time resolution of about 70 ps when collecting 20 photons per kaon, which is significantly better than the design expectations. The K^+ is identified requiring at least 5 (of the 8) sectors in coincidence to remove background from π^+, as shown in Figure 3-right. The efficiency is approximately 99 % while the probability of misidentifying a pion as a kaon while operating at the kaon pressure is estimated to be of $\mathcal{O}(10^{-4})$.

In conclusion, the performance of the KTAG kaon tagger has been estimated and meets the requirements of the NA62 experiment.

References

[1] The NA62 Collaboration 2010 *NA62: Technical Design Document, NA62-10-07*, available [online] `https://cds.cern.ch/record/1404985`
[2] Ruggiero G, in KAON16 proceedings, this conference
[3] Goudzovski E *et al. Development of the kaon tagging system for the NA62 experiment at CERN*, 2015 Nucl. Instrum. Meth. **A801** 86
[4] Bovet C *et al. The CEDAR counters for particle identification in the SPS secondary beams*, 1982 CERN-1982-013

The CHarged particle ANTIcounter for the NA62 experiment at CERN

Marco Mirra[1]
Università degli studi di Napoli "Federico II" e Sezione INFN di Napoli, Italy
E-mail: mirra@na.infn.it

Abstract.
The design and performance of a charged particle detector named CHANTI is described. The detector, which operates in vacuum, it is used as a veto counter in the NA62 experiment at CERN.

1. The NA62 experiment

The NA62 experiment at CERN[1] aims at measuring the Branching Ratio (BR) of the ultra-rare kaon decay $K^+ \to \pi^+ \nu\bar{\nu}$ with a 10% uncertainty. It is a theoretically very clean Flavour Changing Neutral Current process and represents a valid test for the flavor structure of the Standard Model (SM). The SM prediction is BR($K^+ \to \pi^+ \nu\bar{\nu}$)=$(0.84 \pm 0.10) \times 10^{-10}$[2], while the current best measurement, performed by the E787/E949 collaboration, is BR($K^+ \to \pi^+ \nu\bar{\nu}$)=$(1.73^{+1.15}_{-1.05}) \times 10^{-10}$[3]. NA62 uses a 75 GeV/c unseparated hadron beam from the SPS with a 6% kaon component. The apparatus is designed to collect kaon decays in flight within a 65 m long evacuated fiducial volume. A critical point of the experiment is the rejection power of the kaon decays that can mimic the signal: kinematical constraints (measuring both K^+ and π^+ momenta), particle identification and veto detectors have to be used.

2. CHANTI detector

The interaction of the beam with the kaon spectrometer (called GTK) represents a possible source of background. Indeed inelastic scattering events can mimic the signal if a produced pion falls into the decay product spectrometer acceptance and is badly reconstructed inside the fiducial volume while no other track is detected. The CHANTI detector is placed just after the GTK to help the rejection of this background by tagging the particle emitted at large angles with respect to the beam (from 49 to 1340 mrad). The detector is composed of 6 stations, of square shape (300×300 mm^2), placed inside the vacuum tube starting at a distance of 27 mm from the last station of the GTK along the beam axis. The hole inside each station is 65 mm in Y and 95 mm in X. Each station is made up of two layers (X and Y) and each layer is composed of 24 scintillator bars arranged parallel to the X(Y) direction and individually shaped at the appropriate length. The bars consists of an extruded polystyrene core doped with blue-emitting fluorescent compounds with a TiO_2 coating. The cross-section of the bar is an isosceles triangle

[1] on behalf of the NA62 CHANTI group: F. Ambrosino, T. Capussela, D. Di Filippo, P. Massarotti, M. Mirra, M. Napolitano, L. Roscilli, G. Saracino, G. Corradi, C. Paglia, D.Tagnani

Content from this work may be used under the terms of the Creative Commons Attribution 3.0 licence. Any further distribution of this work must maintain attribution to the author(s) and the title of the work, journal citation and DOI.
Published under licence by IOP Publishing Ltd

Figure 1. Single scintillator bar (left) and CHANTI stations inside the vacuum chamber (right)

with a base of 33 mm and a height of 17 mm; a hole (1.7 mm diameter) is placed at 8.5 mm from the base to host a WaveLenght Shifting (WLS) fibre Bicron BCF92 (Fig.1 left). The fibre has one side mirrored and is read from the other side by a Hamamatsu SiliconPhotoMultiplier (SiPM). Figure 1 shows the first five CHANTI stations assembled in the vacuum vessel (right). More details about CHANTI can be found in [4].

A VME 9U board (CHANTI-FE) has been developed to power and read the SiPMs. It also gives fast amplification of the signals before passing them to a Time over Threshold (ToT) board [5] which gives a LVDS output of duration equal to the time a signal is above a given threshold. Actually the ToT board is able to give two LVDS outputs for each analog input corresponding to different adjustable level thresholds. This feature allows to correct for time-walk effects with a linear extrapolation when both thresholds are crossed by the signal and to calibrate an average time-walk vs ToT correction, to be used when only the lower thresholds is crossed. The LVDS signal is then processed by a fast TDC. A careful calibration procedure for the CHANTI-FE has been setup in order to precisely control and readout the value of the voltage bias and to readout the current; the calibration constants for DAC and ADC are obtained by measuring the voltage-current curve of a set of high impedance calibrated input resistors. A threshold scan is done for each of the readout SiPMs of the CHANTI system by counting the dark rate at different values of the threshold. Since the SiPM dark rate diminishes by roughly one order of magnitude as the threshold is raised by an amount equivalent to one photoelectron amplitude, a fit of this threshold scan with a proper step function can detemine the amplitude (in mV) of the single photoelectron peak for each channel (Fig.2). This procedure allows to set the thresholds in terms of a quantity directly related to the fundamental process in the scintillator/SiPM (number of photoelectrons) rather than in terms of signal amplitude in mV that depends on the gain of each SiPM and on the electronics channel amplification gain.

The CHANTI took part in the data taking of NA62 in 2014, 2015 and 2016. The efficiency and spatial resolution have been measured by means of muon halo tracks, which provide a clean sample of straight and penetrating Minimum Ionizing Particles (MIP). Single station efficiency has been measured to be above 99.9% as expected from tests performed on the detector prototype using cosmic rays. A CHANTI hit position is reconstructed in each view as the average of the positions of at least two bars weighted by the energy release in each bar. Indeed the triangular shape of each bar results in a linear relation between the amount of energy released by a straight particle crossing the bar and the distance of its interaction from the bar center. The center-of-charge technique allows to reach a resolution of 1.8 mm (Fig.3 left), better than the intrinsic one resulting from the pitch of the WLS fibres used in the detector. Given the high rate of spurious tracks crossing the detector (muon halo, pion inelastic interactions, etc.), a good time resolution is important to reduce the random vetoes fraction to a few percent, while keeping a

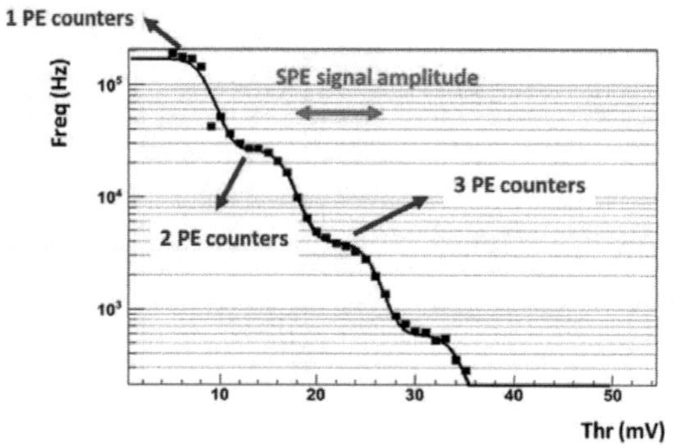

Figure 2. A typical threshold scan curve where the flat part corresponds to the counters of a certain number of photoelectron signal (PE). The distance between two consecutive inflection points of the curve is the single photoelectron (SPE) amplitude. For this particular channel it is found to be about 10 mV.

high efficiency in detecting kaon inelastic interactions. Reaching a few ns single hit resolution is therefore mandatory for CHANTI. Once time-walk correction is taken into account, the typical single hit resolution is found to be 1.2 ns (Fig.3 right).

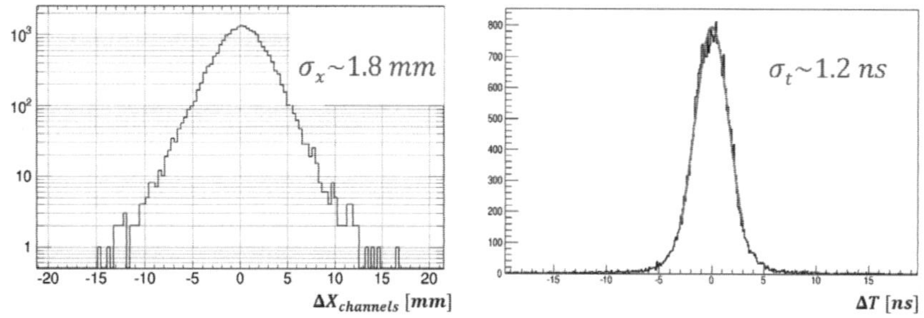

Figure 3. Difference between X position of two hit points on different stations for straight MIP (left) and difference in time between bar hits, after time-walk correction (right).

3. Conclusion

In this work the CHANTI detector was presented. The achieved performance has shown that CHANTI can safely operate in vacuum with a time and spatial resolution of 1.2 ns and 1.8 mm, respectively, in line with the design requirements.

References

[1] A. Ceccucci et al.: *Proposal to Measure the Rare Decay at CERN SPS*, CERN-SPSC-2005-013, SPSC-P-326.
[2] A.J. Buras, D. Buttazzo, J. Girrbach-Noe and R. Knegjens, *$K^+ \to \pi^+ \nu\bar{\nu}$ and $K_L \to \pi^0 \nu\bar{\nu}$ in the Standard Model: status and perspectives*, JHEP 11 (2015) 033 [arXiv:1503.02693].
[3] Anisimovsky et al. [E949 Collaboration], Phys. Rev. Lett. 93, (2004).
[4] F. Ambrosino, et al.: *CHANTI: a fast and efficient charged particle veto detector for the NA62 experiment at CERN.*, JINST 11 P03029, http://dx.doi.org/10.1088/1748-0221/11/03/P03029
[5] A. Antonelli et. al, arXiv:1111.5768

The RICH Detector of the NA62 experiment at CERN

Patrizia Cenci[1,2] and Mauro Piccini[1]

[1]INFN Perugia – Via A Pascoli – 06123 Perugia (I)

E-mail: patrizia.cenci@pg.infn.it

Abstract. The Ring Imaging Cherenkov (RICH) detector of the NA62 experiment at CERN is a key element for particle identification. This paper describes the detector design and some measurements of its performance obtained from the analysis of the data collected so far by NA62. The results fully match the detector requirements.

1. Introduction

The NA62 experiment [1] is the lastest generation kaon experiment at the CERN SPS, exploiting an intense beam of charged kaons. It aims to measure the Branching Ratio (BR) of the golden $K^+ \rightarrow \pi^+ \nu\bar{\nu}$ ultra-rare decay with a 10% precision. This process is highly suppressed in the Standard Model (SM): the BR prediction is of the order of 10^{-10} with minimal theoretical uncertainty [2]. Only one measurement is available so far, compatible with the SM within a large uncertainty [3]. As a result, the precision measurement of the BR will be a remarkable achievement for the SM flavour sector.

The most challenging aspect of the NA62 experiment is the suppression of decay channels with BR up to 10 orders of magnitude higher than the signal and similar experimental signature, such as the main background mode $K^+ \rightarrow \mu^+ \nu$, with BR = 63.55%. To this purpose a high kinematic rejection power, an effective μ and γ reduction and very good particle identification (PID) are required. The Ring Imaging Cherenkov detector (RICH) is a key elements of PID in NA62. Its purposes are: to identify pions from muons in the momentum range 15 to 35 GeV/c with a muon suppression factor better than 10^2; to provide precise measurements of the Cherenkov angle and of the pion crossing time, below 100 *ps*, to minimize wrong matching with the parent particle; to provide a fast signal to the trigger system. The NA62 detector[3] was commissioned in 2014 and 2015. The experiment has been taking data since 2015 and will run until the year 2018. The RICH performance has been studied in detail and some results will be described in this paper.

2. The NA62 RICH detector design

The NA62 RICH consists of a steel vacuum-tight cylindrical vessel around 17.5 m long, constructed from 4 longitudinal sections with decreasing diameter (4 to 3.4 m) enclosing a central beam pipe. The vessel is filled with about 200 m^3 of Neon gas slightly above atmospheric pressure, suitable as radiator medium thanks to appropriate refractive index, good light transmission and minimal chromatic

---

[2] On behalf of the RICH Working Group of the NA62 experiment: Anzivino G, Aisa D, Barbanera M, Bizzarri M, Bizzeti A, Bucci F, Campeggi C, Cenci P, Checcucci B, Ciaranfi R, Duk V, Iacopini E, Imbergamo E, Latino G, Lenti M, Lollini R, Pepe M, Piccini M, Piluso A, Volpe R.

[3] The NA62 detector details can be found in http://na62.web.cern.ch/NA62/Documents/TechnicalDesign.html.

Content from this work may be used under the terms of the Creative Commons Attribution 3.0 licence. Any further distribution of this work must maintain attribution to the author(s) and the title of the work, journal citation and DOI.
Published under licence by IOP Publishing Ltd

dispersion. Gas recirculation is not required. Due to the low emission rate of Cherenkov photons per unit length in Neon, the RICH exploits the maximum space available along the NA62 beam line. Hamamatsu R7400-U03 PM have been chosen to collect the Cherenkov light due to their compactness and excellent timing properties. Figure 1 shows the RICH vessel with the two aluminium flanges that each hold ~1000 photomultipliers (PM) and are outside the detector acceptance, and the mirror layout.

The mirror system consists of a mosaic made of 18 spherical mirror segments with a 35 cm side hexagonal shape and two semi-hexagonal mirrors at the RICH center to fit the beam pipe. The mosaic covers a surface of about 3 m in diameter. Each mirror segment consists of a 2.5 cm thick glass with 17 m focal length, coated with aluminium that is protected with a thin dielectric film to provide the suitable optical quality. The mirror support panel consists of two halves of aluminium honeycomb. Each mirror segment is suspended with a dowel inserted in the back and oriented to point toward the two PM flanges in order to avoid the beam pipe shadow on the reflected Cherenkov photons. High precision piezo-motors, out of the detector acceptance, pull two thin aluminum ribbons installed at ±45 degree with respect to the vertical direction, providing an independent two-axis alignment for each mirror. A third ribbon prevents on-axis rotations. A laser system is used for the initial mirror alignment, performed with the vessel open. Individual mirror positions are monitored during the data taking and corrected via software or adjusted remotely with piezo-motors. Thin entrance and exit beam windows, made of an aluminium alloy, minimize the material thickness in the decay volume. The mirror installation, with details of the movement system, is illustrated in figure 1 (right).

The PM signals feed custom-made current amplifiers with differential output. Customized front-end boards use the NINO ASIC chip [4] as discriminator operating in time-over-threshold mode, with LVDS output and about 50 *ps* intrinsic time resolution. The RICH exploits an integrated trigger and data acquisition system developed to read-out most of the NA62 detector subsytems. The key element is the FPGA-based TEL62 board [5] which houses four custom-made TDC boards (TDCB), with 512 channels in total. The TDCB sends the signal leading and trailing times to the TEL62 processors, which buffer data, produce trigger primitives in parallel with the read-out and send them to the Level 0 (L0) trigger processor. A three-level trigger system reduces the detector rate in the decay region from order of 10 MHz to few kHz. The RICH signals are well suited to be used as reference time and to select charged particles in the NA62 trigger system. A fast selection based on PM hit multiplicity is exploited by the L0 trigger processor. A selection based on multi-ring reconstruction can be implemented at higher trigger levels to increase the NA62 background rejection power.

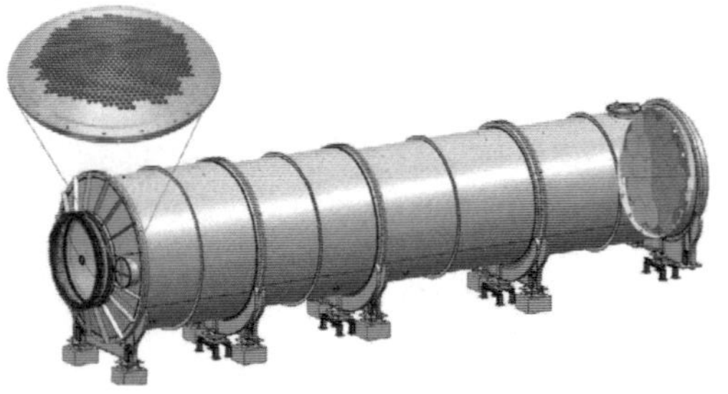

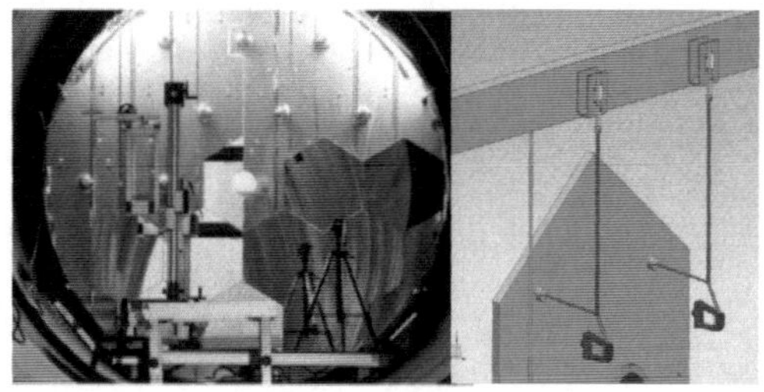

Figure 1. Left: Schematic drawing of the NA62 RICH detector showing vessel, PM assembly and mirror layout. Right: the installation of the mirror system and a detail of the mirror movement system.

3. The NA62 RICH Performance

A full-length RICH prototype was built and tested with hadron beams at CERN in 2007 and 2009 to demonstrate the performance of the proposed layout [6][7]. The construction was completed by mid-2014 and the detector, commissioned in 2014 and 2015, is successfully working since then.

The RICH time resolution has been first measured in 2014 with charged pion samples from the $K^+ \rightarrow \pi^+\pi^0$ decay mode. The hits of a Cherenkov ring have been split into two halves to compute the average time of each group. The RICH intrinsic time resolution is given by the difference of those average times. The distribution, shown in figure 2 (left), has a sigma width of 140 *ps*, resulting in an event time resolution of about 70 *ps*. This result has been confirmed with the 2015 and 2016 data.

In the 2016 run the complete mirror alignment has been accomplished for the first time after a special maintenance intervention operated during the winter shutdown. This maintenance aimed to restore the proper operation of the mirror movement system after repairing the small amount of damage suffered during the detector installation. All the mirrors were first individually aligned with a laser when the vessel was open. The alignment has been later fine-tuned with particle data by piezo-motor remote adjustment getting a mirror orientation within ±60 μrad with respect to the reference semi-hexagonal mirrors.

The PID performance of the RICH has been assessed with the physics data collected by NA62 in 2015. The Cherenkov ring radius has been reconstructed from the hit information and the center of the rings has been compared to the extrapolated track position measured with the spectrometer. In this way different particle types have been well identified, as clearly seen in figure 2 (center), illustrating the ring radius as a function of the particle momentum measured by the spectrometer. Positrons, muons, pions and a kaon component coming from beam particles scattered in the upstream beam tracker can be clearly distinguished in different bands.

In order to separate charged pions from muons a cut on the reconstructed squared mass m^2 is used. The mass is defined using the track momentum p and some RICH information as $m^2 \approx p^2(\theta_{max}{}^2 - \theta_C{}^2)$, where θ_{max} is the maximum Cherenkov angle and θ_C the Cherenkov angle measured by the fitted ring radius divided by the mirror focal length. The pion identification efficiency as a function of the muon misidentification probability has been measured by selecting clean pion and muon samples and scanning over the cut on m^2. The preliminary result obtained with the 2016 data is shown in figure 2 (right): a pion identification efficiency of 90% corresponds to some 0.8% muon misidentification probability. This result improves the previous measurement done in 2015 due to the better alignement of the mirror sytem achieved after the special maintenance intervention of the winter shutdown.

The performance of the RICH detector measured with the data collected so far by the NA62 experiment matches the expectations and meets the main detector requirements: time resolution better than 100 *ps* and a 90% pion identification efficiency for about 1% muon misidentification probability.

References

[1] Anelli G *et al.* (NA62 Collaboration) 2005 CERN-SPSC-2005-013 SPSC-P-326
[2] Buras A J, Buttazzo D, Girrbach-Noe J and Knegjens R 2015 *J. High Energy Phys.* 033 1511
[3] Artamonov A V *et al.* (E949 Collaboration) 2009 *Phys. Rev.* D **79** 092004
[4] Anghinolfi F *et al.* 2004 *Nucl. Instrum. Method* A **533** 183
[5] Avanzini C *et al.* 2010 *Nucl. Instrum. Method* A **623** 543
[6] Anzivino G *et al.* 2008 *Nucl. Instrum. Method* A **593** 314
[7] Angelucci B *et al.* 2010 *Nucl. Instrum. Method* A **621** 205

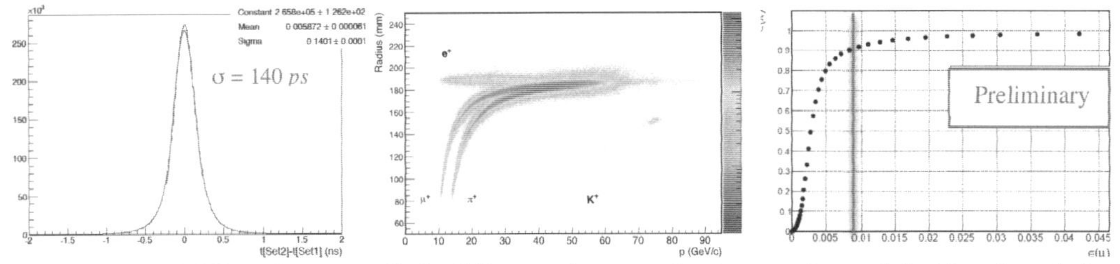

Figure 2. The RICH performance. Left: difference between the average times of the hits of two halves of a Cherenkov ring. Center: the Cherenkov ring radius as a function of the particle momentum. Different particle types are clearly identified. Right: Pion identification efficiency (a line is drawn at 90% value) versus muon misidentification probability (preliminary result, 2016 data).